AF678278

Fuzzy Logic and the Semantic Web

CAPTURING INTELLIGENCE

Series Editors

F. van Harmelen
D. Dubois
J. Taylor
S. Thrun

ELSEVIER

AMSTERDAM • BOSTON • HEIDELBERG • LONDON • NEW YORK • OXFORD
PARIS • SAN DIEGO • SAN FRANCISCO • SINGAPORE • SYDNEY • TOKYO

Fuzzy Logic and the Semantic Web

Edited by

Elie Sanchez

LIF, Université de la Méditerranée
Marseille, France

ELSEVIER

AMSTERDAM • BOSTON • HEIDELBERG • LONDON • NEW YORK • OXFORD
PARIS • SAN DIEGO • SAN FRANCISCO • SINGAPORE • SYDNEY • TOKYO

Elsevier
Radarweg 29, PO Box 211, 1000 AE Amsterdam, The Netherlands
The Boulevard, Langford Lane, Kidlington, Oxford OX5 1GB, UK

First edition 2006

Copyright © 2006 Elsevier B.V. All rights reserved

No part of this publication may be reproduced, stored in a retrieval system or transmitted in any form or by any means electronic, mechanical, photocopying, recording or otherwise without the prior written permission of the publisher

Permissions may be sought directly from Elsevier's Science & Technology Rights Department in Oxford, UK: phone (+44) (0) 1865 843830; fax (+44) (0) 1865 853333; email: permissions@elsevier.com. Alternatively you can submit your request online by visiting the Elsevier web site at http://elsevier.com/locate/permissions, and selecting *Obtaining permission to use Elsevier material*

Notice
No responsibility is assumed by the publisher for any injury and/or damage to persons or property as a matter of products liability, negligence or otherwise, or from any use or operation of any methods, products, instructions or ideas contained in the material herein. Because of rapid advances in the medical sciences, in particular, independent verification of diagnoses and drug dosages should be made

Library of Congress Cataloging-in-Publication Data
A catalog record for this book is available from the Library of Congress

British Library Cataloguing in Publication Data
A catalogue record for this book is available from the British Library

ISBN-13: 978-0-444-51948-1
ISBN-10: 0-444-51948-3
Series ISSN: 1574-9576

For information on all Elsevier publications
visit our website at books.elsevier.com

Printed and bound in The Netherlands

06 07 08 09 10 10 9 8 7 6 5 4 3 2 1

This volume is dedicated to Prof. Lotfi A. Zadeh

Preface

It is with great delight that I write the preface for this, the very first volume in Elsevier's new book series "Capturing Intelligence".

This series aims at publishing books on research from all disciplines dealing with and affecting the issue of understanding and reproducing intelligent artificial systems. The series will cast its net wide, aiming for contributions from diverse areas such as symbolic AI, biological approaches, self-organisation and emergence, and physically embodied systems.

Symbolic knowledge representation has a long tradition in the quest to understand the notion of "intelligence", and it is fitting that the first book in our series is rooted in this tradition.

However, the current volume is not simply a treatise on classical topics. Instead, it confronts a classical, rich and well-established area, namely the representation of fuzzy, non-crisp concepts, with a new and highly exciting challenge, namely the vision of the Semantic Web.

The vision of a more Semantic Web has been attracting much attention by scientists and industrialists alike since early publications on the topic, notably Tim Berners-Lee's book "Weaving the Web" and his Scientific American publication jointly with Jim Hendler and Ora Lasilla in 2001. Higher precision of search engines, searching with semantic instead of syntactic queries, information integration among different web-resources, personalisation, and semantic web services, are all part of the promises of this vision.

The insight that any realistic approach to the Semantic Web will have to take into account the lessons from fuzzy logic approaches is gaining ground in a wide community, and slowly but surely, the Semantic Web community is waking up to this fact.

For this reason, the current volume is extremely timely. The different papers in this book propose a number of ways in which current Semantic Web technology can be "fuzzified", and they discuss a number of ways in which such fuzzy representations can be used for tasks such as search and information retrieval.

I am convinced that this book is a very good and timely start of this new book series, and I am looking forward to future volumes in this series, of equally high quality and relevance.

Frank van Harmelen
(Series Editor)

Foreword

Fuzzy Logic in the Semantic Web: Covering a Missing Link

1. Introduction

During recent years, important initiatives have led to reports of connections between Fuzzy Logic and the Internet. For example, at the initiative of Lotfi Zadeh and Masoud Nikravesh, FLINT meetings ("Fuzzy Logic and the Internet") have been organized by BISC ("Berkeley Initiative in Soft Computing"). The proceedings of the FLINT 2001 workshop [1] and of the NAFIPS-FLINT 2002 Joint International conference [2] have been published. Meanwhile, scattered papers were published on Fuzzy Logic and the Semantic Web. A special session, "Fuzzy Logic in the Semantic Web, a New Challenge", was organized during the IPMU 2004 conference, at Perugia, Italy [3]. The idea for the present book came from enthusiastic authors and participants at that well-attended session, including: Trevor Martin, Ernesto Damiani, Umberto Straccia and Henrik Legind Larsen. More recently, in February 2005, the first workshop on Fuzzy Logic and the Semantic Web [4] at Marseille was attended by European experts in the field (they are all contributors to this book, too). This volume, the first in the new series *Capturing Intelligence*, shows the positive role Fuzzy Logic, and more generally Soft Computing, can play in the development of the Semantic Web, filling a gap and facing a new challenge. It covers concepts, tools, techniques and applications exhibiting the usefulness, and the necessity, of using Fuzzy Logic in the Semantic Web.

2. The Semantic Web and Fuzzy Logic

Most of today's Web content is suitable for human consumption. The Semantic Web is presented as an extension of the current web in which information is given well-defined meaning, better enabling computers and people to work in cooperation. For example, within the Semantic Web, computers *will understand* the meaning of semantic data on a web page by following links to specified ontologies. But while the vision of the Semantic Web and associated research attracts attention, as long as two-value-based logical methods are used, no progress can be expected in handling ill-structured, uncertain or imprecise information encountered in real world knowledge. It is also worth noting that Fuzzy Logic (in a simple form) is used in almost all major search engines. Fuzzy Logic will not be the basis for the Semantic Web but its related concepts and techniques will certainly reinforce the systems classically developed within W3C. In fact, Fuzzy Logic cannot be ignored if we are to bridge the gap between human-understandable *soft logic* and machine-readable *hard logic*. None of the usual logical requirements can be guaranteed: there is no centrally defined format for data, no guarantee of truth for assertions made, no guarantee of consistency. To support these arguments, this book shows how components of the Semantic Web (such as XML, RDF, OWL, Description Logics, Conceptual Graphs, Ontologies) can be covered, with in each case a Fuzzy Logic focus.

The Semantic Web, as presented under W3C recommendations, deals with hard semantics in the description and manipulation of crisp data. RDF based languages do not have the ability to represent soft semantics. The Semantic Web allows relational knowledge to be embedded as metadata in web pages enabling machines to use ontologies and inference rules in retrieving and manipulating data. Ontologies bridge an effective communication gap between users and machines. The construction of ontologies is crucial in the development of the Scientific Web. The key ingredients to build up an ontology are a vocabulary of basic terms and, when possible, a precise specification of the meaning of these terms. In fact, computers require precise definitions but humans normally work better without precise definitions and, mostly due to the nature of the information in world knowledge, there is a need for a collection of tools drawn from Fuzzy Logic, for example Zadeh's PNL (Precisiated Natural Language). A Fuzzy Ontology structure can be defined as consisting of concepts, of fuzzy relations among concepts, of a concept hierarchy or taxonomy, and of a set of ontology axioms, expressed in an appropriate logical language. Then, a lexicon for a fuzzy ontology can consist of lexical entries for concepts (knowledge about them can be given by fuzzy attributes, with context-dependent values), of lexical entries for fuzzy relations, coupled with weights expressing the strength of associations, and of reference functions linking lexical entries to concepts or relations they refer to. DLs (Description Logics) are a logical reconstruction of frame-based knowledge representation languages that can be used to represent the knowledge of an application domain in a structured and formally well-understood way. They are considered as a good compromise between expressive power and computational complexity. DLs are essentially the theoretical counterpart of the Web Ontology Language OWL DL, the state of the art language to specify ontologies. DLs can be used to define, integrate and maintain ontologies. DLs have been extended with fuzzy capabilities, yielding FDLs (Fuzzy Description Logics) in which concepts are interpreted as fuzzy sets. These are only a few examples of using Fuzzy Logic in the Semantic Web and many more extended developments are presented in this volume.

3. Structure of the book

The book covers such different topics as overviews, proposals for language definitions, ontology construction, information retrieval and search. It is composed of seven sections. These sections do not constitute a crisp partition. There is some overlap between the chapters (multiple introductions to RDF, OWL, protoforms, etc.). However, this makes the chapters readable independently. Our grouping of chapters was a choice among many other possible ones, a choice to give the book more structure for the reader and to make the material more easily accessible.

The first section, *Introduction*, starts with the chapter "On the Expressiveness of the Languages for the Semantic Web – Making a Case for 'A Little More' " by Christopher Thomas and Amit Sheth. It introduces the need for fuzzy-probabilistic formalisms on the Semantic Web, in particular within OWL. It is followed by the chapter "Fuzzy Ontologies for Information Retrieval on the WWW" by David Parry, which deals with the concept of fuzzy ontology, and presents a broad survey of relevant techniques, leading up to the notions of fuzzy search and fuzzy ontologies. The third chapter "Capturing Basic Semantics Exploiting RDF-oriented Classification" by Vincenzo Loia and Sabrina Senatore, presents a system designed to support searching and indexing tools oriented towards semantic-based information discovery. It proposes a semantic RDF-oriented classification of documents to characterize relationships between concepts and web resources.

The second section, *Fuzzy Description Logics for Ontology Construction*, deals with fuzzy description logics in theoretical aspects and applications. The first chapter is "A Fuzzy Description Logic for the Semantic Web" by Umberto Straccia. It describes a fuzzy version of SHOIN(D), the corresponding Description Logic of the ontology description language OWL DL, showing that its representation and reasoning capabilities go clearly beyond classical SHOIN(D). The next chapter "What Does Mathematical Fuzzy Logic Offer to Description Logic?" by Petr Hájek, begins with a survey on continuous t-norm based fuzzy predicate logic. Then, it proposes a fuzzy description logic based on fuzzy predicate logic, to deal with vague (imprecise) concepts. The following chapter "Possibilistic Uncertainty and Fuzzy Features in Description Logic. A Preliminary Discussion" by Didier Dubois, Jérôme Mengin, and Henri Prade, is another approach to injecting fuzzy features in Description Logics, this time based on fuzzy and possibilistic logic. The chapter "Uncertainty and Description Logic Programs over Lattices" by Umberto Straccia, presents a Description Logic framework for the management of uncertain information. In this approach, sentences are certain to some degree, where certainty values are taken from a certainty lattice. The last chapter of this section, "Fuzzy Quantification in Fuzzy Description Logics", by Daniel Sanchez and Andrea Tettamanzi, introduces reasoning procedures for a fuzzy description logic with fuzzy quantifiers.

The third section, *Search and Protoforms*, contains chapters with concepts introduced in "From Search Engines to Question Answering Systems – The Problems of World Knowledge, Relevance, Deduction and Precisiation", by Lotfi A. Zadeh. This chapter deals with question-answering systems – systems with deduction capabilities – with perception-based information, and especially perceptions of probabilities, which are intrinsically imprecise. It introduces and discusses new tools to deal effectively with world knowledge, relevance, deduction and precisiation. These tools, drawn from fuzzy logic, are: Precisiated Natural Language (PNL), Protoform Theory (PFT), and the Generalized Theory of Uncertainty (GTU). The following chapter, "A Perception-based Search with Fuzzy Semantic" by Chris Tseng and Toan Vu, is a proposal for a perception-based search methodology established on fuzzy semantic for improving the websearch. It contains an implementation and case-study in the search domain of health and food. Finally, the chapter "Using Knowledge Trees for Semantic Web Querying" by Ronald Yager, proposes a formalism, involving knowledge trees, for querying the web. Protoforms are also discussed to aid in deduction and manipulation of knowledge.

The fourth section, *XML based Approaches, Building Ontologies, and (Conceptual) Graphs*, starts with the chapter "Fuzzy Data Mining for the Semantic Web: Building XML Mediator Schemas" by A. Laurent, P. Poncelet and M. Teisseire. It considers the problem of mining XML mediator schemas from a database perspective. XML documents are modeled by labeled ordered trees, Fuzzy Tree Inclusion is addressed and it is proposed to extend the knowledge on the mined semantic structures by providing fuzzy links within frequent trees. Then the chapter "Bottom-up Extraction and Maintenance of Ontology-based Metadata" by Paolo Ceravolo, Angelo Corallo, Ernesto Damiani, Gianluca Elia, Marco Viviani and Antonio Zilli, presents an approach to build fuzzy ontologies in a bottom-up fashion, by clustering documents, based on a fuzzy representation of XML documents structure and content. This section ends with the chapter "Approximate Knowledge Graph Retrieval: Measures and Realization" by T.H. Cao and Dat T. Huynh. It is a proposal for degrees of subsumption, and an engineering work on top of well-established Semantic Web infrastructure software (Sesame). Conceptual graphs are employed for a user-friendly interface and easily readable query expressions.

The fifth section, *Integration and Processing of Fuzzy Information*, begins with the chapter "Soft Integration of Information with Semantic Gaps" by Trevor Martin and Ben Azvine. The work aims to identify sources of information referring to a same entity, enabling merging attributes and integrating attributes to provide a fused information source. Then, the chapter "Processing Fuzzy Information in Semantic Web Applications" by Sebastian Kloeckner, Klaus Turowski and Uwe Weng, is a discussion on several approaches for processing fuzzy information, to improve the actual concepts of the semantic web. Finally, the chapter "Fuzzy Logic Aggregation for Semantic Web Search for the Best (Top-k) Answers" by Peter Vojtáš, presents an application to find best (top-k) answers depending on scoring multicriterial user requirements on the web.

The sixth section, *Ontologies, Information Retrieval*, starts with "A Fuzzy Logic Approach to Information Retrieval using an Ontology-based Representation of Documents", by Mustapha Baziz, Mohand Boughanem, Gabriella Pasi and Henri Prade. This chapter is related to Information Retrieval using fuzzy ontologies and it includes a benchmark evaluation. It is followed by the chapter "Towards a Semantic Portal for Oncology using a Description Logic with Fuzzy Concrete Domains" by Mathieu d'Aquin, Jean Lieber and Amedeo Napoli. It is a work on encoding medical guidelines in fuzzy description logics and using that for a portal. The three systems that are presented are fully implemented (KASIMIR oncology project). The last chapter of this section, "Fuzzy Relational Oncological Model in Information Search Systems" by Rachel Pereira, Ivan Ricarte and Fernando Gomide, is another approach to Information Retrieval using fuzzy ontologies encoded by fuzzy relations, with a test case on scientific articles about 'Computational Intelligence'.

The last and seventh section, *Soft Computing in Various Domains*, starts with the chapter "Evolving Ontologies for Intelligent Decision Support', by Paulo Gottgtroy, Nikola Kasabov and Stephen MacDonell. It integrates soft computing techniques and ontology engineering. It deals with the rather different (but important) topic of evolving ontologies and it presents biomedical case studies. Finally, the chapter "Enhancing the Power of the Internet using Fuzzy Logic-based Web Intelligence: Beyond the Semantic Web" by Masoud Nikravesh, proposes the notion of "Concept-based Databases for Intelligent Decision Analysis" and it discusses several soft computing concepts and tools to search the web.

4. Concluding thoughts

These are exciting times in the fields of Fuzzy Logic and the Semantic Web, and this book will add to the excitement, as it is the first volume to focus on the growing connections between these two fields. This book will be a valuable aid to anyone considering the application of Fuzzy Logic to the Semantic Web, because it contains a number of detailed accounts of these combined fields, written by leading authors in several countries. The field of Fuzzy Logic has been maturing for forty years. These years have witnessed a tremendous growth in the number and variety of applications, with a real-world impact across a wide variety of domains with humanlike behavior and reasoning. And we believe that in the coming years, the Semantic Web will be a major field of applications of Fuzzy Logic.

Paraphrasing Lotfi A. Zadeh [5], *we can say that "in moving further into the age of machine intelligence and automated reasoning, we have reached a point where we can speak, without exaggeration, of systems which have a high machine IQ* (*MIQ*) . . . *In the context of the Semantic Web, MIQ becomes Semantic Web IQ, or SWIQ, for short"*

References

[1] "New Directions in Enhancing the Power of the Internet" (Proceedings UCB/ERL, Berkeley, Memo N^{O} M01/28, August 2001) and "Enhancing the Power of the Internet", M. Nikravesh, B. Azvine, R. Yager, L.A. Zadeh (Eds.), Springer-Verlag, 2004.
[2] NAFIPS-FLINT 2002 Int. Conf., IEEE SMC Proceedings 02TH8622, New-Orleans, 2002.
[3] IPMU 2004, Special Session "Fuzzy Logic in the Semantic Web: a New Challenge", IPMU04@dipmat.unipg.it, Perugia, Italy, 2004, Proceedings, pp. 1017–1038.
[4] "Fuzzy Logic and the Semantic Web" Workshop, Extended abstracts available at http://www.lif.univ-mrs.fr/FLSW, Marseille, France, 2005.
[5] L.A. Zadeh, "Web Intelligence and Fuzzy Logic – The concept of Web IQ (WIQ)", Invited talk at the 2003 IEEE/WIC Int. Conference on Web Intelligence (WI 2003), Halifax, Canada, available at www.comp.hkbu.edu.hk/IAT03/InvitedTalk1.htm.

Acknowledgements

I would like to take this opportunity to give special thanks to all the contributors and also the reviewers for their commitment, hard work and time spent writing the various chapters. Without their efforts, this volume could not have come to fruition.

Elie Sanchez
Laboratoire d'Informatique Fondamentale
Biomathématiques et Informatique Médicale
Faculté de Médecine, Université de la Méditerranée
Marseille, France
September 2005

Contents

Introduction

CHAPTER 1

On the Expressiveness of the Languages for the Semantic Web – Making a Case for 'A Little More'

Christopher Thomas and Amit Sheth*

Large Scale Distributed Information Systems Lab, University of Georgia, Athens, GA, USA
E-mail: chaos@uga.edu; amit@cs.uga.edu

Abstract

The best pair of shoes you can find are the ones that fit perfectly. No inch too short, no inch too wide or long. The same, of course, holds for applications in all fields of computer science. It should serve our needs perfectly. If it does more, it usually comes with a tradeoff in performance or scalability. On top of that, for logic based systems, the maintenance of a consistent knowledge base is important. Hence a decidable language is needed to maintain this consistency computationally. Recently, the restriction of the Semantic Web standard OWL to bivalent logic has been increasingly criticized for its inability to semantically express uncertainties. We will argue for the augmentation of the current standard to close this gap. We will argue for an increased expressiveness at different layers of the cake and we want to show that only a spiced up version of some of the layers can take the blandness out of it. We want to show that it is possible to have a mixture that can account for reasoning based on uncertainties, possibilistic measures, but also for other epistemologically relevant concepts, such as belief or trust.

Keywords

semantic web, fuzzy logic, probability theory, knowledge representation

> "So far as the laws of mathematics refer to reality, they are not certain. And so far as they are certain, they do not refer to reality."
>
> Albert Einstein

* Corresponding author.

FUZZY LOGIC AND THE SEMANTIC WEB
Edited by Elie Sanchez
© 2006 Elsevier B.V. All rights reserved

1. Introduction

During the 2004 WWW conference, Tim Berners-Lee, the spiritual leader of the semantic web community, made a statement that might impact further research on semantic web foundations. Asked whether it is necessary to have a representation for uncertainty on the layered semantic web stack, his answer was basically "no". Knowing about the impact of a person like Berners-Lee, this view might have great impact on the community and hence on unfortunate students and researchers funded for the development of uncertainty calculi for the semantic web. But aside from personal considerations, this answer can be challenged on many levels. After all, history has shown many times that "the revolution eats its children".

When talking about extending the kinds of logical inference used in local or global knowledge bases, we have to be aware of the different ways in which this can be done. The semantic web standard assumes a monotonic bivalent logic. Bivalent means that statements are either true or false; no third possibility, such as unknown, exists, nor anything in between true and false. In the remainder of the paper we will refer to these bivalent logics as FOL (first-order logics).

Research in logic has produced a plethora of different formalisms which are more or less used in real-world applications. There are many-valued logics, continuous-valued logics with different semantics of their value assignments, such as fuzzy logics and probabilistic logics [13,15], non-monotonic logics [38], paraconsistent logics [5] and combinations thereof [10]. One can make a case for the use of any of these logics for different applications. The question for semantic web researchers is, whether to augment the famous layer cake by adding a different layer for non-bivalent and/or non-monotonic logics or whether these augmentations should exist locally, for groups of agents that agree to their own representation of non-FOL statements.

Berners-Lee's main argument for the rejection of a representation of uncertainty at the very basis of Semantic Web formalisms was their lacking scalability. This is a strong argument. For example, in order to do complete probabilistic inference on dependent objects in a Bayesian Network or any other technology for probabilistic inference, potentially every path through the network has to be taken into account. The same holds for any kind of non-monotonic logic. Given that, even in the case of FOL, such as SHIQ [18] or similar Description logics, inference cannot be done in a computationally efficient way, it seems mandatory to keep the evaluation of statements local.

One major problem that monotonic knowledge bases face is that of inconsistency. In a monotonic logic, such as each of the various flavors of bivalent Description Logics and hence OWL, it is assumed that if a true statement can be derived from a set S of facts and rules, then it can also be derived from every larger set S' that contains S. This is an appealing assumption to make, because it allows reasoning to be local and to only take into account the rules and facts that are immediately necessary to deduce a new statement. But it is also an unrealistic assumption to make, because the world, even the formalized one, is full of contradictions. And the more this world of formal statements grows, the more likely it is that a contradiction occurs. For this reason, the CYC® Ontology is partitioned into consistent micro theories or contexts [31]. Inconsistencies with other micro theories are explicitly stated. To give an example of how easily inconsistencies can occur when we get to know more about a domain, we cite an example that is widely used in AI textbooks. It involves birds. We will state the knowledge base informally:

- Birds fly ($\forall x$: bird(x) $\rightarrow$ fly(x))
- Penguins are birds ($\forall x$: penguin(x) $\rightarrow$ bird(x))
- Joe is a penguin (penguin(Joe))

We can deduce now, that Joe is a bird and hence can fly. But there's one more rule that, for the sake of accuracy, could be added to the domain representation.

- Penguins don't fly ($\forall x$: penguin(x) $\rightarrow$ $\neg$fly(x))

This statement causes an inconsistency in the knowledge base. If we follow the path Joe is a penguin, penguins are birds, birds fly, then Joe flies. If we follow the path Joe is a penguin, penguins don't fly, then Joe doesn't. The addition of a rule made the knowledge base inconsistent, even though it was a sensible rule to add. And locally, the knowledge base was also correct before this rule was added. It was just missing some information and it still is missing information. Complete formalized knowledge of a domain is an illusion.

It is no question that inconsistencies arise when combining knowledge from multiple sources. It is also no question that most real-world phenomena cannot be sufficiently described in a bivalent logic. This holds in two cases. We can look at this from a scientific point of view, where we have to deal with uncertainty, randomness and partial knowledge. In this case even extremely accurate measurements result in uncertainties about the implications of the obtained data. The second case is that of knowledge representation and computation based on human perception as described in [40]. Here, non-bivalence and inconsistencies occur because of a fundamental inability to conduct completely accurate measurements on the one hand but the corresponding human ability to accurately reason with this inaccurate knowledge. The question for the Semantic Web research is, though, whether the basis of Semantic Web standards should be the bivalent monotonic case and the uncertain and/or non-monotonic cases are a specialization thereof, or whether the basis should be the uncertain case with bivalent monotonic logics as a special case.

1.1. *Science*

In essence, it is also a question about the extent to which the use of ontologies will have an impact on how humans gain and evaluate knowledge. The time we are living in is partially signified by the specialization of the scientific fields. The time in which an Aristotle could have all scientific knowledge of his cultural realm is long gone. It is even impossible to know about all current developments in adjacent fields. Over time, science evolved into a hydra who's heads are autonomous and quite ignorant about what happens in most of it's other heads, a giant distributed apparatus that is striving towards progress without the ability to take a comprehensive snapshot of the current state. Hypotheses are generated, but are mostly evaluated against the knowledge of a small field. A formalization of these hypotheses could allow the scientific communities to extensively evaluate their findings. Agents could roam the web and cross check hypotheses with data obtained by other labs and compare the results and hypotheses generated from the data to find similarities and contradictions.

In a scientific environment or in general in fields that are mainly driven by empirically closing the gaps of incomplete knowledge of their domain, it is necessary that the knowledge bearing agent is aware of the partiality of its knowledge. In a localized system, that is easily accomplished. If we want to share information across the Internet, then the formalism that encapsulates this knowledge needs to reflect this partiality, so it can be propagated to other agents. Otherwise it is like telling rumors. I heard from my not very trustworthy acquaintance Paul that Peter and Mary broke up recently. While I might still have propagated it as a rumor that I did not really trust to Cathy, she might have told it as a fact to Marc who was Peter's best friend but always had his eyes on Mary. After asking her out on a date, his Friendship with Peter was shattered. If every link in this gossip chain had kept the information that it was just a rumor, Marc and Peter would be drinking beer now.

1.2. *Information retrieval*

In information retrieval, we are dealing with all sorts of noisy data. Web search engines and proposed question answering systems that are based on information available on the Internet or in local ontologies will on average yield results that are not totally relevant or reliable. The annotation of data with metadata will improve this situation, but not fundamentally change it. Hence, an estimate of relevance or truth, delivered together with the answer or answer set, would be desirable.

Recently, more statistical techniques have emerged in Internet search engines. While Google's page rank [8] measures the popularity of a page and makes the implicit assumption that popularity is somewhat linked to importance, other search engines use clustering to order their search results. The Vivisimo search engine [42] builds a hierarchy for the search results using a hierarchical clustering technique.

The Vivisimo approach is only the first of many search engines that give a hierarchical structure of their search results. We will also encounter situations in which we want to build ontologies or topic maps on the fly from sets of documents that have not yet been classified[1] [21,29]. Any automatic approach to ontology generation, be it statistical, using pattern based, NLP or other techniques, must be uncertain in nature and the map must reflect its own uncertainty about the inferred knowledge.

In the Semantic Web context, this uncertainty cannot just be stored within the generating or retrieving agent, but must be propagated through the deductions or other reasoning steps that are performed with the information. In a future situation, when a derived fact is annotated with the trace of its derivation (i.e. its provenance information), this will become clear. In community based knowledge accumulation, such as an augmented "Semantic Wikipedia" project [41], the community could assign degrees of belief to the formalized facts. Each agent in turn can assign degrees of belief to selected ontologies or single facts within ontologies. The latter is an individual assignment and doesn't require a formalization, but the former needs to be made available to general purpose reasoners, if we want the Semantic Web vision of connectivity between heterogeneous knowledge sources become a reality.

For the next generation of web search engines, question answering machines, it will be crucial to have a powerful representation of vagueness, because the terms humans use to describe the world will be vague and ambiguous [28,30,40].

1.3. *Semantic Web*

Semantic heterogeneity is an inherent reality of the Internet. It is possible to encompass all types of syntactic, schematic and semantic heterogeneity that have been discussed in

[1] The term ontology usually implies a rigid knowledge representation which meets formal criteria of inheritance or soundness/completeness issues. The term topic map has a less formal connotation.

research on heterogeneous databases [22,24,33]. This diversity makes it the most useful, ubiquitous and unbiased collection of knowledge that has ever existed. The challenge to a computational use of this smorgasbord of opinions is to keep the diversity while still having a sound basis for computation. Inability of first order to model heterogeneity even in relational databases was conclusively discussed [26]. In the real world realization of the Semantic Web, very few ontologies will be consistent with each other. These inconsistencies can arise for multiple reasons. The easiest resolution is that of ambiguity in terms and actual formalization while still agreeing on the conceptualization as such. But ontologies can also be fundamentally inconsistent, because the ontological and metaphysical basis is different, i.e. the ontologies don't share the same view of the world or their domain, or the factual knowledge used to populate them come from different sources whose observations are inconsistent with each other.

This paper is meant to show that, despite computational disadvantages, many signs point towards incorporating the representation of more powerful semantics into the Semantic Web layer cake. While it is computationally impossible to logically track every statement on the web and use it in every logical inference, a restriction to deduction within local ontologies based on FOL will only exploit the local knowledge, but never help acquiring new knowledge. Hence the Semantic Web formalisms need to account for the possibilities that inductive and abductive inferences offer alongside with the ability to combine ontologies that, in FOL, would result in inconsistent and hence unusable knowledge.

The discussion about having representations of uncertainty, fuzzy reasoning and continuous representations in general is less concerned about whether this representation is necessary, but where it is necessary. Looking at the problems that non-classical logics pose to inference mechanisms, the placement has to be done carefully. The easy way out is to argue that each community that relies on a non-classical representation, does so in their own ontology with their own agents processing this information. There is a lot of merit to this argument, but we are going to show that it will not be sufficient in the long run, if the Semantic Web is to be exploited to its full potential.

In a current scenario, in which we see the mere beginning of the application of Semantic Web technologies, ontologies tend to be a convenient way of porting database information. This is a large step by itself, but while it allows machines to read and maybe process data from multiple sources, it doesn't give us a way of reliably putting derived knowledge from these multiple sources back on the web. With human intervention, we filter information that we think wasn't derived in a scientifically viable way and we have the ability to trust and distrust sources. In most current applications, there is only one step of computational inference between two tasks accomplished by humans. If now we are trying to have an automated chain of inference, the agents involved need to tell each other about the confidence they have in the truth of their derived information, which is in turn a function of the reliability of their sources and their inference methods. An internal procedure, even a very sophisticated one that allows an agent to filter out information that according to its criteria does not exceed a certain threshold of reliability will not be of any use, if it propagates the remaining inferred information as completely reliable information or knowledge.

The Semantic Web has the potential to make an immense impact on human society in general and on science in particular. The decisions made in the beginning will be difficult to revise at a later stage. We see, for example, how hard it is to change from a 32 bit IP address to a 128 bit address. For these reasons we need to be careful what we design as the foundational formalisms of this Semantic Web. We need to make sure that we promote knowledge, deepen and broaden it. But alongside knowledge goes doubt. Every careful representation of knowledge needs to be aware of the fallibility of each statement. The Semantic Web offers a great opportunity, but it might also just offer the chance to augment existing junk by a whole lot of meta-junk. We want to improve, not worsen the situation that T.S. Eliot depicts: "Where is the wisdom we have lost in knowledge? Where is the knowledge we have lost in information?" (T.S. Eliot)

This paper attempts to name fundamental issues with the use of non-classical representations on the Semantic Web. It does not intend to solve them. Much research has been done in single fields, which we are going to address and correlate with the problems that the semantic web is facing.

The remainder of the paper is organized as follows. First, we give two more detailed examples of real-world situations in which we need a representation of uncertainty. In Section 3, we will briefly introduce several kinds of logics that exceed the bivalent paradigm. In Section 4 we will outline the envisioned framework of logics for the semantic web. Section 5 is concerned with issues of soundness, completeness, consistency and computability. Section 6 sketches the shortcomings of OWL as a basis for the framework and Section 7 introduces related work from the fields of Artificial Intelligence, Computer Science, Philosophy and Psychology. Section 8 finally concludes the sketch of this framework and addresses open questions. The discussion of the logical formalisms is mostly held on an informal level. Essentially we are presenting a vision rather than a theory or its implementation.

2. Some motivating examples

2.1. *Glycomics*

In a project funded by the *National Center for Research Resources* [4], the LSDIS lab and the Complex Carbohydrate Research Center at UGA are building an ontology to model the interaction of complex carbohydrates with genes, proteins, enzymes and cells.

Human cells have complex carbohydrates on their surfaces, and various experiments have shown that the surface carbohydrates of some invasive cancer cells have unusual structures. Therefore, biochemists are very interested in the relationship between these atypical cell surface carbohydrates and the process of metastasis (invasion by cancer cells).

The complex carbohydrates are synthesized by enzymes that are encoded by genes, and the amount of each enzyme depends on the level to which the gene is activated in the cell. Although it is relatively straightforward to determine whether a gene is "turned on" in a tissue (i.e. whether it is actively making the RNA message that leads to production of a specific enzyme), characterization of complex carbohydrates at the cell surface is technically challenging.

Nevertheless, one can find correlations between the presence of certain complex carbohydrates on the cell surface and the activation of specific genes (such as glycosyl transferases). There is a general but well-defined sequence of events starting from gene activation to carbohydrate synthesis, but the players (specific genes and enzymes) involved in the production of a particular carbohydrate are often not known with certainty. Correlative evidence (e.g., gene homologies and relationships between the amount of carbohydrate and the level of gene activation) is often available, but not sufficient to say with certainty that "activation of gene X results in the production of enzyme Y, which synthesizes carbohydrate Z". Similarly, the hypothesis that the carbohydrate is involved in causing a cancer cell to become invasive is based on correlative evidence. Hence, observation of a measurable quantity – the level of gene activation – only gives us a likelihood of the presence of the abnormal glycoprotein and that only provides the likelihood that the cancer cell is invasive. Additional evidence for these relationships must be accumulated piecemeal. Every bit of evidence modifies the likelihood of the initial hypothesis, which may eventually have to be rejected if the probability of its accuracy becomes too low.

That in cases like this the representation of likelihoods is absolutely necessary is not disputed. But since BioInformatics data is usually obtained and used by experts within the BioInformatics domain, it is not necessary to have a representation of these uncertainties at the lowest possible level, viz. the level of the semantic web layer cake. BioInformatics could agree on an upper ontology that is used for all representations of uncertainties within this field. On the other hand, the next example shows a more common scenario that will arise when we are using advanced Semantic Web technologies for information retrieval. If situations like the following shows the need for a representation of uncertainty at a lower level, we will have a reason to argue for its implementation in the layer cake.

2.2. *Information retrieval*

Imagine an advanced information retrieval situation. A question we could ask the question answering system would be "what is the population of New York City?" The system could either return one value or it could ask several other question answering systems and return all answers ranked by some sort of confidence value. It could also return the one that was given with the highest confidence or an average value. In any case, it can be given directly to the user. Hence no formal representation is necessary.

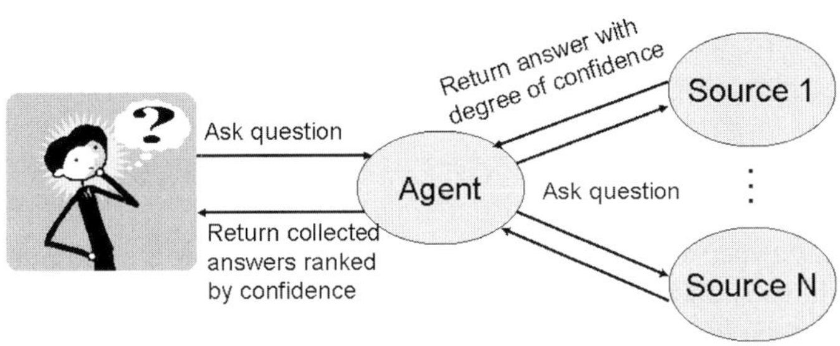

Fig. 1. Simple scenario, one step between user and question answering systems.

A more complicated case arises when we ask an agent to deduce an answer. We could, for example, ask the agent whether the largest city in the USA has more inhabitants than the largest city in Mexico. Now the system has to deal with several levels of uncertainty. First of all, it has to retrieve the name of the largest cities of both countries. Depending on the systems it consults, this could potentially return different answers with varying degrees of certainty. Then it has to take those results and ask again for the population of the cities. In the deductive step, the agent takes either all or the most confident results of the queries

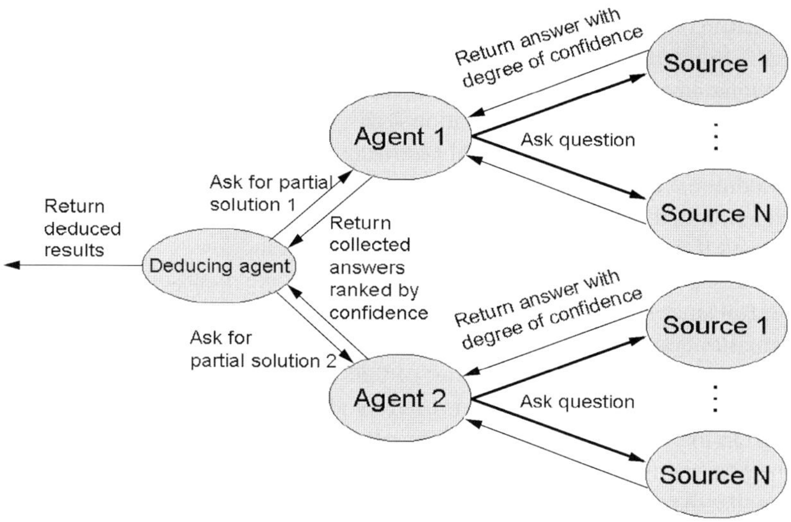

Fig. 2. Complex scenario, involving deductive steps between user and QA systems.

and performs the comparison. Depending on the settings, the user sees either the answer that was deduced with the highest confidence or all answers ranked by their degrees of confidence.

In this case, the representation of uncertainty needs to be on a fairly low level, preferably directly within the Semantic Web layer cake. We need to assume that the different agents that are consulted by the deducing agent do not agree on the same domain ontology, but use all the same knowledge exchange format. Hence uncertainty does not only arise due to an initial uncertainty returned by the QA systems, but also due to different ontological representations that need to be merged.

3. Challenges for multi-valued logics

3.1. *The fuzzy case*

Fuzzy logic has a special place amongst those logics that allow the assignment of continuous values. In its simplest case, it exhibits monotonicity. This means that the scalability argument cannot be used against an application of a fuzzy logic. Fuzzy logic is based on the notion of fuzzy sets. In classical set theory a set has crisp boundaries, i.e. an object is either a member of a set A or it is member of the set $\neg A$ (the complement of A). In fuzzy set theory sets have fuzzy boundaries. This notion is closely associated with prototype theory. A set has prototypical members that best describe the set. Then there are things that closely resemble the prototype, but are not considered to meet the properties of the prototype completely. A 7 feet tall person is prototypical for a tall person, while a 5 foot 8 inches tall person is less prototypical, but can still be considered as somewhat tall. In fuzzy logic, how much or little prototypical an object is, is identified by the set-membership function, which maps an attribute value to the range [0..1], which is the degree of membership. These set membership function can potentially take arbitrary forms, for example linear or

Gaussian. In the special case of a step function the fuzzy set takes the form of a classical set.

Umberto Straccia [35] has developed a fuzzy description logic that can easily be expressed in the form of OWL statements. SWRL [19] can then be used to map values to degrees of membership in fuzzy sets.

3.2. *Probabilities and possibilities*

Probabilistic reasoning gives us a way of finding out how likely something is the case. If 75% of the students in a class passed the exam, then we have a 75% chance to randomly pick one that passed. However, it doesn't make a claim about every particular student. Each single one either passed or failed. There is a fundamental difference in the semantics of fuzzy logic and probabilistic logic. In fuzzy logic, a statement can be true to a certain extent or an entity belongs to a class to a certain degree. This degree is assumed to be known with certainty. In probabilistic reasoning, there is a probability that a statement is true or false, but the statement itself is either true or false, but neither both nor something in between. Hence fuzzy logic sees the world as continuous instead of binary, while probabilistic logics make a claim about the randomness of the world or the observer's state of certainty.

For the task of knowledge acquisition, the statistical analysis of data allows the exploration of relationships that are not explicitly stated. Statistical techniques give us great insight into a corpus of documents or a large collection of data in general, when a program exists that can actually "pose the right questions to the data", i.e. analyze the data according to our needs. All derived relationships are statistical in nature and we only have an idea or a likelihood of their validity.

In the traditional bivalent-logic based formalisms we, that is the users or the systems, have to make a decision. Once two contradictory statements are identified, one has to be chosen as the right one. While this is possible in domains that are axiomatized, fully explored or in which statements are true by definition, it is not possible for most scientific domains. In the life sciences for instance, hypotheses have to be evaluated, contradicting statements have promoting data, etc. Decisions have to be deferred until enough data is available that either verifies or falsifies the hypothesis. Nevertheless, it is desirable to express these hypotheses formally to have means to computationally evaluate them on the one hand and to exchange them between different systems on the other.

3.3. *Belief and trust*

As the number of ontologies available grows and maybe in the future also their interoperability, we will be facing the problem of inference across ontologies. We will have to take into account that ontologies evolve and that facts change. In the paradigm of FOL and FOL-based formalisms, this can completely throw off the inference engine. It is very likely that a set of related ontologies is inconsistent. Assume a good-case scenario in which the interoperability issue is solved insofar as a common thesaurus has been used to name concepts in a set of ontologies and the merging of the ontologies has been accomplished already. Imagine further a set of news related ontologies O. And for simplicity, assume that all ontologies follow a bivalent paradigm and that each ontology by itself is consistent. It is very likely that some subset $O' \subset O$ entails a fact a, while some other subset $O'' \subset O$, $O'' \cap O' = \varnothing$, $O'' \cup O' = O$, entails $\neg a$. Each entailment is logically valid and sound. But the merged set is inconsistent and will lead to false inferences. A simple modal solution is to put a big epistemological **B** in front of each ontology o_i to show that "it is believed by the creator(s) of o_i, that the facts in o_i are true".

With such a framework for distributed belief, the problem of inconsistency in merged ontologies is somewhat solved, but on the other hand, it does not make us any wiser. There exists no calculus that, based on these beliefs, tells us, what is more likely to be a true belief.

The first possibility is to assign a belief value to every belief. Each knowledge bearing agent, i.e. each ontology and its creator(s), can assign confidence values to their beliefs [6]. A calculus allows the evaluation of the overall distributed belief. This, however, requires epistemic honesty on the side of each agent and, in a distributed environment with unreliable sources, may result in wrong conclusions, since "The whole problem with the world is that fools and fanatics are always so certain of themselves, but wiser people so full of doubts" (Bertrand Russell).

Even though it might not be epistemologically correct, we are making the simplified assumption that the complement to belief on the side of the knowledge providing agent is trust on the side of the knowledge consuming agent. Trust, as much as belief, is a state of mind and hence subjective and biased. However, the consumer is in charge of it. As opposed to relying on the opposite side's honesty, the consumer can assign a trust value to sets of concepts. Each of these sets might encompass a complete ontology, a subset of the ontology or single classes or facts. The consumer can now assign trust values of 0.8 to the New York times and the Economist, 0.6 to CNN and BBC, 0.01 to the Enquirer.[2] The same calculus as used for belief in [6] can now be applied for trust. Combining both is also possible and is in the hands of the consumer.

The statement that trust is purely subjective can be relaxed at this point. The statements of other sources can be compared to a trusted source of choice and the trust value can be a function of the jointly held beliefs vs. contradicting beliefs.

4. A framework for logical extensions on the Semantic Web

This section will outline the parts of a possible framework for several multi-valued logics. We will present one example of a combined fuzzy and probabilistic logic and a partial translation into RDF as an extension to OWL.

4.1. *A fuzzy uncertain logic*

Cao and Rossiter [10] describe a fuzzy and probabilistic language for database systems, implemented in the language FRIL++ [3]. In this language, a property is of the form $\varphi[l, u]$, where φ is a fuzzy atom and $[l, u]$ describes the probability range that the proposition of the atom is the case. A class *Dinner* could have a fuzzy probabilistic attribute

[2] One might actually imagine a negative assignment such as in "I trust with certainty 0.5 that everything what the government says is false". Consider as well, that is it not the same as claiming that the opposite is true or believed.

time_of_day(*evening*)[0.8, 0.95], indicating, that there is an 80% to 95% chance that diners are held in the evening. The label *evening* of course identifies a fuzzy set, indicating the times of day roughly between late afternoon and nightfall.

For practical purposes, it is assumed, that a class totally subsumes its subclasses, even though fuzzy logic would allow an overlap. Hence the knowledge engineer needs to be careful when designing a taxonomy of fuzzy concepts.

4.2. *OWL-FP: combining OWL and a fuzzy-probabilistic logic*

From a theoretical point of view it is easy to conceive of a language just like OWL that has a meaningful continuous representation for all sorts of relationships. The beauty of it is, that it can handle standard bivalent facts as a special case in which the class membership value 1.0 and the probability range is [1, 1]. Hence, a reasoner for a fuzzy and probabilistic logic can translate an OWL property into a fuzzy-probabilistic-OWL property. We will exemplify this on material implication:

$$\frac{\begin{array}{l} a \supset b \\ a \end{array}}{b} \equiv \frac{\begin{array}{l} \Pr(b \mid a) = 1;\ \Pr(b \mid \neg a) = [0, 1] \\ \Pr(a) = 1 \end{array}}{\Pr(b) = \Pr(b \mid a)\Pr(a) + \Pr(b \mid \neg a)\Pr(\neg a) = 1}$$

The reverse translation is much harder, as it is always more difficult to map a description in a more powerful language to a less powerful one. If the ontology is meant to be translated, the burden lies on the ontology designer who has to be careful to design the ontology in such a way that a less powerful inference engine than the one the ontology was designed for can still use the contained information. For example, when an ontology that has been designed for a reasoner that processes a fuzzy and probabilistic logic is given to a traditional OWL reasoner, it is desirable that the OWL part of the ontology still be consistent. There are different ways to solve this problem; a simple solution is to just disregard the non-OWL constructs, similar to an HTML parser that just skips non-HTML tags. If a fuzzy assignment made the instance a fuzzy member of two classes that are according to the OWL representation supposed to be disjoint, then it can be made a member of the lowest common super class in the class hierarchy. An ontology designer can make sure that this happens by having this class be the primary class of the instance. For example

```
<X:instance_1>
  <fuzzyOWL:type rdf:resource = "X.1" membership_value = "0.4"/>
  <fuzzyOWL:type rdf:resource = "X.2" membership_value = "0.6"/>
</X:instance_1>
```

In this case, *instance_1* is still an instance of *X*. The assignment of *instance_1* with a membership degree to the subsets can either be discarded or treated as a simple named relationship within OWL.

Given the power that a language like FRIL++ offers for fuzzy and probabilistic reasoning, it is easy to conceive of an extension to the syntax of OWL that allows the representation of fuzzy and probabilistic constraints:

This is an example of a fuzzy and probabilistic attribute. It describes the classes OWL-FP:Property, OWL-FP:Fuzzy_Set as well as the slots OWL-FP:fuzzy_value, OWL-FP:upper_probability_bound and OWL-FP:lower_probability_bound, which, together

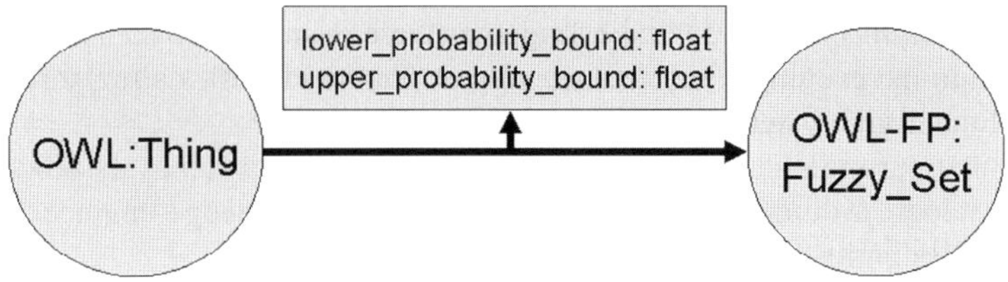

Fig. 3. Example of a fuzzy and probabilistic attribute.

with the domain of Property, describe φ, l and u. Since the second-order attributes of the relationship are limited, it is still computable. The RDF syntax for it is:

```
<rdfs:Class rdf:ID = "Fuzzy-Set">
  <rdfs:label>Datatype</rdfs:label>
  <rdfs:comment>
    The class of all fuzzy sets
  </rdfs:comment>
  <rdfs:subClassOf rdf:resource = "http://www.w3.org/2000/01/rdf-schema#Class"/>
</rdfs:Class>

<rdf:Property rdf:ID = "upper_probability_bound">
  <rdfs:label>upper probability bound</rdfs:label>
  <rdfs:comment>
    upper probability bound, u in [l, u]
  </rdfs:comment>
  <rdfs:domain rdf:resource = "#Restriction"/>
  <rdfs:range rdf:resource = "http://www.w3.org/2000/10/XMLSchema#float"/>
</rdf:Property>

<rdf:Property rdf:ID = "lower_probability_bound">
  <rdfs:label>lower probability bound</rdfs:label>
  <rdfs:comment>
    lower probability bound, l in [l, u]
  </rdfs:comment>
  <rdfs:domain rdf:resource = "#Restriction"/>
  <rdfs:range rdf:resource = "http://www.w3.org/2000/10/XMLSchema#float"/>
</rdf:Property>

<rdf:Property rdf:ID = "has_fuzzy_value">
  <rdfs:label>has fuzzy value</rdfs:label>
  <rdfs:comment>
    for a fuzzy set F and (a, b) in Property P, b is in F
  </rdfs:comment>
  <rdfs:domain rdf:resource = "#Restriction"/>
  <rdfs:range rdf:resource = "#Fuzzy-Set"/>
</rdf:Property>
```

Because of the ugliness of all XML, we forego a further description of the OWL-FP syntax. This example is just meant to show, that it is possible to easily define a language more powerful than OWL, if the underlying semantics are already well defined. Definition is one step. In order to have any sort of semantics, the notion of the fuzzy set and the probability bounds have to be computationally used, of course.

5. Soundness and completeness

Soundness and completeness are seen as crucial properties of a logical inference. Soundness refers to the fact that given true premises, the inferred conclusions are always true within the system. Completeness means that every conclusion that is true in the system can be derived. For many applications however, the property of completeness of the inference mechanism will be secondary to its soundness. Hence the first step will have to be the proof of the soundness of different combinations of logics.

The relaxation of the requirement of completeness also gives rise to the design of more efficient algorithms. Domain independent and domain specific heuristics can be applied to the inference task. Some domain independent heuristics could be "follow the sequences with the higher likelihoods first", "consider the most prototypical members of fuzzy sets first" or the converse rules.

The language proposed by Cao and Rossiter [10] was proven to be sound but not complete. Since for now, we have based our formalism on this language, reasoning with OWL-FP will also be sound, but not complete. Soundness always means soundness with respect to the semantics of the language, not with respect to the "real world". We have seen above that non-monotonic logics need to do global inference in order to give complete results. This, in our case, is equivalent to finding the tightest probability bounds. However, the derived probability bounds are not inconsistent with the tightest ones. This allows us to restrict the reasoning to local inference instead of global. The danger in not finding the tightest intervals is, though, that we might find many meaningless statements, such as $P(a) = [0, 1]$, which means that the probability of a is somewhere between zero and one. This is apparently true, but does not convey much information. It would be an interesting exercise to see whether, if the knowledge base grows indefinitely, all inference would result in meaningless statements.

Relaxing the completeness requirement also raises the question of relaxing the notion of soundness. While it is absolutely necessary to have a sound inference leveling a deductive proof, many tasks cannot be sound by definition. Any sort of inductive or abductive reasoning faces the danger of producing incorrect results. Here, the focus is merely to have a statistical bound on the percentage of errors. Any tool that is meant to assist with the derivation of new knowledge needs human evaluation. It can merely suggest the results of its reasoning and add a confidence measure. In essence, it again depends on the area of application. The difference to traditional models of inductive and abductive reasoning is, however, that the Semantic Web technologies are used for the exchange of data amongst agents and the reasoning across local domains or knowledge bases. A general purpose reasoner need not consider the abductive and inductive cases.

6. Fundamental limitations of the current standard

While it is one problem to clearly define these different modal and uncertain properties, the challenge is to define their actual function by defining operations on these properties.

With the latest versions of SWRL [19], this is possible in the context of the Semantic Web layers themselves. And even though this is not the most efficient way of computing, the ontologies for uncertain reasoning can carry their own rules and functions. In the case of OWL-FP, the use of FRIL++ improves performance.

While for every local use, any sort of extension to OWL can be used to formalize and a proprietary inference mechanism can be used to meaningfully interpret the data, in a global scenario these extensions become meaningless. With "Contextualizing Ontologies" and the C-OWL formalism, [7] point in the right direction. The C-OWL language allows assigning certain parts of remote ontologies as consistent or equivalent, other parts as inconsistent. This allows us to use parts of remote ontologies and to just leave out the inconsistent statements. However, a lot of information that could have been gained from the remote ontologies is lost, because the approach is too rigid. One can picture overlapping concepts A and B, such that $|A \cap B| > 0$, $|A \cup B| < |A| + |B|$, that in this case, since neither $A \supseteq B$, nor $A \subseteq B$, nor $A \equiv B$, the concepts would be inconsistent. For knowledge discovery it is desirable to know about the concept overlap. $A \approx B$ to a degree λ.

In a recent Weblog [37], basically the same question was raised that we raise here. The author's conclusion as to why we need a representation of uncertainty is also very much along our lines: "*Fundamentally because the semantic web is about finding connections*." And most likely we want to find connections across ontologies that have not been found before and declared to be consistent. OWL and bivalent DLs in general lack the basic capabilities of expressing degrees of consistency in a meaningful way.

7. Related work

Even though, with the exception of a probabilistic extension to OWL [11], no XML-style serialization of a more powerful formalism than that of OWL has been published, AI, KR and IR research has brought forward many formalisms that extend FOL in one or the other direction. The extremely powerful Knowledge Interchange Format (KIF) language [43] is an example for a higher-order language that is actually used in ontology creation, but due to its complexity on the one hand and its requirement of locality of the knowledge on the other hand not considered for the semantic web stack. It is known that increasing the expressive power of a KR language causes problems relating to computability and decidability. This has been the main reason for limiting the expressive power of KR languages. The real power behind human reasoning however is the ability to do so in the face of imprecision, uncertainty, inconsistencies, partial truth, and approximation. There have been attempts made in the past at building KR languages that allow such expressive power.

As previously outlined, major approaches to reasoning with imprecision are (1) probabilistic reasoning, (2) possibilistic reasoning [12] and (3) fuzzy reasoning. In [40], Lotfi Zadeh proposed a formalism that combines fuzzy logic with probabilistic reasoning to exploit the merits of both approaches. Zadeh stresses the different semantics of fuzzy sets. For background on this, see also Dubois and Prade [13]. With the *Fuzzy Logic for the Internet* (*FLINT*) initiative, the BISC group presents fuzzy logic as a part of a larger framework including the notions of precisiated natural language (PNL), fuzzy knowledge bases for background world knowledge, etc. as a basis for future question answering systems. [28] present a survey of applications of fuzzy logic in contemporary search engines and challenges for future applications of fuzzy logic in perception based querying and question answering.

Other formalisms have focused on resolving local inconsistencies in knowledge bases; see for instance the works by Blair, Kifer, Lukasiewicz, Subrahmanian and others in annotated logic and paraconsistent logic [5,23]. In [27], Thomas Lukasiewicz proposes a weak probabilistic logic and addresses the problem of inheritance. In [9], T.H. Cao proposed an annotated fuzzy logic approach that is able to handle inconsistencies and imprecision.

The easiest transformations to a Semantic Web language can be done from a Description Logic. Many formalisms have been proposed that immediately extend description logics with uncertain or fuzzy reasoning support. Umberto Straccia has done extensive work on Fuzzy Description Logics, see for example [35] or the excellent survey of existing approaches in [36]. With P-CLASSIC, Daphne Koller and others presented an early approach to probabilistic description logics implemented in Bayesian Networks [25]. Other probabilistic description logics have been proposed by Heinsohn [17] and Jaeger [20]. Early research on Bayesian-style inference on OWL was done by Zhongli Ding [11]. In her formalism, OWL is augmented to represent prior probabilities. However, the problem of inconsistencies arising through inheritance of probability values [27] is not taken into account. Jonathan Rossiter addresses this issue with regard to uncertain class hierarchies in [32]. An interesting probabilistic extension to the Description Logic SHOQ(D), which is very close to the OWL-DL standard, is described in Guigno and Lukasiewicz [15]. The Authors can prove soundness and completeness. An assessment of the computational complexity of P-SHOQ(D) is proposed for future research. A different turn was taken by Wang et al. [38], who provided a non-monotonic extension to the SHOQ(D) Description Logic to be able to process incomplete knowledge on the Semantic Web.

In our previous discussion in [34], we give examples of the applications of soft computing techniques in classical database, IR and Internet tasks. We claim that only with the processing of different types of semantics, the Semantic Web vision can be accomplished. It is concluded, that: "[...] the focus of effort in pursuit of the Semantic Web vision needs to move towards an approach that encompasses all three types of semantics both in representation of the knowledge and the computational methods that support the creation of such knowledge." The three types of semantics considered are semantics derived using statistical inference (implicit), semantics based on formal bivalent logics (formal) and formal soft representations (powerful).

In a very comprehensive essay, Wheeler and Pereira [39] analyze the epistemological problems that were only glanced at in this survey and relate them to non-monotonic and default-reasoning. While we have grossly simplified the semantics of fuzzy and probabilistic logic in our approach of a fuzzy and probabilistic OWL, the authors investigate different semantics of the expression of uncertainty, with regards to truth and belief. Dubois and Prade [14] also stress the different semantics that such modal operators can have and the resulting differences for their computation.

Other researchers look at approaches that are not based on logic. Hence the criteria of rigid soundness and completeness do not apply. The paradigm is rather that the computer should assist the human investigator to arrive at conclusions by showing results that are relevant to previously defined formulas and constraints. Golbeck and Hendler [16], true to the slogan "A little semantics goes a long way", coined by James Hendler, define a model of trust that is logically less rigid. An agent assigns trust values to other agents. In a network of agents, it can be inferred by a transitivity function, how much the first agent would trust an agent that he does not immediately know, but that is a few hops down the network path.

Anyanwu et al. [2,1] present an algorithm to find and rank complex relationships in RDF graphs. Any sound deduction in a logic paradigm can be seen as a path in a graph. In addition to the valid deduction, there are most likely several paths in the graph that would produce invalid deductions. According to a classical paradigm, this is, because parts of the paths just do not entail the rest of the path. But this can also arise due to inconsistencies between two RDF schemes. A path finder does not have to worry about these inconsistencies and can simply follow all paths between two entities. The presented ranking mechanism then decides which paths are more likely to be informative and which are less. In this order the paths are then presented to the user. This, like many graph search algorithms, does not really take edge semantics into account that go beyond specificity and other purely structural semantics. However, what at first sight seems to be a drawback proves to be extremely useful, because it allows fast access to the relevant part of the knowledge base and hence allows us to narrow down the number of statements we have to consider in a rigid logical proof.

8. Conclusions

As so often when it comes to practical applications of great ideas, we find ourselves in a conundrum between possibility and usability. Many of the great technologies that could make everything so easy and perfect cannot feasibly be applied. On the other hand, not taking at least the ideas behind those technologies into account will cause the Semantic Web formalisms to be too weak to actually do what we envision it to do.

In our work we are looking towards combining faster, but incomplete graph-based algorithms with more rigid logic formalisms. For future work we want to use the former to identify important regions of a graph that potentially represents multiple ontologies to narrow down the number of statements we have to do logical inference on. We believe that combining different kinds of reasoning on different kinds of semantics is crucial to make reasoning on the Semantic Web a successful and computationally feasible endeavor [34].

Undoubtedly, nothing can solve all our problems. Not even a powerful formalism for the representation of logic on the Semantic Web. But it is not supposed to. The secret is, as so often, the application of the right tools at the right time and the right place. Whether we need a fuzzy description of our concepts depends on the domain and the degree of accuracy we need in modeling it. Whether we need a probabilistic description or not depends on our certainty and on the randomness or indeterminacy of the domain. As it happens, it all depends on the scope of our application. When two companies want to use ontologies to speed up their legal contracts, it is mandatory that no fuzziness or uncertainties remain. In this case the user wants to be aware of every inconsistency so that s/he can assess the legal implications.

For the task of finding knowledge across different ontologies, it is necessary to allow for the resolution of inconsistencies and partial belief or knowledge. The question whether the representation of this is internal to the knowledge-acquiring agent or external and explicitly available is crucial. If the acquired knowledge is to be used instantaneously and then thrown away, it is up to each agent to deal with it. If it is used to formalize new knowledge that is available to users and artificial agents on the web, it better reflects the degree of uncertainty under which it was gained and the sources from which it was obtained. Otherwise, errors in knowledge representation can propagate through the web. In a situation in which knowledge is still mostly curated by humans, this is a minor problem. In a scenario in which thousands of ontologies are available and used mainly by machines, the restriction to purely bivalent non-monotonic logics will become a major obstacle for knowledge retrieval and inference.

Acknowledgements

This work is funded primarily by Bioinformatics of Glycan Expression, a component of the "Integrated Technology Resource for Biomedical Glycomics", funded by the *National Center for Research Resources*, July 1, 2003–June 30, 2008. We gratefully acknowledge the collaboration with Prof. William S. York in the BioChemistry domain which is the basis of one of the examples discussed in this chapter.

References

[1] K. Anyanwu, A. Maduko, A. Sheth, *SemRank: ranking complex relationship search results on the Semantic Web*, The 14th International World Wide Web Conference (WWW2005), Chiba, Japan, May 10–14, 2005.

[2] K. Anyanwu, A. Sheth, *The ρ operator: discovering and ranking associations on the Semantic Web*, SIGMOD Record **31** (4) (2002), 42–47.

[3] J.F. Baldwin, T.H. Cao, T.P. Martin, J.M. Rossiter, *Towards soft computing object-oriented logic programming*, in: Proc. 9th IEEE Internat. Conf. on Fuzzy Systems, San Antonio, USA, 2000, pp. 768–773.

[4] Bioinformatics for Glycan Expression. Integrated Technology Resource for Biomedical Glycomics: Technological Research and Development Project IV, A Project funded by National Center for Research Resources, http://lsdis.cs.uga.edu/projects/Glycomics/.

[5] H.A. Blair, V.S. Subrahmanian, *Paraconsistent logic programming*, Theoret. Comput. Sci. **68** (1989), 135–154.

[6] L. Boldrin, A. Saffiotti, *A modal logic for merging partial belief of multiple reasoners*, J. Logic Comput. **9** (1) (1999), 81–103.

[7] P. Bouquet, F. Giunchiglia, F. van Harmelen, L. Serafini, H. Stuckenschmidt, *Contextualizing ontologies*, J. Web Semantics **1** (4) (2004), 325–343.

[8] S. Brin, L. Page, *The anatomy of a large-scale hypertextual web search engine*, in: Proceedings of the Seventh International World Wide Web Conference, 1998.

[9] T.H. Cao, *Annotated fuzzy logic programs*, Internat. J. Fuzzy Sets and Systems **113** (2000), 277–298.

[10] T.H. Cao, J.M. Rossiter, *A deductive probabilistic and fuzzy object-oriented database language*, Fuzzy Sets and Systems **140** (2003), 129–150.

[11] Z. Ding, Y. Peng, *A probabilistic extension to ontology language OWL*, Proceedings of the Hawai'i International Conference on System Sciences, January 5–8, 2004, Big Island, Hawaii.

[12] D. Dubois, J. Lang, H. Prade, *Possibilistic logic*, in: D.M. Gabbay, et al. (Eds.), Handbook of Logic in Artificial Intelligence and Logic Programming, vol. 3, Oxford University Press, Oxford, 1994, pp. 439–514.

[13] D. Dubois, H. Prade, *The three semantics of fuzzy sets*, Fuzzy Sets and Systems **90** (1997), 141–150.

[14] D. Dubois, H. Prade, *Possibility theory, probability theory and multiple-valued logics: a clarification*, Ann. Math. Artificial Intelligence **32** (2001), 35–66.

[15] R. Giugno, T. Lukasiewicz, *P-SHOQ(D): a probabilistic extension of SHOQ(D) for probabilistic ontologies in the Semantic Web*, in: Proceedings of the European Conference on Logics in Artificial Intelligence, 2002.

[16] J. Golbeck, J. Hendler, *Accuracy of Metrics for Inferring Trust and Reputation*, in: Proceedings of 14th International Conference on Knowledge Engineering and Knowledge Management, October 5–8, 2004, Northamptonshire, UK.

[17] J. Heinsohn, *Probabilistic description logics*, in: UAI, 1994, pp. 311–318.

[18] I. Horrocks, P.F. Patel-Schneider, F. van Harmelen, *From SHIQ and RDF to OWL: the making of a Web Ontology Language*, J. Web Semantics **1** (1) (2003), 7–26.

[19] I. Horrocks, P.F. Patel-Schneider, H. Boley, S. Tabet, B. Grosof, M. Dean, *SWRL: a Semantic Web Rule Language combining OWL and RuleML*, W3C Member Submission 21 May 2004. http://www.w3.org/Submission/2004/SUBM-SWRL-20040521/.

[20] M. Jaeger, *Probabilistic reasoning in terminological logics*, in: KR, 1994, pp. 305–316.

[21] V. Kashyap, C. Ramakrishnan, C. Thomas, A. Sheth, *TaxaMiner: an experimental framework for automated taxonomy bootstrapping*, Internat. J. Web and Grid Services, Special Issue on Semantic Web and Mining Reasoning, September 2005, submitted for publication.

[22] V. Kashyap, A. Sheth, *Semantic heterogeneity in global information systems: the role of metadata, context and ontologies*, Cooperative Information Systems (1996).

[23] M. Kifer, V.S. Subrahmanian, *Theory of generalized annotated logic programming and its applications*, Logic Programming **12** (1992), 335–367.

[24] W. Kim, I. Choi, S.K. Gala, M. Scheevel, *On resolving schematic heterogeneity in multidatabase systems*, Distributed and Parallel Databases **1** (3) (1993), 251–279.

[25] D. Koller, A. Levy, A. Pfeffer, *P-CLASSIC: a tractable probabilistic description logic*, in: Proc. of the 14th Nat. Conf. on Artificial Intelligence (AAAI-97), 1997, pp. 390–397.

[26] R. Krishnamurthy, W. Litwin, W. Kent, *Language features for interoperability of databases with schematic discrepancies*, in: J. Clifford, R. King (Eds.), Proceedings of ACM SIGMOD Conference, ACM, May 1991, pp. 40–49.

[27] T. Lukasiewicz, *Weak nonmonotonic probabilistic logics*, Principles of Knowledge Representation and Reasoning (KR), 2004.

[28] M. Nikravesh, V. Loia, B. Azvine, *Fuzzy logic and the Internet (FLINT): Internet, World Wide Web, and search engines*, Soft Computing – A Fusion of Foundations, Methodologies and Applications **6** (5) (Aug. 2002), 287–299.

[29] B. Omelayenko, *Learning of ontologies for the Web: the analysis of existent approaches*, in: Proceedings of the International Workshop on Web Dynamics, 2001.

[30] G. Ramakrishnan, S. Chakrabarti, D. Paranjpe, *Is question answering and acquired skill*? in: Proceedings of the 13th International Conference on World Wide Web, 2004.

[31] S. Reed, D.B. Lenat, *Mapping ontologies into CYC*, in: AAAI 2002 Conference Workshop on Ontologies for the Semantic Web, Edmonton, Canada, July 2002.

[32] J. Rossiter, *Fusing partially inconsistent expert and learnt knowledge in uncertain hierarchies*, in: Proceedings of the Third International Conference on Intelligent Data Engineering and Automated Learning, 2002.

[33] A. Sheth, *Changing focus on interoperability in information systems: from system, syntax, structure to semantics*, in: M.F. Goodchild, M.J. Egenhofer, R. Fegeas, C.A. Kottman (Eds.), Interoperating Geographic Information Systems, Kluwer Academic Publishers, 1999, pp. 5–30.

[34] A. Sheth, C. Ramakrishnan, C. Thomas, *Semantics for the Semantic Web: the implicit, the formal and the powerful*, Internat. J. on Semantic Web & Information Systems **1** (1) (Jan–Mar 2005), 1–18.

[35] U. Straccia, *A Fuzzy description logic*, in: Proceedings of AAAI-98, 15th Conference of the American Association for Artificial Intelligence.

[36] U. Straccia, *Uncertainty and description logic programs: a proposal for expressing rules and uncertainty on top of ontologies*, Technical Report 2004-TR-14.

[37] SWAD-Europe Weblog 05/30/04: The semantic web and uncertainty: http://esw.w3.org/mt/esw/archives/000055.html.

[38] S. Wang, S. Zhou, F. Wang, A. Zhou, *N-SHOQ(D): a nonmonotonic extension of description logic SHOQ(D)*, in: J.X. Yu, X. Lin, H. Lu, Y. Zhang (Eds.), APWeb 2004, Lecture Notes in Comput. Sci., vol. 3007, 2004, pp. 912–915.

[39] G. Wheeler, L.M. Pereira, *Epistemology and logical artificial intelligence*, J. Appl. Logic **2** (4) (2004), 469–493.

[40] L.A. Zadeh, *Toward a perception-based theory of probabilistic reasoning with imprecise probabilities*, J. Statist. Plann. Inference **105** (2002), 233–264.

[41] http://platypuswiki.sourceforge.net/, http://en.wikipedia.org/wiki/Semantic_Wikipedia

[42] http://www.vivisimo.com

[43] http://logic.stanford.edu/kif/dpans.html

CHAPTER 2

Fuzzy Ontologies for Information Retrieval on the WWW

David Parry

Auckland University of Technology, New Zealand
E-mail: Dave.parry@aut.ac.nz

Abstract

Fuzzy logic and soft computing have important roles to play in the development of the semantic web. Ontologies represent a means of representing and sharing knowledge between users and systems, and are important tool for information retrieval. However many different ontologies have been established, and communication between them can be difficult. In particular, many terms are located in multiple locations within large ontologies. Different stakeholders in the information retrieval process, "Authors", "Readers" and "Librarians", are identified. This work deals with the concept of the "Fuzzy Ontology", which is designed to allow the representation of differing viewpoints within a single framework. The means of constructing, refinement and use of the fuzzy ontology are described. Fuzzy ontology membership values can be learned from documents or users, based around existing ontologies such as the Unified Medical Language System (UMLS). Use of the fuzzy ontology approach for information retrieval may allow the efficient sharing of knowledge between users and groups.

Keywords

ontology, fuzzy set theory, information retrieval, semantic web, collaborative filtering

1. Introduction

The explosion in availability of documents on the World Wide Web (WWW) has made issues of information retrieval even more important than previously. With the next-generation vision of the "semantic web" [6] the need to support the user in successful searching despite the environment of low-cost, unregulated publishing, that the current WWW provides

FUZZY LOGIC AND THE SEMANTIC WEB
Edited by Elie Sanchez
© 2006 Elsevier B.V. All rights reserved

is more vital than ever. A vision of the semantic web that is able to provide appropriate, reliable information for users requires new techniques to be developed to support users and producers of information.

This chapter describes the concepts associated with "fuzzy ontologies", and argues that such improvements in the representation of knowledge are essential if individuals and groups of users are to gain the benefit of improved indexing and searching tools. Because the web is dynamic, and uncontrolled by any one group, this approach focuses primarily on benefits that may be obtained via representations that do not require modification of the source documents of the WWW. Any scheme that requires that all authors and producers of documents on the WWW subscribe to a strict set of standards in terms of labelling, indexing and structure ignores one of the fundamental basis for the success of the WWW. Currently on the WWW the standards are easy to adhere to and easily parsed by simple browser programs, and there remains a great deal of flexibility in the structure and mark-up of documents, even when they are written to more demanding standards. Even more fundamentally it is a characteristic of human artefacts that their meaning and appropriate classification is not determined entirely by the author, but depends at least partly on the user and context of use. One does not have to be a fully-fledged "deconstructionist" to realise that documents are used to prove the opposite of what the author intended, or that the significance of particular documents may change with interpretation, time and the requirements of the reader.

This chapter is organised to introduce to the reader some issues of information retrieval on the WWW, along with some of the approaches taken to improve the searching process, in particular the use of fuzzy searching techniques and the issues of probabilistic relevance. It then outlines the advantages of ontologies for information retrieval. The concept of the fuzzy ontology is then introduced, and mechanisms for fuzzy ontology development are outlined. The examples given generally relate to the medical information domain and in particular the field of obstetric information.

A "Fuzzy Ontology" can be defined as an ontology organised such that each object within the ontology only occurs once, but with each object having a relation to every other object in the ontology. Each relation has a membership value associated with it.

Each object relationship has a fuzzy membership value (μ) between 0 and 1 for each of the objects involved where the sum of each μ adds up to 1 for each object. Most potential relations will have a $\mu = 0$.

$$\sum_{i=1}^{n} \mu_i = 1.$$

This compares to a crisp ontology where relations either exist ($\mu = 1$) or do not ($\mu = 0$) but the sum of μ can be greater than 1, that is, objects may have multiple relations between each other with equal weight. This makes the mapping between terms and locations in an ontology ambiguous when using the ontology for query expansion and refinement.

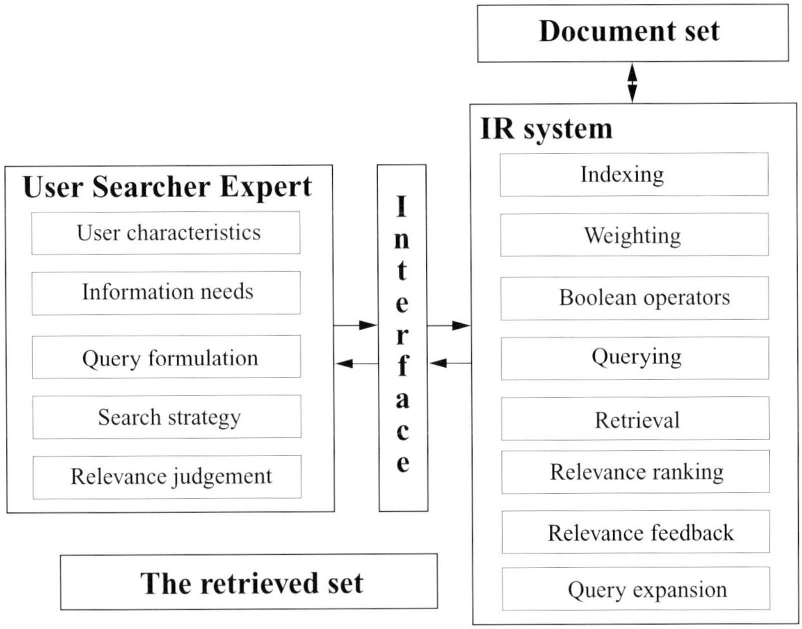

Fig. 1. Aspects of the IR process from [29].

2. Information retrieval and the semantic web

2.1. *Information retrieval issues*

An excellent sourcebook for issues surrounding information retrieval problems is the textbook "Modern Information Retrieval" [4].

Searching for information can be an extremely complex task. Many searches performed are inadequate or cursory as [24] show in the context of the ERIC database. Successful searching relies on a high level of cognitive effort, for example by using the techniques of critical reflection [12]. However there is data to support the idea that users are loath to spend more than 30 minutes learning a system such as a catalogue [9].

In order to build a successful system for information retrieval it is necessary to have an understanding of the process from both the systems, and the users point of view. Classic texts such as [25] and more modern works such as [5] have covered both areas. Logs alone will not provide this information or guidance.

Kagolovsky's [29] view of the IR process is given in Figure 1.

The large number of attributes assigned to both the user and the system, emphasise the fact that determining the success or otherwise of a particular searching system does not rely entirely on relevance scores. [25] does give a useful summary of some parameters that may be important for assessing an information retrieval system. These are shown, with comments in Table 2.

For a given set of documents, if the retrieved set of documents is given by B, and the relevant set of documents is given by A then the following Table 1 is used:

Table 1
Parameters for calculation (from [58]).

	Relevant	Non-relevant	
Retrieved	$A \cap B$	$\overline{A} \cap B$	B
Not retrieved	$A \cap \overline{B}$	$\overline{A} \cap \overline{B}$	$\overline{B}$
	A	$\overline{A}$	

Table 2
Attributes of information sources, adapted from [58].

Attribute	Description	Comments
Coverage	The degree to which the collection being searched includes material of use	With multiple information sources being queried, this is now extremely broad
Time lag	The time taken for the search to take place	Largely a function of bandwidth available and activity at search sites
Presentation	The form of the output	The use of HTML and PDF has standardised the format, although not the appearance of returned results
User effort	The work expended by the user in obtaining the results	Covers the time spent formulating queries, as well as the effort in thinking about them, the filtering of long lists of retrieved items is especially trying
Recall	The proportion of relevant material in the collection actually retrieved in answer to a search request	$= \dfrac{\|A \cap B\|}{\|A\|}$
Precision	The proportion of retrieved material that is actually relevant	$= \dfrac{\|A \cap B\|}{\|B\|}$

This problem is not trivial. The last of Swanson's "Postulates of Impotence" [55] says that "In sum, the first eight postulates imply that consistently effective fully automatic indexing and retrieval is not possible. The conceptual problems of IR – the problems of meaning – are no less profound than thinking or any other form of intelligent behaviour" (page 558). Assuming that he is correct, then the correct role of any system that aims to contribute to IR is one that supports the user, and in particular the cognitive and metacognative processes, rather than trying to replace them.

2.2. *Relevance*

Relevance is a key concept in information retrieval. The complete definition of relevance continues to elude the discipline, but one definition of relevance is the degree to which the retrieved document set matches the users requirement for information. Obviously, there are some imponderable issues here, the following diagram (Figure 2) may be helpful:

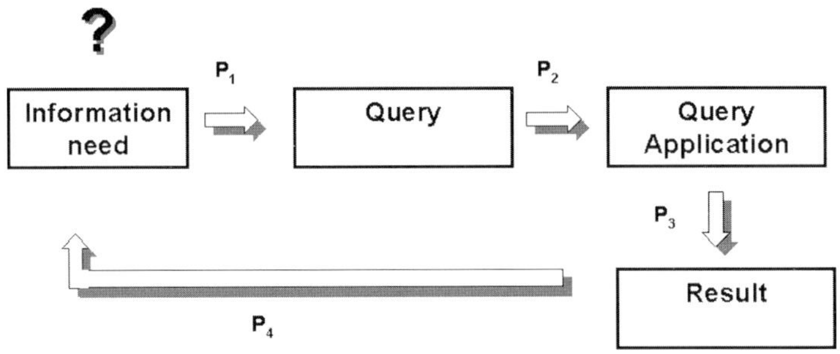

Fig. 2. Information need satisfaction.

Each process, $(P_1..P_4)$, contributes to the overall efficiency of the search for information.

P_1 – The query formulation process.
P_2 – The parsing of the query by the searching device.
P_3 – The database selection and searching process.
P_4 – Result presentation.

A large number of different methods have been proposed for determining the quality of the relevance of a querying system. The most commonly used method in the information retrieval community may be the F-measure, described by [58]. The F-measure is calculated using the precision (P) and the recall (R):

$$\text{F-measure} = \frac{2 \times P \times R}{(P + R)}.$$

This approach is extremely useful when assessing the quality of information retrieval schemes on tagged test data (for example the TREC dataset [39]). However, this is a purely binary view of relevance, either a document is relevant or it is not, degrees of relevance have no meaning in this scheme. Difficulties occur when the relevance is not known in advance for a testing set or when the degree of relevance for a particular document is required.

Kagolovsky and others have proposed that rather than considering the user as a person with an information need to be satisfied, they may be thought of as being in a "anomalous state of knowledge" (ASK) [30]. This approach is shown diagrammatically in Figure 3.

This reflects the fact that the extent of the information need may not be clear to the user when they begin their search. This concept is related to Swanson's notion (see above) that the information need is hard to measure at the start of the process, and only becomes clear at the end, when it has been satisfied. This approach is particularly relevant when considering the actions of users searching for information where there is a lack of information as to what is a reliable or authoritative source, or even an understanding of the organisation of the information. It is important to realise that information seeking is an aspect of learning, rather than an isolated task, and thus approaches that increase relevance quality for particular queries, may not be as successful as those that support the information retrieval process as a whole.

An outline of some of the roles traditionally associated with documents is shown in Figure 4.

Early database systems used exact matching techniques for queries, based on Boolean queries. Issues with this approach include the difficulty of formulating vague queries, and

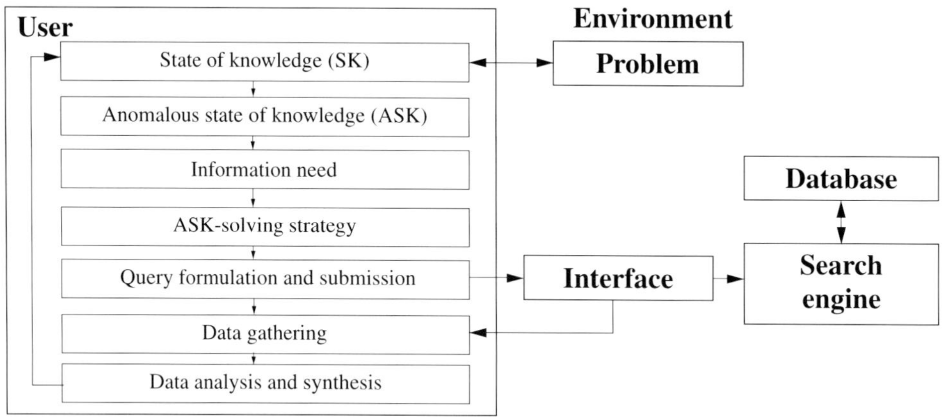

Fig. 3. The anomalous state of knowledge in IR from [30].

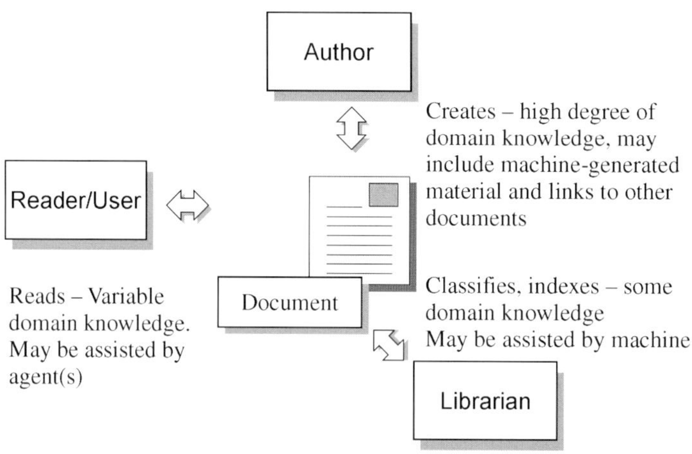

Fig. 4. Roles in document production, storage and use.

issues with dealing with terms which are not indexed. The necessity for the user to know in advance the index terms used by the system, even in cases where "Full-text" search is available adds another hurdle. In systems, which are searching an open space of discovered documents such as the WWW, rather than a constrained digital library with a limited set of index terms, such approaches are often not successful. Similarly, database queries and information retrieval queries may appear similar, but the difference often lies in the greater control database managers have over the content of their database. User queries based on Boolean searches are often poorly formulated, with little understanding of the use of so-called "advanced search" pages, and little desire to expend effort using them [18]. In these approaches, the librarian of Figure 4 effectively controls the searching process.

More sophisticated models are more commonly used for matching queries to documents. The vector space approach [49] calculates a vector for each document according to its content of weighted index terms contained within it. Generally the weight of each term reflects its relative rarity, and so-called "stop words" that occur very commonly are not included. The query is then represented in vector space in a similar fashion, and the degree of similarity between the query vector and each of the document vectors can produce a relevance score.

Other approaches that have more recently developed include citation-based systems designed to use the presence of hyperlinks between hypertext documents. The most famous of these is probably the "Pagerank" algorithm [44], the basis for the successful "GOOGLE" search engine. Citation analyses, assign weights to documents on the basis of citations pointing to them including index terms, effectively using the reputation of the document as a measure of its authority and hence relevance. This approach begins to include the authors of documents, as shown in Figure 4, and even the readers, when they become authors themselves and include links in their publication. The low cost of publishing on the WWW encourages large numbers of authors to publish who would have been unlikely to publish in traditional media.

2.3. *Query formulation*

Query formulation is often particularly difficult for users that do not have training in information retrieval. This is an old problem [9] that has continued despite improvements in interfaces and "computer literacy" [10]. Essentially the message of Borgman is that understanding of the underlying principles of the organisation of knowledge databases including for example the use of index terms and Boolean logic is necessary to use them efficiently. However users are reluctant to undergo the necessary training for this, indeed training for longer than half an hour is seen as excessive. In addition, the rapid rate of change of interfaces, and the wide variety of interfaces used have made the users job harder. Problems that continue to occur include:

- Choice of index term
- Expansion of terms
- Combination of terms

Of particular interest are systems like Inquirus2 [19,20], that are designed to answer particular clinical queries rather than act as a general-purpose search tool. The drawback of these systems is the degree to which constant editing and updating is required.

Another approach uses the traditional IR approach of allowing complex Boolean statements to be entered into the system. However, examination of the logfiles of search engines reveals that the use of "advanced" search tools comprises less than 10% of all queries [52], and some of these are thought to be errors. Indeed most queries are made using one or two words with an implicit "AND". [61] has found that the mean number of words in each query did not change between 1997 and 1999 and remains at 2.4. Some systems such as "Ask Jeeves" and GOOSE [34] attempt to parse natural language queries and construct queries from them in order to avoid the problems users appear to have in constructing efficient queries. GOOSE is interesting because it attempts to build a model of what the user 'means' – by using the repository of 'commonsense' as part of its parsing. Other methods use quite simple parsing for the identification of potential keywords and in the case of "Ask Jeeves", by attempting to match the user's query to a library of preformulated questions. However, the argument that users are to blame in effect for the perceived poor performance of search engines and that better education would improve this appears to be contradicted by the results of the work of [16]. These authors took a selection of queries that had been logged in the Excite search engine that used Boolean operators. They then ran these queries on a number of search engines (AOL, MSN and Google), with or without the Boolean operators included. They then examined the retrieved documents for relevance and ranking. Somewhat surprisingly the results showed little difference between queries using Boolean linkages, and those that did not. Before discarding the use of information retrieval training however, it should be remembered that search engines often parse queries before executing them – for example GOOGLE may include implicit "AND" statements. Crucially for this work the authors acknowledge that this work has been done on general-purpose search engines, with quite commonly used, and certainly non-technical, terms. Obviously any successful search engine in general use will be optimised to satisfy likely queries, and may indeed use precalculated results for queries that it judges to be similar to the current one, rather than performing a Boolean operation at all. The case of technical terms, and rarely used query terms may be very different, and sources of information other than search engines – such as PubMed – still use Boolean approaches.

MEDLINE currently contains various filters for terms such as 'clinical trial' and 'review'. These filters have become increasingly sophisticated and a study published recently explains some of the background to this process [60]. In order to assess the best way of returning results related to the concept of 'prognosis', the authors examined large numbers of journal articles by hand. Articles were then rated for quality and relevance to the goal concept. Searches were then performed on MEDLINE and the specificity, sensitivity, accuracy and precision calculated. The combination of search terms that were required to get high, equal levels of sensitivity and specificity were derived from candidate terms and combinations of terms obtained from experienced MEDLINE users. The aim of this work was to allow a search filter to be constructed, using search terms, that would identify methodologically high quality articles to be retrieved. In practise the filters should be used in conjunction with other, clinically relevant terms in order to obtain better accuracy. A similar article was published in the [23], outlining the changing precision, sensitivities and specificities of searches, from MEDLINE using various terms. The information in Table 3 is derived from this information.

If U are the articles in the database, and R are those retrieved by the chosen search strategy where U_{high} are high quality articles and U_{low} are low quality articles and R_{high}, and R_{low} are corresponding classes for the retrieved articles then – the following table describes the values calculated.

Table 3
Derivation of terms regarding effectiveness of searching.

Measure	Meaning	Calculation
Sensitivity	Proportion of high quality articles retrieved	$= N(R_{\text{high}})/N(U_{\text{high}})$
Specificity	Proportion of low quality articles not retrieved	$= (N(U_{\text{low}}) - N(R_{\text{low}}))/N(U_{\text{low}})$
Precision	Proportion of retrieved articles of high quality	$= N(R_{\text{high}})/N(R)$

However, although sensitivities and specificities of particular searches can go into the 80%+ range, the low precision of such searches (around 10%) may present the greatest barrier to usage. The reason for this is neatly explained in the article [3], which introduces the concept of "number needed to read" (NNR), in analogy to the concept of "number needed to treat" (NNT). NNT is a familiar concept from Evidence Based Medicine [48] and refers to the number of patients you would need to treat with a particular intervention in order to gain a certain benefit for one patient. It is used particularly in the evaluation of preventative or screening medical interventions [2]. The NNR is calculated as 1/precision, and calculates the number of articles or abstracts that would be read in order to obtain one good one. For 10% precision this number is 10. This raises particular problems for clinical users of information systems. The typical "first page" presentation of abstracts in a system such as MEDLINE will have between 1–3 abstracts that are of high quality, even when a filtering system is in place, with reasonable sensitivity. Bachmann's work emphasises that with less sensitive searching, the precision rises, but the important point is made that such a filtering system may well not be randomly excluding articles, but may do so with a particular bias, which implies that the result set may not reflect the true picture. One can imagine such a situation in particular, where a term begins to gain a more or less specific meaning over time, a strategy based around searching for such a term may emphasise documents produced at the point of the terms most popular use. An example may be the term WEB, where the combination of WEB + computer may be a particularly useful one before 1994 for discovering computer simulations of spider web constructions, but not so after that date. A similar case may exists with obsolete diagnoses or deprecated terms in medical searching. One of the interesting aspects of the work of [60] is that search techniques based around formal traversing of the MeSH tree may not be as successful as the use of text words combined in a more empirical way. Thus the work done in order to avoid systematic bias by the use of specific search terms (such as the specialist lexicon), while valuable, may become somewhat irrelevant if only applied to keyword-based query expansion and refinement.

2.4. *Ontologies to support information retrieval*

Ontologies have been defined as "a way to express formally a shared understanding of information" [41, p. 60]. Another definition has been given by Gruber "An Ontology is a specification of a conceptualisation" [21] – this paper also first described "Ontoligua", one of the first approaches to the specifications of ontologies.

Web-based ontologies can be seen as a part of implementing the "semantic web". The Stamford medical informatics group suggest their language – PROTÉGÉ-2000 [41] as a means of representing ontologies on the web. The central role of ontologies in knowledge management is emphasised by [53].

Ontology acts as guide to the relationship between concepts in a domain, which may or may not themselves represent physical objects. By the definitions above, it may be a little uncertain as to what an ontology is useful for but Gruber would assert that an ontology is useful for purposes of knowledge sharing and reuse. That is, by understanding the relationships between objects then the interaction between objects, the operations that can be performed on them, and the appropriate position of new objects becomes easier. Tacit ontologies seem to exist in everyday life, as people generally appear able to share knowledge of relations in normal activities.

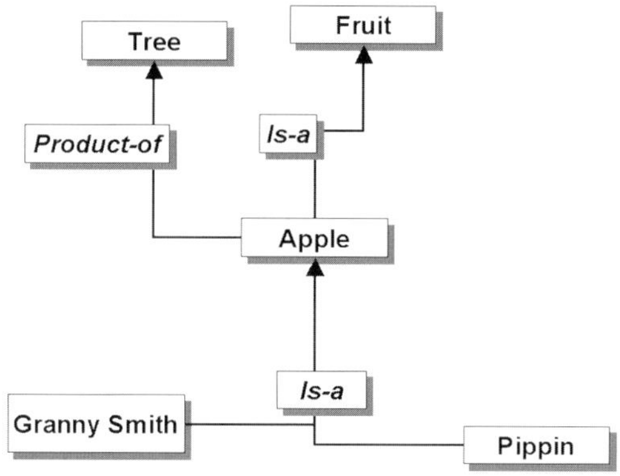

Fig. 5. A simple ontology.

A simple example of an ontology is shown in Figure 5. In terms of information retrieval, the ontology can be used to assign relevance scores to documents on the basis of the relations described. For example, a query for "Apples" may recover items that include "Granny Smith" or "Pippin" even if the term "Apple" does not appear in the document or document index at all.

Programming computer-based systems to understand and share these relations in a flexible and efficient manner appears to be more difficult. However, the problem of whether different sources or domains of knowledge should have their own ontology, or whether all knowledge can be fitted into one ontology is still debated. In practise, multiple ontologies exist, and much current work is devoted to combining them, for example the work of [43]. Many ontologies have extremely rich sets of relations available, e.g., Unified Medical language system (UMLS) [40], WordNet [50] and PROTÉGÉ-II [38]. In some domains, for example medical informatics, ontology construction and use is seen as a central task [37]. In many ways an ontology is similar to an XML document, or a class in an object-orientated programming language [17]. The representation encodes a hierarchical structure, with inheritance of properties from root to branch, with additional attributes at each level. The advantage over the fixed hierarchy of objects is that such ontologies can be modified with the inclusion of new branches or leaves, and the location of such a new item can give information about it even to users that are not aware of the new terms used. In addition, the wide range of relations available allow a deeper representation of the arrangement of knowledge.

The ontology approach has obvious similarities to the concepts of object-orientated programming (OOP), both use hierarchical collections of classes for example, but as Noy points out [42], ontologies concentrate on the structure of a class, whereas OOP concentrates on the operation of a class – OOP is about what a class does, ontology is about what it is. Both have the emphasis on constructing structures that can be reused and crucially, are understandable, translatable and limited in scope.

Systems for implementing ontologies in order to implement the "semantic web" have gained increasing prominence recently. In should be emphasised that with the increasing use of document management and knowledge management systems, the "semantic web" implicitly includes intranet as well as Internet applications. The problems of identification of the meaning of documents and other objects exist within systems as well as between systems and this explains some of the interest in tools such as XML for system development.

The UMLS system includes an approach to semantic indexing. In the UMLS concepts are identified which may be expressed as strings or words. Synonyms for these terms are stored, so for example a particular concept – "Asthma" has a concept code – strings such as asthma Asthmas etc. are assigned to the same concept. This approach is obviously useful to identify roots, and allow systems to be used in different languages. The ontological relationships from UMLS then allow these concepts to be related to other concepts.

A standard for the expression of ontologies on the WWW has recently been developed. This standard – Ontology Web Language (OWL) [51] allows ontology relations to be coded in a machine and human readable form within files than can be located on the WWW.

An example of an OWL statement is shown below in Figure 6 – the example is taken from [51].

```
<owl:Class rdf:ID=''TexasThings''>
 <owl:equivalentClass>
  <owl:Restriction>
   <owl:onProperty rdf:resource=''#locatedIn''/>
   <owl:someValuesFrom rdf:resource=''#TexasRegion''/>
  </owl:Restriction>
 </owl:equivalentClass>
</owl:Class>
```

Fig. 6. Owl example.

This example shows the definition of an equivalent class – items that satisfy the conditions given in the equivalent class construction are regarded as equivalent. In this particular example, things from Texas are assigned to be "Texas things". This project is still at a fairly early stage, but obviously represents a potentially beneficial approach to standardisation that may be useful for the deployment of the fuzzy ontology described later.

Systems for encoding semantic information do not have to rely on XML or other general-purpose systems. An extremely general-purpose approach to adding semantic information to documents is the WorldNet approach described in [36]. In WorldNet, there are a number of keywords, each of which has a large potential set of attributes, such as synonyms, hyponyms, related terms, etc., which are used for searching. The authors describe a successful test of this approach on the Cranfield data set. Unfortunately this approach still includes difficult problems – such as disambiguation and word root identification. As an approach focused on one particular problem – retrieval of text documents – it may be that it is initially simpler than efforts to encode and transmit meaning such as XML and RDF.

A number of systems have been designed to allow communication between ontologies – for example XOL, an ontology exchange language implemented in XML and developed from Ontolingua [31]. The Ontolingua server [54], allows online collaborative editing of ontologies, and the Knowledge system laboratory is also the home of a library of ontologies – for example the enterprise ontology and others.

In terms of the information retrieval process, ontologies allow controlled query expansion and refinement, a means of assigning users preferences, and establishes links between retrieved documents as described by [14].

In a more general sense, the use of ontologies allows a matching between the ontological structure of the users information need and the ontological structure of the documents that may satisfy that need [22]. However, this still assumes that the structure of the users information need can be assumed, and this may not be correct.

2.5. *Uncertainty in information retrieval*

Aside from relevance scores derived from the vector space model, other probabilistic approaches have been used. A wide variety of these techniques are described in [15]. Much of this work deals with methods for calculating relevancy scores, and most use some application of Bayes theorem [47]. If $\vec{x}$ represents a particular document vector, and R is the relevance then:

$$P(R \mid \vec{x}) = \frac{P(R) \cdot P(\vec{x} \mid R)}{P(\vec{x})}$$

where $P(R)$ is the prior probability of a relevant document being recovered from a particular set, $P(\vec{x})$ is the probability of a document being represented by the document vector $\vec{x}$, and $P(\vec{x} \mid R)$ is the probability of a particular document having relevance R. Thus the probability that a particular document is relevant to a particular query is related to the closeness of the document and query in vector space, and the degree to which such document vectors commonly occur.

These models have shown a good deal of success. Other approaches have refined these models by introducing weightings related to the location of terms – for example the work [32].

Because of the nature of the information retrieval process it is likely that the degree of uncertainty or vagueness in queries will alter as the process continues.

2.6. *Fuzzy search*

Fuzzy search in databases and information retrieval has been popular for some time, with extensions to the SQL command set being discussed in [11,26], and in more detail in later publications such as [27]. The advantages of fuzzy search include; more natural mapping of term combinations in queries ("somewhat related to", "strongly related to") and greater meaning assigned to relevance scores ("a little relevant", "very relevant"). Zadrozny and Kacprzyk have also implemented fuzzy searching in an access environment as the Fquery interface [64]. The challenge of using fuzzy querying on the web is also gaining increasing attention [28].

A recent approach is described in [59]. In this paper, the authors use a fuzzy ontology to represent the relationship between terms. Effectively it uses fuzzy search techniques, as opposed to the more traditional Boolean or crisp logic, but the authors realise that such an approach can also be represented as an ontology. This can be compared to the MeSH system described above, where the basic relation is a crisp "is-a", but other relations such as contains, is a synonym etc. are also included. In fact, the full UMLS relationship model is more complex and this is described below. The Widyantoro method envisages two relations, *Narrower than* (NT) and *Broader than* (BT).

The advantage of this method is that the membership functions can be calculated by using the frequencies of the occurrences of the terms being used. The relations are simple, but they can be calculated automatically. In addition, by calculating degrees of membership automatically, this method allows automated trimming of the trees produced, by combining the membership values of the elements, calculating the overall membership value and then rounding, so that membership values greater than 0.5 become true and those 0.5 or less become false. False relations are then removed from the final tree.

More sophisticated representation than merely NT and BT are also possible. Takagi [56] describes the use of associative memories to allow assessment of objects by potentially large numbers of different criteria. However this has the disadvantage that calculation of the membership function is not obvious and may become complex. In particular the practical difficulties involved in defining which aspects should have fuzzy values, and which characteristics of the objects should be chosen to contribute to these membership values remains problematic. Essentially, it is very important that the end-user should be able to understand how the system has assigned values, and be able to modify the result by using their own criteria.

3. Fuzzy ontology

The fact that ontologies is a plural raises the major difference between the philosophical and computer science approach to the term. A philosophical ontology would encompass the whole of the universe, but computer scientists allow the existence of multiple, overlapping ontologies, each focused on a particular domain. Indeed an understanding of the ontology of a particular domain may be crucial to any understanding of the domain. The combination of ontologies, and communication between them, is therefore, a major issue within computer science, although such issues are problematic with the philosophical use of the term. At the limit, an ontology that perfectly expresses one persons understanding of the world is useless for anyone else with a different view of the world. Communication between ontologies is necessary to avoid this type of solipsism.

Similar issues also influences the scope and coverage of ontologies. As summarised in a recent discussion in IEEE expert [13], ontologies can attempt to represent the maximum amount with the minimum complexity – a "Newtonian" approach, or take the "Liebenizian" route where the ontology represents the complexity inherent in the different objects covered by it. Moving to a practical level, much of the complexity of relations in systems such as Ontolingua is due to the need for machines to be able to reliably use the systems for classification or comparison. If an ontology is primarily used by humans, then some of the more complex and specific relations may not be required. Ultimately, if an ontology is dealing with human-defined or subjective terms – for example particular sensations or observations, then a great deal of precision and complexity in the ontology relations may give a misleading view of the precision of the description of the objects included in the ontology. At the same time, ontologies may be able to staddle the differences between the Newtonian and Liebenizian views by having different levels of representation for different purposes, this may involve the merging of classes and relations for a broad

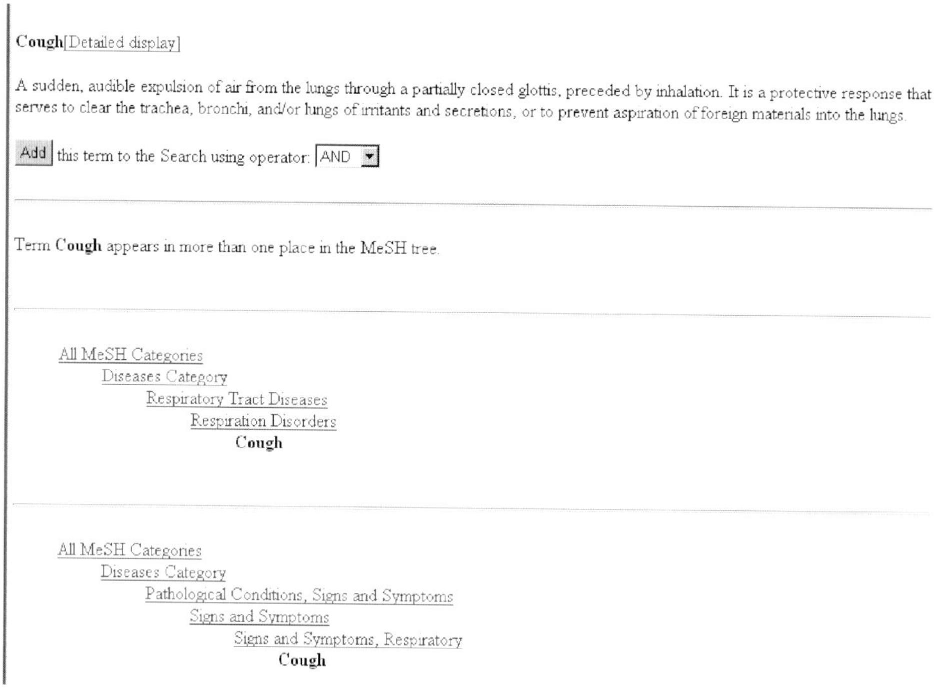

Fig. 7. The UMLS entry for "cough".

view and the dissection and sub-classification of the same domain when greater precision is required. This issue represents one of the main drivers for the development of the fuzzy ontology, by setting the membership values appropriately such a transformation can be made.

3.1. *Rationale*

Useful ontologies can include a very large number of terms. If one considers UMLS MeSH as an ontology, then it is a particularly useful one for the medical domain. However, of the 21836 terms within it, 10072 appear in more than one place. An example of the UMLS entry for "cough" – a term that occurs in multiple locations is shown in Figure 7. Thus the "overloading" of terms in a mature ontology can be seen to be significant. In addition, as ontologies are extended – for example outside their home domain this problem is likely to get worse.

In his fascinating textbook "Linguistic Categorization" [57], Taylor discusses another complicating factor, that of polysemy. According to WordNet polysemy, also known as lexical ambiguity, refers to "the ambiguity of an individual word or phrase that can be used (in different contexts) to express two or more different meanings" [50]. Taylor gives a number of examples that show the subtlety of differences of meaning when polysemous terms are used in everyday sentences. This is mostly the domain of natural language processing and as such largely out of the domain of this work, but a number of issues are relevant.

Table 4
Multiple Occurrence Example – "Pain".

Term	Concept ID	Parent	Depth	Root term
Pain	G11.561.796.444	Sensation	4	Musculoskeletal, neural, and ocular physiology
Pain	F02.830.816.444	Sensation	4	Psychological phenomena and processes
Pain	C23.888.646	Signs and symptoms	3	Pathological conditions, signs and symptoms
Pain	C23.888.592.612	Neurologic manifestations	4	Pathological conditions, signs and symptoms
Pain	C10.597.617	Neurologic manifestations	3	Nervous system diseases

Firstly there are many common polysemous terms, and they have considerable impact on the meaning of sentences. Secondly, dictionary definitions do not always distinguish effectively between these meanings, and the same may apply to ontological use. For example using the term "pain", the sentences "Using a computer can be a real pain", and "If I use my computer without a break, I end up in pain" have different meanings as sentences. However, the appropriate location of the term "pain" in the ontology – (see Table 4) – is probably G11.561.796.444 in both cases. The ontology does not reveal whether the term is being used metaphorically or not.

This makes searching difficult as any broadening of an initial particular term may expand an inappropriate set of related terms, which occur in a different part of the ontology. WordNet [50] is a good example of a system that has attempted to overcome this issues by including definitions, synonyms, hypernyms ("x is a kind of", i.e. objects higher on the ontology), hyponyms ("is a kind of x", i.e. objects lower in the ontology), holonyms ("x is a part of", i.e. dependent child of another term), meronyms ("parts of x", i.e. dependent children of this term). Along with examples of the way the term is used – contextualisation – WordNet allows users to identify the meaning of a particular word. For terms in a language with an accepted syntax this is fine. The problem occurs when the term has a number of meanings for different people, this is especially prevalent in the case of people searching for new information, where their own understanding, and knowledge of key terms in a field is limited, or coloured by their understanding of previously encountered terms.

Table 4 demonstrates how these issues arise in existing ontologies. "Pain" occurs in five locations in the MeSH Ontology – see Table 4. Because the term is located in a number of different places, query expansion for this term is difficult, because there are wide numbers of "related" terms. In the case of Pain for example, a fairly standard expansion using the immediate parent, and the immediate "offspring", i.e. terms below Pain in the ontology yields the following potential expansion of 5 Parent terms + 19 Child terms, giving a total of 24 potential expansions, where an average of 1 parent term and 7 child terms would be needed if a correct term location was known at the start (there is some overlap between children of the different original terms). A simple expansion that does not understand the intended location of the query term may lead to many irrelevant results being returned, as the user presumably has one particular meaning of "Pain" in mind.

3.2. *Description of fuzzy ontology*

The concept of a fuzzy ontology has been introduced in [46]. The fuzzy ontology is based around the concept that each index term or object is related to every other term (or object) in the ontology, with a degree of membership assigned to that relationship based on fuzzy logic as introduced by [63]. The fuzzy membership value μ is used where $0 < \mu < 1$, and μ corresponds to a fuzzy membership relation such as "strongly", "partially", "somewhat", "slightly" where for each term;

$$\sum_{i=1}^{n} \mu_i = 1$$

where n is the number of relations a particular object has, where $n = (N - 1)$, with N representing the total number of objects in the ontology. That is, each term used in the system has the total membership value of its relations as a value of 1 summed over each dependant relation. This rule is not commutative, for the relationship between two objects, A, B:

$$A \xrightarrow{related\_to} B \ldots \mu_{AB}$$

$$B \xrightarrow{related\_to} A \ldots \mu_{BA}$$

$\mu_{AB} > \mu_{BA}$ or $\mu_{AB} < \mu_{BA}$ or $\mu_{AB} = \mu_{BA}$ are all possible.

An example, Figure 8, may make this clearer, each μ represents the membership value of the relationship from the Apple to tree, fruit and Computer Company. Any relationships not shown, e.g., between IPOD and pippin, are assumed to have a $\mu = 0$, relationships directed to the Apple do not have μ values shown for clarity.

The apple can be a product of a tree (to distinguish an apple tree from another kind of tree); it can be a fruit (if the query relates to distinguishing between fruit, for example

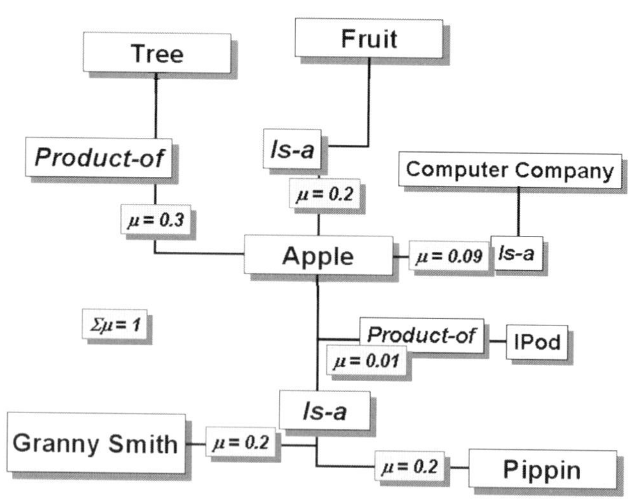

Fig. 8. A fuzzy ontology scheme.

apple pie, cherry pie) or a computer company. The fuzzy ontology applies membership values to each of these possibilities; depending on how likely it is that a particular relation is required. These different relations will have different membership values depending on the context of the query, and particularly the user's view of the world.

The ontology can be seen as an acyclic directed graph, with nodes representing index terms or objects, and edges representing the relations between them. In conventional ontologies, particular objects may occur in multiple locations, leading to ambiguity when being used for query construction. In addition [7], unwanted cyclic relationships may occur. In the fuzzy ontology all nodes can be linked to all other nodes, but with the addition of a membership value for that edge, where the "offspring to parent" type of relations are assigned membership values associated with the offspring – noted as arrows away from the object (Apple) in question in Figure 8. Unwanted relationships can be assigned zero membership values.

Ontologies have been devised for many domains and a number of systems have been designed to allow communication between ontologies – for example XOL, an ontology exchange language implemented in XML and developed from Ontolingua [31]. The Ontolingua server [54], allows online collaborative editing of ontologies, and the Knowledge system laboratory is also the home of a library of ontologies – for example the enterprise ontology and others. Ontologies or hierarchies have also been constructed from sets of documents [32], and used for recommender systems [1,35]. Particular issues arise from the use of multiple ontologies, and the fuzzy ontology is suggested in order to allow a common framework, or base ontology, with different membership values associated with different users and groups. It should be noted that because of the learning methods involved, only "is-a" type relations are currently used, based on the currently existing MeSH hierarchy.

4. Methods of learning fuzzy ontologies

4.1. *Learning from documents*

Documents are particularly useful sources for learning large numbers of relations in a fairly short time. Before the process begins, each term which exists in multiple locations, is given an equal and proportionate membership value (i.e. if there are two locations, each would have an initial membership value of 0.5). In this experiment the support for each term at each location within each document $S(C)$ was calculated by:

$$S(C) = K \sum_{i=1}^{n} \frac{1}{|D|_i} \Big/ L$$

where K is a scaling constant, n is equal to the number of relatives of the particular location being tested, D is the distance in character space from the start of each relative term and L is the total length in characters of the document. This approach was taken in order to assign more weight to terms that are close to the target term in the document, and avoid long documents automatically weighting terms more heavily. Support is related to the membership value of a relationship by calculating the total support S_{tot} of a particular term in the entire set along with the support S_{AB} of a particular relationship A with B.

$$\mu_{AB} = \frac{S_{AB}}{S_{\text{tot}}(A)}.$$

Table 5
"Public health" membership values.

Concept ID	Support	Root term
N01.400.550	4.6 × 10-3	Population Characteristics
G02.403.776.630.560	4.7 × 10-9	Health Occupations
G03.850	4.6 × 10-6	Environment and Public Health

If and only if $S_{\text{tot}}(A) = S_{\text{tot}}(B)$, then as $S_{AB} = S_{BA}$, then $\mu_{AB} = \mu_{BA}$.

With suitable scaling the fuzzy ontology membership value can then be modified using:

$$\mu_{\text{New}} = \mu_{\text{Old}} \pm \left(\sqrt{(\mu_i - \mu_{\text{Old}})}^2 / Q_{\text{Hist}}\right)$$

where

$$\mu_i \approx S(C).$$

Where the new membership (μ_{New}) is determined by the old membership (μ_{Old}) the membership calculated for this query (μ_i), and the number of queries that have confirmed the intended meaning of this term (Q_{Hist}). Q is used as a denominator in order to reduced the effect of later learning in order to stabilise the values. For this experiment, only the support values were calculated which are equivalent to (μ_i), depending on the choice of K. Normalisation or the membership value is necessary to avoid multiply located terms being given undue weight.

Relatives of the location include parents and children of the location in question but not siblings, as these would be likely to refer to distinct entities.

The algorithm is represented by:

For each ambiguous term in the collection
For each location of this term
For each document in the collection
Discover relatives
Calculate distance
Next document
Calculate average support for this location
Next location
Next term

This approach is in some way similar to the approach described in [62]. The authors of this paper use the UMLS Specialist lexicon and elements from the metathesaurus in order to build a dynamic taxonomy. A dynamic taxonomy can be used to support browsing operations by allowing only those terms that are related to the current term or document to be highlighted. However this approach is most useful for individual searching episodes rather than inter-searcher support over a longer timeframe. The results for this approach, described in [45], include wide variations in the membership value for terms that are multiply located as shown in Table 5.

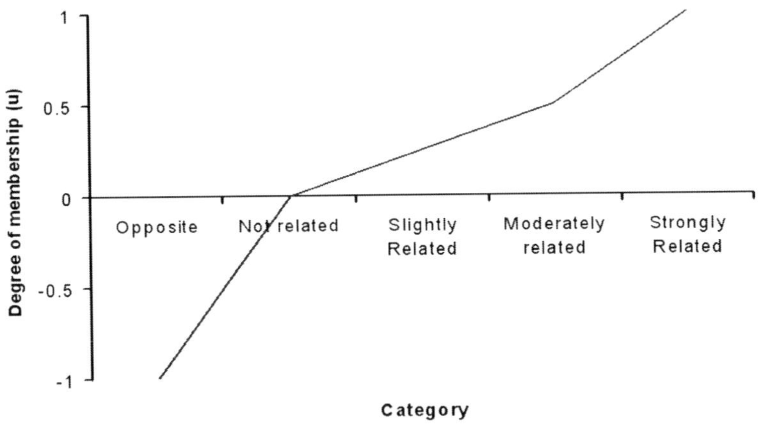

Fig. 9. Fuzzy ontology membership.

4.2. *Learning from users*

The terms used for the fuzzification are shown in Figure 9, along with the degree of membership associated with each relation.

These terms were chosen as fuzzy labels because they are easy to understand, and refer to relations between terms rather than a general quality of a term. This is important because the system requires relations to be investigated to construct the fuzzy ontology. The "opposite" label is used in order to allow otherwise useless terms or documents containing these terms to be excluded. Its role is similar to that of the "NOT" operator, if for example one was searching using a term with a number of different meanings. An example would be the use of "Labour" NOT "Unions" to exclude organised labour pages when searching for obstetric labour.

A system for eliciting these membership values was described in [46]. A specialised browser is used that allows queries to be sent to search engines, and the results displayed. For each document in the recovered set, terms that exist within the ontology are extracted automatically, affixes removed by stemming where possible using a lookup table. The user can then drag these terms into "related boxes" according to the degree of relatedness to the original query term. There are boxes for "Opposite", "Not Related", "Slightly Related", "Moderately Related" and "Strongly Related". A value for "Usefulness" of the document is also recorded via a slider. "Opposite" is used as a shorthand for this term is not related to the desired target, and documents containing this term are unlikely to be relevant – so for example in searching for "Cold" the term "Frigid" would be assigned to the opposite category, if the user was looking for documents about the common cold. The ontology update then takes place in two stages:

1. The intended ontology location of the search term is calculated. Only documents that receive a usefulness value greater than a certain threshold are included in the process. The query term is then compared to the terms in the "related terms boxes".
2. A score is calculated for each potential meaning of each query term by summing the membership values of terms in the "related terms boxes" that are related to each potential location of the query term.

3. The location of the query term with the largest score is assumed to be the location the user intended.

To calculate the membership value of the query term in a particular location, as indicated by the users response to the retrieved document then the following formula is used:

$$\mu_{\text{result}} = \frac{\sum_{i=0}^{n} \mu_i}{n}$$

where μ_i is the membership value for each term the user has put into in the "related boxes". Only terms that are parents or children of the preferred meaning of the query term are included in this part of the calculation. If a term from the retrieved document occurs more than once, then each instance of the term is included. The value n is given by the number of such terms, including duplicates. If the calculation yields a membership value of <0 then the value is reset to 0.

A fuzzy ontology membership value can therefore be used to identify the most likely location in the ontology of a particular term. Each user would have their own values for the membership assigned to terms in the ontology, reflecting their likely information need and world view. However, this still requires the appropriate membership values to be assigned to each occurrence of a term for a particular user. This process can be performed in a number of ways, but the most direct approach – of assigning values during the searching process is unlikely to succeed. This is because of the reluctance of users to use existing term browsers – as noted above. The use of relevance feedback is potentially more fruitful, but this has limited application in the context of queries dealing with previously unused terms. A collective searching system may be more successful in this case.

The key element in the creation of the collective searching systems is the creation of groups of users based on common professional group, status and task. These groups can then share a base set of membership values for each term in the Fuzzy Ontology, with modifications via an incremental learning algorithm when queries take place and the relevance score noted. As part of the data structure of a Fuzzy Ontology the number of queries using a term is recorded along with the membership value for that term. Recalling the formula for updating of the group Fuzzy Ontology above:

$$\mu_{\text{New}} = \mu_{\text{Old}} \pm \left(\sqrt{(\mu_i - \mu_{\text{Old}})^2 / Q_{\text{Hist}}}\right).$$

The membership value of any other equivalent terms are decreased or increased in proportional amounts in order to maintain normality. For example, consider 3 locations for a particular term, (L_1, L_2, L_3) with membership values $(\mu_1 = 0.6, \mu_2 = 0.3, \mu_3 = 0.1)$. If L_1's membership value is decreased to 0.5 then the extra membership values are split in the, so that the new value for μ_2 is given by:

$$\mu_2(\text{New}) = \mu_2(\text{Old}) \pm \mu_{\text{Change}} x \left(\mu_2(\text{Old}) / \left(\mu_2(\text{Old}) + \mu_3(\text{Old})\right)\right)$$

where μ_{Change} is the change in μ_1 in this case -0.1.

The new values are then $(\mu_1 = 0.5, \mu_2 = 0.375, \mu_3 = 0.125)$. The process is shown graphically in Figure 10.

This scheme is very simple and other incremental learning schemes could be adopted. A similar algorithm performs the individuals' Fuzzy Ontology updating, however in this case the number of queries represents the number of queries by that particular user. In the future a "user authority" factor may need to be introduced, related to the degree of

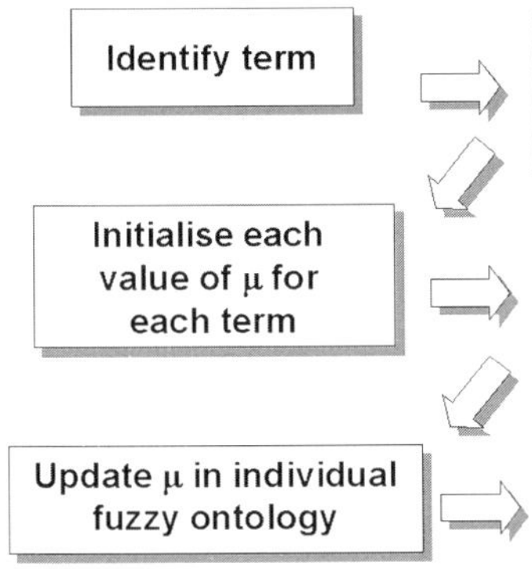

Fig. 10. Group fuzzy ontology updating.

experience or the degree of similarity between a user's Fuzzy Ontology and the collective one in other cases.

It is important to note that the updating process will produce different values for the individual and the collective membership values within the Fuzzy Ontology for the same term. Therefore a collective Fuzzy Ontology, and an individual Fuzzy Ontology both need to be maintained. The important aspects of this approach are:

- The assignment of terms is done in the context of a document.
- Terms are located into the ontology by assigning a degree of membership, using the term, rather than having a slot in an existing ontology requiring a term.
- Only terms that the user believes are important are included in the analysis.
- The ontology assignment process is voluntary, and is linked to the user's assessment of the usefulness of the document.

5. Example of use of fuzzy ontology

As with Boolean searching techniques, fuzzy identifiers can be combined. Traditional fuzzy searching and probabilistic relevance measures have used a number of these approaches – for example some of those described in [15].

In the case of the fuzzy ontology proposed, there are three main operators used:

1. AND – conjunction
2. OR – disjunction
3. NOT – negation

The fuzzy logic approach to these combinations is very similar to the crisp one, except that the membership value is included in the calculation. The reason this is important is in the retrieval part of the process once the fuzzy ontology is generated. Similar work has been done on this already for example [8], this section outlines some of the methods that can be used to utilise the fuzzy ontology once constructed.

Table 6
AND truth table.

A	B	A AND B ($A \wedge B$)	Membership value of combined terms
1	1	1	Min(μ_A, μ_B)
1	0	0	0
0	1	0	0
0	0	0	0

5.1. *Fuzzy AND operator*

For this operator, two or more groups of results are combined in order to provide the most selective coverage of the topic.

The truth table for the AND conjunction is shown in Table 6.

The fuzzy AND operator used in this paper is the *min* operator. In this approach, the membership values of each piece are calculated and the minimum membership value is assigned to the combination. For example, assume that there are two terms included in a query; "complications" and "forceps". "Forceps" occurs twice in the MeSH ontology and "complications" occurs 11 times. Simplifying the situation considerably it can be imagined that there are two possible domains that may be required when searching for "Forceps AND Complications":

(a) General Surgical complications – associated with the use or loss of forceps during operations.
(b) Complications of labour and delivery that require the use of forceps.

It can be seen that in case (a) the forceps cause the complication, but in case (b) they are used in an attempt to avoid it. Assuming that the membership value of relations involving "forceps" remains at the initial value of 0.5, and the value for each of the relations involving "complications" is around .09, then the fuzzy Min relation returns a value of 0.09 for documents containing "forceps" AND "complications" in the obstetric portion of the ontology. This is the case where the ontology membership values have not been altered since instantiation. However, if the ontology being used has been modified, so that forceps is considered to have a membership value of .8 for the obstetric portion of the hierarchy, and complications has .5 for the obstetric portion then the membership value of the conjunction is now .5, for the obstetric domain. If the fuzzy ontology approach is not used, then it is still possible that a query of "forceps AND complications" will return many obstetric results, however if the desired results are not obtained, it is not certain as to which sets of related terms should be used for query expansion and refinement. In the case of the fuzzy ontology approach, the query expansion goes to related terms – such as assisted delivery, vacuum extraction, etc.

5.2. *Fuzzy OR operator*

The fuzzy "OR" relation takes the maximum value of the membership value of its elements, as the combined membership value. In an information retrieval system, the use of fuzzy operators can be combined with a threshold and/or a weighting system in order to assign relevance scores. The effect of this is to increase the importance of "certain", i.e. high membership value terms in a compound query. This means that, any automatic query expansion is in danger of being biased towards these terms if memberships are translated as weights for relevance scores.

Table 7
Truth table for OR.

A	B	A OR B $(A \vee B)$	Membership value of combined terms
1	1	1	$\text{Max}(\mu_A, \mu_B)$
1	0	1	μ_A
0	1	1	μ_A
0	0	0	0

The fuzzy ontology approach uses these membership values. As an example, consider a query with 3 terms with different membership values associated with them in different locations – "obstetric" ($\mu = 0.5, 0.5$), "delivery" ($\mu = 0.3, 0.3, 0.3$) and "theatre" ($\mu = 0.2, 0.8$). Firstly, the terms are examined to discover the likely region of the ontologies. This is done by examining the various possible locations of the terms in the ontologies. For each term-location pair the parents, siblings and children of the term in each location are examined. Siblings are included in this calculation because even if siblings represent orthogonal concepts they still relate to the same area. The system then selects an area one level further out – e.g., grandparents, grandchildren aunts/uncles and nieces/nephews (i.e. siblings of parents and children). This process continues until all of the terms in the query are identified.

The score for a particular location in the ontology is then given by the membership value of the original term plus the sum of the membership values of the nearest terms divided by the network distance to each of those terms. This value is given as the support for this location.

$$\text{Support}(t, i) = \mu_i \sum_{1}^{n} \mu_j \Big/ D(i, j)$$

where μ_i is the membership value of the term t at that location i, μ_j is the membership value of the other terms $1..n$ at the closest location to the location i, and $D(i, j)$ is the network traversal distance, i.e. number of edges within the ontology that are needed to travel from i to j. The support is calculated for each term in the query, and the maximum value selected. This maximum value is known as the *centre of the query*, and the query expansion takes place on the basis that expansion terms are chosen which are adjacent to locations nearest to this centre where terms are ambiguously located. To return to our example, it may be discovered that the combination that has the greatest support is centred on the term "Theatre" which had the membership value (0.2).

In practise this means that non-ambiguous terms in queries will tend to have the greatest influence on the location of the centre of the query and that distant modifiers will not have much influence at all. Rather than this "winner-takes-all" approach another method would allow a number of centres to be used, with the relevance score – however calculated – for each recovered document to be multiplied by the support for each centre.

6. Future research directions

Many learning schemes can be proposed for updating the membership values in a fuzzy ontology. Browser-based approaches, driven by individual users are attractive, but may not provide the width and depth of coverage necessary. However, such approaches, if combined with the benefits of collaborative searching, may be successful. The more specialised the domain, the more likely it is that many relationships will in fact have a very low or zero membership value. The use of a "forgetting algorithm" to reduce the support for terms when they are deemed unrelated may also assist in this process.

7. Structural fuzzy ontology

Fuzzy ontologies can have a wider use outside term definition and selection. In the present case the fuzzy ontology approach could be used to relate different sorts of documents in terms of their structure. The combination of fuzzy ontologies as described above, could be used for the identification of documents likely to be of suitable structure. An example is given below:

Consider a user who is a member of a group, for example (English speaking, physiotherapist, advanced user). If we represent their document structure preferences as a set of fuzzy relations, then we can use the combination rules described above, if documents are assigned to a set of similar groups. In this example the document structure preferences of this example user may be as shown in Table 8.

The hierarchy is established by, e.g., structural analysis of pages or vector space modelling using the content or even Kolmogorov distance as described in [33]. The combination of the user weighting and the genre score in each category is then represented as an overall "quality factor" between 0 and 1. This means that a new document can then be located in a structural fuzzy ontology where it is assigned to a particular genre with a particular membership.

Figure 11 shows the construction of such an ontology, where an hierarchical clustering process using Kolmogorov distance produces the base relations between the genres. A sample of documents belonging to each genre is then identified and the mean value for each parameter calculated. A new document is then assigned to one or more different genres with different membership values for each.

Table 8
Simulated preferences.

Parameter	Value	Membership
Official source	1	High (1)
Graphic content	0.8	Low (0.2)
Number of links	0.7	Medium (0.5)
Non-English	1	Unwanted (−1)

CHAPTER 3

Capturing Basic Semantics Exploiting RDF-oriented Classification

Vincenzo Loia and Sabrina Senatore

Dipartimento Matematica e Informatica, Universita' di Salerno, via Ponte Don Melillo, 84084 Fisciano (SA), Italy

E-mail: {loia,ssenator}@unisa.it

Abstract

Semantic Web, as extension of the current one, promises to provide more automated services based on machine-processable semantics of data and heuristics that make use of metadata. A key challenge of the Semantic Web is the adaptation of Web contents according to web parameters, ontologies, dictionaries. Processing metadata scalability involves infrastructures to interpret, translate and convert different metadata languages automatically. In addition, contextual information and user exigencies should also be considered to assure flexible and proactive interaction. The emergence of sharing semantic knowledge moves towards comfortable reuse of knowledge (thanks to the availability of metadata dictionaries) and machine-processable semantics of data, (through agent-based services, middleware applications, ontology-based system). In this sense, benefits of garnering information come from middle-way services (brokering, matchmaking assistance) aimed to enable more interoperability among heterogeneous approaches.

This work is based on a web service infrastructure for semantic-based information discovery. Through techniques of fuzzy clustering, human comprehensible information (relevant metadata) are captured from an RDF-oriented collection of machine-readable documents, in order to support semantic navigation into the Web that is very heterogeneous space, in terms of languages, dictionaries and ontologies. This approach proposes a semantic RDF-oriented classification of documents as prerequisite for a suitable characterization of relationships between concepts and web resources.

Keywords

fuzzy clustering, rule-based classification, linguistic modifiers, Semantic Web, RDF model

FUZZY LOGIC AND THE SEMANTIC WEB
Edited by Elie Sanchez
© 2006 Elsevier B.V. All rights reserved

1. Introduction

The increasing volume of exchanged information, supported by ubiquitous networks has amplified the complexity of searching and arranging the retrieved information. Web searching is considered a task of social and financial relevance, in view of the fact that hundreds of millions of worldwide users, with different profiles and backgrounds, look for pages related to their requests. Typical queries submitted to Web search engines are sequences of keywords, with the same weight and without context relation. As a result, lists of web pages are ranked according to the occurrence of terms inside the pages retrieved by the query, but often, quite distant to user requests. Last trends go on additional semantics, in order to guide searching in the WWW space, although currently most search features are based on raw lexical content. It is difficult for machines to meaningfully process and integrate heterogeneous information in structures, format, subjects and content. Famous directories and search engines guarantee some assistance but are still far away from getting the level of knowledge discovery needed to perform accurate, context-sensitive and semantic responses. As extension of Word Wide Web, the Semantic Web provides enhanced data processing capabilities aimed to aid users in getting required information. In order to achieve these worthwhile capabilities, innovative approaches need to interpret the meaning of information on the Web, extract the embedded semantics, make deduction and organize information for the user. The strong exigency of semantics models machine-interpretable resources and intermediary services which provide environments where data can be shared and processed by automated tools as well as by people. Beyond supporting automation procedures, the Semantic Web improves the way people directly access meaning rather than simple keywords, through the clean separation between content and presentation. In the Semantic Web, information is semantically marked-up instead of just being formatted for displaying, providing a universally accessible infrastructure which can guarantee more effective discovery, automation, integration and reuse across heterogeneous, end-user applications [10].

This paper presents a framework for the semantic-based classification of information from an RDF-oriented collection. The approach seeks to capture the unbiased content (in form of metadata) of each one of RDF documents and to group them according to similar meanings.

After a brief introduction about the actual semantic-oriented scenery of the Web, a global overview of system performance and its functional objectives are presented. Then the architectural approach follows, providing specific details about the main components and evidencing the relative theoretical approach. Experiments complete the overall description of proposed framework. Conclusions and further developments close the paper.

2. A look at the Semantic Web panorama

Recent web technologies evidence the different views of the information, from targeting human consumption to machine readable format. The absence of a regulated standard in the construction of a HTML document has facilitated the human interpretation associating Web content with explicit meaning, but some etymological problems has been met by keyword-based Web search tools about the polysemy or synonyms of words in

```
<?xml version="1.0"?>
  <rdf:RDF
    xmlns:rdf="http://www.w3.org/TR/WD-rdf-syntax#"
     xmlns:dc="http://purl.org/dc/elements/1.1/">
    <rdf:Description about="http://www.dmi.unisa.it/people/loia/www/LASAlab/home.htm">
    <dc:title>LASA Home</dc:title>
        <dc:description>The LASA Laboratory, founded in 2004, is
            a new research group at the University of Salerno, Italy.
            The laboratory is run under direction of Professor Vincenzo LOIA
            and is composed of computer scientists from a wide variety of backgrounds.
   </dc:description>

   <dc:subject>Agent Techonlogy, Web Knowledge Discovery,
               Fuzzy Control, Ambient Intelligence
   </dc:subject>
   <dc:language>IT</dc:language>
  </rdf:Description>
</rdf:RDF>
```

Fig. 1. RDF/XML syntax.

different contexts, considering that keyword indices do not preserve the notion of relationships among words, too. XML signs the passage from data displaying to data description; through a more structured, syntax rule-based formalism, it suggests smarter skills to make data machine processable. The introduction of XML namespaces allows differentiation of the meaning of identical element and attribute names and permit different markup languages to be mixed together without ambiguity. Although XML requires textual contents are wrapped by tags, appreciable flexibility on the machine interpretation is guaranteed by XML schema, which constrains occurred tags in XML document to comply to given vocabularies and hierarchical structures. It follows XML semantics are implicit: the interpretation of the XML elements derives from human inference or from an explanation in natural language. However, XML schema presents some semantic drawbacks:

1. difficult mapping of concepts defined using different vocabularies;
2. conflicts of word meanings;
3. lack of interpretation of contextual information.

The Resource Description Framework [1] defines a syntactical data model for representing knowledge in more structured form; restrictions of classical user-oriented HTML format, where the human expressiveness, meaning vagueness and complexity of data sources make more complicate to provide a right interpretation of user intentions. At the same time, it revises XML-based formalism, creating and modeling semistructured metadata about a document as a stand-alone entity, seeing RDF proposes a scalable model, which can be used for exchange of information among automated activities or knowledge-management applications through a recommended XML serialization syntax (RDF/XML).

The RDF data model uses URIs to unambiguously denote objects, and properties to describe relationships between objects. The structure of any expression in RDF is a collection of triple-based statements, each consisting of a subject, a predicate and an object. Each triple is a statement which represents a relationship, indicated by the predicate, holds between the things denoted by subject and object of the triple. The RDF data model can be thought as a semantic network of nodes connected by labeled arcs where the nodes represent Web resources and the arcs represent properties which links these resources [1]. The key syntax of RDF consists of a *Description* element, which depicts the resource; in particular, Figure 1 shows a sketched example (in XML format), where the *about* attribute identifies the main resource by a specific uniform resource identifier (URI); other property elements complete the resource definition. Furthermore a peculiar characteristic of RDF model is to describe, like resources not only Web pages or textual documents, but also media objects such as images or audio files. Specifically, RDF presents salient characteristics:

- It is designed to defined *independently* resources, without any constraints about imaginary names, producing a waste and heterogeneous semantic networks.
- RDF statements can be serialized or converted into XML, so they are easy to *interchange*.
- On the light of incessant growth of Web, RDF proposes a simple, *scalable* triple-based model, which is easy to handle and, through add-in tools it is quick to consult and look things up.
- The simplicity of RDF assertion, based on the triple resource–property–values permits to build nested assertions: properties might be resources (with own properties and they can be found and manipulated like any other resource), at the same way, values can be resources (most web pages will have a property named Home-Page which points at the home page of their site). So an assertion could be enclosed other inner assertions.

Annotation methodologies [12] move in this direction: they create a semantic infrastructure needed to handle and associate metadata with content of traditional web documents; RDF-based annotations are manually added to web documents in order to better describe related resources. In particular Magpie [24] is a tool supporting the interpretation of web pages; rather than defining manual annotation, it associates a semantic ontology-based layer to a web resource in order to assist users with finding the web pages by means of automatic conceptual annotations. Even though the concepts are high-level described, in order to invoke relevant services within a standard web browser, Mapgie is not able to express users intentions through the management of the user experiences during the web navigation.

The RDF vocabularies can be formally defined using sets of rules, which define the domain elements and describe the relationships which blend these elements to each other. RDF schema (RDFS) specification [3] defines sets of RDF resources and properties, enabling users to create their own RDF vocabularies. Although RDF-based model represents an important goal for formally representing the data and describing data about this data (i.e. metadata), it exhibits a limited expressive power and weak reasoning capabilities only suitable for constraint checking, besides thanks to its potential flexibility should be exploited to support for different natural languages, so a consequent conceptual and lexical mapping is necessary. Community is actively developing and improving tools which meet user's needs and provide semantic-based assistance. Several automated reasoning applications and inference engines are based on knowledge representation languages and taxonomies: markup languages and ontologies languages have recently been developed; indeed DAML + OIL (DARPA Agent Markup Language and Ontology Inference Layer) [7,6] represents the result of W3C research to create a Web Ontology language (OWL) [23], in order to create arbitrary domain ontologies which support the unambiguous description of Web resources. In this sense, the ontology-based approach provides the way to express the semantics of data, document collections through resource definition which is machine interpretable, reusable, extensible and sharable. Several standardized ontologies, which share and annotate information have been developed: SUMO [15,27] which is online accessible, enabling the navigation of the ontology; two interesting ontology editing tools Onto linga/Chimaera [8] and Portégé [14] present skills to extend ontologies to the desktop of domain experts. In particular Portégé is an open platform for ontology modelling and knowledge acquisition, that moves towards the definition of infrastructure for ontology-based Semantic Web applications. Further noteworthy approach is SEAL (SEmantic PortAL) [22], a framework which has been developed to exploit ontologies for managing community Web sites and web portals. It is based on OntoWeb, an European thematic network for sharing of research related ontology technologies, so SEAL represents a methodology for integrating heterogeneous information from distributed resources, through ontology-based semantic integration, for acquiring, structuring and sharing information between humans or machines.

Based on the RDF/RDFS model, DAML+OIL and OWL represent the standard markup languages for describing ontologies, classifying of resources, providing facilities for agent interaction on the Web. In particular, the diffusion of standards for Web services infrastructure, based on multiagent coordination mechanisms like matchmaking and brokering, sees the Web Ontology language for Semantic Web Service (OWL-S) [25], the responding language construct for supporting dynamic service discovery via matchmakers [28]. OWL-S, formerly DAML-S [5] describes Web Services, allowing software agents to interpret descriptions about how to interact with the services they describe. Due to the proliferation of ontologies to be used for describing different Semantic Web Services, a lot of interoperability problems arise. Thus, interoperability also requires semantic translation between the representation which clients employ to entail service goals and the representation the services apply to satisfy requests. Traditionally, middle agents have been employed for translating messages sent between clients and servers, but this approach results unsuitable when dynamically unfamiliar services (i.e. associated to different ontologies) are invoked: client should embed reasoning about data relationship and knowledge background, in order to appropriately characterize their request message [4].

3. Framework

Nowadays, what is missing in classification systems of Web documents is a mechanism for recording machine readable statements that semantically describe resources or more generally, concepts. Previous our works tackled these themes: a partially supervised classification system has been described in [19,16] where, the user expressed a personal judgment about how two web (HTML-based) pages were similar, according to his subjective perception. In this direction, further developments have been carried out, aimed at providing assisted web navigation based on the proximity of web documents [21,20].

Then, the increasing need of capturing the actual web content of heterogeneous documents moves towards approaches oriented to the semantics of (RDF-based) pages, as previously discussed. The current work is the result of an incremental modeling that, through a knowledge discovery process which, grabs the enclosed, semantically meaningful concepts inside RDF-oriented pages. This approach extends a minimal frame composed of the *clustering* and *fuzzy rule-based classification* modules introduced in [18]. This minimal frame represents a starting point for modeling wider and complex architectures. In particular, in [17] we described a system for searching semantic-based information, through a user-driven classification, on the basis of the topics given as input to the system. Rather than returning all the pages related to a requested topic, it arranges a collection of knowledge based on the user's expectation, organized in accordance with concept-based features instead of simple keywords.

The current approach instead, aims at providing a background framework for supporting enhanced activities of semantic interpretation of web contents; although it is based on the same core frame, it gets the information discovery by a different viewpoints, emphasizing the conceptual aspect of information, in terms of semantic relationships defined between metadata. Finally, it builds further enhancements through linguistic modifiers which, applied to fuzzy rules, provide more flexibility to the interpretation domain of the extracted concepts. In summary, this work focus to achieve the following goals: (1) a model to capture semantics information, (2) services-based facilities for managing and providing support to interaction (agent-based) mechanism, (3) presentation of information based on RDF classification (by fuzzy rules) and (4) use of linguistic label to extend the lexical domain among semantically related concepts. Therefore, the system prototype described in the sequel, lends itself to being exploited as basic repository layer to accomplish a starting screening of irrelevant information from salient concepts in terms of RDF-oriented metadata. Further ontology-based extensions have been taken into account in order to enlarge the semantic domain of action.

3.1. *Logical view*

Figure 2 shows the system architecture pointing out the main interaction activities (depicted as rounded rectangles). They are described in the following list.

- *Extraction and selection of candidate metadata*

  This task carries out the extraction of meaningful metadata from the analysis of the RDF-based corpus, given as input. A preliminary parsing activity is achieved, in order to validate the XML/RDF documents. The parsing exploits the SiRPAC [29] package, created by the W3C; it is a reference implementation of an RDF parser which executes a compilation process from XML encoding syntax to the triple-based statements of RDF model. Final goal of this activity is the selection of the metadata (associated to the basic concepts), which are more recurring, meaningful and, hence, the most representative in given RDF collection; further details are dealt with Section 4.
- *Building metadata-page matrix*

  The metadata, which are elicited in the previous task, represent the features for building the metadata-page matrix. Indeed, each row of this matrix describes an RDF page through vector-based representation and each element of vector is a weight-value in the range [0, 1] associated to each correspondent metadata.
- *Clustering*

  This is the clustering task. It takes as input the built matrix, performs the clustering algorithm on it and returns groupings (i.e. clusters) of pages, assembled according to related argumentation. The clustering approach is the FCM algorithm [2] which a-priori fixes the number of clusters. A service mediator attends to set the number of clusters, through the interaction with semantic dictionary/ontology, according to the number of concepts identified by the selected metadata.
- *Misclassified pruning*

  This task analyzes the obtained clusters. Considering the fuzzy clustering, it could happen some pages are misclassified, because they badly belong to each one of attained clusters (recall the FCM clustering assigns to each pattern a membership degree for each cluster). The uncertain pages are marked. If the number of pages overcomes

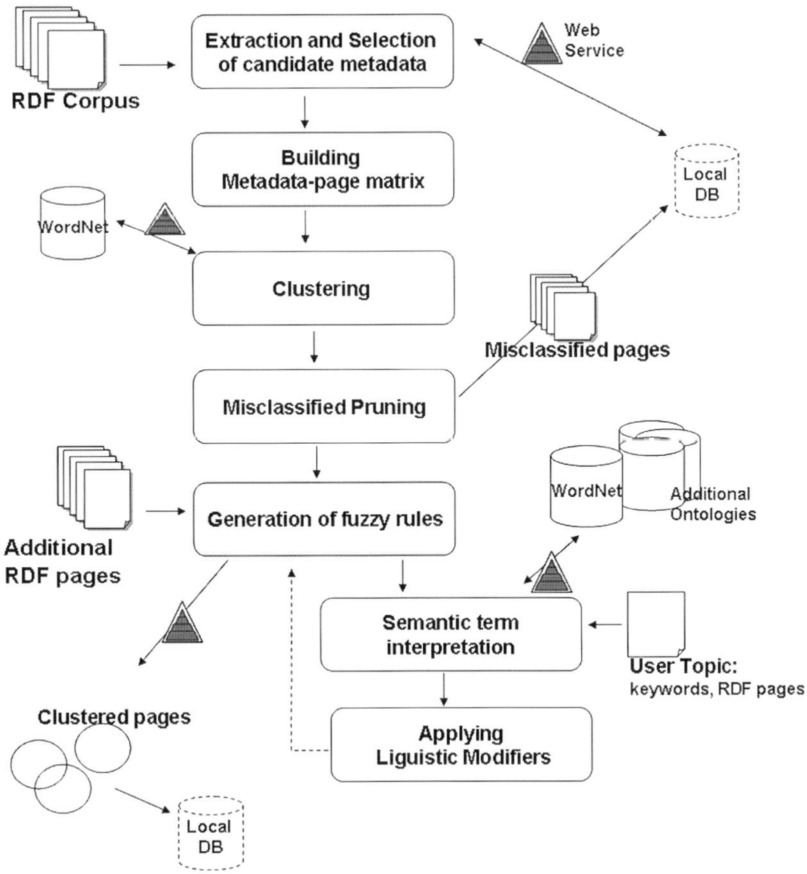

Fig. 2. The system architecture.

a prefixed threshold (for instance, the 25% of all), these pages are discarded from clusters and the clustering activity is performed again.

- *Generation of fuzzy rules*

  In this step, a higher level description of clusters is achieved through fuzzy rules generation. In other word, the obtained clustering is translated in meaningful fuzzy rules, whose arguments are just the selected metadata. Moreover, thanks to fuzzy rule-based approach, new incoming RDF pages (discovered for instance, during additional spidering activities) can be classified into an existing cluster.
- *Semantic term interpretation*

  This module contemplates the human interaction: a user request can occur about a specific topic in terms of keywords or RDF page. A service mediator attends this phase: it establishes a communication aimed at analyzing and interpreting the user input (through POS activities). Then, it interacts with a stored ontology in order to find out additional information and hierarchic relationships about the formulated argumentation.

- *Applying linguistic modifiers*
  According to the relationships or semantically correlate words, linguistic modifiers are defined on the basis of fixed constraints. This task evaluates possible semantic ties between the metadata-word of fuzzy rules and the words of user input. Major details are provided in the following.

3.2. *Web semantic technologies*

The infrastructure consists of modular components, which interact to each other through service-based exchange. Although the prototype presents some weaknesses in terms of project choices, it is designed to be independent on the implementation details, just thought to flexible extensions towards more structured languages, ontologies, services. Main intervened technologies are:

- Service-based invocation: The web research has developed ontologies and languages for describing web services. The system uses service mediators, which interposed between ontological databases and other information sources, hide the underlying infrastructure. They interpret the system requests, perform semantic mapping to solve ambiguous data and synchronously reply to invocations. The system components are self-contained web applications which exhibit their functionality, provide interfaces described by WSDL (Web Services Description Language) and communicate using SOAP (Simple Object Access Protocol). Enhanced developments based on OWL-S are still in progress. Anyway extensions on SOAP messages with additional information (i.e. security, reliability, expiration time) guarantee basic WS-Security specification.
- Ontological databases: Ontologies provide a common language for representing and sharing knowledge. Often, even in the same domain, data models used for ontologies are different, due to the encoding of the semantics or conceptualization of the domain. Such cases require translation, even though supporting to interoperability is guaranteed by using OWL ontologies. Ontological mapping become a crucial requirement, although it is often partial (not complete correspondence between concepts). The system prototype employs WordNet which, although it is not an ontological standard for the semantic Web, represents an effective standard for general-purpose lexical databases. The RDF version of WordNet is exploited to detect taxonomic and similarity relations among concepts (defined by metadata), then generate a domain concept forest (where hierarchic-related words provide additional information to better depict the concept).
- RDF approach: The system takes as input an RDF corpus of documents. Analyzed documents contain homogeneous data, represented in RDF/XML context. RDF provides a simple model for describing relationship between web resource in terms of properties and their values. As wider explained before, the RDF/XML syntax easily permits to serialize RDF triples-based content. The current approach describes some abstraction levels of RDF documents:

  RDF reference: Namespaces declared by URI references inside an RDF document enable to represent web information, through formal semantics specification, dictionaries, ontologies.

RDF metadata: RDF-data tags (surrounded by angle parenthesis) defined into the RDF reference, enable to describe resources and relationships. The parsing activity performs at this level.

RDF content: Textual information enclosed into RDF tags (metadata). This level provides an high-level view of described metadata.

- Spidering activity: The system employs a crawling component to harvest RDF files from the Web and resolve cycles. Spidering is carried out simply in a breadth-first way (for instance, by following the links related to *rdfs:seeAlso* tags).

3.3. *System behavior*

The system acquires the knowledge (concepts, information, etc.), extracting the basic and more recurring metadata from *RDF corpus* of documents. Indeed, the ingoing RDF-based pages are parsed: for each pages, the system selects, on the basis of specific relevance criteria (discussed later, in Section 4) the candidate metadata. The *local database* stores each RDF page and its relative metadata.

The whole corpus of RDF documents represents the embedded collective knowledge: elicited metadata embody the most representative information, that is, the main concepts which describe the collection. This way the metadata-page matrix synthetically describes the global knowledge, exploiting as features, the computed values associated to the metadata for each RDF page.

On the basis of the given data matrix, the *Clustering* activity generates K clusters (in correspondence to number of concepts). The parameter setting for the clustering activity is entrusted to a service mediator; on the basis of the SOAP protocol, information is interchanged with WordNet to provide taxonomic relationship among metadata, in order to invariably identify metadata related with other metadata (i.e. property or resource names) or domain concepts.

Clusters returned by clustering activity are analyzed: misclassified pages are reconsidered for a possible removal from the RDF corpus (with the consequent re-clustering). The obtained clusters are translated into fuzzy rules (more approachable to human beings). In this phase, additional RDF pages can be classified on the basis of he performed clustering.

At this stage, the training activity is completed: the system has the knowledge about the thematic of RDF corpus and has a synthetic formalism for expressing the main argumentations, in terms of metadata.

The module *Semantic term interpretation* comprises the interaction with the user who can specifies own search topic (regarding keywords or RDF pages). The user interaction is handled by the service mediator which semantically interprets terms (through a request of service to Wordnet) and discovers words (synonyms, meronyms, hypernyms, hyponyms, etc.) related to the grabbed metadata (sensed words).

The service mediator identifies terminology used in the domain of user input, semantically interprets terms, associating unambiguous concept identifiers with the extracted terms. Semantic interpretation is based on given ontology (the current system prototype exploits WordNet) semantically tagged corpus to perform this function.

A computed semantics relation between metadata and user-topic terms provides more flexibility to the fuzzy rules: through linguistic modifiers, they enable to adapt the fuzzy interpretations (formally associated fuzzy sets) of concepts to semantics related words (of user topics).

This way the system reaches the following functionalities:

1. User requests are translated in rule-compliant form and classified according to the semantic of RDF collection.
2. Furthermore, wider range of new ingoing RDF pages (which should be unclassified through original fuzzy rules), can be further classified by means of linguistic modifiers.

As output, the system produces the classification of given RDF corpus, according to the discovered main concepts.

4. Matrix building

The building of metadata-page matrix is achieved through a preliminary selection of relevant metadata, extracted from RDF pages and then, for each RDF page, the vector-based representation is constructed.

In detail, the *Extraction and Selection of candidate metadata* activity produces ranked list of relevant metadata extracted in the RDF collection. The most relevant ones (in the top of the list) will represent the column headings of data matrix, whose values, estimated according to each parsed RDF document, will form the feature space.

4.1. *Relevance criteria for eliciting the metadata/features*

The selection of relevant metadata represents, at semantic level, the content of the collection of the RDF documents.

Recall that the RDF model describes each resource through triple-based statements: classes, properties and values. RDF pages describe data exploiting pre-constructed *definition* of classes and properties (through dictionaries and ontologies whose references are declared inside the RDF documents). In the following, a measure of relevance is computed, for each metadata in RDF-based collection. Let us begin by considering the following assumptions:

Collection of RDF documents: it is a set P of n RDF pages $P = \{P_1, P_2, \ldots, P_n\}$.

Collection of dictionaries: it represents a set D of RDF schemas and – without loss of generality – dictionaries and/or ontologies referenced in the collection P $D = \{D_1, D_2, \ldots, D_m\}$.

Collection of dictionaries related to a page: for each page $P_i \in P$, $1 \leqslant i \leqslant n$, $DP_i = \{D_{i_1}, D_{i_2}, \ldots, D_{i_m}\}$ is a set of dictionaries associated to P_i where $D_{i_h} \in D$ for $1 \leqslant h \leqslant m$.

Each RDF page P_i consists of triple-based statements; each statement describes a resource, represented by an instance of a class (defined into the dictionaries DP_i), a named property plus the value of that property for that instance (that can be a literal or another resource). Besides, more instances of the same class can exist inside the same RDF document and a statement can be composed of some nested statements.

Bearing in mind these considerations, in the calculus of relevant metadata, the level of nesting is considered; besides, this analysis only considers RDF metadata (class and property) names, without care of the relative assumed values.

In the following, two measures of relevance (named *instance* and *property* relevance) will be defined on instances of classes and relative properties, for each RDF page P_i. Then, these measurements will be applied to all class and property names of the whole RDF collection P. A collective relevance measure will be computed for the same metadata in the pages. It will represent the global relevance of a metadata in the whole collection. The ranked list of these values will describe the more relevant metadata, candidate for representing the features on the matrix building.

The following sketched sequence of steps describes the selection procedure:

Step 1: *for* each class name C in an RDF page P_i
for all statements A_s where C appears, compute:
the *Instance Relevance* measure on each instance of the class C
for each property name p defined to the instance of class C, compute:
the *Property Relevance* measure on the instance of C with respect to p

Step 2: Consider a metadata M (class C or property p names) of the RDF page: get a summarized, unique relevance value for all the statements including the metadata M.

Step 3: Repeat Steps 1–2 for all the pages of the given RDF corpus.

Step 4: Compute the average of all the relevance measures, related to the metadata M names, on the whole corpus.

4.1.1. *Instance & property relevance values*

This section depicts details about the Step 1. For each RDF instance, in a statement, the nesting degree and the *accuracy* and *relevance* measures are evaluated for each instance.

Fixed a class C defined into dictionary DP_i, select the sequence of (nested) statements in the current RDF page, that describe (or, more specifically, have as subject) instances of C. More specifically:

DEFINITION 1. Given a class name C of the dictionary DP_i, let $A_1, A_2, \ldots, A_h$ be the statements in an RDF page P_i, that include instance of C. Let $\pi(A_s, C)$ be the function which computes the number of distinct properties associated to the class C, within the statement A_s. The *accuracy* is described as:

$$Accuracy(A_s, C) = \frac{\pi(A_s, C)}{\sum_{j=1}^{h} \pi(A_j, C)}. \tag{1}$$

This value $\pi(A_s, C)$ indicates the detail (or granularity) degree by which each instance of class is described in that RDF page. In other words, the accuracy measure represents the influence of a statement with respect of all the others representing that class.

DEFINITION 2. Given a class name C of the dictionary DP_i and given a sequence of statements $A_1, A_2, \ldots, A_h$ of an RDF page P_i including instance of C. Let $nlev_s$ be the

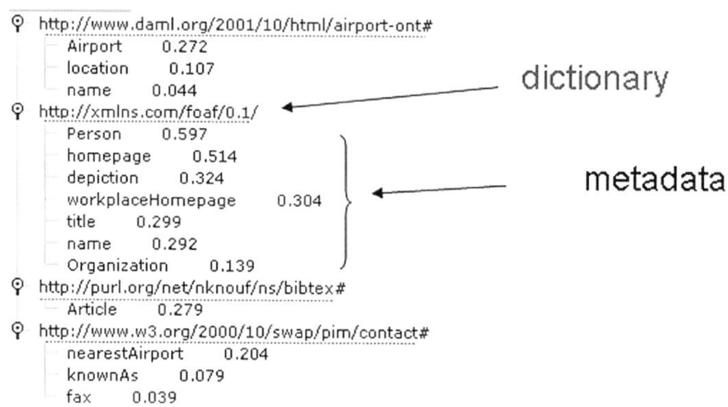

Fig. 3. Ranked list of metadata associated to each dictionary.

nesting level of the statement A_s with respect to the external statement (which includes it). The *instance relevance* of class C within the statement A_s is:

$$Inst\_Relevance(A_s, C) = \frac{Accuracy(A_s, C)}{nlev_s}. \tag{2}$$

This expression highlights the influence of the nesting level of the statement A_s, describing the instance of class C, with respect to the accuracy by which the instance is described. This measure computes the relevance associated to each statements of an RDF page. An analogous evaluation is attained for the property names associated to the instances of these classes, in each statement of an RDF page. Each resource (instance) can have different properties or more occurrences of same property. Let us recall that a property represents a predicate, or a relationship in the statement. Naturally, it follows the definition:

DEFINITION 3. Given a class name C of the dictionary DP_i, let $Inst\_Relevance(A_s, C)$ be the relevance value of class C in the statement A_s, let p be the property name associated to an instance of class C; the *property relevance* value is:

$$Prop\_Relevance(p, C) = Inst\_Relevance(A_s, C) \cdot \#p \tag{3}$$

where $\#p$ represents the number of occurrences of property p in the statement A_s.

The *property* relevance tightly depends on the *instance relevance* and assumes value greater or at least equal. Specifically, if an instance of a class has more occurrences of the same property, the relative *property relevance* value will be greater or at least equal to *instance relevance* value, for a given instance of C, where the property is defined. In summary, two relevance measures have been previously defined: the former describes the relevance of the resources into the statements for each page, another one evaluates the relevance of the properties, associated to the given resources.

4.1.2. *Candidate metadata*

As anticipated into Steps 2–4 of Section 4.1, in order to characterize the feature space of the metadata-page matrix, a digest (summarized) relevance value is computed for each selected metadata (class and property name) with regard to each RDF page. Fixed the class C, the sum of all relevance measures $Inst\_Relevance(A_s, C)$ is computed for all instances A_s of C. Similarly, the sum of relevance values $Prop\_Relevance(p, C)$ for each property name p is computed.

In general, let us name $Meta\_Relevance(M)$ the summarized relevance measure where M can represent class C or property p names. This way, a collection of relevance-based metadata is associated to each RDF page. Finally, in order to evaluate a global relevance of each RDF metadata M, with regard to the whole RDF collection, the averages on all $Meta\_Relevance(M)$ for each distinct M, are computed. The ranked list of relevance values corresponds to a list of metadata in the dictionary DP of the given RDF collection P.

The first r metadata (with arbitrarily fixed r, according to the fixed number of features or equivalently the number of columns defined for the matrix) associated to a ranked list represent the default selection which the system proposes.

Figure 3 gives an idea about the returned ranked list of metadata, candidate to represent the features, grouped for relative dictionary.

Once the metadata are selected, the metadata-page matrix is computed: each row describes an RDF page, by means of a vector-based representation and each cell of vector contains a normalized $Meta\_Relevance(M)$ value, for the metadata M.

5. Clustering

The clustering phase is accomplished by the well-known FCM algorithm [2], particularly useful for flexible data organization. Generally, FCM algorithm takes as input a data matrix representing the given patterns and it tries to get an "optimal" partitioning of the feature space, given a collection of data samples. It is based on the maximization of homogeneity, grouping into the same cluster the samples that are closer. FCM recognizes spherical clouds of points (clusters) in a multi-dimensional data space and each cluster is represented by its center point (prototype).

The FCM algorithm aims at minimizing the weighted sum of the distances between data points $\underline{x}_i$ and the centers $\underline{v}_j$, according to this formula:

$$J(U, k) = \sum_{i=1}^{k} \sum_{j=1}^{n} u_{i,j}^{m} d_{i,j}^{2} \tag{4}$$

where $k \geqslant 2$ is the number of classes, $u_{i,j} \in [0,1]$ is the membership degree of $\underline{x}_j$ $(j = 1, \ldots, n)$ in the ith cluster A_i $(i = 1, \ldots, k)$, $d_{i,j} = d(\underline{x}_i, \underline{v}_j)$ represents the Euclidean distance between the data $\underline{x}_i$ and the center of the cluster $\underline{v}_j$ and finally $m > 1$ controls the quantity of fuzziness in the classification process (usually $m = 2$).

In this approach, each row of *term-document* matrix is a weight-based vector representing an RDF page $P \Leftrightarrow \underline{x} = (x_1, x_2, \ldots, x_n)$, where each component of n-dimensional vector is a (real) value associated to the metadata/feature (normalized $Meta\_Relevance(M)$, as seen in the previous section). After the FCM execution, partitions of RDF documents are returned, in a prior fixed number k of clusters.

As anticipated before, the number of cluster is computed on the basis of the concepts represented by the metadata. For example, considering Figure 3, there are metadata of four dictionaries: through the semantic (dictionary-based) services, three main concepts are extracted: Person, Airport, Article.

Because the complexity of the FCM algorithm is proportional to data size, it is not practical to handle frequent re-clustering, every time there is an updating of RDF collection. If the case of several misclassified pages occurs, accurate service-based analysis of these pages are executed, in order to decide if discard them and re-apply the clustering.

6. Generation of fuzzy rules

Additional, ingoing web pages can be classified through fuzzy rules [11], generated by the analysis of clusters of the RDF collection. In literature, fuzzy if–then rules are often used for this kind of task, because they adapt well to provide a human description of structures in data [13]. Thanks to the fuzzy rules generated by the FCM clustering, ingoing RDF pages (discovered, for instance, during a spidering activity) can be assigned to a cluster/class, without running through the whole clustering algorithm again.

Recall the FCM-algorithm takes as input a data matrix (metadata-page matrix in our case). Each row of matrix (an RDF page P) is a vector $\underline{x} = (x_1, x_2, \ldots, x_n) \in [0, 1]^n$, where each x_j represents the weight-value (or normalized relevance) of the jth feature (or metadata) of $\underline{x}$ (evaluated on the page P). The FCM produces as output prototypes and a partitioning matrix of the data patterns. This matrix contains the membership of the elements of the given data set in each of the k clusters.

The fuzzy if–then rules define a way to determine the membership of additional data, not considered in the data matrix. To get this, the discrete membership matrix is extended to a continuous membership function. One way to obtain the continuous functions is through the linear interpolation of memberships; in order to easily assign linguistic label (with no higher-dimensional domain), the membership function is computed by linear interpolation of the projected membership to each axis (according to the n-dimension space), for each cluster. Then, this method, called the cylindrical extension [11], defines a continuous membership function taking as argument the projection of n-dimensional argument vector, in the n-dimensional projection space. So, each fuzzy cluster K_i with $i = 1, \ldots, k$ can be represented through n functions $\mu_{i_1}, \mu_{i_2}, \ldots, \mu_{i_n}$, obtained by the projected and interpolated memberships.

The goal is to obtain a representation (or interpretation) of the clusters through fuzzy sets, whose membership functions are associated linguistic labels, in order to make fuzzy sets easy to interpret by humans. This way, the membership of a n-dimensional vector $\underline{x} = (x_1, x_2, \ldots, x_n)$ is evaluated in the fuzzy sets "x_1 *is* μ_1", "x_2 *is* μ_2", ..., "x_n *is* μ_n", where each μ_j ($j = 1, \ldots, n$) represents a linguistic label.

Exploiting the conjunction (evaluated in *and*) of the cylindrical extensions, it is possible to provide a whole interpretation of the clusters. The linguistic interpretability and the assignment of label to membership functions enable clusters to be interpreted as classes [11]; so it follows that the fuzzy rule associated for the ith class is so described:

$$R_i\colon\ \textit{If } (x_1 \textit{ is } \mu_{i_1}) \text{ and } (x_2 \textit{ is } \mu_{i_2}) \text{ and} \ldots \text{and } (x_n \textit{ is } \mu_{i_n}) \textit{ then}$$
$$\underline{x} \textit{ belong to class } K_i \textit{ with } \mu_i(\underline{x})$$

where each component of the n-dimensional vector is evaluated in the fuzzy sets, through the membership functions. Furthermore, the conjunction (of membership degrees in the antecedent rule) is evaluated by the minimum, in fact considering a specific case, the evaluation of membership degree $\mu_i(\underline{x})$ of the vector $\underline{x}$ in the ith cluster K_i with $i = 1, \ldots, K$,

Fig. 4. Rules extraction by features analysis.

is characterized as:

$$\mu_{i_1}(x_1) \wedge \mu_{i_2}(x_2) \wedge \cdots \wedge \mu_{i_n}(x_n)$$

where the conjunction of membership functions corresponds to the minimum $\wedge$ of the computed membership degrees.

The consequent part of the rule provides a unique value evaluated on the membership functions (evaluated in the antecedent rule) as global membership value for the ith class.

For fuzzy classifier systems, a unique final assignment of a (vector-based) object to a class is required; it is the defuzzification process which selects the class with the highest membership degree. In our approach the generic ingoing page, depicted as vector $\underline{x}$ is assigned to class K_j with the rule activation:

$$\mu_j(\underline{x}) = \max_{1 \leqslant i \leqslant n} \mu_i(\underline{x}).$$

Figure 4 provides a graphical example of fuzzy rules where the tendency of membership functions (associated to the features) are shown, for each cluster. The features are generically identified as *Feature 1*, *Feature 2*, ...

In particular, the figure describes an example of data which deals with argumentations related to airports (their coordinates, vicinity to contact people, etc.), people (foaf format information), articles (bibtex format). Just as idea, a fuzzy if–then rule herein generated is so sketched:

$$R_i: \textit{ If } (\textit{Airport is } \mu_{i_1}) \text{ and } (\textit{NearestAirport is } \mu_{i_2}) \text{ and } (\textit{Location is } \mu_{i_3}) \text{ and } (\textit{Person is } \mu_{i_4}) \text{ and } (\textit{homepage is } \mu_{i_5}) \text{ and } (\textit{Article is } \mu_{i_6}) \text{ and } (\textit{hasAuthor is } \mu_{i_4}) \textit{ then } \underline{x} \textit{ belong to class } K_i \textit{ with } \mu_i(\underline{x}). \quad (5)$$

In summary, fuzzy rules enable incoming RDF pages to be assigned to the existing (FCM-based) classes; sometimes, some misclassifications can occur because new pages could not accurately be assigned to classes. There are two possibilities to consider: the provided information is not sufficient to classify them (low membership values for each class) or the classification introduces an approximation error. These pages become "undecided" due to the fact that much information (with further features) occurs to correctly classify them, then new classification phase needs.

6.1. *Semantic interpretation through linguistic modifiers*

As shown above, the generated fuzzy rules are based on an antecedent part consisting of fuzzy sets associated to the feature components.

An interesting characteristic of fuzzy systems is based on the linguistic modifiers of fuzzy values, which represent an effort to maintain close ties to the natural language, and to allow for the generation of fuzzy statements through mathematical calculations. Linguistic modifiers produce a semantic alteration of the original fuzzy set: for example, traditional linguistic modifiers as "very" or "more or less" produce respectively concentration or dilation. In other words they might both increase and decrease the uncertainty. A classical example transforms the statement "Jane is old" to "Jane is very old". The linguistic modifier "very" is usually defined as follows:

$$\mu_{\mathrm{VERY}}(x) = \mu(x)^2.$$

Thus, if $\mu_{\mathrm{OLD}}(\textit{Jane}) = 0.8$, then $\mu_{\mathrm{VERY\,OLD}}(\textit{Jane}) = 0.64$.

The point of this example is to emphasize the role of linguistic modifier as further enhancement for the description of the fuzzy concept (fuzzy set).

The fuzzy rules characterize classes of related information, in terms of the relevant concepts (through features/metadata), defined in the clusters. The current goal is to exploit the linguistic modifiers, in order to relax the classification constraints to enclosed *additional* objects. Our approach aims at the classification of RDF-based pages, according to the semantics of relevant metadata: in general, further ingoing RDF pages are classified on the basis of occurrences of given features in these pages. If a new page presents no occurrences of the defined features, it will not be classified. Exploiting the linguistic modifiers, each concept, associated to a fuzzy set, can be modified (with a slight change in the meaning of the basic terms), getting additional expressivity that enables the classification of further metadata, that although they do not directly match with the computed features, could describe related or similar concepts.

This way, new concepts can be obtained by applying linguistic modifiers to the basic metadata which have been computed in the previous steps.

Therefore, every page can be analyzed to find out semantic relationships with the selected features of the fuzzy rules; dictionaries and ontologies are the sources, which are consulted by ad-hoc services, to get contextual information. On the other hand, basic metadata (exploited in a rule) which are related to same ontology or describing the same concept, enable the characterization of the context of concepts and disambiguate polysemic words.

6.1.1. *A sketch about practical use*

In order to provide some practical details about how the linguistic modifiers can be applied, let us consider a new RDF-based page to classify by means of the fuzzy if–then rules: (see rule (5) in Section 6):

$$\begin{aligned}&\textit{If } (\textit{Airport is } \mu_{i_1}) \text{ and } (\textit{NearestAirport is } \mu_{i_2}) \text{ and } (\textit{Location is } \mu_{i_3})\\ &\text{and } (\textit{Person is } \mu_{i_4}) \text{ and } (\textit{homepage is } \mu_{i_5}) \text{ and } (\textit{Article is } \mu_{i_6})\\ &\text{and } (\textit{hasAuthor is } \mu_{i_4}) \textit{ then } \underline{x} \textit{ belong to class } K_i \textit{ with } \mu_i(\underline{x}).\end{aligned}$$

The new parsed page presents, among the relevant metadata, the term *Paper* (defined by DAML dictionary); querying WordNet about this term, one of the senses of *Paper* is an hypernym (*a kind of* or a generalization) of *Article*. This way, the fuzzy rule can be interpreter as:

$$\begin{aligned}&R_i\text{: } \textit{If } (\textit{Airport is } \mu_{i_1}) \text{ and } (\textit{NearestAirport is } \mu_{i_2}) \text{ and } (\textit{Location is } \mu_{i_3})\\ &\quad\text{and } (\textit{Person is } \mu_{i_4}) \text{ and } (\textit{homepage is } \mu_{i_5}) \text{ and } (\textit{Paper is } \mu_{\text{SORT OF}})\\ &\quad\text{and } (\textit{hasAuthor is } \mu_{i_4}) \textit{ then } \underline{x} \textit{ belong to class } K_i \textit{ with } \mu_i(\underline{x}).\end{aligned}$$

This way, the original rule is modified changing the meaning of concept *Article*, that through the application of the linguistic modifier SORT OF, can represent the concept *Paper*. Similarly, let us suppose another ingoing page deals with flight description of a certain airway company; the service mediator can interrogate a specific ontology, declared inside to new page, in order to better center the context of topic. The OWL-based ontology travel definition [26] of airport and flight, is so sketched in the following:

```
< owl : Class rdf : ID = "Airport" >
< rdfs : subClassOf >
< owl : Class rdf : about = "#FlightDomain"/ >
< /rdfs : subClassOf >
< /owl : Class >
< owl : Class rdf : ID = "Flight" >
< rdfs : subClassOfrdf : resource = "#FlightDomain"/ >
< /owl : Class >
```

The terms *Airport* and *Flight* have in common the same parent *FlightDomain*, so they are related to the same concept, and in this case, they can be seen as synonyms. It is natural to exploit linguistic modifiers such as *MORE OR LESS* or *SOMEWHAT*; the fuzzy rule becomes, in this case:

$$\begin{aligned}&\textit{If } (\textit{Flight is } \mu_{\text{MORE OR LESS}}) \text{ and } (\textit{NearestAirport is } \mu_{i_2}) \text{ and } (\textit{Location is } \mu_{i_3})\\ &\text{and } (\textit{Person is } \mu_{i_4}) \text{ and } (\textit{homepage is } \mu_{i_5}) \text{ and } (\textit{Article is } \mu_{i_6}) \text{ and } (\textit{hasAuthor is } \mu_{i_4})\\ &\textit{then } \underline{x} \textit{ belong to class } K_i \textit{ with } \mu_i(\underline{x}).\end{aligned}$$

Table 1
Parameters configuration evaluated on some experiments.

Ex. No.	Pages	Features	Clusters	Concepts	Ingoing pages	Classif. pages /wrong pages	Classif. pages with LM* /wrong pages
1	31	8	3	3	7	4/0	0/0
2	31	10	3	5	7	4/0	2/0
3	60	12	3	n.c.	13	2/0	1/0
4	60	12	4	n.c.	13	4/1	4/1
5	55	9	3	3	22	12/1	7/2
6	80	12	3	3	18	8/2	5/3
7	80	8	3	3	18	12/1	3/1
8	80	10	3	3	18	5/0	3/2

*Considering the unclassified remaining pages.

7. Experimental results

Some experiments have been carried out, exploiting the RDF-based pages from the FOAF site [9], though some pages have been created ad hoc, based on some existing RDF schemas (foaf, vs, bibtex, etc.). In particular, Table 1 describes the system performance in terms of the global parameter configuration. Each row presents an experiment and provides details about the number of RDF-based pages (given as input), the number of metadata chosen as features, the number of salient concepts enclosed into RDF corpus, the associated number of clusters, the additional ingoing pages to classify through the fuzzy rules, those ones classified, exploiting the original fuzzy rules and finally, on the remaining unclassified ones, the further pages classified applying the linguistic modifier on the fuzzy rules. The *wrong pages* represent the classification error (pages associated to a wrong class), evaluated on the rule-based classification (with or without the linguistic modifiers).

Let us observe some sketched results, applying the whole working flow, illustrated in Section 3.

A suitable selection of RDF pages (dealing with some elicited topics) have been carried out for describing some topic-based categories, in accordance with the number of clusters.

Each experiment is composed of two phases: the first phase evaluates the rule-based classification performance on the ongoing RDF-based pages; the other one, instead, applies the linguistic modifiers on the fuzzy rules, for estimating the remaining unclassified pages.

Experiments 1 and 2 are based on the same set of (31) pages, but defining a different number of concepts. Let us note the number of features is strongly related to the concepts identified by the service-based system. Indeed, the concepts are usually recognized considering the metadata/features belonging to the same dictionary or to different dictionaries but describing a common piece of information.

Starting from the same configuration setting (experiments 1–2), a better qualification of main concepts (enriched through additional features) corresponds to a better characterization of information classes and consequently, an improved final classification, through the linguistic modifiers, is obtained. Anyway these experiments are supported by a human assistance to identify the meaningful features to describe the main correspondent concepts. Further proof is the absence of classification error.

Otherwise, an unsupervised execution (without considering the number of clusters with reference to concepts) has been evaluated in the experiments 3 and 4, where automatic clustering and then classification have been performed on the same configuration, but with a different number of clusters. In particular in this specific case, the choice of considering 4 clusters instead of 3 guarantees a better classification of pages.

Further comparing results are examined considering the three experiments (6–8) on a corpus of 80 pages, varying the number of features. Although the experiments begin by considering a common intersection of features, the response of system for classifying new pages apparently returns an unexpected situation.

The common eight features are meaningful, in fact in the relative experiment, only 9 pages out of 18 remains to classify (one of them is misclassified); applying the rules with the linguistic modifiers, further 2 pages have been classified (the same misclassified page has come up). The other two cases (with 10 and 12 features) show results with lower performance. This is due to the features that represent RDF metadata with similar or general meanings, getting confusing or very similar rules in terms of fuzzy sets associated to features. The application of linguistic modifiers remarks this unclear situation. The experiment 5 has been accomplished on an ad-hoc (human-defined) corpus, identifying proper concepts in correspondence of clusters. At the same way, the ingoing pages are accurately selected. Globally, this case shows a well-behaved performance. Additional experimental results are in progress; anyway, further improvements are taken into account, in order to improve the feature space representation which represents the focal point for increasing the system performance.

In conclusion, although the system performance presents interesting characteristics, some weaknesses should be re-considered:

1. sometimes the number of clusters chosen according to the number of principal concepts does not bring toward a good solution: clusters could not represent the concepts, elicited by system services. Rather, as seen before, the classification shows some structural similarity of the pages that are far to purely concept-based arranging;
2. the analysis of system highlights the strong importance of the use of the ontologies, rather than only general-purpose dictionaries, sometimes not able to elicit proper concept through contextual information. Combined use of domain-independent ontologies such as Dublin Core (http://www.dublincore.org), domain-specific ontologies and vocabularies are required.

8. Conclusions

Over the past few years, human view of web interaction has been altering from processes which consume data to the semantic power of data itself. Indeed, recently, it has been passing from a protect approach to the data, according to the proprietary restrictions, to a novel paradigm in which data assumes the prominent role to characterize semantic awareness. Markuped languages such as XML and the new semantic Web technologies (RDF, ontologies, OWL, etc.) suggest enhanced capabilities to make data machine processable and, therefore, understandable.

The most important advantages of this approach are greater flexibility and interoperability. In the future, the Semantic Web and Web Services will happily coexist and will enrich each other. This makes Web Services an ideal choice for machine-to-machine communication. The approach herein described, presents a system for RDF-based classification of documents as basic layer for identifying concepts and the inherent relationships. Although the system is defined to interact with the final user (through a query interface for the search of RDF information), it is designed to support searching and indexing tools oriented to semantic-based classification. The system, presented as prototype, exhibits potential functionalities for future extensions that on the basis of existing general-purpose ontologies. In particular this approach has been advancing towards advanced translation techniques, such as the employment of ontologies (both local, that is embedded into RDF page itself, and public standard ones) as parameters to estimate semantic similarity between pages, through mapping and equivalences, required to identify synonyms and compare similar concepts. Besides, the Semantic Web aims at providing persistent addresses for XML namespaces, RDF schemas and other systems that use URIs as global identifiers, versus the classical Web, stuck in breaking links and in disappearing of web pages. By getting a more stable Web, able to maintaining resources over a time, meets the idea to design systems, able to provide useful functions, destined to represent resources, built to last.

References

[1] D. Beckett, B. McBride, *RDF/XML Syntax Specification (Revised)*, W3C Recommendation, 10 February 2004. Available http://www.w3.org/TR/rdf-syntax-grammar/.

[2] J.C. Bezdek, *Pattern Recognition and Fuzzy Objective Function Algorithms*, Plenum Press, New York, 1981.

[3] D. Brickley, R.V. Guha, *RDF Vocabulary Description Language 1.0: RDF Schema*, W3C Recommendation, 10 February 2004. Available: http://www.w3.org/TR/2004/REC-rdf-schema-20040210/.

[4] M.H. Burstein, *Dynamic invocation of semantic web services that use unfamiliar ontologies*, IEEE Intelligent Systems **19** (4) (2004), 67–73.

[5] DAML Services Coalition, *DAML-S: Semantic Markup for Web Services*, 2001. Available: http://www.http://www.daml.org/services/daml-s/0.9/ Draft Release.

[6] DAML + OIL, *Annotated DAML + OIL Ontology Markup*, W3C Note, 2001. Available: http://www.w3.org/TR/2001/NOTE-daml+oil-walkthru-20011218/.

[7] DARPA-DAML, *DARPA Agent Markup Language*. Available: http://www.daml.org/.

[8] A. Farquhar, R. Fikes, J. Rice, *The ontolingua server: a tool for collaborative ontology construction*, Int. J. Hum.-Comput. Stud. **46** (6) (1997), 707–727.

[9] FOAF, *The Friend of a Friend (FOAF) project*, http://www.foaf-project.org/.

[10] J. Hendler, T. Berners-Lee, E. Miller, *J. Integrating applications on the semantic web*, J. Institute of Electrical Engineers of Japan **122** (10) (October 2002), 676–680.

[11] F. Hoppner, F. Klawonn, R. Kruse, T. Runkler, *Fuzzy Cluster Analysis – Methods for Image Recognition*, Wiley, New York, 1999.

[12] J. Kahan, M. Koivunen, E. Prud'Hommeaux, R.R. Swick, *Annotea: an open RDF infrastructure for shared web annotations*, in: Proceedings of the 10th International World Wide Web Conference, 2001, pp. 623–632.

[13] A. Klose, *Extracting fuzzy classification rules from partially labeled data*, in: Soft Computing, Springer-Verlag, 2003, pp. 417–427.

[14] H. Knublauch, *An AI tool for the real world – knowledge modeling with protege*, Java World, June 2003. Walkthrough of Protege.

[15] W. Lewis, S. Farrar, T. Langendoen, *Building a knowledge base of morphosyntactic terminology*, in: P. Buneman, S. Bird, M. Liberman (Eds.), Proceedings of the IRCS Workshop on Linguistic Databases, 2001.

[16] V. Loia, W. Pedrycz, S. Senatore, *P-FCM: a proximity-based fuzzy clustering*, Fuzzy Sets and Systems **148** (2004), 21–41.

[17] V. Loia, S. Senatore, *Personalized knowledge models using RDF-based fuzzy classification*, in: E. Herrera-Viedma, G. Pasi, F. Crestani (Eds.), Studies in Fuzziness and Soft Computing, Springer-Verlag, 2005, submitted for publication.

[18] V. Loia, S. Senatore, M.I. Sessa, *Knowledge modeling through RDF classification*, in: 18th International FLAIRS Conference, Clearwater Beach, Florida, May 15–17, 2005, AAAI Press.

[19] V. Loia, W. Pedrycz, S. Senatore, *P-FCM: a proximity-based fuzzy clustering for user-centered Web applications*, Int. J. Approximate Reasoning (IJAR) **34** (23) (November 2003), 121–144.

[20] V. Loia, W. Pedrycz, S. Senatore, M.I. Sessa, *Interactive knowledge management for agent-assisted Web navigation*, Int. J. Intelligent Systems (IJIS), Special Issue Intelligent Approaches to Internet Usage and Applications, 2004, in press.

[21] V. Loia, W. Pedrycz, S. Senatore, M.I. Sessa, *Support web navigation by means of cognitive proximity-driven assistant agents*, J. Amer. Soc. Inform. Sci. Technol. 2004, in press.

[22] A. Maedche, S. Staab, N. Stojanovic, R. Studer, Y. Sure, *SEAL – a framework for developing SEmantic Web PortALs*, in: Lecture Notes in Comput. Sci., vol. 2097, 2001, pp. 1–22.

[23] D.L. McGuinness, F. van Harmelen, *Web ontology language (OWL) overview*, W3C Recommendation, February 2004. Available: http://www.w3.org/TR/owl-features/.

[24] E. Motta, J.B. Domingue, M. Dzbor, *Magpie: supporting browsing and navigation on the semantic web*, in: Proceedings of the 9th International Conference on Intelligent User Interface, ACM Press, 2004, pp. 191–197.

[25] OWL Services Coalition, *OWL-S: Semantic Markup for Web Services*, 2004. Available: http://www.daml.org/services/owl-s/.

[26] SchemaWeb, *Travel Ontology*, 2005. Available: http://www.schemaweb.info/-schema/SchemaDetails.aspx?id=236.

[27] SUMO WG, *Sumo ontology site*, 2003. Available URL: http://ontology.teknowledge.com/#FOIS.

[28] K.P. Sycara, M. Paolucci, J. Soudry, N. Srinivasan, *Dynamic discovery and coordination of agent-based semantic web services*, IEEE Internet Computing **8** (3) (2004), 66–73.

[29] W3C SiRPAC, *A simple RDF parser and compiler*, http://www.w3.org/ RDF/Implementations/SiRPAC/.

Fuzzy Description Logics for Ontology Construction

CHAPTER 4

A Fuzzy Description Logic for the Semantic Web

Umberto Straccia
ISTI-CNR, Via G. Moruzzi 1, I-56124 Pisa, Italy
E-mail: straccia@isti.cnr.it

Abstract

In this paper we present a fuzzy version of $\mathcal{SHOIN}(\mathrm{D})$, the corresponding Description Logic of the ontology description language OWL DL. We show that the representation and reasoning capabilities of fuzzy $\mathcal{SHOIN}(\mathrm{D})$ go clearly beyond classical $\mathcal{SHOIN}(\mathrm{D})$. Interesting features are: (i) concept constructors are based on t-norm, t-conorm, negation and implication; (ii) concrete domains are fuzzy sets; (iii) fuzzy modifiers are allowed; and (iv) entailment and subsumption relationships may hold to some degree in the unit interval [0, 1].

Keywords

description logics, ontologies, fuzzy logics

1. Introduction

In the last decade a substantial amount of work has been carried out in the context of *Description Logics* (DLs) [1]. DLs are a logical reconstruction of the so-called frame-based knowledge representation languages, with the aim of providing a simple well-established Tarski-style declarative semantics to capture the meaning of the most popular features of structured representation of knowledge.

Nowadays, DLs have gained even more popularity due to their application in the context of the *Semantic Web* [3,16]. Semantic Web has recently attracted much attention both from academia and industry, and is widely regarded as the next step in the evolution of the World Wide Web. It aims at enhancing content on the World Wide Web with meta-data, enabling *agents* (machines or human users) to *process*, *share* and *interpret* Web content.

Ontologies [9] play a key role in the Semantic Web and major effort has been put by the Semantic Web community into this issue. Informally, an ontology consists of a hierarchical

FUZZY LOGIC AND THE SEMANTIC WEB
Edited by Elie Sanchez
© 2006 Elsevier B.V. All rights reserved

description of important concepts in a particular domain, along with the description of the properties (of the instances) of each concept. DLs play a particular role in this context as they are essentially the theoretical counterpart of the *Web Ontology Language OWL DL*, the state of the art language to specify ontologies. Web content is then annotated by relying on the concepts defined in a specific domain ontology.

However, OWL DL becomes less suitable in all those domains in which the concepts to be represented have not a precise definition. If we take into account that we have to deal with Web content, then it is easily verified that this scenario is, unfortunately, likely the rule rather than an exception. For instance, just consider the case we would like to build an ontology about flowers. Then we may encounter the problem of representing concepts like[1] "Candia is a creamy white rose with dark pink edges to the petals", "Jacaranda is a hot pink rose", "Calla is a very large, long white flower on thick stalks". As it becomes apparent such concepts hardly can be encoded into OWL DL, as they involve so-called *fuzzy* or *vague concepts*, like "creamy", "dark", "hot", "large" and "thick", for which a clear and precise definition is not possible (another issue relates to the representation of terms like "very", which are called fuzzy concepts *modifiers*, as we will see later on).

The problem to deal with imprecise concepts has been addressed several decades ago by Zadeh [33], which gave birth in the meanwhile to the so-called *fuzzy set and fuzzy logic theory* and a huge number of real life applications exists. Unfortunately, despite the popularity of fuzzy set theory, relative little work has been carried out in extending DLs towards the representation of imprecise concepts, notwithstanding DLs can be considered as a quite natural candidate for such an extension [4,12,13,24,26,28,29,31,32] (see also [7, Chapter 6]).

In this paper we consider a fuzzy extension of $\mathcal{SHOIN}(\mathrm{D})$, the corresponding DL of the ontology description language OWL DL, and present its syntax and semantics. The main feature of fuzzy $\mathcal{SHOIN}(\mathrm{D})$ is that it allows us to represent and reason about vague concepts. None of the approaches to fuzzy DLs deal with the expressive power of the fuzzy extension of $\mathcal{SHOIN}(\mathrm{D})$ we present here. Our purpose is also to integrate most of these contributions within an unique setting and, thus, hope to define a reference language for fuzzy $\mathcal{SHOIN}(\mathrm{D})$. Interesting features are: (i) concept constructors are based on t-norm, t-conorm, negation and implication; (ii) concrete domains are fuzzy sets; (iii) fuzzy modifiers are allowed; and (iv) entailment and subsumption relationships may hold to some degree in the unit interval [0, 1].

We will proceed as follows. In the following section we recall the description logic $\mathcal{SHOIN}(\mathrm{D})$. In Section 3 we extend $\mathcal{SHOIN}(\mathrm{D})$ to the fuzzy case and discuss some properties of it. Section 4 presents related work, while Section 5 concludes and presents some topics for further research.

2. Preliminaries

The ontology language OWL DL is strictly related to the DL $\mathcal{SHOIN}(\mathrm{D})$ [16]. Although several XML and RDF syntaxes for OWL-DL exist, in this paper we use the traditional description logic notation. For explicating the relationship between OWL DL and DLs syntax, see, e.g., [15,16]. The purpose of this section is to make the paper self-contained.

[1] Taken from a text book on flowers.

More importantly it helps in understanding the differences between classical $\mathcal{SHOIN}(\mathrm{D})$ and fuzzy $\mathcal{SHOIN}(\mathrm{D})$. The reader confident with the $\mathcal{SHOIN}(\mathrm{D})$ terminology may skip directly to Section 3.

2.1. *Syntax*

$\mathcal{SHOIN}(\mathrm{D})$ allows to reason with concrete data types, such as strings and integers using so-called *concrete domains* [2,21,20,22].

Concrete domains. A *concrete domain* D is a pair $\langle \Delta_{\mathrm{D}}, \Phi_{\mathrm{D}} \rangle$, where Δ_{D} is an interpretation domain and Φ_{D} is the set of concrete domain predicates d with a predefined arity n and an interpretation $d^{\mathrm{D}} \subseteq \Delta^n_{\mathrm{D}}$. For instance, over the integers $\geqslant_{20}$ may be an unary predicate denoting the set of integers greater or equal to 20. For instance, $\mathtt{Person} \sqcap \exists \mathtt{age}. \geqslant_{20}$ will denote a person whose age is greater or equal to `20`.

Alphabets. Let C, R_a, R_c, I_a and I_c be non-empty finite and pair-wise disjoint sets of *concepts names*, *abstract roles names*, *concrete roles names*, *abstract individual names* and *concrete individual names*.

RBox. An *abstract role* is an abstract role name or the inverse S^- of an abstract role name S (concrete role names do not have inverses). An *RBox* $\mathcal{R}$ consists of a finite set of transitivity axioms $\mathtt{trans}(R)$, and role inclusion axioms of the form $R \sqsubseteq S$ and $T \sqsubseteq U$, where R and S are abstract roles, and T and U are concrete roles. The reflexive–transitive closure of the role inclusion relationship is denoted with $\sqsubseteq^*$. A role not having transitive sub-roles is called *simple role*.

Concepts. The set of $\mathcal{SHOIN}(\mathrm{D})$ *concepts* is defined by the following syntactic rules, where A is an atomic concept, R is an abstract role, S is an abstract simple role, T_i are concrete roles, d is a concrete domain predicate, a_i and c_i are abstract and concrete individuals, respectively, and $n \in \mathbb{N}$:

$$
\begin{aligned}
C \longrightarrow\ & \top \mid \bot \mid A \mid C_1 \sqcap C_2 \mid C_1 \sqcup C_2 \mid \neg C \mid \\
& \forall R.C \mid \exists R.C \mid (\geqslant n S) \mid (\leqslant n S) \mid \{a_1, \ldots, a_n\} \mid \\
& (\geqslant n T) \mid (\leqslant n T) \mid \forall T_1, \ldots, T_n.D \mid \exists T_1, \ldots, T_n.D \\
D \longrightarrow\ & d \mid \{c_1, \ldots, c_n\}
\end{aligned}
$$

For instance, we may write the concept

$$\mathtt{Flower} \sqcap (\exists \mathtt{hasPetalWidth}.(\geqslant_{\mathtt{20mm}} \sqcap \leqslant_{\mathtt{40mm}})) \sqcap \exists \mathtt{hasColour.Red})$$

to informally denote the set of flowers having petal's dimension within `20mm` and `40mm`, whose color is red. Here $\geqslant_{\mathtt{20mm}}$ (and $\leqslant_{\mathtt{40mm}}$) is a concrete domain predicate. We use $(= 1\ S)$ as an abbreviation for $(\geqslant 1\ S) \sqcap (\leqslant 1\ S)$.

TBox. A *TBox* $\mathcal{T}$ consists of a finite set of concept inclusion axioms $C \sqsubseteq D$, where C and D are concepts. For ease, we use $C = D \in \mathcal{T}$ in place of $C \sqsubseteq D$, $D \sqsubseteq C \in \mathcal{T}$.

An abstract simple role S is called *functional* if the interpretation of role S is always functional (see later for the semantics). A functional role S can always be obtained from an abstract role by means of the axiom $\top \sqsubseteq (\leqslant 1\ S)$. Therefore, whenever we say that a role is functional, we assume that $\top \sqsubseteq (\leqslant 1\ S)$ is in the TBox.

ABox. An *ABox* $\mathcal{A}$ consists of a finite set of concept and role assertion axioms and individual (in)equality axioms $a{:}C$, $(a, b){:}R$, $(a, c){:}T$, $a \approx b$ and $a \not\approx b$, respectively.

Knowledge base. A $\mathcal{SHOIN}(\mathrm{D})$ *knowledge base* $\mathcal{K} = \langle \mathcal{T}, \mathcal{R}, \mathcal{A} \rangle$ consists of a TBox $\mathcal{T}$, an RBox $\mathcal{R}$, and an ABox $\mathcal{A}$.

2.2. *Semantics*

Interpretation. An *interpretation* $\mathcal{I}$ with respect to a concrete domain D is a pair $\mathcal{I} = (\Delta^{\mathcal{I}}, \cdot^{\mathcal{I}})$ consisting of a non-empty set $\Delta^{\mathcal{I}}$ (called the *domain*), disjoint from Δ_{D}, and of an *interpretation function* $\cdot^{\mathcal{I}}$ that assigns to each $C \in \mathrm{C}$ a subset of $\Delta^{\mathcal{I}}$, to each $R \in \mathrm{R}_a$ a subset of $\Delta^{\mathcal{I}} \times \Delta^{\mathcal{I}}$, to each $a \in \mathrm{I}_a$ an element in $\Delta^{\mathcal{I}}$, to each $c \in \mathrm{I}_c$ an element in Δ_{D}, to each $T \in \mathrm{R}_c$ a subset of $\Delta^{\mathcal{I}} \times \Delta_{\mathrm{D}}$ and to each n-ary concrete predicate d the interpretation $d^{\mathrm{D}} \subseteq \Delta^{n}_{\mathrm{D}}$.

The mapping $\cdot^{\mathcal{I}}$ is extended to concepts and roles as usual:

$$
\begin{aligned}
\top^{\mathcal{I}} &= \Delta^{\mathcal{I}} \\
\bot^{\mathcal{I}} &= \emptyset \\
(C_1 \sqcap C_2)^{\mathcal{I}} &= C_1^{\mathcal{I}} \cap C_2^{\mathcal{I}} \\
(C_1 \sqcup C_2)^{\mathcal{I}} &= C_1^{\mathcal{I}} \cup C_2^{\mathcal{I}} \\
(\neg C)^{\mathcal{I}} &= \Delta^{\mathcal{I}} \setminus C^{\mathcal{I}} \\
(S^{-})^{\mathcal{I}} &= \{\langle y, x\rangle{:}\ \langle x, y\rangle \in S^{\mathcal{I}}\} \\
(\forall R.C)^{\mathcal{I}} &= \{x \in \Delta^{\mathcal{I}}{:}\ R^{\mathcal{I}}(x) \subseteq C^{\mathcal{I}}\} \\
(\exists R.C)^{\mathcal{I}} &= \{x \in \Delta^{\mathcal{I}}{:}\ R^{\mathcal{I}}(x) \cap C^{\mathcal{I}} \neq \emptyset\} \\
(\geqslant n\ S)^{\mathcal{I}} &= \{x \in \Delta^{\mathcal{I}}{:}\ |S^{\mathcal{I}}(x)| \geqslant n\} \\
(\leqslant n\ S)^{\mathcal{I}} &= \{x \in \Delta^{\mathcal{I}}{:}\ |S^{\mathcal{I}}(x)| \leqslant n\} \\
\{a_1, \ldots, a_n\}^{\mathcal{I}} &= \{a_1^{\mathcal{I}}, \ldots, a_n^{\mathcal{I}}\}
\end{aligned}
$$

and similarly for the other constructs, where $R^{\mathcal{I}}(x) = \{y{:}\ \langle x, y\rangle \in R^{\mathcal{I}}\}$ and $|X|$ denotes the cardinality of the set X. In particular,

$$
(\exists T_1, \ldots, T_n.d)^{\mathcal{I}} = \{x \in \Delta^{\mathcal{I}}{:}\ [T_1^{\mathcal{I}}(x) \times \cdots \times T_n^{\mathcal{I}}(x)] \cap d^{\mathrm{D}} \neq \emptyset\}.
$$

Satisfiability. The *satisfiability* of an axiom E in an interpretation $\mathcal{I} = (\Delta^{\mathcal{I}}, \cdot^{\mathcal{I}})$, denoted $I \models E$, is defined as follows: $I \models C \sqsubseteq D$ iff $C^{\mathcal{I}} \subseteq D^{\mathcal{I}}$, $I \models R \sqsubseteq S$ iff $R^{\mathcal{I}} \subseteq S^{\mathcal{I}}$, $I \models T \sqsubseteq U$ iff $T^{\mathcal{I}} \subseteq U^{\mathcal{I}}$, $I \models \mathtt{trans}(R)$ iff $R^{\mathcal{I}}$ is transitive, $I \models a{:}C$ iff $a^{\mathcal{I}} \in C^{\mathcal{I}}$, $I \models (a, b){:}R$ iff $\langle a^{\mathcal{I}}, b^{\mathcal{I}} \rangle \in R^{\mathcal{I}}$, $I \models (a, c){:}T$ iff $\langle a^{\mathcal{I}}, c^{\mathcal{I}} \rangle \in T^{\mathcal{I}}$, $I \models a \approx b$ iff $a^{\mathcal{I}} = b^{\mathcal{I}}$, $I \models a \not\approx b$ iff $a^{\mathcal{I}} \neq b^{\mathcal{I}}$.

For a set of axioms $\mathcal{E}$, we say that I *satisfies* $\mathcal{E}$ iff I satisfies each element in $\mathcal{E}$. If $I \models E$ (resp. $I \models \mathcal{E}$) we say that I is a *model* of E (resp. $\mathcal{E}$). I *satisfies* (is a *model* of) a knowledge base $\mathcal{K} = \langle \mathcal{T}, \mathcal{R}, \mathcal{A} \rangle$, denoted $I \models \mathcal{K}$, iff I is a model of each component $\mathcal{T}$, $\mathcal{R}$ and $\mathcal{A}$, respectively.

Logical consequence. An axiom E is a *logical consequence* of a knowledge base $\mathcal{K}$, denoted $\mathcal{K} \models E$, iff every model of $\mathcal{K}$ satisfies E. According to [15], the entailment and subsumption problem can be reduced to knowledge base satisfiability problem (e.g., $\langle \mathcal{T}, \mathcal{R}, \mathcal{A} \rangle \models a{:}C$ iff $\langle \mathcal{T}, \mathcal{R}, \mathcal{A} \cup \{a{:}\neg C\} \rangle$ unsatisfiable), for which decision procedures and reasoning tools exist (e.g., RACER [10] and FACT [14]).

EXAMPLE 1. Let us consider the following excerpt of a simple ontology (TBox $\mathcal{T}$) about cars, with empty RBox ($\mathcal{R} = \emptyset$):

$$\mathtt{Car} \sqsubseteq (= 1\,\mathtt{maker}) \sqcap (= 1\,\mathtt{passenger}) \sqcap (= 1\,\mathtt{speed})$$

$$(= 1\,\mathtt{maker}) \sqsubseteq \mathtt{Car} \qquad \top \sqsubseteq \forall \mathtt{maker}.\mathtt{Maker}$$
$$(= 1\,\mathtt{passenger}) \sqsubseteq \mathtt{Car} \qquad \top \sqsubseteq \forall \mathtt{passenger}.\mathbb{N}$$
$$(= 1\,\mathtt{speed}) \sqsubseteq \mathtt{Car} \qquad \top \sqsubseteq \forall \mathtt{speed}.\mathtt{Km/h}$$

$$\mathtt{Roadster} \sqsubseteq \mathtt{Cabriolet} \sqcap \exists \mathtt{passenger}.\{2\}$$
$$\mathtt{Cabriolet} \sqsubseteq \mathtt{Car} \sqcap \exists \mathtt{topType}.\mathtt{SoftTop}$$
$$\mathtt{SportsCar} = \mathtt{Car} \sqcap \exists \mathtt{speed}.\geqslant_{\mathtt{245km/h}}$$

In $\mathcal{T}$, the value for `speed` ranges over the concrete domain of kilometers per hour, `Km/h`, while the value for `passengers` ranges over the concrete domain of natural numbers, $\mathbb{N}$. The concrete predicate $\geqslant_{\mathtt{245km/h}}$ is true if the value is greater or equal than to `245km/h`.

The ABox $\mathcal{A}$ contains the following assertions:

$$\mathtt{mgb{:}Roadster} \sqcap (\exists \mathtt{maker}.\{\mathtt{mg}\}) \sqcap (\exists \mathtt{speed}.\leqslant_{\mathtt{170km/h}})$$
$$\mathtt{enzo{:}Car} \sqcap (\exists \mathtt{maker}.\{\mathtt{ferrari}\}) \sqcap (\exists \mathtt{speed}.>_{\mathtt{350km/h}})$$
$$\mathtt{tt{:}Car} \sqcap (\exists \mathtt{maker}.\{\mathtt{audi}\}) \sqcap (\exists \mathtt{speed}.=_{\mathtt{243km/h}})$$

Consider the knowledge base $\mathcal{K} = \langle \mathcal{T}, \mathcal{R}, \mathcal{A} \rangle$. It is then easily verified that, e.g.,

$$\mathcal{K} \models \mathtt{Roadster} \sqsubseteq \mathtt{Car} \qquad \mathcal{K} \models \mathtt{mg{:}Maker}$$
$$\mathcal{K} \models \mathtt{enzo{:}SportsCar} \qquad \mathcal{K} \models \mathtt{tt{:}\neg SportsCar}.$$

The above example illustrates an evident difficulty in defining the class of sport cars. Indeed, it is highly questionable why a car whose speed is `243km/h` is *not* a sport car any more. The point is that essentially, the higher the speed the more likely a car is a sports car, which makes the concept of sports car rather a *fuzzy* concept, i.e. *vague* concept, rather than a crisp one. In the next section we will see how to represent such concepts more appropriately.

3. Fuzzy OWL DL

Fuzzy sets have been introduced by Zadeh [33] as a way to deal with vague concepts like `low pressure`, `high speed` and the like. Formally, a *fuzzy set* A with respect to a universe X is characterized by a *membership function* $\mu_A : X \to [0, 1]$, assigning an A-membership degree, $\mu_A(x)$, to each element x in X. $\mu_A(x)$ gives us an estimation of the belonging of x to A. Typically, if $\mu_A(x) = 1$ then x definitely belongs to A, while $\mu_A(x) = 0.8$ means that x is "likely" to be an element of A.

When we switch to fuzzy logics, the notion of degree of membership $\mu_A(x)$ of an element $x \in X$ w.r.t. the fuzzy set A over X is regarded as the *degree of truth* in [0, 1] of the statement "x *is* A". Accordingly, in our fuzzy DL, (i) a concept C, rather than being interpreted as a classical set, will be interpreted as a fuzzy set and, thus, concepts become *imprecise*; and, consequently, (ii) the statement "a is C", i.e. a:C, will have a truth-value in [0, 1] given by the degree of membership of being the individual a a member of the fuzzy set C.

In the following, we present first some preliminaries on fuzzy set theory (for a more complete and comprehensive presentation see, e.g., [6,11,17,8]) and then define fuzzy $\mathcal{SHOIN}(\mathrm{D})$.

3.1. *Preliminaries on fuzzy set theory*

Let X be a crisp set and let A be a fuzzy subset of X, with membership function $\mu_A(x)$, or simply $A(x) \in [0, 1]$, $x \in X$. The *support* of A, $\mathrm{supp}(A)$, is the crisp set $\mathrm{supp}(A) = \{x \in X\colon A(x) \neq 0\}$. The *scalar cardinality* of A, $|A|$, is defined as $|A| = \sum_{x \in X} A(x)$. The *fuzzy powerset* of X, $\mathcal{F}(X)$, is the set of all the fuzzy sets over X. Let $A, B \in \mathcal{F}(X)$. We say that A and B are *equivalent* iff $A(x) = B(x)$, $\forall x \in X$. A is a *crisp subset* of B iff $A(x) \leqslant B(x)$, $\forall x \in X$. Note that either A is a subset of B or it is not. We will see later on a different notion of subset, in which A is a subset of B to some degree in [0, 1]. We next give the basic definitions on fuzzy set operations (complement, intersection and union).

The *complement* of A, $\neg A$, is given by membership function $(\neg A)(x) = n(A(x))$, for any $x \in X$. The function $n : [0, 1] \to [0, 1]$, called *negation*, has to satisfy the following conditions and extends Boolean negation:

- $n(0) = 1$ and $n(1) = 0$;
- $\forall a, b \in [0, 1]$, $a \leqslant b$ implies $n(b) \leqslant n(a)$;
- $\forall a \in [0, 1]$, $n(n(a)) = a$.

Several negation functions have been given in the literature, e.g., Lukasiewicz negation $n_L(a) = 1 - a$ (syntax, $\neg_L$) and Gödel negation $n_G(0) = 1$ and $n(a) = 0$ if $a > 0$ (syntax, $\neg_G$).

The *intersection* of two fuzzy sets A and B is given $(A \wedge B)(x) = t(A(x), B(x))$, where t is a *triangular norm*, or simply *t-norm*. A t-norm, called *conjunction*, is a function $t : [0, 1] \times [0, 1] \to [0, 1]$ that has to satisfy the following conditions:

- $\forall a \in [0,1], t(a,1) = a$;
- $\forall a, b, c \in [0,1], b \leqslant c$ implies $t(a,b) \leqslant t(a,c)$;
- $\forall a, b \in [0,1], t(a,b) = t(b,a)$;
- $\forall a, b, c \in [0,1], t(a, t(b,c)) = t(t(a,b), c)$.

Examples of t-norms are: $t_L(a,b) = \max(a+b-1, 0)$ (Lukasiewicz t-norm, syntax $\wedge_L$), $t_G(a,b) = \min(a,b)$ (Gödel t-norm, syntax $\wedge_G$), and $t_P(a,b) = a \cdot b$ (product t-norm, syntax $\wedge_P$). Note that $\forall a \in [0,1], t(a,0) = 0$.

The *union* of two fuzzy sets A and B is given $(A \vee B)(x) = s(A(x), B(x))$, where s is a *triangular co-norm*, or simply *s-norm*. A s-norm, called *disjunction*, is a function $s : [0,1] \times [0,1] \to [0,1]$ that has to satisfy the following conditions:

- $\forall a \in [0,1], s(a,0) = a$;
- $\forall a, b, c \in [0,1], b \leqslant c$ implies $s(a,b) \leqslant s(a,c)$;
- $\forall a, b \in [0,1], s(a,b) = s(b,a)$;
- $\forall a, b, c \in [0,1], s(a, s(b,c)) = s(s(a,b), c)$.

Examples of s-norms are: $s_L(a,b) = \min(a+b, 1)$ (Lukasiewicz s-norm, syntax $\vee_L$), $s_G(a,b) = \max(a,b)$ (Gödel s-norm, syntax $\vee_G$), and $s_P(a,b) = a + b - a \cdot b$ (product s-norm, syntax $\vee_P$). Note that if we consider Lukasiewicz negation, then Lukasiewicz, Gödel and product s-norm are related to their respective t-norm according to the De Morgan law: $\forall a, b \in [0,1], s(a,b) = n(t(n(a), n(b)))$.

Another important operator is *implication*, denoted $\to$, that gives a truth-value to the formula $A \to B$, when the truth of A and B are known. A fuzzy implication is a function $i : [0,1] \times [0,1] \to [0,1]$ that has to satisfy the following conditions:

- $\forall a, b, c \in [0,1], a \leqslant b$ implies $i(a,c) \geqslant i(b,c)$;
- $\forall a, b, c \in [0,1], b \leqslant c$ implies $i(a,b) \leqslant i(a,c)$;
- $\forall a \in [0,1], i(0,b) = 1$;
- $\forall a \in [0,1], i(a,1) = 1$;
- $i(1,0) = 0$.

In classical logic, $a \to b$ is a shorthand for $\neg a \vee b$. A generalization to fuzzy logic is, thus, $\forall a, b \in [0,1], i(a,b) = s(n(a), b)$. For instance, $\forall a, b \in [0,1], i_{KD}(a,b) = \max(1-a, b)$ is the so-called Kleene–Dienes implication (syntax, $\to_{KD}$). Another approach to fuzzy implication is based on the so-called *residuum*. His formulation starts from the fact that in classical logic $\neg a \vee b$ can be re-written as $\max\{c \in \{0,1\}: a \wedge c \leqslant b\}$. Therefore, another generalization of implication to fuzzy logic is

$$\forall a, b \in [0,1], \quad i(a,b) = \sup\{c \in [0,1]: t(a,c) \leqslant b\}.$$

For residuum based implication, $i(a,b) = 1$ if $a \leqslant b$. If $a > b$ then, according to the chosen t-norm, we have that, e.g., $i_L(a,b) = 1 - a + b$ for Lukasiewicz implication (syntax, $\to_L$), $i_G(a,b) = b$ for Gödel implication (syntax, $\to_G$) and $i_P(a,b) = a/b$ for product implication (syntax, $\to_P$). Note that, for Lukasiewicz implication, s-norm and negation, we have $i_L(a,b) = s_L(n_L(a), b)$. The same holds using Kleene–Dienes implication, Lukasiewicz negation and Gödel s-norm. On the other hand $i_P(a,b) \neq s_P(n_G(a), b)$ (for instance, for $0 < a \leqslant b < 1$, $i_P(a,b) = 1$, while $s_P(n_G(a), b) = b < 1$).

Another interesting question is when $\forall a, b \in [0,1], i(a,b) = n(t(a, n(b)))$ holds, which in formulae is formulated as $a \to b \equiv \neg(a \wedge \neg b)$. It turns out that, e.g., in Zadeh's logic [33] (i.e. using $\to_{KD}, \wedge_G, \neg_L$) this relation holds. It holds as well in the so-called Lukasiewicz logic (i.e. using $\to_L, \wedge_L, \neg_L$), while it does neither hold for Gödel logic (i.e. using $\to_G, \wedge_G, \neg_G$) nor for the product logic (i.e. using $\to_P, \wedge_P, \neg_G$). For them, just consider the case $1 > a > b > 0$ to verify the inequality. We will see later on that whenever $i(a, b) \neq n(t(a, n(b)))$ then under the fuzzy semantics, $\forall R.C$ is not equivalent to $\neg\exists R.\neg C$.

Fuzzy implication can also be used to determine the degree of subset relationship between two fuzzy subsets A and B over X. Indeed, we define the *degree of subsumption* between A and B, denoted $A \to B$, as $\inf_{x \in X} i(A(x), B(x))$, where i is an implication function. Note that if $\forall x \in [0, 1]$, $A(x) \leqslant B(x)$ holds then $A \to B$ evaluates to 1. Of course, it may be that $A \to B$ evaluates to a value $0 < v < 1$ as well.

We conclude the discussion on fuzzy implication by noting that we have the following inferences: assume $a \geqslant n$ and $i(a, b) \geqslant m$. Then

- under Kleene–Dienes implication we infer that if $n > 1 - m$ then $b \geqslant m$. Indeed, from $i(a, b) = \max(1 - a, b) \geqslant m$, either $1 - a \geqslant m$ or $b \geqslant m$. But $a \geqslant n$, so $1 - a \geqslant m$ implies $1 - m \geqslant a \geqslant n > 1 - m$, a contradiction. Therefore, $b \geqslant m$ must hold. This has been used in [28];
- under residuum based implication w.r.t. a t-norm t, we infer that $b \geqslant t(n, m)$. Indeed, from $i(a, b) = \sup\{c\colon t(a, c) \leqslant b\} \geqslant m$ and $a \geqslant n$ we have $t(n, m) \leqslant t(n, c) \leqslant t(a, c) \leqslant b$.

A (binary) *fuzzy relation* R over two crisp sets X and Y is a function $R : X \times Y \to [0, 1]$. The *inverse* of R is the function $R^{-1} : Y \times X \to [0, 1]$ with membership function $R^{-1}(y, x) = R(x, y)$, for every $x \in X$ and $y \in Y$. The *composition* of two fuzzy relations $R_1 : X \times Y \to [0, 1]$ and $R_2 : Y \times Z \to [0, 1]$ is defined as $(R_1 \circ R_2)(x, z) = \sup_{y \in Y} t(R_1(x, y), R_2(y, z))$, where t is a t-norm. A fuzzy relation R is said to be *transitive* iff $(R \circ R)(x, z) \leqslant R(x, z)$.

We conclude this part with fuzzy modifiers. Fuzzy modifiers applies to fuzzy sets to change their membership function. Well known examples are modifiers like `very`, `more_or_less`, `slightly`, etc. These allow us to define fuzzy sets like `very(High)` and `slightly(Mature)`. Formally, a *modifier*, f_m, is a function $f_m : [0, 1] \to [0, 1]$. For instance, we may define `very`$(x) = x^2$, while define `slightly`$(x) = \sqrt{x}$.

In the following, we use $\wedge$, $\vee$, $\neg$ and $\to$ in infix notation, in place of a t-norm t, s-norm s, negation n and implication operator i.

3.2. *Fuzzy* $\mathcal{SHOIN}(\mathrm{D})$

In this section we give syntax and semantics of fuzzy $\mathcal{SHOIN}(\mathrm{D})$, using the fuzzy operators defined in the previous section. We generalize the semantics given in [13,28,31].

3.2.1. *Syntax*

We have seen that $\mathcal{SHOIN}(\mathrm{D})$ allows to reason with concrete data types, such as strings and integers using so-called concrete domains. In our fuzzy approach, concrete domains may be based on fuzzy sets as well.

Concrete fuzzy domain. A *concrete fuzzy domain* is a pair $\langle \Delta_{\mathrm{D}}, \Phi_{\mathrm{D}} \rangle$, where Δ_{D} is an interpretation domain and Φ_{D} is the set of concrete fuzzy domain predicates d with a predefined arity n and an interpretation $d^{\mathrm{D}} : \Delta_{\mathrm{D}}^n \to [0, 1]$, which is a n-ary fuzzy relation over Δ_{D}.

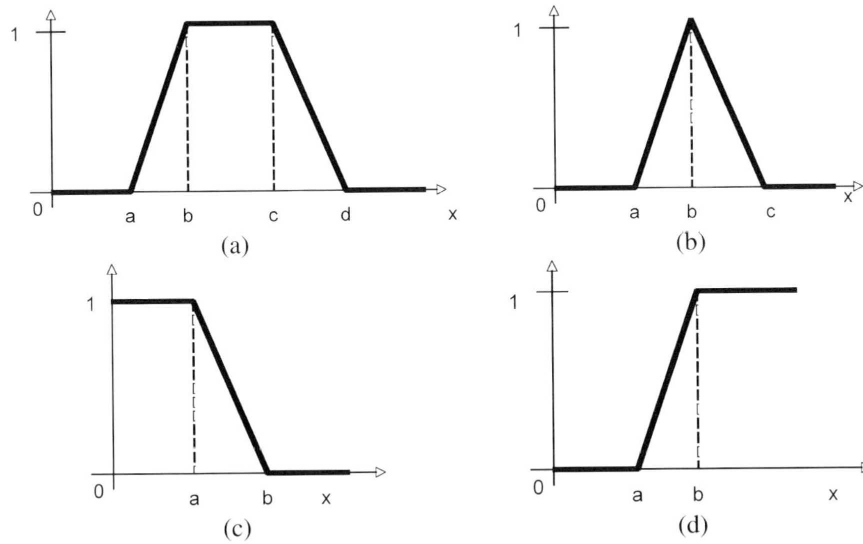

Fig. 1. (a) Trapezoidal function; (b) triangular function; (c) L-function; (d) R-function.

For instance, as for $\mathcal{SHOIN}(\mathrm{D})$, the predicate $\leqslant_{18}$ may be an unary crisp predicate over the natural numbers denoting the set of integers smaller or equal to 18, i.e. $\leqslant_{18}$: Natural $\to [0, 1]$ and

$$\leqslant_{18}(x) = \begin{cases} 1 & \text{if } x \leqslant 18, \\ 0 & \text{otherwise.} \end{cases}$$

So,

$$\texttt{Minor} = \texttt{Person} \sqcap \exists \texttt{age}. \leqslant_{18} \tag{1}$$

defines a person, whose age is less or equal 18, i.e. it defines a minor.

On the other hand, concerning non crisp fuzzy domain predicates, we recall that in fuzzy set theory and practice there are many membership functions for fuzzy sets membership specification. However, the triangular, the trapezoidal, the L-function and the R-function are simple, yet most frequently used to specify membership degrees. The functions are defined over the set of non-negative reals $\mathbb{R}^+ \cup \{0\}$. The *trapezoidal function*, $trz(x, a, b, c, d)$, is defined as follows: let $a < b \leqslant c < d$ be rational numbers then (see Figure 1)

$$trz(x; a, b, c, d) = \begin{cases} 0 & \text{if } x \leqslant a, \\ (x-a)/(b-a) & \text{if } x \in [a, b], \\ 1 & \text{if } x \in [b, c], \\ (d-x)/(d-c) & \text{if } x \in [c, d], \\ 0 & \text{if } x \geqslant d. \end{cases}$$

A *triangular function*, $tri(x; a, b, c)$, is such that (see Figure 1)

$$tri(x; a, b, c) = \begin{cases} 0 & \text{if } x \leqslant a, \\ (x-a)/(b-a) & \text{if } x \in [a, b], \\ (c-x)/(c-b) & \text{if } x \in [b, c], \\ 0 & \text{if } x \geqslant c. \end{cases}$$

Note that $tri(x; a, b, c) = trz(x; a, b, b, c)$.

The L-function is defined as (see Figure 1)

$$L(x; a, b) = \begin{cases} 1 & \text{if } x \leqslant a, \\ (b - x)/(b - a) & \text{if } x \in [a, b], \\ 0 & \text{if } x \geqslant b. \end{cases}$$

Finally, the R-function is defined as (see Figure 1)

$$R(x; a, b) = \begin{cases} 0 & \text{if } x \leqslant a, \\ (x - a)/(b - a) & \text{if } x \in [a, b], \\ 1 & \text{if } x \geqslant b. \end{cases}$$

Using these functions, we may then define, for instance, $\mathtt{Young}:\mathtt{Natural} \to [0, 1]$ to be a fuzzy concrete predicate over the natural numbers denoting the degree of youngness of a person's age. The concrete fuzzy predicate `Young` may be defined as

$$\mathtt{Young}(x) = L(x; 10, 30).$$

So,

$$\mathtt{YoungPerson} = \mathtt{Person} \sqcap \exists \mathtt{age}.\mathtt{Young} \tag{2}$$

will denote a young person.

Modifiers. We allow modifiers in fuzzy $\mathcal{SHOIN}(\mathrm{D})$. Fuzzy modifiers, like `very`, `more_or_less` and `slightly`, apply to fuzzy sets to change their membership function. Formally, a *modifier* is a function $f_m : [0, 1] \to [0, 1]$. For instance, we may define $\mathtt{very}(x) = x^2$, while define $\mathtt{slightly}(x) = \sqrt{x}$. Modifiers has been considered, for instance, in [13,31]. From a syntax point of view, if $\mathtt{M}$ is a new alphabet for modifier symbols, $m \in \mathtt{M}$ is a modifier and C is a $\mathcal{SHOIN}(\mathrm{D})$ concept, then $m(C)$ is fuzzy $\mathcal{SHOIN}(\mathrm{D})$ concept as well. For instance, by referring to Example 1, we may define the concept of sports car as the concept

$$\mathtt{SportsCar} = \mathtt{Car} \sqcap \exists \mathtt{speed}.\mathtt{very}(\mathtt{High}), \tag{3}$$

where `very` is a concept modifier, with membership function $\mathtt{very}(x) = x^2$, and `High` is a fuzzy concrete predicate over the domain of speed expressed in kilometers per hour and may be defined as

$$\mathtt{High}(x) = R(x; 80, 250).$$

Similarly, we may represent "Calla is a very large, long white flower on thick stalks" as

$$\begin{aligned} \mathtt{Calla} = \mathtt{Flower} \sqcap \big(\exists \mathtt{hasSize}.\mathtt{very}(\mathtt{Large})\big) \sqcap \\ \sqcap(\exists \mathtt{hasPetalWidth}.\mathtt{Long}) \sqcap \\ \sqcap(\exists \mathtt{hasColour}.\mathtt{White}) \sqcap (\exists \mathtt{hasStalks}.\mathtt{Thick}), \end{aligned}$$

where `Large`, `Long` and `Thick` are fuzzy concrete predicates.

In summary, the syntax of fuzzy $\mathcal{SHOIN}(\mathrm{D})$ concepts is as follows:

$$\begin{aligned} C \longrightarrow & \top \mid \bot \mid A \mid C_1 \sqcap C_2 \mid C_1 \sqcup C_2 \mid \neg C \mid m(C) \\ & \forall R.C \mid \exists R.C \mid (\geqslant n\ S) \mid (\leqslant n\ S) \mid \{a_1, \ldots, a_n\} \mid \\ & (\geqslant n\ T) \mid (\leqslant n\ T) \mid \forall T_1, \ldots, T_n.D \mid \exists T_1, \ldots, T_n.D \\ D \longrightarrow & d \mid \{c_1, \ldots, c_n\} \end{aligned}$$

Concerning axioms and assertions, similarly to [28], we introduce fuzzy axioms. Let be $n \in (0, 1]$.

Fuzzy RBox. A fuzzy *RBox* $\mathcal{R}$ is a finite set of $\mathcal{SHOIN}(\mathrm{D})$ transitivity axioms $\mathtt{trans}(R)$ and *fuzzy role inclusion axioms* of the form $\langle \alpha \geqslant n \rangle$, $\langle \alpha \leqslant n \rangle$, $\langle \alpha > n \rangle$ and $\langle \alpha > n \rangle$, where α is a $\mathcal{SHOIN}(\mathrm{D})$ role inclusion axiom.

Fuzzy TBox. A fuzzy *TBox* $\mathcal{T}$ consists of a finite set of *fuzzy concept inclusion axioms* of the form $\langle \alpha \geqslant n \rangle$, $\langle \alpha \leqslant n \rangle$, $\langle \alpha > n \rangle$ and $\langle \alpha < n \rangle$ where α is a $\mathcal{SHOIN}(\mathrm{D})$ concept inclusion axiom.

Fuzzy ABox. A fuzzy *ABox* $\mathcal{A}$ consists of a finite set of *fuzzy concept* and *fuzzy role assertion axioms* of the form $\langle \alpha \geqslant n \rangle$, $\langle \alpha \leqslant n \rangle$, $\langle \alpha > n \rangle$, or $\langle \alpha < n \rangle$, where α is a $\mathcal{SHOIN}(\mathrm{D})$ concept or role assertion. As for the crisp case, $\mathcal{A}$ may also contain a finite set of individual (in)equality axioms $a \approx b$ and $a \not\approx b$, respectively.

For instance, $\langle a{:}C \geqslant 0.1 \rangle$, $\langle (a, b){:}R \leqslant 0.3 \rangle$, $\langle R \sqsubseteq S \geqslant 0.4 \rangle$, or $\langle C \sqsubseteq D \leqslant 0.6 \rangle$ are fuzzy axioms. Informally, from a semantics point of view, a fuzzy axiom $\langle \alpha \leqslant n \rangle$ constrains the membership degree of α to be less or equal to n (similarly for $\geqslant, >, <$). As a consequence,

$$\langle \mathtt{jim{:}YoungPerson} \geqslant 0.2 \rangle,$$

i.e. $\langle \mathtt{jim{:}Person} \sqcap \exists \mathtt{age.Young} \geqslant 0.2 \rangle$, dictates that `jim` is a `YoungPerson` with degree at least 0.2.

On the other hand, a fuzzy concept inclusion axiom of the form

$$\langle C \sqsubseteq D \geqslant n \rangle$$

dictates that the subsumption degree between C and D is at least n.

Fuzzy knowledge base. A $\mathcal{SHOIN}(\mathrm{D})$ *fuzzy knowledge base* $\mathcal{K} = \langle \mathcal{T}, \mathcal{R}, \mathcal{A} \rangle$ consists of a fuzzy TBox $\mathcal{T}$, a fuzzy RBox $\mathcal{R}$, and a fuzzy ABox $\mathcal{A}$.

3.2.2. *Semantics*

The semantics extends [28]. The main idea is that concepts and roles are interpreted as fuzzy subsets of an interpretation's domain. Therefore, $\mathcal{SHOIN}(\mathrm{D})$ axioms, rather being satisfied (true) or unsatisfied (false) in an interpretation, become a degree of truth in [0, 1].

Fuzzy interpretation. A *fuzzy interpretation* $\mathcal{I}$ with respect to a concrete domain D is a pair $\mathcal{I} = (\Delta^{\mathcal{I}}, \cdot^{\mathcal{I}})$ consisting of a non empty set $\Delta^{\mathcal{I}}$ (called the *domain*), disjoint from Δ_{D}, and of a *fuzzy interpretation function* $\cdot^{\mathcal{I}}$ that assigns

- to each abstract concept $C \in \mathrm{C}$ a function $C^{\mathcal{I}} : \Delta^{\mathcal{I}} \to [0, 1]$;
- to each abstract role $R \in \mathrm{R}_a$ a function $R^{\mathcal{I}} : \Delta^{\mathcal{I}} \times \Delta^{\mathcal{I}} \to [0, 1]$;
- to each abstract functional role $R \in \mathrm{R}_a$ a partial function $R^{\mathcal{I}} : \Delta^{\mathcal{I}} \times \Delta^{\mathcal{I}} \to [0, 1]$ such that for all $x \in \Delta^{\mathcal{I}}$ there is an unique $y \in \Delta^{\mathcal{I}}$ on which $R^{\mathcal{I}}(x, y)$ is defined;
- to each abstract individual $a \in \mathrm{I}_a$ an element in $\Delta^{\mathcal{I}}$;
- to each concrete individual $c \in \mathrm{I}_c$ an element in Δ_{D};
- to each concrete role $T \in \mathrm{R}_c$ a function $R^{\mathcal{I}} : \Delta^{\mathcal{I}} \times \Delta_{\mathrm{D}} \to [0, 1]$;

- to each concrete functional role $T \in \mathrm{R}_c$ a partial function $t^{\mathcal{I}} : \Delta^{\mathcal{I}} \times \Delta_{\mathrm{D}} \to [0, 1]$ such that for all $x \in \Delta^{\mathcal{I}}$ there is an unique $v \in \Delta_{\mathrm{D}}$ on which $T^{\mathcal{I}}(x, v)$ is defined;
- to each modifier $m \in \mathtt{M}$ the modifier function $f_{\mathtt{m}} : [0, 1] \to [0, 1]$;
- to each n-ary concrete fuzzy predicate d the fuzzy relation $d^{\mathrm{D}} : \Delta_{\mathrm{D}}^{n} \to [0, 1]$.

The mapping $\cdot^{\mathcal{I}}$ is extended to concepts and roles as specified in the following table (where $x, y \in \Delta^{\mathcal{I}}$, $v \in \Delta_{\mathrm{D}}$):

$$
\begin{aligned}
\top^{\mathcal{I}}(x) &= 1 \\
\bot^{\mathcal{I}}(x) &= 0 \\
(C_1 \sqcap C_2)^{\mathcal{I}}(x) &= C_1^{\mathcal{I}}(x) \wedge C_2{}^{\mathcal{I}}(x) \\
(C_1 \sqcup C_2)^{\mathcal{I}}(x) &= C_1^{\mathcal{I}}(x) \vee C_2{}^{\mathcal{I}}(x) \\
(\neg C)^{\mathcal{I}}(x) &= \neg C^{\mathcal{I}}(x) \\
\big(m(C)\big)^{\mathcal{I}}(x) &= f_{\mathtt{m}}\big(C^{\mathcal{I}}(x)\big) \\
(\forall R.C)^{\mathcal{I}}(x) &= \inf_{y \in \Delta^{\mathcal{I}}} R^{\mathcal{I}}(x, y) \to C^{\mathcal{I}}(y) \\
(\exists R.C)^{\mathcal{I}}(x) &= \sup_{y \in \Delta^{\mathcal{I}}} R^{\mathcal{I}}(x, y) \wedge C^{\mathcal{I}}(y) \\
(\geqslant n\ S)^{\mathcal{I}}(x) &= \sup_{\substack{\{y_1, \ldots, y_n\} \subseteq \Delta^{\mathcal{I}} \\ |\{y_1, \ldots, y_n\}| = n}} \bigwedge_{i=1}^{n} S^{\mathcal{I}}(x, y_i) \\
(\leqslant n\ S)^{\mathcal{I}}(x) &= \neg(\geqslant n + 1\ S)^{\mathcal{I}}(x) \\
\{a_1, \ldots, a_n\}^{\mathcal{I}}(x) &= \bigvee_{i=1}^{n} a_i{}^{\mathcal{I}} = x \\
d(v) &= d^{\mathrm{D}}(v) \\
\{c_1, \ldots, c_n\}^{\mathcal{I}}(v) &= \bigvee_{i=1}^{n} c_i{}^{\mathcal{I}} = v \\
(\forall T_1, \ldots, T_n.D)^{\mathcal{I}}(x) &= \inf_{y_1, \ldots, y_n \in \Delta_{\mathrm{D}}{}^{\mathcal{I}}} \left(\bigwedge_{i=1}^{n} T_i{}^{\mathcal{I}}(x, y_i) \right) \to D^{\mathcal{I}}(y_1, \ldots, y_n) \\
(\exists T_1, \ldots, T_n.D)^{\mathcal{I}}(x) &= \sup_{y_1, \ldots, y_n \in \Delta_{\mathrm{D}}^{\mathcal{I}}} \left(\bigwedge_{i=1}^{n} T_i{}^{\mathcal{I}}(x, y_i) \right) \wedge D^{\mathcal{I}}(y_1, \ldots, y_n) \\
(S^{-})^{\mathcal{I}}(x, y) &= S^{\mathcal{I}}(y, x).
\end{aligned}
$$

We comment briefly some points. The semantics of $\exists R.C$

$$
(\exists R.C)^{\mathcal{I}}(d) = \sup_{y \in \Delta^{\mathcal{I}}} R^{\mathcal{I}}(x, y) \wedge C^{\mathcal{I}}(y)
$$

is the result of viewing $\exists R.C$ as the open first order formula $\exists y.F_R(x, y) \wedge F_C(y)$ (where F is the obvious translation of roles and concepts into First-Order Logic – FOL) and the existential quantifier $\exists$ is viewed as a disjunction over the elements of the domain. Similarly,

$$(\forall R.C)^{\mathcal{I}}(x) = \inf_{y \in \Delta^{\mathcal{I}}} R^{\mathcal{I}}(x, y) \to C^{\mathcal{I}}(y)$$

is related to the open first-order formula $\forall y.F_R(x, y) \to F_C(y)$, where the universal quantifier $\forall$ is viewed as a conjunction over the elements of the domain. However, as we already pointed out in Section 3.1, unlike the classical case, in general we do not have that $(\forall R.C)^{\mathcal{I}} = (\neg\exists R.\neg C)^{\mathcal{I}}$. For instance, it holds in Lukasiewicz logic, but not in Gödel logic. Also interesting is that (see [12]) the axiom $\top \sqsubseteq \neg(\forall R.A) \sqcap (\neg\exists R.\neg A)$ has no classical model. However, in [12] it is shown that in Gödel logic it has no finite model, but has an infinite model.

Another point concerns the semantics of number restrictions. The semantics of the concept $(\geqslant n\ S)$

$$(\geqslant n\ S)^{\mathcal{I}}(x) = \sup_{\substack{\{y_1,\dots,y_n\} \subseteq \Delta^{\mathcal{I}} \\ |\{y_1,\dots,y_n\}| = n}} \bigwedge_{i=1}^{n} S^{\mathcal{I}}(x, y_i)$$

is the result of viewing $(\geqslant n\ S)$ as the open first-order formula

$$\exists y_1, \dots, y_n. \bigwedge_{i=1}^{n} S(x, y_i) \wedge \bigwedge_{1 \leqslant i < j \leqslant n} y_i \neq y_j.$$

That is, there are at least n distinct elements that satisfy to some degree $S(x, y_i)$. This guarantees us that $\exists S.\top \equiv (\geqslant 1\ S)$. The semantics of $(\leqslant n\ S)$ is defined in such a way to guarantee the classical relationship $(\leqslant n\ S) \equiv \neg(\geqslant n+1\ S)$.

An alternative definition for the $(\geqslant n\ S)$ and the $(\leqslant n\ S)$ constructs may rely on the scalar cardinality of a fuzzy set. However, we prefer to stick on the formulation, which derives directly from its FOL translation.

Finally, the mapping $\cdot^{\mathcal{I}}$ is extended to non-fuzzy axioms as specified in the following table (where $a, b \in \mathtt{I}_a$):

$$\begin{aligned}
(R \sqsubseteq S)^{\mathcal{I}} &= \inf_{x,y \in \Delta^{\mathcal{I}}} R^{\mathcal{I}}(x, y) \to S^{\mathcal{I}}(x, y) \\
(T \sqsubseteq U)^{\mathcal{I}} &= \inf_{x,y \in \Delta^{\mathcal{I}}} T^{\mathcal{I}}(x, y) \to U^{\mathcal{I}}(x, y) \\
(C \sqsubseteq D)^{\mathcal{I}} &= \inf_{x \in \Delta^{\mathcal{I}}} C^{\mathcal{I}}(x) \to D^{\mathcal{I}}(x) \\
(a{:}C)^{\mathcal{I}} &= C^{\mathcal{I}}(a^{\mathcal{I}}) \\
((a, b){:}R)^{\mathcal{I}} &= R^{\mathcal{I}}(a^{\mathcal{I}}, b^{\mathcal{I}}).
\end{aligned}$$

Note here that, e.g., the semantics of a concept inclusion axiom $C \sqsubseteq D$ is derived directly from its FOL translation, which is of the form $\forall x.F_C(x) \to F_D(x)$. This definition is novel and is clearly different from the approaches in which $C \sqsubseteq D$ is viewed as $\forall x.C(x) \leqslant D(x)$. This latter approach has the effect that the subsumption relationship is a classical $\{0, 1\}$ relationship, while the former has the advantage that subsumption is determined up to a certain degree in $[0, 1]$.

Satisfiability. The notion of *satisfiability* of a fuzzy axiom E by a fuzzy interpretation $\mathcal{I}$, denoted $I \models E$, is defined as follows: $\mathcal{I} \models \mathtt{trans}(R)$, iff $\forall x, y \in \Delta^{\mathcal{I}}.R^{\mathcal{I}}(x, y) = \sup_{z\in\Delta^{\mathcal{I}}} R^{\mathcal{I}}(x, z)\wedge R^{\mathcal{I}}(z, y)$. $I \models \langle\alpha \geqslant n\rangle$, where α is a role inclusion or concept inclusion axiom, iff $\alpha^{\mathcal{I}} \geqslant n$. Similarly, for the other relations $\leqslant$, $<$ and $>$. $I \models \langle\alpha \geqslant n\rangle$, where α is a concept or a role assertion axiom, iff $\alpha^{\mathcal{I}} \geqslant n$. Similarly, for the other relations $\leqslant, <, >$. Finally, $\mathcal{I} \models a \approx b$ iff $a^{\mathcal{I}} = b^{\mathcal{I}}$ and $\mathcal{I} \models a \not\approx b$ iff $a^{\mathcal{I}} \neq b^{\mathcal{I}}$.

For a set of fuzzy axioms $\mathcal{E}$, we say that I *satisfies* $\mathcal{E}$ iff I satisfies each element in $\mathcal{E}$. If $I \models E$ (resp. $I \models \mathcal{E}$) we say that I is a *model* of E (resp. $\mathcal{E}$). I *satisfies* (is a *model* of) a fuzzy knowledge base $\mathcal{K} = \langle\mathcal{T}, \mathcal{R}, \mathcal{A}\rangle$, denoted $I \models \mathcal{K}$, iff I is a model of each component $\mathcal{T}$, $\mathcal{R}$ and $\mathcal{A}$, respectively.

Logical consequence. A fuzzy axiom E is a *logical consequence* of a knowledge base $\mathcal{K}$, denoted $\mathcal{K} \models E$ iff every model of $\mathcal{K}$ satisfies E.

The interesting point is that according to our semantics, e.g., a minor is a young person to a certain degree and is obtained without explicitly mentioning it. This inference can not be achieved in classical $\mathcal{SHOIN}(\mathrm{D})$. Similarly, by referring to Example 1, we will have that the car `tt` will be a sports car to a certain degree. Therefore, unlike Example 1, `tt` is now likely a sport car, *as it should be*. The following two examples highlight these points.

EXAMPLE 2. Let us consider Example 1, where all axioms of the TBox and ABox are asserted with degree 1, i.e. are of the form $\langle\alpha \geqslant 1\rangle$. We replace the definition of `SportsCar` with definition (3). Then we have that (interpreting conjunction as min)

$$\mathcal{K} \models \langle \mathtt{SportsCar} \sqsubseteq \mathtt{Car} \geqslant 1\rangle \qquad \mathcal{K} \models \langle \mathtt{mgb{:}SportsCar} \leqslant 0.25\rangle$$
$$\mathcal{K} \models \langle \mathtt{enzo{:}SportsCar} \geqslant 1\rangle \qquad \mathcal{K} \models \langle \mathtt{tt{:}SportsCar} \geqslant 0.82\rangle.$$

Note how the maximal speed limit of the `mgb` car ($\leqslant_{\mathtt{170km/h}}$) induces an upper limit, 0.25, of the membership degree. Neither this inference is possible in classical $\mathcal{SHOIN}(\mathrm{D})$, nor the one involving `tt`.

EXAMPLE 3. Consider the knowledge base $\mathcal{K}$ with definitions (1) and (2). Then under Lukasiewicz logic we have that (see [30])

$$\mathcal{K} \models \langle \mathtt{Minor} \sqsubseteq \mathtt{YoungPerson} \geqslant 0.2\rangle,$$

which is a relationship not captured with classical $\mathcal{SHOIN}(\mathrm{D})$.

Best truth value bound. Finally, given $\mathcal{K}$ and an axiom α, where α is neither a transitivity axiom, nor an individual (in) equality axiom, it is of interest to compute α's best lower and upper degree value bounds. The *greatest lower bound* of α w.r.t. $\mathcal{K}$ (denoted $glb(\mathcal{K}, \alpha)$) is

$$glb(\mathcal{K}, \alpha) = \sup\{n \colon \mathcal{K} \models \langle\alpha \geqslant n\rangle\},$$

while the *least upper bound* of α with respect to $\mathcal{K}$ (denoted $lub(\mathcal{K}, \alpha)$) is

$$lub(\mathcal{K}, \alpha) = \inf\{n \colon \mathcal{K} \models \langle\alpha \leqslant n\rangle\},$$

where $\sup\emptyset = 0$ and $\inf\emptyset = 1$. Determining the *lub* and the *glb* is called the *Best Degree Bound* (BDB) problem. For instance, the entailments in Examples 2 and 3 are the best possible degree bounds. Furthermore, note that,

$$lub(\Sigma, a{:}C) = \neg glb(\Sigma, a{:}\neg C),$$

i.e. the *lub* can be determined through the *glb* (and vice-versa). Similarly, $lub(\Sigma, (a, b){:}R) = \neg glb(\Sigma, a{:}\neg\exists R.\{b\})$ holds. Also, note that, $\Sigma \models \langle\alpha \geqslant n\rangle$ iff $glb(\Sigma, \alpha) \geqslant n$, and similarly $\Sigma \models \langle\alpha \leqslant n\rangle$ iff $lub(\Sigma, \alpha) \leqslant n$ hold.

Concerning the entailment problem, it is quite easily verified that, as for the crisp case, the entailment problem can be reduced to the unsatisfiability problem:

$$\langle\mathcal{T}, \mathcal{R}, \mathcal{A}\rangle \models \langle\alpha \geqslant n\rangle \quad \text{iff} \quad \big\langle\mathcal{T}, \mathcal{R}, \mathcal{A} \cup \{\langle\alpha < n\rangle\}\big\rangle \text{ is not satisfiable,}$$
$$\langle\mathcal{T}, \mathcal{R}, \mathcal{A}\rangle \models \langle\alpha \leqslant n\rangle \quad \text{iff} \quad \big\langle\mathcal{T}, \mathcal{R}, \mathcal{A} \cup \{\langle\alpha > n\rangle\}\big\rangle \text{ is not satisfiable.}$$

3.3. *Reasoning*

Unfortunately, from a computational point of view, no calculus exists yet checking satisfiability of fuzzy $\mathcal{SHOIN}(\mathrm{D})$ knowledge bases. [13,31] report a calculus for the case of $\mathcal{ALC}$ [25] (with concept constructors $\top, \bot, \neg, \sqcap, \sqcup, \forall, \exists$) with modifiers and simple TBox, with min, max and $\rightarrow_{KD}$ connectives. No indication for the BDB problem is given. [26,28] reports a calculus for $\mathcal{ALC}$ and simple TBox, with min, max and $\rightarrow_{KD}$ connectives and addresses the BDB problem and, [29] shows how the satisfiability problem and the BDB problem can be reduced to classical $\mathcal{ALC}$ and, thus, can be resolved by means of a tools like FACT and RACER. However, despite these negative results, recently [30] reports a calculus for $\mathcal{ALC}(\mathrm{D})$ whenever the connectives, the modifiers and the concrete fuzzy predicates are representable as a bounded Mixed Integer Program. For instance, Lukasiewicz logic satisfies these conditions as well as the membership functions for concrete fuzzy predicates we have presented in this paper. Additionally, modifiers should be a combination of linear functions. In that case the calculus consists of a set of constraint propagation rules and an invocation to an oracle for bounded Mixed Integer Programming. But, indeed, the computational aspect is definitely a point that has to be addressed in forthcoming works.

4. Related work

Several ways of extending DL using the theory of fuzzy logic have been proposed in the literature. The first work is due to Yen [32] who considered a sub-language of $\mathcal{ALC}$, $\mathcal{FL}^-$ [5]. However, it already informally talks about the use of modifiers and concrete domains. Though, the unique reasoning facility, the subsumption test, is a crisp yes/no question. Tresp [31] considered fuzzy $\mathcal{ALC}$ extended with a special form of modifiers, which are a combination of two linear functions. min, max and $1 - x$ membership functions has been considered and a sound and complete reasoning algorithm testing the subsumption relationship has been presented. Similar to our approach, a linear programming oracle is needed. Assertional reasoning has been considered by Straccia [26–28], where fuzzy assertion axioms have been allowed in fuzzy $\mathcal{ALC}$ (with min, max and $1-x$ functions), concept modifiers are not allowed however ([27] reports a four-valued variant of fuzzy $\mathcal{ALC}$). He also introduced the BDB problem and provided a sound and complete reasoning algorithm based on completion rules ([29] provides a translation of fuzzy $\mathcal{ALC}$ into classical $\mathcal{ALC}$).

For an application see [23]. In the same spirit [13] extend Straccia's fuzzy $\mathcal{ALC}$ with concept modifiers of the form $f_m(x) = x^\beta$, where $\beta > 0$. A sound and complete reasoning algorithm for the graded subsumption problem, based on completion rules, is presented. Finally, [24] start addressing the issue of alternative semantics of quantifiers in fuzzy $\mathcal{ALC}$ (without the assertional component). No reasoning algorithm is given. Concerning [30], we already addressed it in the previous section. Finally, [12] considers $\mathcal{ALC}$ under arbitrary t-norm and reports, among others, a procedure deciding $\models \langle C \sqsubseteq D \geqslant 1\rangle$ and deciding whether $\langle C \sqsubseteq D \geqslant 1\rangle$ is satisfiable, by a reduction to the propositional BL logic.

5. Conclusions and outlook

We have presented a fuzzy extension of $\mathcal{SHOIN}(\mathrm{D})$ showing that its representation and reasoning capabilities go clearly beyond classical $\mathcal{SHOIN}(\mathrm{D})$. Interestingly, we allow modifiers, fuzzy concrete domain predicates and fuzzy axioms to appear in a $\mathcal{SHOIN}(\mathrm{D})$ knowledge base and the entailment and the subsumption relationship hold to a certain degree. To the best of our knowledge, no other work has yet extended the semantics to $\mathcal{SHOIN}(\mathrm{D})$ in such a way. The argument supporting the necessity of such an extension relies on the fact that vague concepts are abundant in human knowledge and, thus, appear *likely* in Web content.

The main direction for future work involves the computational aspect. Currently, we are addressing the fundamental issue to develop a calculus for reasoning within $\mathcal{SHOIN}(\mathrm{D})$, extending [30].

Another direction is in extending fuzzy $\mathcal{SHOIN}(\mathrm{D})$ with *fuzzy quantifiers* (see [18,19] for an overview on fuzzy quantifiers), where the $\forall$ and $\exists$ quantifiers are replaced with fuzzy quantifiers like `most`, `some`, `usually` and the like (see [24] for a preliminary work in this direction). This allows to define concepts like

$$\mathtt{TopCustomer} = \mathtt{Customer} \sqcap (\mathtt{Usually})\mathtt{buys.ExpensiveItem}$$

$$\mathtt{ExpensiveItem} = \mathtt{Item} \sqcap \exists \mathtt{price.High}.$$

Here, the fuzzy quantifier `Usually` replaces the classical quantifier $\forall$ and `High` is a fuzzy concrete predicate. Fuzzy quantifiers can be applied to inclusion axioms as well, allowing to express, e.g.,

$$(\mathtt{Most})\mathtt{Bird} \sqsubseteq \mathtt{FlyingObject}.$$

Here the fuzzy quantifier `Most` replaces the classical universal quantifier $\forall$ assumed in the inclusion axioms. The above axiom allows to state that most birds fly.

Ultimately, we believe that the fuzzy extension of $\mathcal{SHOIN}(\mathrm{D})$ is of great interest to the Semantic Web community, as it allows to express naturally a wide range of concepts of actual domains, for which a classical $\mathcal{SHOIN}(\mathrm{D})$ representation is unsatisfactory.

References

[1] F. Baader, D. Calvanese, D. McGuinness, D. Nardi, P.F. Patel-Schneider (Eds.), The Description Logic Handbook: Theory, Implementation, and Applications, Cambridge University Press, 2003.

[2] F. Baader, P. Hanschke, *A schema for integrating concrete domains into concept languages*, in: Proceedings of the 12th International Joint Conference on Artificial Intelligence (IJCAI-91), Sydney, 1991, pp. 452–457.
[3] T. Berners-Lee, J. Hendler, O. Lassila, *The semantic web*, The Scientific American **284** (5) (2001), 34–43.
[4] P. Bonatti, A. Tettamanzi, *Some complexity results on fuzzy description logics*, in: A. Petrosino, V. Di Gesù, F. Masulli (Eds.), WILF 2003 International Workshop on Fuzzy Logic and Applications, Lecture Notes in Comput. Sci., vol. 2955, Springer-Verlag, Berlin, 2004.
[5] R.J. Brachman, H.J. Levesque, *The tractability of subsumption in frame-based description languages*, in: Proceedings of AAAI-84, 4th Conference of the American Association for Artificial Intelligence, Austin, TX, 1984, pp. 34–37.
[6] D. Dubois, H. Prade, *Fuzzy Sets and Systems*, Academic Press, New York, 1980.
[7] J.Z. Pan, et al., *Specification of coordination of rule and ontology languages*, Technical Report, Knowledgeweb Network of Excellence, EU-IST-2004-507482, 2004. Deliverable D2.5.1.
[8] S. Gottwald, *A Treatise on Many-Valued Logics*, A Research Studies Press Book, 2000.
[9] N. Guarino, R. Poli, *Formal ontology in conceptual analysis and knowledge representation*, Internat. J. Human and Computer Studies **43** (5/6) (1995), 625–640.
[10] V. Haarslev, R. Möller, *RACER system description*, in: Proceedings of International Joint Conference on Automated Reasoning (IJCAR-01), Siene, Italy, Lecture Notes in Artificial Intelligence, vol. 2083, Springer, 2001, pp. 701–705.
[11] P. Hájek, *Metamathematics of Fuzzy Logic*, Kluwer Academic Publishers, 1998.
[12] P. Hájek, *Making fuzzy description logics more expressive*, Fuzzy Sets and Systems (2005), submitted for publication.
[13] S. Hölldobler, H.-P. Störr, T.D. Khang, *The subsumption problem of the fuzzy description logic ALC_{FH}*, in: Proceedings of the 10th International Conference on Information Processing and Management of Uncertainty in Knowledge-Based Systems (IPMU-04), 2004.
[14] I. Horrocks, *Using an expressive description logic: fact or fiction?*, in: Proc. of the 8th Int. Conf. on the Principles of Knowledge Representation and Reasoning (KR-98), 1998.
[15] I. Horrocks, P. Patel-Schneider, *Reducing OWL entailment to description logic satisfiability*, J. Web Semantics (2004), in press.
[16] I. Horrocks, P.F. Patel-Schneider, F. van Harmelen, *From SHIQ and RDF to OWL: the making of a web ontology language*, J. Web Semantics **1** (1) (2003), 7–26.
[17] G.J. Klir, B. Yuan, *Fuzzy Sets and Fuzzy Logic: Theory and Applications*, Prentice-Hall Inc., Upper Saddle River, NJ, 1995.
[18] Y. Liu, E.E. Kerre, *An overview of fuzzy quantifiers. (i). Interpretations*, Fuzzy Sets and Systems **95** (1) (1998), 1–21.
[19] Y. Liu, E.E. Kerre, *An overview of fuzzy quantifiers. (ii). Reasoning and applications*, Fuzzy Sets and Systems **95** (2) (1998), 135–146.
[20] C. Lutz, *Reasoning with concrete domains*, in: Proceedings of the Sixteenth International Joint Conference on Artificial Intelligence, Morgan Kaufmann Publishers Inc., 1999, pp. 90–95.
[21] C. Lutz, *Description logics with concrete domains – a survey*, in: Advances in Modal Logics, vol. 4, King's College Publications, 2003.
[22] C. Lutz, *Nexp time-complete description logics with concrete domains*, ACM Trans. Comput. Logic **5** (4) (2004), 669–705.
[23] C. Meghini, F. Sebastiani, U. Straccia, *A model of multimedia information retrieval*, J. ACM **48** (5) (2001), 909–970.
[24] D. Sánchez, G.B. Tettamanzi, *Generalizing quantification in fuzzy description logics*, in: Proceedings 8th Fuzzy Days in Dortmund, 2004.
[25] M. Schmidt-Schauß, G. Smolka, *Attributive concept descriptions with complements*, Artificial Intelligence **48** (1991), 1–26.
[26] U. Straccia, *A fuzzy description logic*, in: Proc. of the 15th Nat. Conf. on Artificial Intelligence (AAAI-98), Madison, USA, 1998, pp. 594–599.
[27] U. Straccia, *A framework for the retrieval of multimedia objects based on four-valued fuzzy description logics*, in: F. Crestani, G. Pasi (Eds.), Soft Computing in Information Retrieval: Techniques and Applications, Physica-Verlag (Springer-Verlag), Heidelberg, 2000, pp. 332–357.
[28] U. Straccia, *Reasoning within fuzzy description logics*, J. Artificial Intelligence Research **14** (2001), 137–166.

[29] U. Straccia, *Transforming fuzzy description logics into classical description logics*, in: Proceedings of the 9th European Conference on Logics in Artificial Intelligence (JELIA-04), Lisbon, Portugal, Lecture Notes in Comput. Sci., vol. 3229, Springer-Verlag, 2004, pp. 385–399.

[30] U. Straccia, *Fuzzy description logics with concrete domains*, Technical Report 2005-TR-03, Istituto di Scienza e Tecnologie dell'Informazione, Consiglio Nazionale delle Ricerche, Pisa, Italy, 2005.

[31] C. Tresp, R. Molitor, *A description logic for vague knowledge*, in: Proc. of the 13th European Conf. on Artificial Intelligence (ECAI-98), Brighton (England), August 1998.

[32] J. Yen, *Generalizing term subsumption languages to fuzzy logic*, in: Proceedings of the 12th International Joint Conference on Artificial Intelligence (IJCAI-91), Sydney, Australia, 1991, pp. 472–477.

[33] L.A. Zadeh, *Fuzzy sets*, Inform. and Control **8** (3) (1965), 338–353.

CHAPTER 5

What Does Mathematical Fuzzy Logic Offer to Description Logic?[1]

Petr Hájek

Institute of Computer Science, Academy of Sciences of the Czech Republic, 182 07 Prague, Czech Republic
E-mail: hajek@cs.cas.cz

Abstract

Continuous t-norm based fuzzy predicate logic is surveyed as a generalization of classical predicate logic; then a kind of fuzzy description logic based on our fuzzy predicate logic is briefly described as a powerful but still decidable formal system of description logic dealing with vague (imprecise) concepts.

Description logic is a flourishing domain of research (see [2]) and has been for long developed in such a way that it is naturally embeddable into the classical (two-valued, Boolean) predicate logic. Early papers on a possible fuzzy description logic, notably [14,11,12] work with a rather minimalistic system of fuzzy logic. In my paper [7] I develop a system (ALC-like) of fuzzy description logic based on the formal system of fuzzy logic from my monograph [6]. The presented paper is a *companion* of [6], not containing any proofs but concentrating to a presentation of fuzzy predicate logic as a natural and rich generalization of classical predicate logic (Section 1), a presentation of fuzzy description logic as a natural and powerful generalization of "classical" description logic (Section 2) and some examples and some discussion (Section 3). For description logic see also [1,3,4,15].

1. From classical logic to fuzzy logic

We start with quickly surveying the basic notions of classical predicate logic (which is undoubtedly the queen of all logics). The reader is assumed to know them and hoped to accept a slightly non-traditional presentation prepared for generalization to fuzzy logic.

[1] This paper is a part of the project number 1ET100300419 of the Program of the Information Society – Thematic Program II of the national Research Program of the Czech Republic. The work was partly supported by the Institutional Research Plan AV0Z10300504.

FUZZY LOGIC AND THE SEMANTIC WEB
Edited by Elie Sanchez
© 2006 Elsevier B.V. All rights reserved

A *language* is given by *predicates* $P, Q, \ldots$, each with its arity (number of arguments – unary, binary, ..., n-ary) and *object constants* $a, b, c, \ldots$. *Logical symbols* are *object variables* $x, y, \ldots$, *connectives* (conjunction $\wedge$, disjunction $\vee$, implication $\rightarrow$, equivalence $\equiv$, negation $\neg$), *quantifiers* (universal $\forall$, existential $\exists$) and (possibly) truth constants $\top$ (truth), $\bot$ (falsity). Formulas are built from these in the obvious way (atomic having the form $P(t_1, \ldots, t_n)$ where t_i's are variables or constants; $\top$ and $\bot$ are formulas; other formulas are built using connectives and quantifiers). For example, $(\forall x)(P(x, c) \vee (\exists y)Q(x, y))$ is a formula; it is closed, contains no free variables (all variables are quantified).

There are two truth values: 1 – true, 0 – false. An *interpretation* of such a language is a structure $\mathbf{M} = (M, (r_P)_{P \text{ predicate}}, (v_c)_{c \text{ constant}})$ where M is a non-empty set (the *domain*), for each predicate P of arity n, r_P is the characteristic function of an n-ary relation on M named by P, i.e., $r_P : M^n \rightarrow \{0, 1\}$. For each constant c, v_c is an element of M.

Given $\mathbf{M}$, for each formula $\varphi(x_1, \ldots, x_n)$ with free (non-quantified) variables $x_1, \ldots, x_n$ and for each n-tuple $u_1, \ldots, u_n \in M$, $\|\varphi(u_1, \ldots, u_n)\|_{\mathbf{M}}$ is the truth value of $\varphi(x_1, \ldots, x_n)$ in $\mathbf{M}$ for the elements $u_1, \ldots, u_n$; it is 1 ($u_1, \ldots, u_n$ satisfy φ in $\mathbf{M}$) or 0 (they do not satisfy). This is defined inductively, e.g., for an atomic formula $P(x, y)$ and $u, v \in M$, (u, v) satisfy $P(x, y)$ if $r_P(u, v) = 1$, thus $\|P(u, v)\|_{\mathbf{M}} = r_P(u, v)$. For connectives one uses the well-known truth tables, e.g. (φ stands for $\varphi(u_1, \ldots)$ etc.)

$\|\varphi \wedge \psi\|_M = 1$ iff $\|\varphi\|_M = \|\psi\|_M = 1$, thus $\|\varphi \wedge \psi\|_{\mathbf{M}} = \min(\|\varphi\|_{\mathbf{M}}, \|\psi\|_{\mathbf{M}})$;
$\|\varphi \rightarrow \psi\|_{\mathbf{M}} = 1$ iff $\|\psi\|_{\mathbf{M}} = 1$ or $\|\varphi\|_{\mathbf{M}} = 0$, i.e., iff $\|\varphi\|_{\mathbf{M}} \leqslant \|\psi\|_{\mathbf{M}}$;
$\|\neg\varphi\|_{\mathbf{M}} = 1$ iff $\|\varphi\|_{\mathbf{M}} = 0$; $\|\top\|_{\mathbf{M}} = 1$, $\|\bot\|_{\mathbf{M}} = 0$ for all $\mathbf{M}$.
$\|(\forall x)\varphi(x, \ldots)\|_{\mathbf{M}} = 1$ iff for all $v \in M$, $\|\varphi(v, \ldots)\|_{\mathbf{M}} = 1$,
$\|(\exists x)\varphi(x, \ldots)\|_{\mathbf{M}} = 1$ iff some $v \in M$, $\|\varphi(v, \ldots)\|_{\mathbf{M}} = 1$, thus
$\|(\forall x)\varphi(x, \ldots)\|_{\mathbf{M}} = \min\{\|\varphi(v, \ldots)\|_{\mathbf{M}} \mid v \in M\}$,
$\|(\exists x)\varphi(x, \ldots)\|_{\mathbf{M}} = \max\{\|\varphi(v, \ldots)\|_{\mathbf{M}} \mid v \in M\}$.

Clearly, if φ is closed (has no free variables) then $\|\varphi\|_{\mathbf{M}}$ is just the truth value of φ in $\mathbf{M}$; if $\|\varphi\|_{\mathbf{M}} = 1$ we say that φ is *true* in $\mathbf{M}$ and denote by $\mathbf{M} \models \varphi$.

A formula φ is a *tautology* if it is true in all interpretations. Just mention the existence of *axioms* and *deduction rules* giving the notion of a formula *provable* in the predicate calculus. The *completeness theorem* says that a formula is provable in the predicate calculus iff it is a tautology.

To close this telegraphic summary of the classical predicate calculus let us mention that one can choose some connectives to be basic (or starting) and define the others from them. For example, one can take $\wedge$ and $\neg$ for starting and define $\varphi \vee \psi$ to be $\neg(\neg\varphi \wedge \neg\psi)$, define $\varphi \rightarrow \psi$ to be $\neg\varphi \vee \psi$ and define $\varphi \equiv \psi$ to be $(\varphi \rightarrow \psi) \wedge (\psi \rightarrow \varphi)$. Or take $\rightarrow$ and $\neg$ to be starting and define $\varphi \wedge \psi$ to be $\neg(\varphi \rightarrow \neg\psi)$, define $\varphi \vee \psi$ to be $\neg\varphi \rightarrow \psi$ etc. This is OK since the semantics of the defining and defined formula is the same.

*

Fuzzy logic is the logic of vague (imprecise) notions; formulas of fuzzy logic may be not just true of false but may be partially true, true in a degree. Mathematical fuzzy logic (see [6] and, e.g., [5,8,9]) is a formal system like classical logic but having more than two truth values that are ordered (possibly partially ordered), a logic with a comparative notion of truth. (Think of the truth degree of a sentence like "John is young" etc.) As such, fuzzy logic is a kind of multiple-valued logic (many-valued logic) with some specific properties and aims. The *standard domain of truth degrees* is the real unit interval $[0, 1]$. The definition of a *language* is the same as in classical logic; a (standard) *interpretation* of a language is a structure $\mathbf{M} = (M, (r_P)_{P \text{ predicate}}, (v_c)_{c \text{ constant}})$ where M and v_c are as above but for each P, r_P is a *fuzzy relation* on M, i.e., a mapping $r_P : M^n \to [0, 1]$ assigning to each n-tuple $(v_1, \ldots, v_n)$ of elements of M the *degree* $r_P(v_1, \ldots, v_n)$ in which the tuple is in the relation. A trivial example: $M = \{1, 2, 3\}$, one binary predicate L (read $L(x, y)$ "x likes y") and r_P is given by the following table:

L	1	2	3
1	1	0.2	0
2	0.7	0.8	0.8
3	0	0.1	0.5

Starting connectives are *conjunction, implication* (binary) and the truth constants $\top$ (truth), $\bot$ (falsity). Since we shall have two different conjunctions we start with the conjunction denoted by &; implication will be $\to$. Since early days of fuzzy logic, for the truth function of & one takes a *continuous t-norm*, which is a binary operation on $[0, 1]$ continuous as a real-valued function and satisfying the following for each $x, y, z \in [0, 1]$:

associativity	$x * (y * z) = (x * y) * z$,
commutativity	$x * y = y * z$,
monotony	$x \leqslant y$ implies $x * z \leqslant y * z$,
zero and unit	$0 * x = 0,\ 1 * x = 1$.

Three most important continuous t-norms are

Łukasiewicz	$x * y = \max(0, x + y - 1)$,
Gödel	$x * y = \min(x, y)$,
product	$x * y = x \cdot y$ (real product).

Each continuous t-norm is "composed" from a linearly ordered at most countable system of copies of these three t-norms (Łukasiewicz, Gödel, product) – in a precise well defined sense (Mostert–Shields theorem). For a detailed formulation see, e.g., [6]; here we only mention that the system may and may not have a least (first) element.

Thanks to continuity each continuous t-norm has its *residuum* $x \Rightarrow y = \max\{z \mid x * z \leqslant y\}$ (left continuity suffices). The operation $\Rightarrow$ is the *truth function of implication* given by the t-norm. This implication has very good properties, in particular, for each continuous t-norm $*$, $x \Rightarrow y = 1$ iff $x \leqslant y$. For $x > y$ different t-norms give different residua, notably: for $x > y$,

Łukasiewicz	$x \Rightarrow y = 1 - x + y$,
Gödel	$x \Rightarrow y = y$,
product	$x \Rightarrow y = y/x$.

Given this, we may define the truth degree of each formula $\varphi(x_1, \ldots, x_n)$ given by elements $u_1, \ldots, u_n \in M$ denoted $\|\varphi(u_1, \ldots, u_n)\|^*_{\mathbf{M}}$ (since it depends on the choice of our t-norm) in analogy to the classical logic as follows:

For an atomic formula $P(x_1, \ldots, x_n)$, $\|P(u_1, \ldots, u_n)\|^*_{\mathbf{M}} = r_P(u_1, \ldots, u_n)$, and similarly if the atomic formula contains some constants, e.g., for $P(x, c, y)$ and $u_1, u_2 \in M$,

$\|P(u_1, c, u_2)\|^*_{\mathbf{M}} = r_P(u_1, v_c, u_2)$ (where v_c is the interpretation of c in $\mathbf{M}$).

$$\|\top\|^*_{\mathbf{M}} = 1, \qquad \|\bot\|^*_{\mathbf{M}} = 0, \qquad \|\varphi \mathbin{\&} \psi\|^*_{\mathbf{M}} = \|\varphi\|^*_{\mathbf{M}} * \|\psi\|^*_{\mathbf{M}},$$
$$\|\varphi \to \psi\|^*_{\mathbf{M}} = \|\varphi\|^*_{\mathbf{M}} \Rightarrow \|\psi\|^*_{\mathbf{M}}.$$

(In our example above, $\|L(2,2)\|^*_{\mathbf{M}} = 0.8$, $\|L(2,1)\|^*_{\mathbf{M}} = 0.7$, thus for $*$ being Ł (Łukasiewicz) $\|L(2,2) \to L(2,1)\|^{Ł}_{\mathbf{M}} = 1 - 0.8 + 0.7 = 0.9$; for Gödel you get $\|L(2,2) \to L(2,1)\|^*_{\mathbf{M}} = 0.7$ etc.)

Some defined connectives:

$$\neg\varphi \text{ is } \varphi \to \bot,$$
$$\varphi \wedge \psi \text{ is } \varphi \mathbin{\&} (\varphi \to \psi),$$
$$\varphi \vee \psi \text{ is } \big((\varphi \to \psi) \to \psi\big) \wedge \big((\psi \to \varphi) \to \varphi\big),$$
$$\varphi \equiv \psi \text{ is } (\varphi \to \psi) \mathbin{\&} (\psi \to \varphi).$$

Here $\neg$ is *negation*, $\wedge$ is min-conjunction, $\vee$ is max-disjunction, $\equiv$ is equivalence. For any choice of $*$, the truth function of $\wedge$ is minimum (i.e., $\|\varphi \wedge \psi\|^*_{\mathbf{M}} = \min(\|\varphi\|^*_{\mathbf{M}}, \|\psi\|^*_{\mathbf{M}})$, the truth function of $\vee$ is maximum; negation depends on $*$, in particular:

$$\|\neg\varphi\|^{Ł}_{\mathbf{M}} = 1 - \|\varphi\|^{Ł}_{\mathbf{M}}$$

but for $*$ being G of Π (product) we get Gödel negation: if $\|\varphi\|^G_{\mathbf{M}} = 0$ then $\|\neg\varphi\|^G_{\mathbf{M}} = 1$, but $\|\varphi\|^G_{\mathbf{M}} > 0$ implies $\|\neg\varphi\|^G_{\mathbf{M}} = 0$. (Negation of 0 is 1, negation of a positive value is 0.) Clearly, $\|\varphi \equiv \psi\|^*_{\mathbf{M}} = 1$ iff $\|\varphi\|^*_{\mathbf{M}} = \|\psi\|^*_{\mathbf{M}}$.

For quantifiers the definition is as follows:

$$\big\|(\forall x)\varphi(x,\ldots)\big\|^*_{\mathbf{M}} = \inf\big\{\|\varphi(v,\ldots)\|^*_{\mathbf{M}} \mid v \in M\big\},$$
$$\big\|(\exists x)\varphi(x,\ldots)\big\|^*_{\mathbf{M}} = \sup\big\{\|\varphi(v,\ldots)\|^*_{\mathbf{M}} \mid v \in M\big\},$$

thus the truth degree of a universally quantified formula is the infimum of truth degrees of its instances and similarly for existential quantification and supremum. If M is finite, we may replace "infimum" by "minimum" and "supremum" by "maximum"; for infinite M this may not be the case. For example let M be the set of positive natural numbers $1, 2, 3, \ldots$ and let $r_{Sm}(n) = \frac{1}{n}$ (read Sm "small"); then (for any $*$) $\|(\forall x)Sm(x)\|^*_{\mathbf{M}} = \inf_n \frac{1}{n} = 0$, but $\|Sm(n)\|^*_{\mathbf{M}}$ is positive for each n. (This is related to the so-called Sorites paradox, solved in fuzzy logic.) A formula φ is a $*$-tautology if $\|\varphi\|^*_{\mathbf{M}} = 1$ for each $\mathbf{M}$; φ is a *standard tautology* if it is a $*$-tautology for each continuous t-norm $*$. Analogously to classical logic, some standard tautologies are taken for axioms of the basic fuzzy predicate logic BL∀; deduction rules are as in classical logic (modus ponens and generalization). This gives the notion of *provability* in BL∀. It is complete with respect to a more general semantics over so-called BL-algebras which are some algebras of truth functions, algebras given by continuous t-norms being particular BL-algebras. (BL-algebras are particular *residuated lattices*, we shall not go into any details here, see [6].) Note that the set of all standard predicate tautologies is not (effectively) axiomatizable.

The following are some few examples of standard tautologies:

$$\big(\varphi \to (\psi \to \chi)\big) \equiv \big(\psi \to (\varphi \to \psi)\big)$$
$$\big(\varphi \to (\psi \to \chi)\big) \equiv \big((\varphi \mathbin{\&} \psi) \to \chi\big)$$

$$(\varphi \to \psi) \to (\neg\psi \to \neg\varphi)$$

$$(\varphi \to \psi) \to \big((\varphi \,\&\, \chi) \to (\psi \,\&\, \chi)\big)$$

$$(\forall x)(\varphi \to \psi) \to \big((\forall x)\varphi \to (\forall x)\psi\big)$$

$$(\forall x)(\varphi \to \psi) \to \big((\exists x)\varphi \to (\exists x)\psi\big)$$

$$(\exists x)\varphi \to \neg(\forall x)\neg\varphi$$

Now let us present examples of formulas that are *not* standard tautologies, i.e., for some continuous t-norm $*$ they are not $*$-tautologies. (All of them are tautologies of classical logic.) In brackets, Ł, G, Π means that the formula *is* a $*$-tautology for $*$ being Łukasiewicz, Gödel or product t-norm.

$\neg\neg\varphi \equiv \varphi$	(Ł)
$(\varphi \,\&\, \varphi) \equiv \varphi$	(G)
$\neg(\varphi \wedge \neg\varphi)$	(G, Π)
$\varphi \vee \neg\varphi$	(none)
$(\exists x)\varphi \equiv \neg(\forall x)\neg\varphi$	(Ł)
$(\forall x)\varphi \equiv \neg(\exists x)\neg\varphi$	(Ł)

(The reader may verify easily that these formulas are tautologies of the logics indicated remembering the definition of Łukasiewicz negation and of Gödel negation as well as the fact that in Gödel logic the conjunctions & and $\wedge$ have the same semantics.)

Similarly as in classical logic, using fuzzy logic one gets some feeling (or practice) in recognizing well-known tautologies and well-known non-tautologies (but, I repeat, there is no algorithm to decide on any given formula if it is a standard tautology; similarly for (non-)tautologies of a fixed continuous t-norm).

We shall also use the following terminology: an interpretation **M** is a $*$-*model* of a closed formula φ if $\|\varphi\|^*_{\mathbf{M}} = 1$. **M** is a $*$-model of a set T of closed formulas if $\|\varphi\|^*_{\mathbf{M}} = 1$ for each $\psi \in T$. Finally, T $*$-*entails* φ if each $*$-model of T is an T-model of φ.

To close this section let us mention a "minimalistic" fuzzy logic KD used early papers in fuzzy logic (and much later in early papers on fuzzy description logic). It uses only connectives $\wedge, \vee$ (min-conjunction, max-disjunction), Łukasiewicz negation $\|\neg\varphi\|^{\mathrm{KD}}_{\mathbf{M}} = 1 - \|\varphi\|^{\mathrm{KD}}_{\mathbf{M}}$ and so-called Kleene–Dienes implication $\|\varphi \to \psi\|^{\mathrm{KD}}_{\mathbf{M}} = \max(1 - \|\varphi\|^{\mathrm{KD}}_{\mathbf{M}}, \|\psi\|^{\mathrm{KD}}_{\mathbf{M}})$, inspired by mechanical use of the classical tautology $(\varphi \to \psi) \equiv (\neg\varphi \vee \psi)$. Observe that connectives of KD are definable from connectives given by the Łukasiewicz t-norm; but the KD-implication does not have good properties. For example, observe that in our example above (with the predicate "likes"), for any continuous t-norm $*$, the formula $(\forall x, y)(L(x, y) \to L(x, x))$ has the truth value 1 (since for any u, v, $r_L(u, v) \leqslant r_L(u, u)$, i.e., every x likes himself at least as much as he likes anybody else), i.e., $\|(\forall x, y)(L(x, y) \to L(x, x))\|^*_{\mathbf{M}} = 1$, but $\|(\forall x, y)(L(x, y) \to L(x, x))\|^{\mathrm{KD}}_{\mathbf{M}} = \frac{1}{2}$ since in particular $\|L(3, 3) \to L(3, 3)\|^{\mathrm{KD}}_{\mathbf{M}} = (\frac{1}{2} \Rightarrow_{\mathrm{KD}} \frac{1}{2}) = \frac{1}{2}$. This seems to be rather counter-intuitive.

2. From description logic to fuzzy description logic

We shall restrict ourselves to the description logic $\mathcal{ALC}$ and its fuzzy counterpart. Recall that in $\mathcal{ALC}$ concepts are built from finitely many unary predicates (atomic concepts), and finitely many binary predicate (roles) using connectives $\wedge$, $\vee$, $\neg$ and quantifier constructs: if C, D are concepts then $C \wedge D$, $C \vee D$, $\neg C$ are concepts, if C is a concept and R is role then $(\forall R.C)$, $(\exists R.C)$ are concepts. From the point of view of classical predicate logic, concepts correspond to particular formulas with one free variable: if A is an atomic concept, take $A(x)$; if $C(x)$ and $D(x)$ are defined then $C(x) \wedge D(x)$, $C(x) \vee D(x)$, $\neg C(x)$ have clear meaning. The quantifier constructs are understood as follows.

$$(\forall R.C)(x) \quad \text{means} \quad (\forall y)\big(R(x, y) \to C(y)\big),$$
$$(\exists R.C)(x) \quad \text{means} \quad (\exists y)\big(R(x, y) \wedge C(y)\big).$$

Also for each concept C and a constant a, $C(a)$ has clear meaning. Saying that a concept is valid we mean that $(\forall x)C(x)$ is a predicate tautology; saying that it is satisfiable we mean that $C(a)$ has a model (an interpretation $\mathbf{M}$ such that $\mathbf{M} \models C(a)$). Saying that C is subsumed by D (in symbols, $C \sqsubseteq D$) we mean that $(\forall x)(C(x) \to D(x))$ is a tautology. Since in classical logic implication is definable from $\wedge$ and $\neg$, for any concepts C, D we have the concept $C \to D$ equivalent to the concept $\neg C \vee D$ and we see that $C \sqsubseteq D$ (C is subsumed by D) iff the concept $C \to D$ is valid (i.e., the formula $(\forall x)(C(x) \to D(x))$ is a tautology). It is known that the question if a given concept is valid is decidable and the same for the question if a given concept is satisfiable (see, e.g., [10]) consequently, also subsumption $C \sqsubseteq D$ is decidable.

Note also the finite model property: C is valid iff $(\forall x)C(x)$ is true in all *finite* interpretation (having finite domain); and C is satisfiable iff $C(a)$ is true in a finite interpretation.

For simplicity we restrict ourselves just to those problems, not discussing terminological axioms and axioms-facts. Our aim is to show on this simplest fragment how it can be combined with *fuzzy logic* and what are the problems with this.

Thus take the same language of finitely many atomic concepts and finitely many roles. Concepts are built from atomic concepts using $\bot$, $\&$, $\to$ (i.e., $\bot$ is a concept; each atomic concept is a concept; if C, D are concepts then $C \,\&\, D$, $C \to D$ are concepts) and using the quantifier constructs $(\forall R.C)$, $(\exists R.C)$. Translation to predicate formulas is clear, only take $(\exists R.C)(x)$ to be $(\exists y)(R(x, y) \,\&\, C(y))$ (use the $\&$-conjunction). Also here observe that defined connectives can be used to construct concepts, thus if C, D are concepts then so are $C \wedge D$, $C \vee D$, $\neg C$, $C \equiv D$.

Now choose a continuous t-norm $*$; you may ask whether a concept C is $*$-valid (i.e., $\|(\forall x)C(x)\|^*_{\mathbf{M}} = 1$ for each $\mathbf{M}$), whether it is $*$-satisfiable (there is some $\mathbf{M}$ with $\|C(a)\|^*_{\mathbf{M}} = 1$, i.e., $C(a)$ has a $*$-model), whether C is $*$-subsumed by D and similar. We present here (without proofs) the main results of the paper [7] and illustrate them by some examples.

First we shall introduce some important notions. Recall that the truth degree of an universally quantified formula is defined as the infimum of the set of truth values of its instances and that this set need not have a minimal element. Similarly for existentially quantified formulas supremum and maximum. Let us *define*: let $\varphi(x, y_1, \ldots, y_n)$ be a formula $\mathbf{M}$ an interpretation and $u_1, \ldots, u_n \in M$. An object $v \in M$ is a $*$-*witness* in $\mathbf{M}$ for $(\forall x)\varphi(x, y_1, \ldots, y_n)$ and $u_1, \ldots, u_n$ if $\|(\forall x)\varphi(x, u_1, \ldots, u_n)\|^*_{\mathbf{M}} =$

$\|\varphi(v, u_1, \ldots, u_n)\|^*_{\mathbf{M}}$; similarly for $(\exists x)\varphi(x, y_1, \ldots, y_n)$, i.e., $\|(\exists x)\varphi(x, u_1, \ldots, u_n)\|^*_{\mathbf{M}} = \|\varphi(v, u_1, \ldots, u_n)\|^*_{\mathbf{M}}$. An interpretation $\mathbf{M}$ is $*$-*witnessed* if each formula beginning by a quantifier has a $*$-witness in $\mathbf{M}$ for any evaluation of its free variables by elements of M.

Surely, there are non-witnessed interpretations (we saw some above); each finite interpretation (with a finite domain) is witnessed.

For the Łukasiewicz t-norm Ł the following holds true: A formula φ is Ł-true in some interpretation iff it is Ł-true in some witnessed interpretation; and φ is Ł true in all interpretations iff it is Ł-true in all witnessed interpretations (see [7]); but for other t-norms it is not the case. Observe the formula $\neg(\forall x)P(x)\,\&\,\neg(\exists x)\neg P(x)$, which has a G-model (and a Π-model) but no witnessed G-model (Π-model): To get a G-model, take the interpretation of Sm above: $r_{Sm}(n) > 0$ for all $n \in M$ but $\inf_n r_{Sm}(n) = 0$. Thus $\|(\forall x)Sm(x)\|^G_{\mathbf{M}} = 0$, $\|(\exists x)\neg Sm(x)\|^G_{\mathbf{M}} = 0$, hence $\|\neg(\forall x)Sm(x)\,\&\,\neg(\exists x)\neg Sm(x)\|^G_{\mathbf{M}} = 1$. In Section 3 we show that the formula has no witnessed model.

This leads us to the investigation of witnessed models of concepts. (For the aims of description logic non-witnessed models appear to be pathological.) We get the following:

THEOREM 1. *Let C be a concept, $*$ a continuous t-norm.*

(1) *$C(a)$ has a witnessed $*$-model iff it has a finite $*$-model.*

(2) *$(\forall x)C(x)$ is true in all witnessed $*$-interpretations iff it is $*$-true in all finite $*$-interpretations.*

Observe that, e.g., for C being $A \equiv \neg A$ the formula $C(a)$ has a finite Ł-model (in which $\|A(a)\|^{Ł}_{\mathbf{M}} = \frac{1}{2}$) but $C(a)$ has no G-model (Π-model); if C is $A \equiv (A\,\&\,A)$ then $(\forall x)C(x)$ is true in all G-interpretations but not in all Ł-interpretations, neither in all Π-interpretations.

For a full proof of the theorem see [7]; here we only sketch the main idea. Given a formula $C(a)$ (where C is a concept and a is a constant) we construct a finite number of new unary predicates and new object constants, from which we construct effectively (using an algorithm) a finite theory T whose axioms are closed quantifier-free formulas (propositional combinations of atoms) and a closed quantifier-free formula *prop*$(C(a))$ such that for each continuous t-norm $*$ and each $*$-model $\mathbf{M}$ of T, $\|prop(C(a))\|^*_{\mathbf{M}} = \|C(a)\|^*_{\mathbf{M}}$. In particular, due to the very simple (propositional) form of *prop*$(C(a))$, any $*$-model $\mathbf{M}$ of T determines a *finite* $*$-model $\mathbf{M}_0$ of T consisting just of (interpretations of) constants occurring in T and $\|prop(C(a))\|^*_{\mathbf{M}_0} = \|prop\,C(a)\|^*_{\mathbf{M}}$. Moreover, each witnessed $*$-interpretation of the language of $C(a)$ expands to a model of T. Now $C(a)$ has a $*$-model iff $T \cup \{prop(C(a))\}$ has a finite $*$-model from constants as above; this reduces the problem to $*$-satisfiability of a finite set of propositional formulas. $(\forall x)C(x)$ is a $*$-tautology iff T (propositionally) $*$-entails *prop*$(C(a))$, i.e., *prop*$(C(a))$ is true in each $*$-model of T. This gives the result.

We shall not present the algorithm for constructing T here, see [7]. But below we illustrate it by giving some simple illustrative examples. See also [8].

Our construction reduces the problems of witnessed $*$-satisfiability of C (i.e., $C(a)$ having a witnessed $*$-model) and witnessed $*$-validity of C ($(\forall x)C(x)$ being a witnessed $*$-tautology) to problems of propositional $*$-satisfiability and propositional $*$-entailment from a finite set of assumptions. These problems are decidable (see [7] for references), which gives us the following:

THEOREM 2. (1) *The problem of witnessed $*$-satisfiability (equivalently, of finite $*$-satisfiability) of a concept is decidable.*

(2) *The same for the problem of witnessed (finite) $*$-validity of a concept.*

For $*$-satisfiability we can say more. By the structural characterization of continuous t-norms (Mostert–Shields theorem, see [6]), continuous t-norms can be divided into two disjoint groups:

- beginning by Łukasiewicz, i.e., for some $0 < a \leqslant 1$, the restriction of $*$ to $[0, a]^2$ is isomorphic to Łukasiewicz t-norm, and
- having Gödel negation ($\|\neg\varphi\|^*_{\mathbf{M}} = 1$ for $\|\varphi\|^*_{\mathbf{M}} = 0$ and $\|\neg\varphi\|^*_{\mathbf{M}} = 0$ if $\|\varphi\|^*_{\mathbf{M}} > 0$).

THEOREM 3. (1) *For any $*$ beginning by Łukasiewicz and each concept C, C is $*$-satisfiable iff C is Ł-satisfiable (Ł denoting Łukasiewicz t-norm).*

(2) *For any other $*$ (i.e., $*$ having Gödel negation) and each concept C, C is $*$-satisfiable iff C is satisfiable in Boolean logic. (Again see* [7].)

3. Examples and comments

We first illustrate the transformation of $C(a)$ to a formula *prop*$(C(a))$ and the corresponding theory T by giving two examples. They are rather particular since they do not contain nested quantifier constructs. What we show here is just one step sufficient here but needing iteration in the general case.

EXAMPLE 1. Take the concept C defined as $\forall R.(A \equiv \neg A) \to \neg\exists R.A$. Is it witnessedly $*$-valid? Equivalently: is $\forall R(A \equiv \neg A)$ $*$-subsumed by $\neg\exists R.A$? We consider the formula $C(a)$. In each witnessed $*$-interpretation, the two quantifier concepts have witnesses, i.e., for some c, d, the following formulas are $*$-true.

$$\forall R.(A \equiv \neg A)(a) \equiv \big[R(ac) \to \big(A(c) \equiv \neg A(c)\big)\big],$$
$$\exists R.A(a) \equiv \big(R(a, d) \,\&\, A(d)\big).$$

Since c witnesses the formula $(\forall x)(R(a, x) \to (A(x) \equiv \neg A(x)))$, d must satisfy (write Rac for $R(a, c)$ etc.)

$$\big[Rac \to (Ac \equiv \neg Ac)\big] \to \big[Rad \to (Ad \equiv \neg Ad)\big] \tag{1}$$

and similarly, since d witnesses $(\exists x)(R(a, x) \,\&\, A(x))$, c must satisfy

$$(Rac \,\&\, Ac) \to (Rad \,\&\, Ad). \tag{2}$$

The formula $C(a)$ becomes equivalent to

$$\big[Rac \to (Ac \equiv \neg Ac)\big] \to \neg(Rad \,\&\, Ad). \tag{3}$$

$C(a)$ is witnessedly $*$-valid iff (1) and (2) $*$-entail (3). Indeed, if they do then take any $\mathbf{M}$ and introduce witnesses c and d, getting a model of (1) and (2); (3) follows and replacing witnesses by the corresponding quantified formulas you get $*$-truth of $C(a)$. Conversely, if you can evaluate all atoms involved by truth values in such a way that (1) and (2) are $*$-true but (3) not, you get a $*$-interpretation (having just three elements named a, c, d) in which $C(a)$ has value less then 1.

Thus first take Łukasiewicz. Put $\|Rac\| = \|Rad\| = 1$, $\|Ac\| = \|Ad\| = \frac{1}{2}$. Then (1), (2) are Ł-true, but (3) gets value $[1 \to (\frac{1}{2} \equiv \frac{1}{2})] \to \neg(1 \,\&\, \frac{1}{2}) = 1 \to \frac{1}{2} = \frac{1}{2} < 1$.

But for Gödel any formula $\varphi \equiv \neg\varphi$ is always false (has the value 0), thus (1) gives $(Rac \to \bot) \to (Rad \to \bot)$, hence $\neg Rac \to \neg Rad$ and (3) becomes equivalent to $\neg Rac \to \neg(Rad \,\&\, Ad)$. Since $\neg Rad \to \neg(Rad \,\&\, Ad)$ is a tautology (for each $*$), (1) implies $\neg Rac \to \neg(Rad \,\&\, Ad)$ which is (3). $C(a)$ is a (witnessed) G-tautology. (Note that $C(a)$ is even a tautology of Gödel logic with general models.)

EXAMPLE 2. Let C be $\neg\forall R.A \,\&\, \neg\exists R.\neg A$. Let us show that $C(a)$ is not witnessedly $*$-satisfiable, whatever $*$ you take. (But recall that $C(a)$ is G-satisfiable and Π-satisfiable in a non-witnessed model, see above.) Assume witnesses c, d for $\forall R.C$ and $\exists R.\neg C$ respectively. If $\mathbf{M}$ is a witnessed model for $C(a)$, it satisfies

$$\neg(Rac \to Ac), \tag{1}$$

$$\neg(Rad \,\&\, \neg Ad), \tag{2}$$

and by witnessing,

$$(Rac \to Ac) \to (Rad \to Ad), \tag{3}$$

$$(Rac \,\&\, \neg Ac) \to (Rad \,\&\, \neg Ad). \tag{4}$$

From (1) we get $\neg Ac$ (since $\neg(Rac \to Ac) \to \neg Ac$ is a tautology for each $*$); from (4) we get by equivalent transformations

$$\neg Ac \to \big[Rac \to (Rad \,\&\, \neg Ad)\big],$$

hence we get $Rac \to (Rad \,\&\, \neg Ad)$ and by (2) we get $\neg Rac$, hence $Rac \to Ac$, which is a contradiction with (1). $C(a)$ has no *witnessed* model.

*

Now let us return to the title of the paper: *What does mathematical fuzzy logic offer to description logic?* I hope the reader will agree that it offers:

- rich languages, allowing many choices (approaches), but still decidable (whereas of course for each continuous t-norm, the full predicate logic given by $*$ is undecidable)
- precise syntax and semantics (plus deductive systems, not described here, see [6])
- interesting research problems (see below).

In this context let us mention Straccia's paper [13] where the author develops a description logic based on some continuous t-norms and makes several restricting assumptions among them he assumes the inter-definability of quantifiers: $(\forall x)\varphi(x) \equiv \neg(\exists x)\neg\varphi(x)$ and dually. His assumptions are satisfied by Łukasiewicz t-norm and one can show that each continuous t-norm satisfying them is (isomorphic to) Łukasiewicz. (If $*$ is not isomorphic to Ł then either for some $a < 1$ its restriction to $[0, a]^2$ is isomorphic to Ł or it has Gödel negation, then take any $0 < u < 1$. Take an interpretation with $M = \{a\}$ and $r_P(a) = u$. Then $\|(\forall x)P(x)\|^*_{\mathbf{M}} = \frac{1}{2}$, $\|\neg P(a)\|^*_{\mathbf{M}} = 0$, thus $\|(\exists x)\neg P(x)\|^*_{\mathbf{M}} = 0$, $\|\neg(\exists x)\neg P(x)\|^*_{\mathbf{M}} = 1 = \frac{1}{2}$.) Łukasiewicz fuzzy logic is very important but one should know whether one restricts himself to it or one admits more possibilities. This is only a remark illustrating that knowledge of mathematical fuzzy logic may be helpful for developing fuzzy description logic.

What is (still) missing? (Problems.)

- Allowing truth constants into the language. This works well especially for Łukasiewicz logic but can be done in the general. This needs further study.
- Allowing generalized quantifiers as "many", e.g., $(\text{Many}R.C)(a)$ could say "many objects R-related to a have C". This *differs* from the fuzzy $(\forall x)(R(a,x) \to C(x))$. Cf. [6].
- Problems of computational complexity for our fuzzy description logic, in analogy to "classical" description logic.
- Problems of implementation.

This shows that, as said above, our approach to fuzzy description logic is a reasonable and interesting research topic. Any comments are welcome.

References

[1] C. Areces, *Logic engineering – the case of description and hybrid logics*, Thesis, ILLC Univ. Amsterdam, 2000.

[2] F. Baader, et al. (Eds.), The Description Logic Handbook – Theory, Interpretation and Applications, Cambridge University Press, 2003.

[3] P. Bonatti, A. Tettamanzi, *Some complexity results on fuzzy description logics*, in: Proc. Int. Workshop on Fuzzy Logic and Applications (WILF'03), Napoli, 2003.

[4] A. Borgida, *On the relative expressiveness of description logics and predicate logics*, Artificial Intelligence **82** (1996), 353–367.

[5] R. Cignoli, F. Esteva, L. Godo, A. Torrens, *Basic logic is the logic of continuous t-norms and their residua*, Soft Comput. **4** (2000), 106–112.

[6] P. Hájek, *Metamathematics of Fuzzy Logic*, Kluwer Academic Publishers, 1998.

[7] P. Hájek, *Making fuzzy dynamic logic more powerful*, Fuzzy Sets and Systems **154** (2005), 1–15.

[8] P. Hájek, P. Cintula, *On theories and models in fuzzy predicate logic*, submitted for publication.

[9] Z. Haniková, *A note on the complexity of individual t-algebras*, Neural Network World **12** (2002), 453–460.

[10] M. Schmidt-Schauss, G. Smolka, *Attributive concept descriptions with complements*, Artificial Intelligence **48** (1991), 1–26.

[11] U. Straccia, *A fuzzy description logic*, in: Proc. AAAI'98, Madison, WI, on web.

[12] U. Straccia, *Reasoning with fuzzy description logics*, J. AI Research **14** (2001), 137–166.

[13] U. Straccia, *Fuzzy description logics with concrete domains*, Techn. Report 2005-TR-03, Istituto di Sciencza e Tecnologie dell'Informazione CNR Pisa, Italy, 2005.

[14] C.B. Tresp, R. Molitor, *A description logic for vague knowledge*, RWTH-LTCS Report 98-01, Aachen Univ. of Technology.

[15] J. Yen, *Generalizing term subsumption languages to fuzzy logic*, in: IJCAI'91, pp. 472-477.

CHAPTER 6

Possibilistic Uncertainty and Fuzzy Features in Description Logic. A Preliminary Discussion

Didier Dubois, Jérôme Mengin and Henri Prade

Institut de Recherche en Informatique de Toulouse (I.R.I.T.), Université Paul Sabatier, 118 route de Narbonne, 31062 Toulouse cedex 4, France

E-mail:{dubois, mengin, prade}@irit.fr

Abstract

This short paper intends first to emphasize the basic distinction between gradual truth and uncertainty, and its relevance when dealing with classification. Then, the representation capabilities of first-order possibilistic logic are pointed out, before briefly providing some hints, which may be of interest for dealing with uncertainty and handling some fuzzy features in description logic.

1. Introduction

Description logics [2] (initially named "terminological logics") are tractable fragments of first-order logic representation languages that handle the notions of concepts (or classes), of roles (and properties), and of instances or objects, thus directly relying at the semantic level on the notions of set, binary relations, membership, and cardinality. They are especially useful for describing ontologies that consist in hierarchies of concepts in a particular domain, for the semantic web.

Since Yen's pioneering work [30], many proposals have been made for introducing fuzzy features in description logic (Tresp and Molitor [28], Straccia [24,25,27], Hölldobler et al. [16], Hajek [15]) and in semantic web languages (e.g., Mazzieri [20]), since fuzzy sets aim at providing a representation of classes and relations with gradual membership, which may be more suitable for dealing with concepts having a somewhat vague or elastic definition.

Moreover, some authors (Hollunder [17], Straccia [26], Russell [22]) have also expressed concern about handling uncertainty and exceptions in description logic. Hollunder [17] has introduced uncertainty in terminological logics using possibilistic logic [6], which is a direct extension of classical logic that handles uncertainty and partial inconsistency in a qualitative way in the framework of possibility theory [8,32]. A good introductory survey

FUZZY LOGIC AND THE SEMANTIC WEB
Edited by Elie Sanchez
© 2006 Elsevier B.V. All rights reserved

on fuzzy and possibilistic description logics and related works can be found in d'Aquin et al. [4].

This note emphasizes some aspects of fuzzy sets and possibilistic logic that may be relevant for the development of fuzzy description logics. In that respect, it is important to keep in mind that degrees of certainty and degrees of truth are not at all the same [11]. Indeed saying that "you are half certain that a bottle is full" does not mean that "the bottle is half full". Uncertainty calculi can never be fully compositional w.r.t. all connectives. Possibilistic logic handles certainty degrees, while degrees of truth can be combined using a truth functional fuzzy logic.

The paper is organized as follows. Section 2 provides a background on the fuzzy set representation of classes and on related issues, and Section 3 proposes a simplified representation for dealing with the notion of typical elements of a class and provides the basis for a twofold description logic of vagueness. Section 4 surveys first-order possibilistic logic as a tool for handling uncertainty when reasoning about classes and suggests how an extension of possibilistic logic incorporating some fuzzy features could be a basis for a simple extended description logic able to deal with vague classes and exceptions. Section 5 proposes some directions for further extensions of description logics.

2. Classes and instances – the fuzzy set setting

This section aims at restating the fuzzy set notions that are needed for defining a fuzzy description logic able to cope with vagueness. In order to have in mind the required needs, the basic features of standard description logic are first recalled.

2.1. *Basic features of a description logic*

In standard description logic (DL), classes can be represented with concept names from a set $\mathcal{C}$; the members of the classes can be represented with object names from a set $\mathcal{O}$, and binary relations between these objects can be represented with role names from a set $\mathcal{R}$. Complex role expressions can be used to represent the composition or the inversion of relations: the elements of $\mathcal{R}$ are legal role expressions, and if R and S are legal role expressions, so are $R \circ S$ and R^{-1}. The usual set-theoretical operations on classes can be represented using the binary connectives $\sqcap$ (intersection) and $\sqcup$ (union) and the unary connective $\neg$ (complementation); the zero-ary connectives $\top$ and $\bot$ respectively represent the class of all objects and the empty class. Thus, the set of legal concept expressions is the smallest set that contains $\mathcal{C} \cup \{\top, \bot\}$ and such that, if A and B are legal concept expressions and if R is a legal role expression, then $A \sqcap B$, $A \sqcup B$, $\neg A$, $\forall R.A$, $\exists R.A$ are legal too: $\exists R.A$ represents the class of those elements which are in relation R with at least one element of A, it is the set of predecessors of A w.r.t. the relation R; whereas $\forall R.A$ is meant to represent the class of objects which are in relation R with elements of A only, it is the class of objects which are not predecessors of any element not in A. Finally there are counting expressions: $\geqslant nR.A$ represents the class of objects that have at least n successors w.r.t. R in A, and $\leqslant nR.A$ represents the class of objects that have at most n successors w.r.t. R in A.

An interpretation I of concept expressions is defined by a non-empty domain D_I, a set $I(A) \subseteq D_I$ for all $A \in \mathcal{C}$, a set $I(R) \subseteq D_I \times D_I$ for all $R \in \mathcal{R}$, and an element $I(o) \in D_I$ for all $o \in \mathcal{O}$. Then the concept and role expressions are interpreted in the following way:

$$\begin{aligned}
&I(R \circ S) = I(R) \circ I(S); \qquad I\big(R^{-1}\big) = I(R)^{-1};\\
&I(A \sqcap B) = I(A) \cap I(B); \qquad I(A \sqcup B) = I(A) \cup I(B);\\
&I(\neg A) = \overline{I(A)} = D_I - I(A); \qquad I(\top) = D_I; \qquad I(\bot) = \emptyset;\\
&I(\exists R.A) = \big\{x \in D_I \mid \exists y \in D_I \text{ s.t. } (x, y) \in I(R) \text{ and } y \in I(A)\big\}\\
&\qquad = I(R)^{-1}\big(I(A)\big);\\
&I(\forall R.A) = \big\{x \in D_I \mid \forall y \in D_I \text{ if } (x, y) \in I(R) \text{ then } y \in I(A)\big\}\\
&\qquad = \overline{I(R)^{-1}\big(\overline{I(A)}\big)};\\
&I(\geqslant nR.A) = \big\{x \in D_I \mid \big|\big\{y \in D_I \mid (x, y) \in I(R) \text{ and } y \in I(A)\big\}\big| \geqslant n\big\};\\
&I(\leqslant nR.A) = \big\{x \in D_I \mid \big|\big\{y \in D_I \mid (x, y) \in I(R) \text{ and } y \in I(A)\big\}\big| \leqslant n\big\};
\end{aligned}$$

where $\overline{I(A)}$ denotes the complement of $I(A)$ in D_I, $I(R)^{-1}(I(A))$ is the set of predecessors of $I(A)$ w.r.t. the relation $I(R)$, and $|X|$ denotes the cardinality of the set X. In the sequel, $\mathcal{I}^{DL}$ denotes the set of these interpretations.

Knowledge is then represented by a pair $\langle \mathcal{T}, \mathcal{A}\rangle$, where $\mathcal{T}$ is a terminology, that is a set of expressions of the form $A \sqsubseteq B$ or $A = B$, where A and B are legal concept expressions; $\mathcal{A}$ is a set of assertions of the form $A(o)$ or $R(o, o')$, where A is a legal concept expression, R is a role name, and o and o' are object names. The formulas of description logic are these expressions. In the sequel, we denote by $\mathcal{L}^{DL}$ the set of these expressions.

An interpretation I satisfies $\langle \mathcal{T}, \mathcal{A}\rangle$ if it satisfies all formulas in $\mathcal{A} \cup \mathcal{T}$, where:

$$\begin{aligned}
I \models A \sqsubseteq B &\quad \text{if } I(A) \subseteq I(B);\\
I \models A = B &\quad \text{if } I(A) = I(B);\\
I \models A(o) &\quad \text{if } I(o) \in I(A);\\
I \models R(o, o') &\quad \text{if } \big(I(o), I(o')\big) \in I(R).
\end{aligned}$$

2.2. *Concepts and relations interpreted as fuzzy sets*

In order to manipulate vague classes and relations, it is natural to interpret concept and role names by fuzzy sets [31]. A fuzzy set F over a domain D is defined by its membership function, which associates to every $x \in D$ the degree $F(x) \in [0, 1]$ at which x is an element of F. A fuzzy binary relation over D is a fuzzy set of pairs of elements of D.

In a qualitative setting, where only min and max operations and the order-reversing map of the scale are used for defining fuzzy set operations, intersection, union and complementation of fuzzy sets are naturally defined by:

$$\begin{aligned}
&(F \cap G)(x) = \min\big(F(x), G(x)\big); \qquad (F \cup G)(x) = \max\big(F(x), G(x)\big);\\
&\overline{F}(x) = 1 - F(x).
\end{aligned}$$

Given a fuzzy set F and a fuzzy relation G, one can define the image of F by G by its membership function $G(F)$: $G(F)(x) = \sup\{\min(F(y), G(y, x)) \mid y \in D\}$ measures the degree to which x is related to at least one element in F. It is called the upper image of F by G in [9,23], as opposed to the lower inverse image $\overline{G(\overline{F})}$ of F by G which is defined by: $\overline{G(\overline{F})}(x) = \inf\{\max(F(y), 1 - G(y, x)) \mid y \in D\}$ and which measures the degree at which x has no predecessor outside F.

Given a fuzzy set F and a fuzzy relation G, we can also define the fuzzy set of those elements that have at least n successors by G in F: its membership function, denoted by $(\geqslant nG.F)$ is defined by:

$$(\geqslant nG.F)(x) = \sup\bigl\{\min\bigl\{F(y_i), G(x, y_i) \mid 1 \leqslant i \leqslant n\bigr\} \mid (y_1, \ldots, y_n) \in D\bigr\}.$$

Similarly, the fuzzy set of those elements that have at most n successors by R in F: its membership function, denoted by $(\leqslant nG.F)$ is defined by (see, e.g., [10])

$$\begin{aligned}(\leqslant nG.F)(x) &= 1 - \bigl(\geqslant (n+1)G.F\bigr)(x)\\ &= \inf\bigl\{\max\bigl\{1 - F(y_i), 1 - G(x, y_i) \mid 1 \leqslant i \leqslant n+1\bigr\} \mid \\ &\qquad (y_1, \ldots, y_{n+1}) \in D\bigr\}.\end{aligned}$$

Inversion and composition of fuzzy relations F and G are defined by:

$$F^{-1}(x, y) = F(y, x);$$

$$(F \circ G)(x, z) = \sup\bigl\{\min\bigl(F(x, y), G(y, z)\bigr) \mid y \in D\bigr\}.$$

A degree of inclusion of a fuzzy set F in another fuzzy set G can be defined by $\mathrm{IncDeg}(F, G) = \inf\{F(x) \to G(x) \mid x \in D\}$, where $F(x) \to G(x)$ is a measure of the degree at which the membership of a particular x in F implies its membership in G. In a qualitative setting, there are three main ways to define $\to$:

Gödel implication

$$a \to_G b = \sup\bigl\{\alpha \mid \min(a, \alpha) \leqslant b\bigr\} = \begin{cases} 1 & \text{if } a \leqslant b,\\ b & \text{otherwise.}\end{cases}$$

Contraposed of Gödel implication

$$a \to_{cG} b = \begin{cases} 1 & \text{if } a \leqslant b\\ 1 - a & \text{otherwise}\end{cases} = (1 - b) \to_G (1 - a).$$

Dienes implication

$$a \to_D b = \max(1 - a, b).$$

In the sequel, $\mathrm{IncDeg}_X(F, G)$ denotes the degree of inclusion of F in G obtained when considering a particular implication $\to_X$.

Given these extensions to fuzzy sets of classical set-theoretical notions, it is possible to give fuzzy meanings to concepts and roles by associating fuzzy sets to them. An interpretation I is then defined by a non-empty domain D_I, a fuzzy set $I(A)$ on D_I for all $A \in \mathcal{C}$, a fuzzy set $I(R)$ on $D_I \times D_I$ for all $R \in \mathcal{R}$, and an element $I(o) \in D_I$ for all $o \in \mathcal{O}$. Then the concept expressions are interpreted as before, in particular $\forall R.A$ is interpreted as the fuzzy lower inverse image of A by R, and $\exists R.A$ as the fuzzy upper image of A by R (with $I(R)^{-1}(y, x) = I(R)(x, y)$):

- $I(\exists R.A) = I(R)^{-1}(I(A))$ so that

$$I(\exists R.A)(x) = \sup\bigl\{\min\bigl(I(A)(y), I(R)(x, y)\bigr) \mid y \in D\bigr\};$$

- $I(\forall R.A) = \overline{I(R)^{-1}(\overline{I(A)})}$ so that

$$I(\forall R.A)(x) = \inf\{\max(I(A)(y), 1 - I(R)(x, y)) \mid y \in D\}.$$

The counting expressions are interpreted as follows:

$$\begin{aligned} I(\geqslant nR.A)(x) &= (\geqslant nI(R).I(A))(x) \\ &= \sup\{\min\{I(A)(y_i), I(R)(x, y_i) \mid 1 \leqslant i \leqslant n\} \mid \\ &\qquad (y_1, \ldots, y_n) \in D\}; \\ I(\leqslant nR.A)(x) &= (\leqslant nI(R).I(A))(x) \\ &= \inf\{\max\{1 - I(A)(y_i), 1 - I(R)(x, y_i) \mid 1 \leqslant i \leqslant n+1\} \mid \\ &\qquad (y_1, \ldots, y_{n+1}) \in D\}. \end{aligned}$$

When representing knowledge, the terminology $\mathcal{T}$ can now contain expressions of the form $A \sqsubseteq^\alpha B$, meaning that A is included in B to degree α. Assertions of $\mathcal{A}$ can have the form $o \in^\alpha A$, meaning that the degree of membership of o in A is α, or $oR^\alpha o'$. Once a definition of an inclusion degree has been chosen, satisfiability of such expressions can naturally be defined by:

$$\begin{aligned} I \models A \sqsubseteq^\alpha B &\quad \text{if } \mathrm{IncDeg}(I(A), I(B)) = \alpha; \\ I \models o \in^\alpha A &\quad \text{if } I(A)(I(o)) = \alpha; \\ I \models oR^\alpha o' &\quad \text{if } I(R)(I(o), I(o')) = \alpha. \end{aligned}$$

Note that one could also allow expressions of the form $A \sqsubseteq^{X:\alpha} B$, to represent that the degree of inclusion of A in B w.r.t. the fuzzy implication $\rightarrow_X$ is α.

3. A fuzzy DL for dealing with typicality

3.1. *Fuzzy sets as families of classical nested sets*

A fuzzy set F defined over a domain D is equivalent to the collection of its α-level cuts $F_{\geqslant\alpha} = \{x \in D \mid F(x) \geqslant \alpha\}$ for $\alpha \in (0, 1]$ or of its strict α-level cuts $F_{>\alpha} = \{x \in D \mid F(x) > \alpha\}$ for $\alpha \in [0, 1)$: the fuzzy set membership function can be re-obtained from the α-level cuts of F since $F(x) = \sup_\alpha\{\min(\alpha, F_{\geqslant\alpha}(x))\} = \inf_\alpha\{\max(\alpha, F_{>\alpha}(x))\}$. Fuzzy set operations can be defined by their effects on the cuts:

$$\begin{aligned} &(F \cap G)_{\geqslant\alpha} = F_{\geqslant\alpha} \cap G_{\geqslant\alpha}; &&(F \cup G)_{\geqslant\alpha} = F_{\geqslant\alpha} \cup G_{\geqslant\alpha}; \\ &(F \cap G)_{>\alpha} = F_{>\alpha} \cap G_{>\alpha}; &&(F \cup G)_{>\alpha} = F_{>\alpha} \cup G_{>\alpha}; \\ &\overline{F_{\geqslant\alpha}} = \overline{F}_{>1-\alpha}; &&\overline{F_{>\alpha}} = \overline{F}_{\geqslant 1-\alpha}; \\ &(R \circ S)_{\geqslant\alpha} = R_{\geqslant\alpha} \circ S_{\geqslant\alpha}; &&(R \circ S)_{>\alpha} = R_{>\alpha} \circ S_{>\alpha}. \end{aligned}$$

In practice, it may be enough to use a finite collection of level-cuts for interpreting a fuzzy class; it may even be enough to reduce the interpretation of a fuzzy class to a pair $(\mathrm{core}(F), \mathrm{support}(F))$ with $\mathrm{core}(F) = F_{\geqslant 1} = \{x \in D \mid F(x) = 1\}$ and $\mathrm{support}(F) = F_{>0} = \{x \in D \mid F(x) > 0\}$, which enables us to distinguish between typical elements (in the core) and borderline elements in $\mathrm{support}(F) - \mathrm{core}(F)$ [13,21].

This is in agreement with the commonly used, trapezoidal representation of membership functions on continuous unidimensional domains, which are completely defined from their cores and supports.

Clearly, more refined approximations could be easily handled by introducing some intermediary α-level cuts. For instance, using the strict middle level cut $F_{>1/2} = \{x \in D \mid F(x) > 1/2\}$, the middle section $F_{=1/2} = F_{\geqslant 1/2} \cap \overline{F_{>1/2}}$ together with the core and the support we can distinguish between five subsets of elements: core(F), $F_{>1/2} -$ core(F), $F_{=1/2}$, support$(F) - F_{\geqslant 1/2}$, and $\overline{\text{support}(F)}$, i.e., the elements that are completely F, those that are rather F, those that are as much F as $\overline{F}$, those that are rather $\overline{F}$, those that are completely $\overline{F}$. Thus, F is approximated by four nested ordinary subsets, namely core(F), $F_{>1/2}$, $F_{\geqslant 1/2}$ and support(F).

Restricting ourselves to a core/support approximation, we write $c(F)$ for core(F) and $s(F)$ for support(F). We can use core/support approximations for fuzzy relations too. We then have the following properties:

$$\begin{aligned}
&\overline{s(F)} = c(\overline{F}) \quad \text{and} \quad \overline{c(F)} = s(\overline{F});\\
&c\big(G(F)\big) = c(G)\big(c(F)\big) \quad \text{and} \quad s\big(G(F)\big) = s(G)\big(s(F)\big);\\
&c(\geqslant nG.F) = \big(\geqslant nc(G).c(F)\big) \quad \text{and} \quad s(\geqslant nG.F) = \big(\geqslant ns(G).s(F)\big);\\
&c(\leqslant nG.F) = \big(\leqslant ns(G).s(F)\big) \quad \text{and} \quad s(\leqslant nG.F) = \big(\leqslant nc(G).c(F)\big),
\end{aligned}$$

where F is a fuzzy set and G a fuzzy relation, and where $(\geqslant nH.K)$ denotes the set of those elements that have at least n successors by H in K when H is a crisp relation and K a crisp set (and similarly for $(\leqslant nH.K)$). The relationship of the above expressions with the cardinality of fuzzy sets [7,29], and their extensions to fuzzy quantifiers such as "most" or "at least few" is not discussed here for the sake of clarity.

We have described above three ways of defining a degree of inclusion of a fuzzy set F in another fuzzy set G. When we restrict ourselves to simple core/support approximations, only two degrees of inclusion become relevant: we can be interested in checking if F is included in G at degree 1, or at some degree > 0. Depending on the implication we use for defining the degree of inclusion, we obtain several levels of inclusion of $F = (c(F), s(F))$ in $G = (c(G), s(G))$, ordered from the weakest one to the strongest one:

1. $c(F) \subseteq s(G)$, equivalent to $\text{IncDeg}_D(F, G) > 0$ (in the case of finite domains);
2. $c(F) \subseteq c(G)$ or $s(F) \subseteq s(G)$, equivalent to $\text{IncDeg}_G(F, G) > 0$ or $\text{IncDeg}_{cG}(F, G) > 0$;
3. $c(F) \subseteq c(G)$ and $s(F) \subseteq s(G)$, equivalent to $\text{IncDeg}_G(F, G) = 1$ and $\text{IncDeg}_{cG}(F, G) = 1$;
4. $s(F) \subseteq c(G)$, equivalent to $\text{IncDeg}_D(F, G) = 1$.

Note that only the last two inclusion situations define transitive relations. The practical interest of situation 2 is debatable, however it is stronger than situation 1 and weaker than situation 3.

3.2. *Elements for a twofold description logic of vagueness*

The remarks of the above subsections provide the basis for a rather straightforward extension of description logic to vague concepts and vague roles, viewing their extensions as nested pairs of sets that enable us to sort out the typical elements.

An interpretation I of the names in $\mathcal{C}$, $\mathcal{R}$ and $\mathcal{O}$ over a non-empty domain D_I will now be defined by $I(C) = (I_c(C), I_s(C))$ with $I_c(C) \subseteq I_s(C) \subseteq D_I$ for every $C \in \mathcal{C}$, $I(R) = (I_c(R), I_s(R))$ with $I_c(R) \subseteq I_s(R) \subseteq D_I \times D_I$ for every $R \in \mathcal{R}$, and $I(o) \in D_I$ for every $o \in \mathcal{O}$. Then the concept expressions are interpreted in the following way:

$$\begin{aligned}
I(A \sqcap B) &= I(A) \cap I(B) = \big(I_c(A) \cap I_c(B), I_s(A) \cap I_s(B)\big);\\
I(A \sqcup B) &= I(A) \cup I(B) = \big(I_c(A) \cup I_c(B), I_s(A) \cup I_s(B)\big);\\
I(\neg A) &= \overline{I(A)} = \big(\overline{I_s(A)}, \overline{I_c(A)}\big);\\
I(\top) &= (D_I, D_I); \qquad I(\bot) = (\emptyset, \emptyset);\\
I(\exists R.A) &= \big(I_c(R)^{-1}\big(I_c(A)\big), I_s(R)^{-1}\big(I_s(A)\big)\big);\\
I(\forall R.A) &= \big(\overline{I_s(R)^{-1}\big(\overline{I_c(A)}\big)}, \overline{I_c(R)^{-1}\big(\overline{I_s(A)}\big)}\big);\\
I(\geqslant nR.A) &= \big(\big(\geqslant nI_c(R).I_c(A)\big), \big(\geqslant nI_s(R).I_s(A)\big)\big);\\
I(\leqslant nR.A) &= \big(\big(\geqslant nI_s(R).I_s(A)\big), \big(\geqslant nI_c(R).I_c(A)\big)\big);\\
I(R \circ S) &= I(R) \circ I(S) = \big(I_c(R) \circ I_c(S), I_s(R) \circ I_s(S)\big);\\
I\big(R^{-1}\big) &= \big(I_c(R)^{-1}, I_s(R)^{-1}\big).
\end{aligned}$$

With respect to satisfiability, we can now distinguish between four levels of inclusion, here ordered from the weakest to the strongest one:

$$\begin{aligned}
I &\models A \sqsubseteq^{cs} B && \text{if } I_c(A) \subseteq I_s(B);\\
I &\models A \sqsubseteq^{cc/ss} B && \text{if } I_c(A) \subseteq I_c(B) \text{ or } I_s(A) \subseteq I_s(B);\\
I &\models A \sqsubseteq^{ccss} B && \text{if } I_c(A) \subseteq I_c(B) \text{ and } I_s(A) \subseteq I_s(B);\\
I &\models A \sqsubseteq^{sc} B && \text{if } I_s(A) \subseteq I_c(B)
\end{aligned}$$

and two types of membership

$$\begin{aligned}
I &\models o \in^c A && \text{if } I(o) \in I_c(A);\\
I &\models o \in^s A && \text{if } I(o) \in I_s(A);\\
I &\models oR^c o' && \text{if } \big(I(o), I(o')\big) \in I_c(R);\\
I &\models oR^s o' && \text{if } \big(I(o), I(o')\big) \in I_s(R),
\end{aligned}$$

where $o \in^c A$ (resp. $o \in^s A$) expresses that object o is a typical member of the class A (resp. o is an instance of A that is not necessarily typical).

4. Possibilistic uncertainty and first-order possibilistic logic

A fuzzy set F can represent a piece of information stating that the possible values of some variable are fuzzily restricted by F [32]. This gives birth to a possibility distribution $\pi(x) = F(x)$, to be understood in the following way: $\pi(x) = 0$ means that the value x is impossible, while the larger $\pi(x)$, the more x is a possible value for the variable under consideration. Associated with a possibility distribution π is a possibility measure $\Pi(A) = \sup_{x \in A} \pi(x)$ and the dual necessity measure [8] $N(A) = 1 - \Pi(\overline{A})$ (where $\overline{A}$ is the complement of the ordinary set A).

This provides a worth-noticing interpretation of the level cuts of the fuzzy set F taken as representing the available information. Indeed the identity $F(x) = \inf_\alpha\{\max(\alpha, F_{>\alpha}(x))\}$ corresponds to the fact that "it is certain at degree $1 - \alpha$ that $x \in F_{>\alpha}$" for all α, knowing that the possible range of x is restricted by the possibility distribution $\pi(x) = F(x)$, since it can be checked that the necessity measure N, based on π, is such that $N(F_{>\alpha}) \geqslant 1 - \alpha$ (the equality is reached on continuous domains for continuous membership functions).

4.1. *Degrees of truth and degrees of uncertainty*

Degrees of certainty should not be confused with degrees of truth [11]. The degree of membership $F(x)$ of x to the fuzzy class F can be viewed as a degree of truth. It is the degree of truth of the fuzzy statement "the (precisely known) value x is in F". Such a degree is compositional with respect to all logical connectives. For instance, $F(x) \cap G(x) = \min(F(x), G(x))$ and $F(x) \cup G(x) = \max(F(x), G(x))$, expressing that it is all the more true that x is F and G (resp. F or G) as the degrees of membership of x to F and (resp. or) of x to G are higher.

By contrast, necessity measures that express certainty levels are compositional only with respect to conjunction. Indeed, for ordinary sets F and G, the characteristic property of necessity measures is $N(F \cap G) = \min(N(F), N(G))$, while $N(F \cup G) \geqslant \max(N(F), N(G))$, where $N(F)$ estimates the certainty that F restricts the possible values of the variable under consideration. The above equality and inequality still hold for fuzzy sets. The latter inequality agrees with the fact that one may be somewhat certain that the value is in the larger class $F \cup G$ without being certain at all that the value is in the more specific subset F or in the other more specific subset G.

So it is not at all the same to say that a precisely determined object/value is to some degree in the class F, or to say that it is almost certain that the object/value is in class F. Note that the second statement does not presuppose that F is a fuzzy class for making sense.

4.2. *A possibilistic logic refresher* [12,19]

Let $\mathcal{L}$ be the set of formulas of some classical logical language, and let $\mathcal{I}$ be a set of interpretations over $\mathcal{L}$; let us write $I \models \phi$ if the interpretation $I \in \mathcal{I}$ satisfies the formula $\phi \in \mathcal{L}$. We can attach weights from [0, 1] to the formulas of $\mathcal{L}$: we will call possibilistic formula a pair $(\phi, \alpha) \in \mathcal{L} \times [0, 1]$.[1] Such a pair is meant to express a degree of certainty of ϕ: it is understood as $N(\phi) \geqslant \alpha$, where N is a necessity measure. For example, the possibilistic formula $(\forall X \forall Y \mathsf{buy}(X, Y) \wedge \mathsf{expensive}(Y) \to \mathsf{hasmoney}(X), \alpha)$ expresses that it is certain at level α that those who buy an expensive item have money. We denote by $\mathcal{L}_\pi$ the set of possibilistic formulas over $\mathcal{L}$.

Constraints of the form $\Pi(\phi) \geqslant \alpha$ could be handled in the logic too, but they correspond to very poor pieces of information, while $N(\phi) \geqslant \alpha$, which is equivalent to $\Pi(\neg\phi) \leqslant 1-\alpha$ expresses that $\neg\phi$ is somewhat impossible, which is much more informative.

A semantics for possibilistic formulas is defined by the set $\mathcal{P}$ of possibility distributions over the set $\mathcal{I}$: if π is such a possibility distribution, then $\pi \models (\phi, \alpha)$ if $N(\{I \in \mathcal{I} \mid I \models$

[1] Note that any bounded linearly ordered scale could be used instead of [0, 1].

$\phi\}) \geqslant \alpha$, where N is the necessity measure associated with π: $N(\{I \in \mathcal{I} \mid I \models \phi\}) = 1 - \Pi(\{I \in \mathcal{I} \mid I \models \neg\phi\}) = 1 - \sup\{\pi(I) \mid I \models \neg\phi\}$. A possibilistic inference relation $\vdash_\pi$ can be defined on $\mathcal{L}_\pi$ in the usual way by: if $\Phi \cup \{\psi\} \subseteq \mathcal{L}_\pi$ and $\Phi \nvdash \bot$, then $\Phi \vdash_\pi \psi$ if every $\pi \in \mathcal{P}$ that satisfies every possibilistic formula of Φ also satisfies ψ. Then, a fundamental property of this inference relation is the following:

$$\begin{aligned} &\text{If} \quad \{\phi_j \mid j \in J\} \vdash \psi \quad \text{and} \quad \{\phi_j \mid j \in J\} \nvdash \bot \\ &\text{then} \quad \big\{(\phi_j, \alpha_j) \mid j \in J\big\} \vdash_\pi \Big(\psi, \min_j(\alpha_j)\Big) \end{aligned}$$

where $\vdash$ denotes the entailment relation induced by $\mathcal{I}$ on $\mathcal{L}$. In particular, $\{(\forall X p(X), \alpha)\} \vdash_\pi (p(a), \alpha)$ and $\{(\neg p \vee q, \alpha), (p \vee r, \beta)\} \vdash_\pi (q \vee r, \min(\alpha, \beta))$. Besides monotonicity, another important property of $\vdash_\pi$ is the following:

$$\big\{(\phi, \alpha), (\phi, \beta)\big\} \vdash_\pi \big(\phi, \max(\alpha, \beta)\big)$$

Note that in possibilistic logic, truth is two-valued (since p is Boolean), but the validity of (p, α) with respect to classical interpretations is many-valued. Indeed, a possibilistic knowledge base $K = \{(p_i, \alpha_i) \mid i = 1, \ldots, n\}$ is semantically equivalent [6] to the possibility distribution

$$\pi_K(I) = \min_{i=1,\ldots,n} \max\big([p_i](I), 1 - \alpha_i\big)$$

representing the fuzzy set of models of K (with $[p_i](I) = 1$ if I is a model of p_i, and $[p_i](I) = 0$ otherwise). It expresses that I is all the more possibly a model of K as it does not falsify a proposition p_i in K with a high certainty level α_i.

Besides, possibilistic logic can handle inconsistency. Indeed, let $\mathrm{inc}(K) = \max\{\alpha \mid K_\alpha \vdash \bot\}$ be the inconsistency level of K, where $K_\alpha = \{\phi \mid (\phi, \beta) \in K \text{ and } \beta \geqslant \alpha\}$. Then any inference from K_β with $\beta > \mathrm{inc}(K)$ is safe from inconsistency. Then $K \vdash_\pi (\phi, \alpha)$ if and only if $K_\alpha \vdash \phi$ and $\alpha > \mathrm{inc}(K)$. This enables possibilistic logic to encode nonmonotonic reasoning from default rules, once the proper specificity ordering has been determined between the rules [3].

4.3. *Towards a possibilistic description logic*

Applying the ideas described in the previous section to well-formed formulas of description logic, we can now manipulate inclusion and membership expressions with a degree of certainty; that is, formulas of the form $(A \sqsubseteq B, \alpha)$ or $(o \in A, \alpha)$ or (oRo', α), where A and B denote concept expressions, R denotes a role expression, and o and o' are object names. Given a possibility distribution π over the set $\mathcal{I}$ of classical interpretations of concept and role expressions, we have:

$$\begin{aligned} \pi &\models (A \sqsubseteq B, \alpha) \quad \text{if } \sup\big\{\pi(I) \mid I \in \mathcal{I}, I(A) \not\subseteq I(B)\big\} \leqslant 1 - \alpha; \\ \pi &\models (o \in A, \alpha) \quad \text{if } \sup\big\{\pi(I) \mid I \in \mathcal{I}, I \not\models A(o)\big\} \leqslant 1 - \alpha; \\ \pi &\models (oRo', \alpha) \quad \text{if } \sup\big\{\pi(I) \mid I \in \mathcal{I}, I \not\models R(o, o')\big\} \leqslant 1 - \alpha. \end{aligned}$$

In the sequel, we will use shorter syntactic forms for these expressions: we write $A \sqsubseteq_\alpha B$ for $(A \sqsubseteq B, \alpha)$, $o \in_\alpha A$ for $(o \in A, \alpha)$ and $oR_\alpha o'$ for (oRo', α). Note that the index α

denotes here a degree of certainty, whereas the exponent in, e.g., $\sqsubseteq^{ccss}$ indicates a fuzzy inclusion.

Let us now consider a non-monotonic reasoning situation. The classical "birds fly but penguins do not" example can be represented with the terminology

$$\mathcal{T} = \{\mathsf{penguin} \sqsubseteq_1 \mathsf{bird}, \mathsf{penguin} \sqsubseteq_1 \neg\mathsf{fly}, \mathsf{bird} \sqsubseteq_\alpha \mathsf{fly}\}$$

with $\alpha < 1$. From $\mathcal{T}$, one can deduce $\mathsf{penguin} \sqsubseteq_\alpha \mathsf{fly}$, which is clearly inconsistent with the more certain (here fully certain) inclusion $\mathsf{penguin} \sqsubseteq_1 \neg\mathsf{fly}$. However, this inconsistency remains at level $\alpha < 1$, and could be managed by giving priority to formulas strictly above the level of inconsistency of the base. In other words, $\mathsf{penguin} \sqsubseteq_\alpha \mathsf{fly}$ should be rejected from the deductive closure of $\mathcal{T}_\alpha$, since it is directly in contradiction with a consequence from $\mathcal{T}_1$. Moreover, we can also take advantage of the processing of vague concepts as proposed in Section 3.2, for distinguishing between typical subclasses (or instances) and other ones. In the above example, penguins are indeed known as non-typical birds. This is briefly discussed in the next section.

4.4. *Hints for the handling of uncertainty and fuzzy features in description logic*

We are now ready to combine fuzzy sets and degrees of certainty, in order to build a description logic that enables to distinguish between typical and non-typical individuals of a class, and to propagate levels of certainty. For this logic, concept and role expressions are built as before. The formulas in the terminology and the assertions of a knowledge base will now indicate a minimal degree of certainty at which some fuzzy inclusions/memberships hold.

A first way to achieve this is to use the fuzzy inclusions $\sqsubseteq^{cs}$, $\sqsubseteq^{ccss}$, $\sqsubseteq^{sc}$ defined in Section 3.2: for example, we can use $A \sqsubseteq^{cs}_\alpha B$ to express the fact that typical elements of A (the *core* of A) are in the support of B with a degree of certainty at least α. More generally, here is a list of formulas we may use with their intended meaning and a definition of their satisfiability by a possibility distribution π over the set of core/support interpretations $\mathcal{I}$ of concept and role expressions:

$A \sqsubseteq^{cs}_\alpha B$: typical elements of A (the *core* of A) are in the support of B with a degree of certainty at least α;
$\pi \models A \sqsubseteq^{cs}_\alpha B$ if $\sup\{\pi(I) \mid I \in \mathcal{I}, I_c(A) \not\subseteq I_s(B)\} \leqslant 1 - \alpha$.

$A \sqsubseteq^{ccss}_{\alpha,\beta} B$: typical elements of A are typical elements of B with a degree of certainty at least α, and all elements of A are elements of B with a degree of certainty at least β;
$\pi \models A \sqsubseteq^{ccss}_{\alpha,\beta} B$ if $\sup\{\pi(I) \mid I \in \mathcal{I}, I_c(A) \not\subseteq I_c(B)\} \leqslant 1 - \alpha$ and $\sup\{\pi(I) \mid I \in \mathcal{I}, I_s(A) \not\subseteq I_s(B)\} \leqslant 1 - \beta$.

$A \sqsubseteq^{sc}_\alpha B$: all elements of A are typical elements of B with a degree of certainty at least α;
$\pi \models A \sqsubseteq^{sc}_\alpha B$ if $\sup\{\pi(I) \mid I \in \mathcal{I}, I_s(A) \not\subseteq I_c(B)\} \leqslant 1 - \alpha$.

$o \in^c_\alpha A$: o is a typical element of A with a degree of certainty at least α;
$\pi \models o \in^c_\alpha A$ if $\sup\{\pi(I) \mid I \in \mathcal{I}, I(o) \notin I_c(A)\} \leqslant 1 - \alpha$.

$o \in^s_\alpha A$: o is an element of A, possibly non-typical, with a degree of certainty at least α;
$\pi \models o \in^s_\alpha A$ if $\sup\{\pi(I) \mid I \in \mathcal{I}, I(o) \notin I_s(A)\} \leqslant 1 - \alpha$.

$oR^c_\alpha o'$: (o, o') is a typical pair of the relation R with a degree of certainty at least α;
$\pi \models oR^c_\alpha A$ if $\sup\{\pi(I) \mid I \in \mathcal{I}, (I(o), I(o')) \notin I_c(R)\} \leqslant 1 - \alpha$.

$oR^s_\alpha o'$: o is an element of A, possibly non-typical, with a degree of certainty at least α;
$\pi \models oR^s_\alpha o'$ if $\sup\{\pi(I) \mid I \in \mathcal{I}, (I(o), I(o')) \notin I_s(R)\} \leqslant 1-\alpha$.

The lack of compositionality of necessity measures with respect to disjunction makes the handling of $\sqsubseteq^{cc/ss}_\alpha$ more difficult.

Another possibility is to directly manipulate expressions of the form $A \sqsubseteq^{X:\geqslant\alpha}_\beta B$, stating that it is certain at least at level β that the degree of subsumption of A in B is at least α, in the sense of an implication X (e.g., Gödel, Dienes). One may similarly have expressions of the form $o \in^{\geqslant\alpha}_\beta A$ for objects. The interpretation would be as follows:

$$\pi \models A \sqsubseteq^{X:\geqslant\alpha}_\beta B \quad \text{if } \sup\{\pi(I) \mid I \in \mathcal{I}, \mathrm{IncDeg}_X(I(A), I(B)) \ngeqslant \alpha\} \leqslant 1-\beta.$$

If fuzzy sets are represented by their cores and supports only, we have only two types of subsumption relations, namely $\sqsubseteq^{X:\geqslant 1}_\beta$ and $\sqsubseteq^{X:>0}_\beta$ as discussed in Section 3.1. This view could still be used for approximating fuzzy sets by more than two level cuts, as briefly mentioned in Section 3.1.

5. Future work and concluding remarks

It has been noticed (e.g., [14]) that subparts of classical logic formulas may be 'moved' to the weight part of a possibilistic logic formula. Thus, the possibilistic formula $(\forall X p(X) \wedge q(X) \rightarrow r(X), \alpha)$ is semantically equivalent to $(q(X) \rightarrow r(X), \min(P(X), \alpha))$, where $P(X) = 1$ if $p(X)$ is true and $P(X) = 0$ if $p(X)$ is false. It expresses that $q(X) \rightarrow r(X)$ is α-certainly true given the proviso that $p(X)$ is true. This is the basis of the use of possibilistic logic in hypothetical reasoning. A formula such that $(q(X) \rightarrow r(X), \min(P(X), \alpha))$ can be viewed also as an annotated formula. Thus, a formula can be annotated with additional conditions that may be difficult to evaluate but should hold for making sure that the formula is true. These additional conditions are then propagated in the inference when computing the formulas weights.

It is even possible to have variable weights defined through fuzzy set membership functions. Thus, a formula such as $(q(X) \rightarrow r(X), P(X))$ expresses a piece of information of the form "the more X is P, the more certain it is that if $q(X)$ is true then $r(X)$ is true too", where P is a fuzzy set membership function. For instance, $(\mathsf{bird}(X) \rightarrow \mathsf{fly}(X), \min(\mathit{Small}(X), \alpha))$ expresses that the smaller the bird is, the more certain it flies (even if it is not fully certain).

Lastly, fuzzy instantiation can also be defined in possibilistic logic. Let B be a fuzzy set. Then, the fuzzy instantiation of X in $p(X, Y)$ by a fuzzy set B is denoted $p(B, Y)$, and stands for: $\exists X \in B, p(X, Y)$ if B is a classical subset. Then, the following pattern of inference is valid [14]: from $(p(X, Y) \rightarrow q(X, Y), \min(A(X), \alpha))$ and $(p(B, y_0), \beta)$, one can deduce $(q(B, s_0), \min(N_B(A), \alpha, \beta)$ where $N_B(A) = \inf_x\{\max(A(x), 1 - B(x))\}$. See also [1]. This might be useful in description logic only if one would wish to deal with fuzzy objects.

This note is preliminary in many respects. Yet, it advocates two ideas that can be combined. On the one hand, vague concepts can be approximated in practice by pairs of nested sets, thus sorting out the typical elements, in a way that agrees with fuzzy set operations and inclusions. This may be sufficient in practice for describing more expressive ontologies. On the other hand, a possibilistic treatment of uncertainty and exceptions can be performed on top of a description logic. In both cases, the underlying principle is to remain as close as possible to classical logic for preserving computational efficiency as much as possible [18].

Other extensions would be worth investigating, such that allowing for fuzzy or uncertain cardinality information about classes. Then, class inclusions could be associated with interval restrictions on relative cardinalities, such as "between 70% and 80% of P's are Q's". These interval restrictions could be then propagated, even qualitatively [5].

References

[1] T. Alsinet, L. Godo, *A complete calculus for possibilistic logic programming with fuzzy propositional variables*, in: Conf. on Uncertainty in Artificial Intelligence (UAI'2000), 2000, pp. 1–10.

[2] F. Baader, D. Calvanese, D. McGuinness, D. Nardi, P. Patel-Schneider (Eds.), The Description Logic Handbook: Theory, Implementation, and Applications, Cambridge University Press, 2003.

[3] S. Benferhat, D. Dubois, H. Prade, *Representing default rules in possibilistic logic*, in: Proc. of the 3rd Inter. Conf. on Principles of Knowledge Representation and Reasoning (KR'92), Cambridge, 1992, pp. 673–684.

[4] M. d'Aquin, J. Lieber, A. Napoli, *Etude de quelques logiques de descriptions floues et de formalismes apparentés*, Rencontres Francophones sur la Logique Floue et ses Applications (LFA'04), Nantes, 2004. Cépaduès Editions.

[5] D. Dubois, L. Godo, R. Lopez de Mantaras, H. Prade, *A symbolic approach to reasoning with linguistic quantifiers*, J. Intelligent Inform. Systems **2** (1993), 319–363.

[6] D. Dubois, J. Lang, H. Prade, *Possibilistic logic*, in: D.M. Gabbay, C.J. Hogger, J.A. Robinson, D. Nute (Eds.), in: Handbook of Logic in Artificial Intelligence and Logic Programming, vol. 3, Oxford University Press, 1994, pp. 439–513.

[7] D. Dubois, H. Prade, *Fuzzy cardinality and the modeling of imprecise quantification*, Fuzzy Sets and Systems **16** (1985), 199–230.

[8] D. Dubois, H. Prade, *Possibility Theory: An Approach to Computerized Processing of Uncertainty*, Plenum Press, 1988.

[9] D. Dubois, H. Prade, *Upper and lower images of a fuzzy set induced by a fuzzy relation – applications to fuzzy inference and diagnosis*, Inform. Sci. **64** (1992), 203–232.

[10] D. Dubois, H. Prade, *Semantics of quotient operators in fuzzy relational databases*, Fuzzy Sets and Systems **78** (1996), 89–93.

[11] D. Dubois, H. Prade, *Possibility theory, probability theory and multiple-valued logics: a clarification*, Ann. Math. Artificial Intelligence **32** (2001), 35–66.

[12] D. Dubois, H. Prade, *Possibilistic logic: a retrospective and prospective view*, Fuzzy Sets and Systems **144** (2004), 3–23.

[13] D. Dubois, H. Prade, J.P. Rossazza, *Vagueness, typicality and uncertainty in class hierarchies*, Int. J. Intelligent Systems **6** (2) (1991), 167–183.

[14] D. Dubois, H. Prade, S. Sandri, *Possibilistic logic with fuzzy constants and fuzzily restricted quantifiers*, in: T.P. Martin, F. Arcelli Fontana (Eds.), Logic Programming and Soft Computing, Wiley, West Sussex, 1998, pp. 69–90.

[15] P. Hajek, *Making fuzzy description logic more general*, Fuzzy Sets and Systems, submitted for publication.

[16] S. Hölldobler, H.P. Störr, T.D. Khang, N.H. Nga, *The subsumption problem of the fuzzy description logic ALC_FH*, in: Proc. of the 10th International Conference on Information Processing and Management of Uncertainty in Knowledge-Based Systems (IPMU'04), Perugia, 2004, pp. 243–250.

[17] B. Hollunder, *An alternative proof method for possibilistic logic and its application to terminological logics*, in: R.L. de Mantaras, D. Poole (Eds.), Proc. of the 10th Conference on Uncertainty in Artificial Intelligence (UAI'94), San Francisco, CA, USA, 1994, pp. 327–335. Revised version in Int. J. Approximate Reasoning **12** (1995), 85–109.

[18] J. Lang, *Possibilistic logic: complexity and algorithms*, in: J. Kohlas, S. Moral (Eds.), Algorithms for Uncertainty and Defeasible Reasoning, Handbook of Defeasible Reasoning and Uncertainty Management Systems, vol. 5, D. Gabbay, P. Smets, series Eds., Kluwer Academic Publishers, Dordrecht, 2001, pp. 179–220.

[19] J. Lang, D. Dubois, H. Prade, *A logic of graded possibility and certainty coping with partial inconsistency*, in: Proc. of the 7th Conf. on Uncertainty in Artificial Intelligence (UAI'91)–Los Angeles, July 13–15, 1991, Morgan-Kaufmann, 1991, pp. 188–196.

[20] M. Mazzieri, *A fuzzy RDF semantics to represent trust metadata*, in: Proc. of the 1st Italian Semantic Web Workshop on Semantic Web Applications and Perspectives (SWAP), Ancona, Italy, 2004.

[21] J.P. Rossazza, D. Dubois, H. Prade, *A hierarchical model of fuzzy classes*, in: R. De Caluwe (Ed.), Fuzzy and Uncertain Object-oriented Databases. Concepts and Models, World Scientific, 1997, pp. 21–61.

[22] S. Russell, *Readings about the question: it is said that reasoning on the semantic web must be monotonic. Why it is so, when human reasoning, which seems to have served us well, is nonmonotonic*? robustai.net/papers/Monotonic_Reasoning_on_the_Semantic_Web.

[23] E. Sanchez, *Resolution in composite fuzzy relation equations*, Inform. and Control **30** (1976), 38–48.

[24] U. Straccia, *A fuzzy description logic*, in: Proc. 15th National Conf. on Artificial Intelligence (AAAI'98) and 10th Conf. on Innovative Applications of Artificial Intelligence (IAAI'98), AAAI Press, pp. 594–599.

[25] U. Straccia, *Reasoning within fuzzy description logics*, J. Artificial Intelligence Research **14** (2001), 137–166.

[26] U. Straccia, *Uncertainty in description logic: a lattice-based approach*, in: Proc. of the 10th International Conference on Information Processing and Management of Uncertainty in Knowledge-Based Systems (IPMU'04), Perugia, Italy, 2004, pp. 251–258.

[27] U. Straccia, *Towards a fuzzy description logic for the semantic web*, Technical Report, 2004.

[28] C.B. Tresp, R. Molitor, *A description logic for vague knowledge*, in: Proc. of the 13th European Conference on Artificial Intelligence (ECAI'98), Wiley, 1998, pp. 361–365.

[29] M. Wygralak, *Vaguely Defined Objects*, Kluwer Academic Publishers, 1996.

[30] J. Yen, *Generalizing term subsumption languages to fuzzy logic*, in: Proc. of the 12th International Joint Conference on Artificial Intelligence (IJCAI'91), Sidney, August 1991, pp. 472–477.

[31] L.A. Zadeh, *Fuzzy sets*, Inform. and Control **8** (1965), 338–353.

[32] L.A. Zadeh, *Fuzzy sets as a basis for a theory of possibility*, Fuzzy Sets and Systems **1** (1978), 3–28.

CHAPTER 7

Uncertainty and Description Logic Programs over Lattices

Umberto Straccia

ISTI-CNR, Via G. Moruzzi 1, I-56124 Pisa, Italy

E-mail: straccia@isti.cnr.it

Abstract

It is generally accepted that knowledge based systems would be smarter if they can manage uncertainty. In this paper we extend Description Logics, well-known logics for managing structured knowledge, towards the management of uncertainty. We allow (i) to express that a sentence is not just true or false, but certain to some degree, which is taken from a certainty lattice; and (ii) combine the logic with annotated logic programming, in which the management of uncertainty is based on so-called annotation terms.

Keywords

description logics, ontologies, logic programming, lattices

1. Introduction

In the last decade a substantial amount of work has been carried out in the context of *Description Logics* (DLs) [2]. DLs are a logical reconstruction of the so-called frame-based knowledge representation languages, with the aim of providing a simple well-established Tarski-style declarative semantics to capture the meaning of the most popular features of structured representation of knowledge. Nowadays, a whole family of knowledge representation systems has been build using DLs, which differ with respect to their expressiveness and their complexity, and they have been used for building a variety of applications (see the DL community home page http://dl.kr.org/).

Nowadays, DLs have gained even more popularity due to their application in the context of the *Semantic Web* [26]. *Ontologies* play a key role in the Semantic Web and consists of

FUZZY LOGIC AND THE SEMANTIC WEB
Edited by Elie Sanchez
© 2006 Elsevier B.V. All rights reserved

a hierarchical description of important concepts in a particular domain, along with the description of the properties (of the instances) of each concept. Web content is then annotated by relying on the concepts defined in a specific domain ontology. DLs play a particular role in this context as they are essentially the theoretical counterpart of the *Web Ontology Language OWL DL*, a state of the art language to specify ontologies.

On top of the Ontology layer, the Rules layer of the Semantic Web is going to be developed next (see also [24]), which should offer sophisticated representation and reasoning capabilities. A first effort in this direction is RuleML [5], fostering an XML-based markup language for rules and rule-based systems, while the OWL Rules Language [25] is a first proposal for extending OWL by Horn clause rules (see Section 5).

However, despite the growing popularity concerning the integration of rule languages, notably Logic Programs [36] (LPs), with DLs (see, e.g., [12,25,35]), to the best of our knowledge, no work deals with the combination of DLs with LPs towards the management of uncertainty. This is a well-known and important issue whenever the real world information to be represented is of imperfect nature (unfortunately, this rather the rule than an exception). While both uncertainty in DLs as well as uncertainty in LPs has been addressed, a combination of them in this direction has not yet been considered (for DLs, we may mention for probability theory [18,20,28,32,54], for possibility theory [23], metric spaces [40], for many-valued [50,55,62] and for fuzzy theory [6,22,42,52,56–58,63,59,60, 67,71]; concerning logic programming, we may mention for probability theory [39,48], for fuzzy set theory [69], multi-valued logic [8,31,34,38,41], possibilistic logic [13].

The topic of this paper is to combine DLs, LPs and the management of uncertainty into an uniform framework. In particular, we extend DLs towards the management of uncertainty, by allowing to express that a sentence is not just true or false like in classical DLs, but certain to some degree, which is taken from a certainty lattice. The certainty degree dictates to which extend (how certain it is that) a sentence is true. The adopted approach is more general that the fuzzy logic based approach [58], as it subsumes it (just take the lattice over the real unit interval $[0, 1]$ with order $\leqslant$), but is orthogonal to almost all other approaches. A feature of the lattice approach is that it gives us the possibility to address both *quantitative* uncertainty reasoning (by relying, e.g., on $[0, 1]$ or subsets of rational numbers like $\{0, \frac{1}{n-1}, \ldots, \frac{n-2}{n-1}, 1\}$, for natural number n), as well as *qualitative* uncertainty reasoning (by relying, e.g., on {`false`, `likelyfalse`, `unknown`, `likelytrue`, `true`}, in increasing order). From a computational point of view, it is still possible to develop a tableaux calculus in the style of almost all DLs and, under reasonable conditions, the computational complexity does not change, which is especially important as usually, reasoning under uncertainty is more involved than the classical case (see, e.g., [51]). We then extend the DL with rules, where the management of uncertainty is based on so-called annotation terms, inspired by Generalized Annotated Logic Programming framework of Kifer and Subrahmanian [31] (a quite general approach for managing uncertain information).

We proceed as follows. In the next section, we recall some fundamental notions about DLs and extend them with rules. In Section 3 we describe our DL extension to manage uncertain sentences and address the computational aspect of reasoning in it, while in Section 4 we extend it with annotated logic programming. Finally, Section 5 presents related work, while Section 6 concludes the paper.

2. Preliminaries

2.1. *A quick look to $\mathcal{ALC}$*

The specific DL we extend with uncertainty capabilities is $\mathcal{ALC}$, a significant representative of DLs [2]. Consider three alphabets of symbols, *primitive concepts* (denoted A), *primitive roles* (denoted R) and *individuals* (denoted a and b).[1] A *concept* (denoted C or D) of the language $\mathcal{ALC}$ is build out from *primitive concepts* A, the *top concept* $\top$, the *bottom concept* $\bot$ and according the following syntax rule:

$$
\begin{aligned}
C, D \longrightarrow C \sqcap D| &\quad \text{(concept conjunction)}\\
C \sqcup D| &\quad \text{(concept disjunction)}\\
\neg C| &\quad \text{(concept negation)}\\
\forall R.C| &\quad \text{(universal quantification)}\\
\exists R.C &\quad \text{(existential quantification).}
\end{aligned}
$$

An *interpretation* $\mathcal{I}$ is a pair $\mathcal{I} = (\Delta^{\mathcal{I}}, \cdot^{\mathcal{I}})$ consisting of a non-empty set $\Delta^{\mathcal{I}}$ (called the *domain*) and of an *interpretation function* $\cdot^{\mathcal{I}}$ mapping different individuals into different elements of $\Delta^{\mathcal{I}}$ (called *unique name assumption*), primitive concepts into subsets of $\Delta^{\mathcal{I}}$ and primitive roles into subsets of $\Delta^{\mathcal{I}} \times \Delta^{\mathcal{I}}$. The interpretation of complex concepts is defined as usual:

$$
\begin{aligned}
\top^{\mathcal{I}} &= \Delta^{\mathcal{I}}\\
\bot^{\mathcal{I}} &= \emptyset\\
(C \sqcap D)^{\mathcal{I}} &= C^{\mathcal{I}} \cap D^{\mathcal{I}}\\
(C \sqcup D)^{\mathcal{I}} &= C^{\mathcal{I}} \cup D^{\mathcal{I}}\\
(\neg C)^{\mathcal{I}} &= \Delta^{\mathcal{I}} \setminus C^{\mathcal{I}}\\
(\forall R.C)^{\mathcal{I}} &= \{d \in \Delta^{\mathcal{I}} \colon \forall d'.(d, d') \notin R^{\mathcal{I}} \text{ or } d' \in C^{\mathcal{I}}\}\\
(\exists R.C)^{\mathcal{I}} &= \{d \in \Delta^{\mathcal{I}} \colon \exists d'.(d, d') \in R^{\mathcal{I}} \text{ and } d' \in C^{\mathcal{I}}\}.
\end{aligned}
$$

Two concepts C and D are *equivalent* (denoted $C \equiv D$) when $C^{\mathcal{I}} = D^{\mathcal{I}}$ for all interpretations $\mathcal{I}$ (e.g., $\exists R.C \equiv \neg\forall R.\neg C$). An *assertion* (denoted α) is an expression $a{:}C$ ("a is an instance of C"), or an expression $(a, b){:}R$ ("(a, b) is an instance of R"). A *primitive assertion* is either an assertion of the form $a{:}A$, where A is a primitive concept, or an assertion of the form $(a, b){:}R$. An interpretation $\mathcal{I}$ *satisfies* $a{:}C$ (resp. $(a, b){:}R$) iff $a^{\mathcal{I}} \in C^{\mathcal{I}}$ (resp. $(a^{\mathcal{I}}, b^{\mathcal{I}}) \in R^{\mathcal{I}}$). Let A and C be a primitive concept and a concept, respectively. A *terminological axiom* (denoted τ) is either a concept specialization or a concept definition. A *concept specialization* is an expression of the form $A \lessdot C$, while a *concept definition* is an expression of the form $A := C$. An interpretation $\mathcal{I}$ *satisfies* $A \lessdot C$ iff $A^{\mathcal{I}} \subseteq C^{\mathcal{I}}$, while $\mathcal{I}$ *satisfies* $A := C$ iff $A^{\mathcal{I}} = C^{\mathcal{I}}$. A finite set K of assertions and terminological axioms is a *Knowledge Base* (KB). With K_A we denote the set of assertions in K, whereas with K_T we denote the set of terminological axioms in K, also called a *terminology*. A KB K is *purely assertional* if $K_T = \emptyset$. Further, we assume that a terminology K_T is such

[1] Metavariables may have a subscript or a superscript.

that no concept A appears more than once on the left-hand side of a terminological axiom $\tau \in K_T$ and that no cyclic definitions are present in K_T.[2] An interpretation $\mathcal{I}$ *satisfies* (*is a model of*) a KB K iff $\mathcal{I}$ satisfies each element in K. A KB K *entails* an assertion α (denoted $K \models \alpha$) iff every model of K also satisfies α. The problem of determining whether $K \models \alpha$ is called *entailment problem*, while the problem of determining whether K is satisfiable is called *satisfiability problem*. It is well known (see, e.g., [2]) that in $\mathcal{ALC}$: (i) $K \models (a, b){:}R$ iff $(a, b){:}R \in K$; and (ii) $K \models a{:}C$ iff $K \cup \{a{:}\neg C\}$ unsatisfiable. Note that there exists a well known technique based on constraint propagation solving the satisfiability problem [2]. Furthermore, we can restrict our attention to purely assertional KBs, by expanding K to K' and substituting every primitive concept occurring in K, which is defined in K', with its defining term in K'. Informally, the *expansion of a KB K* is as follows [46]: replace each concept specialization $A \lessdot C \in K_T$ with $A := C \sqcap A^*$ (A^* is a new primitive concept); then expand the right-hand side of every concept definition by replacing a primitive concept with its definition until there remain only undefined concepts in the second arguments of concept definitions; and finally, replace in K_A all primitive concepts with their definitions. The transformation has the nice property that $K \models \alpha$ iff $K'_A \models \alpha'$, where α' is obtained by replacing every primitive concept occurring in α, which is defined in K'_T, with its defining term in K'_T. This allows us to restrict our attention to purely assertional KBs only (but, the expansion process can be exponential [46]).

2.2. *A quick look to description logic programs*

For rules, there are mainly three approaches (see [49,16], for an overview), the so-called axiom-based approach (e.g., [25,35]) and the DL-log approach (e.g., [12,14,15]) and the autoepistemic approach (e.g., [11,37]). We are not going to discuss in this section these approaches and defer it to Section 5. We just point out that in this paper we follow the DL-log approach, in which rules may not modify the extension of concepts. Essentially, role names and concept names appearing in K_T *may appear in the body of a rule only* (*except for representing facts*) and are managed as unary and binary predicates, respectively. This because of the underlying assumption that the terminological component *completely* describes the hierarchical structure in the domain, and, therefore, the rules should not allow to make new inferences about that structure.

Syntax. A *term* is either a constant (denoted a, b, c) or a variable (denoted x, y, z). We assume that the description logic component and the rules component share the same alphabet of constants. An *atom* has the form $P(t_1, \ldots, t_n)$, where P is a predicate symbol and t_i are terms. A rule is of the form

$$H \leftarrow B_1, \ldots, B_n,$$

where H and all B_i are atoms. H is called the *rule head*, and the set of all B_i is called the *rule body*. For a rule r, with $H(r)$ and $B(r)$ we indicate the rule head and the rule body or r, respectively. We assume that no rule head atom belongs to the DL signature. For ease

[2] We say that A *directly uses* primitive concept B in K_T, if there is $\tau \in K_T$ such that A is on the left-hand side of τ and B occurs in the right-hand side of τ. Let *uses* be the transitive closure of the relation directly uses in K_T. K_T is *cyclic* iff there is A such that A uses A in K_T.

the readability, DL predicates will have a DL superscript in the rules. An example of rule is

$$\texttt{MatureFruit(x)} \leftarrow \texttt{Fruit}^{\text{DL}}\texttt{(x)}, \texttt{Taste(x, sweet)}$$

Here `Fruit` is a DL predicate, while `MatureFruit` and `Taste` are not.

Note that [14] allows to use DL axioms in place of DL atoms in the rule body (e.g., a concept inclusion may appear in the body of the rule). For easy, we will not deal with this feature.

A *logic program*, denoted $\mathcal{P}$, is a finite set of rules. A *Description Logic Program* (DLP) is a tuple $\mathcal{DP} = \langle K, \mathcal{P} \rangle$, where K is a DL knowledge base and $\mathcal{P}$ is a logic program. For instance, consider the description logic component

$$K = \{\texttt{apple:Fruit}, \texttt{pear:Fruit}, \texttt{a:Fruit} \sqcup \texttt{Cake}\},$$

and the logic program component

$$\begin{aligned}\mathcal{P} = \{&\texttt{MatureFruit(x)} \leftarrow \texttt{Fruit}^{\text{DL}}\texttt{(x)}, \texttt{Taste(x, sweet)}\\ &\texttt{Taste(apple, sweet)} \leftarrow\\ &\texttt{Taste(a, sweet)} \leftarrow\\ \}&,\end{aligned}$$

then $\mathcal{DP} = \langle K, \mathcal{P} \rangle$ is a DLP.

Semantics. Essentially, a DL atom appearing in a rule body acts as a query to the underlying DL knowledge base (see [14]). So, consider a DLP $\mathcal{DP} = \langle K, \mathcal{P} \rangle$. The *Herbrand universe* of $\mathcal{P}$, denoted $H_{\mathcal{P}}$ is the set of constants appearing in $\mathcal{DP}$ (if no such constant symbol exists, $H_{\mathcal{P}} = \{c\}$ for an arbitrary constant symbol c from the alphabet of constants). The *Herbrand base* of $\mathcal{P}$, denoted $\mathcal{B}_{\mathcal{P}}$, is the set of all ground atoms built up from the non-DL predicates and the Herbrand universe of $\mathcal{P}$. With $\texttt{ground}(\mathcal{P})$ we denote the grounding of $\mathcal{P}$ w.r.t. $H_{\mathcal{P}}$. For instance, for the DLP above, $H_{\mathcal{P}} = \{\texttt{apple}, \texttt{pear}, \texttt{a}, \texttt{sweet}\}$, $\mathcal{B}_{\mathcal{P}}$ is the set of all ground atoms that can be built from the predicates `MatureFruit` and `Taste` using the constants in $H_{\mathcal{P}}$, while $\texttt{ground}(\mathcal{P})$ is the set of ground instances of rules in $\mathcal{P}$ that can be built using constants in $H_{\mathcal{P}}$. Thus, e.g.,

$$\begin{aligned}&\texttt{MatureFruit(apple)} \leftarrow \texttt{Fruit}^{\text{DL}}\texttt{(apple)}, \texttt{Taste(apple, sweet)}\\ &\texttt{MatureFruit(a)} \leftarrow \texttt{Fruit}^{\text{DL}}\texttt{(a)}, \texttt{Taste(a, sweet)}\end{aligned}$$

belong to $\texttt{ground}(\mathcal{P})$.

An *interpretation* I w.r.t. $\mathcal{DP}$ is a subset of $\mathcal{B}_{\mathcal{P}}$. We say I is a *model* of a ground atom A w.r.t. K, denoted $I \models_K A$, iff

- $A \in I$, if $A \in \mathcal{B}_{\mathcal{P}}$ (A is a non-DL atom);
- $K \models A$, if A is a DL atom.

We say that I is a *model* of a ground rule r iff $I \models_K H(r)$ whenever $I \models_K B_i$ for all $B_i \in B(r)$. I is *model* of $\mathcal{DP} = \langle K, \mathcal{P} \rangle$ iff $I \models_K r$ for all rules $r \in \texttt{ground}(\mathcal{P})$. We say that a DLP is *satisfiable* if it has a model. Finally, we say that $\mathcal{DP} = \langle K, \mathcal{P} \rangle$ *entails* a ground atom A, denoted $\mathcal{DP} \models A$, iff $I \models_K A$ whenever $I \models \mathcal{DP}$. For instance, given the DLP above,

$$I_1 = \big\{\texttt{MatureFruit(apple)}, \texttt{Taste(apple, sweet)}, \texttt{Taste(a, sweet)}\big\}$$

is a model of $\mathcal{DP}$, while

$$I_2 = I_1 \cup \{\texttt{MatureFruit(a)}\}$$

is not. In the latter case note that $I_2 \not\models_K$ $\texttt{Fruit(a)}$ as $K \not\models$ $\texttt{Fruit(a)}$ and, thus, I_2 is not a model of the ground rule

$$\texttt{MatureFruit(a)} \leftarrow \texttt{Fruit}^{\texttt{DL}}\texttt{(a)}, \texttt{Taste(a, sweet)}$$

Finally, note that $\mathcal{DP} \models$ $\texttt{MatureFruit(apple)}$, $\mathcal{DP} \not\models$ $\texttt{MatureFruit(a)}$, $\mathcal{DP} \not\models$ $\texttt{MatureFruit(sweet)}$ and $\mathcal{DP} \not\models$ $\texttt{MatureFruit(pear)}$ hold.

It is easy to see that the entailment problem of a DL literal in a DLP is independent of the presence of a logic program $\mathcal{P}$. This means that the DL knowledge base is unaffected by the rule system, which can be seen as built on top of the DL knowledge base. With that we mean that if $\mathcal{DP} = \langle K, \mathcal{P}\rangle$ the for any ground DL atom A, $\mathcal{DP} \models A$ iff $K \models A$ (no DL atom can appear in the head of a rule).

Interestingly, it is possible to adapt the standard results of Datalog to our case, which say that a satisfiable description logic program $\mathcal{DP}$ has a minimal model $M_{\mathcal{DP}}$ and entailment can be reduced to model checking in this minimal model.

PROPOSITION 1 [14]. *Let $\mathcal{DP} = \langle K, \mathcal{P}\rangle$ be a DLP. If $\mathcal{DP}$ is satisfiable, then there exists a unique model $M_{\mathcal{DP}} \subseteq \mathcal{B}_{\mathcal{P}}$ such that $M_{\mathcal{DP}} \subseteq I$ for all models $I \subseteq \mathcal{B}_{\mathcal{P}}$ of $\mathcal{DP}$. Furthermore, for any ground atom A, $\mathcal{DP} \models A$ iff $M_{\mathcal{DP}} \models_K A$.*

The minimal model can be computed as the least fixed-point of the following monotone operator. Let $\mathcal{DP} = \langle K, \mathcal{P}\rangle$ be a DLP. Define the operator $T_{\mathcal{DP}}$ on interpretations as follows: for every $I \subseteq \mathcal{B}_{\mathcal{P}}$, let

$$T_{\mathcal{DP}}(I) = \{H(r)\colon r \in \texttt{ground}(\mathcal{P}), I \models_K B_i \text{ for all } B_i \in B(r)\}.$$

It can be shown that $T_{\mathcal{DP}}$ is monotone, i.e. $I \subseteq I' \subseteq \mathcal{B}_{\mathcal{P}}$ implies $T_{\mathcal{DP}}(I) \subseteq T_{\mathcal{DP}}(I')$, and, thus, by the Knaster–Tarski Theorem [66] T_K has a least fixed-point, denoted $lfp(\mathrm{T}_K)$. Moreover, $lfp(\mathrm{T}_K)$ can be computed by finite fixed-point iteration as the signatures and the involved sets in $\mathcal{DP}$ are finite: for every $I \subseteq \mathcal{B}_{\mathcal{P}}$, we define $T^i_{\mathcal{DP}}(I) = I$, if $i = 0$, and $T^i_{\mathcal{DP}}(I) = T_{\mathcal{DP}}(T^{i-1}_{\mathcal{DP}}(I))$, if $i > 0$. Then $lfp(\mathrm{T}_K) = \bigcup_{i=0}^n T^i_{\mathcal{DP}}(\emptyset) = T^n_{\mathcal{DP}}(\emptyset)$, for some $n \geqslant 0$. Finally, $lfp(\mathrm{T}_K) = M_{\mathcal{DP}}$ holds.

From an implementation point of view, given $\mathcal{DP} = \langle K, \mathcal{P}\rangle$, we can compute the set of entailed primitive assertions $F = \{\alpha\colon K \models \alpha\}$ first and then add them as facts to $\mathcal{P}$. Then, we can rely on standard theorem proving in Datalog.

EXAMPLE 1. Consider the following database schema of a company database. Let $\mathcal{DP} = \langle K, \mathcal{P}\rangle$ be such that

$$\begin{aligned} K &= \{\texttt{Employee} \lessdot \forall\texttt{OfficeMate.Employee}, \texttt{Manager} \lessdot \texttt{Employee}\} \\ \mathcal{P} &= \{\texttt{Manager(jim)}, \texttt{OfficeMate(jim, paul)}, \\ &\quad\ \texttt{Q}(x) \leftarrow \texttt{Employee}(x, y)\}. \end{aligned}$$

$\texttt{Q}(x)$ is our query. It turns out that $\mathcal{DP} \models$ $\texttt{Q(jim)}$ and $\mathcal{DP} \models$ $\texttt{Q(paul)}$, i.e. `jim` and `paul` are an answer to our query.

3. The logic $\mathcal{L}$-$\mathcal{ALC}$

Let $\mathcal{L} = \langle \mathcal{T}, \preceq \rangle$ be a *certainty lattice* (a complete lattice), where $\mathcal{T}$ is a set of certainty values and $\preceq$ is a partial order over $\mathcal{T}$. Let $\otimes$ and $\oplus$ be the meet and join operators induced by $\preceq$, respectively. Let f and t be the least and greatest element in $\mathcal{T}$, respectively. We also assume that there is a function from $\mathcal{T}$ to $\mathcal{T}$, called *negation function* (denoted $\neg$) that is anti monotone w.r.t. $\preceq$ and satisfies $\neg\neg\alpha = \alpha, \forall\alpha \in \mathcal{T}$. The main idea is that an assertion $a{:}C$, rather being interpreted as either true or false, will be mapped into a certainty value c in $\mathcal{T}$. The intended meaning is that c indicates to which extend (how certain it is that) 'a is a C'. Typical certainty lattice are: given a set of real values $\mathcal{T}$, consider $\mathcal{L}_{\mathcal{T}} = \langle \mathcal{T}, \leqslant \rangle$. Then $\mathcal{L}_{\{0,1\}}$ corresponds to the classical truth-space, where 0 stands for 'false', while 1 stands for 'true', while $\mathcal{L}_{[0,1]}$, which relies on the unit real interval, is quite frequently used as certainty lattice. In $\mathcal{L}_{[0,1]}$, $\neg\alpha = 1-\alpha$ is quite typical. Another frequent certainty lattice is Belnap's $\mathcal{FOUR}$ [1], where $\mathcal{T}$ is $\{f, t, u, i\}$ with $f \preceq u \preceq t$ and $f \preceq i \preceq t$. Here, u stands for 'unknown', whereas i stands for inconsistency. We denote the lattice as $\mathcal{L}_B$. Additionally, besides $\neg f = t$, we have $\neg u = u$ and $\neg i = i$. Another certainty lattice is $\mathcal{L}_4$, where $\mathcal{T}$ is $\{f, lf, lt, t\}$ with $f \preceq lf \preceq lt \preceq t$. Here, lf stands for 'likely false', whereas lt stands for 'likely true'. Besides $\neg f = t$, we have $\neg lf = lt$. A further popular lattice allows us to reason about *belief and doubt*. Indeed, the idea is to take any lattice $\mathcal{L}$, and to consider the Cartesian product $\mathcal{L} \times \mathcal{L}$. For any pair $(b, d) \in \mathcal{L} \times \mathcal{L}$, b indicates the degree of *belief* a reasoning agent has about a sentence s, while d indicates the degree of *doubt* the agent has about s. The order on $\mathcal{L} \times \mathcal{L}$ is determined by $(b, d) \preceq (b', d')$ iff $b \preceq b'$ and $d' \preceq d$, i.e. belief goes up, while doubt goes down. The minimal element is (f, t) (no belief, maximal doubt), while the maximal element is (t, f) (maximal belief, no doubt). Negation is given by $\neg(b, d) = (d, b)$ (exchange belief with doubt). We indicate this lattice with $\overline{\mathcal{L}}$. The examples above illustrates that the lattice approach gives us the possibility to address both quantitative uncertainty reasoning as well as qualitative uncertainty reasoning.

For a certainty lattice $\mathcal{L} = \langle \mathcal{T}, \preceq \rangle$, an *$\mathcal{L}$-interpretation* is now a pair $\mathcal{I} = (\Delta^{\mathcal{I}}, \cdot^{\mathcal{I}})$, where $\Delta^{\mathcal{I}}$ is, as for the classical case, the *domain*, whereas $\cdot^{\mathcal{I}}$ is an *interpretation function* mapping (i) individuals as for the classical case, i.e. $a^{\mathcal{I}} \neq b^{\mathcal{I}}$, if $a \neq b$; (ii) a concept C into a function $C^{\mathcal{I}} : \Delta^{\mathcal{I}} \to \mathcal{T}$; (iii) a role R into a function $R^{\mathcal{I}} : \Delta^{\mathcal{I}} \times \Delta^{\mathcal{I}} \to \mathcal{T}$. In the following, for ease with interpretation we mean always an $\mathcal{L}$-interpretation, for some certainty lattice $\mathcal{L}$. As anticipated above, if $d \in \Delta^{\mathcal{I}}$ is an object of the domain $\Delta^{\mathcal{I}}$ then $C^{\mathcal{I}}(d)$ gives us the degree of certainty of being the object d an instance of the concept C under the interpretation $\mathcal{I}$. Similarly for roles. The interpretation function $\cdot^{\mathcal{I}}$ has to satisfy the following equations: for all $d \in \Delta^{\mathcal{I}}$,

$$
\begin{aligned}
\top^{\mathcal{I}}(d) &= t \\
\bot^{\mathcal{I}}(d) &= f \\
(C \sqcap D)^{\mathcal{I}}(d) &= C^{\mathcal{I}}(d) \otimes D^{\mathcal{I}}(d) \\
(C \sqcup D)^{\mathcal{I}}(d) &= C^{\mathcal{I}}(d) \oplus D^{\mathcal{I}}(d) \\
(\neg C)^{\mathcal{I}}(d) &= \neg C^{\mathcal{I}}(d) \\
(\forall R.C)^{\mathcal{I}}(d) &= \bigotimes_{d' \in \Delta^{\mathcal{I}}} \{\neg R^{\mathcal{I}}(d, d') \oplus C^{\mathcal{I}}(d')\} \\
(\exists R.C)^{\mathcal{I}}(d) &= \bigoplus_{d' \in \Delta^{\mathcal{I}}} \{R^{\mathcal{I}}(d, d') \otimes C^{\mathcal{I}}(d')\}.
\end{aligned}
$$

Note that the semantics of $\exists R.C$ is the result of viewing $\exists R.C$ as the open first-order formula $\exists y.R(x, y) \wedge \overline{C}(y)$ (where $\overline{C}$ is the translation of C into first-order logic) and $\exists$ is viewed as a disjunction over the elements of the domain. Similarly, the semantics of $\forall R.C$ is related to $\forall y.\neg R(x, y) \vee \overline{C}(y)$, where $\forall$ is viewed as a conjunction over the elements of the domain. The definition of concept *equivalence* is like for $\mathcal{ALC}$. As for classical $\mathcal{ALC}$, dual relationships between concepts hold: e.g., $\top \equiv \neg\bot$, $(C \sqcap D) \equiv \neg(\neg C \sqcup \neg D)$ and $(\forall R.C) \equiv \neg(\exists R.\neg C)$.

An *$\mathcal{L}$-assertion* (denoted ψ) is an expression $\langle\alpha, c\rangle$, where α is an $\mathcal{ALC}$ assertion and $c \in \mathcal{T}$. From a semantics point of view, an $\mathcal{L}$-assertion $\langle\alpha, c\rangle$ constrains the certainty-value of α to be greater or equal to c. An interpretation $\mathcal{I}$ *satisfies* $\langle a{:}C, c\rangle$ (resp. $\langle(a, b){:}R, c\rangle$) iff $C^{\mathcal{I}}(a^{\mathcal{I}}) \succeq c$ (resp. $R^{\mathcal{I}}(a^{\mathcal{I}}, b^{\mathcal{I}}) \succeq c$). Two $\mathcal{L}$-assertions ψ_1 and ψ_2 are *equivalent* (denoted $\psi_1 \equiv \psi_2$) iff they are satisfied by the same set of interpretations. A *primitive $\mathcal{L}$-assertion* is a $\mathcal{L}$-assertion involving a primitive assertion only.

Concerning terminological axioms, an interpretation $\mathcal{I}$ *satisfies* $A < C$ iff $\forall d \in \Delta^{\mathcal{I}}$, $A^{\mathcal{I}}(d) \preceq C^{\mathcal{I}}(d)$, while $\mathcal{I}$ *satisfies* $A := C$ iff $\forall d \in \Delta^{\mathcal{I}}$, $A^{\mathcal{I}}(d) = C^{\mathcal{I}}(d)$. In $\mathcal{L}$-$\mathcal{ALC}$, a *knowledge base* is a finite set of $\mathcal{L}$-assertions and terminological axioms. With K_A we denote the set of $\mathcal{L}$-assertions in K, with K_T we denote the set of terminological axioms in K (the terminology), if $K_T = \emptyset$ then K is *purely assertional*, and we assume that a terminology K_T is such that no concept A appears more than once on the left-hand side of a terminological axiom in K_T and that no cyclic definitions are present in K_T. An interpretation $\mathcal{I}$ *satisfies* (is *a model of*) a knowledge base K iff $\mathcal{I}$ satisfies each element of K. A KB K *$\mathcal{L}$-entails* an $\mathcal{L}$-assertion ψ (denoted $K \models_{\mathcal{L}} \psi$) iff every model of K also satisfies ψ. Finally, given a KB K and an assertion α, it is of interest to compute α's best lower certainty-value bound. To this, the *greatest lower bound* of α w.r.t. K (denoted $glb(K, \alpha)$) is $\bigoplus\{c\colon K \models_{\mathcal{L}} \langle\alpha, c\rangle\}$ ($\bigoplus\emptyset = f$). Determining the *glb* is called the *Best Certainty-Value Bound* (BCVB) problem.

In the following we show an application in the context of logic-based information retrieval.

Logic-based multimedia information retrieval. Let us first roughly present (parts of) the LMIR model of [42,57]. In doing this, we rely on Figure 1 and consider $\mathcal{L}_{[0,1]}$. The model has two layers addressing the multidimensional aspect of multimedia objects $o \in \mathbb{O}$ (e.g., objects o1 and o2 in Figure 1): that is, their *form* and their *semantics* (or *meaning*). The form of a multimedia object is a collective name for all its *media dependent*, typically automatically extracted features, like text index term weights (object of type text), color distribution, shape, texture, spatial relationships (object of type image), mosaiced video-frame sequences and time relationships (object of type video). On the other hand, the semantics (or meaning) of a multimedia object is a collective name for those features that pertain to the slice of the real world being *represented*, which exists independently of the existence of a object referring to it. Unlike form, the semantics of a multimedia object is thus *media independent* (typically, constructed manually perhaps with the assistance of some automatic tool).

Therefore, we have two layers, the *object form layer* and the *object semantics layer*. The former represents media dependent features of the objects, while the latter describes the semantic properties of the slice of world the objects are about. The semantic entities (e.g., Snoopy, Woodstock), which objects can be about are called *semantic index terms* ($t \in \mathbb{T}$). The mapping of objects $o \in \mathbb{O}$ to semantic entities $t \in \mathbb{T}$ (e.g., "object

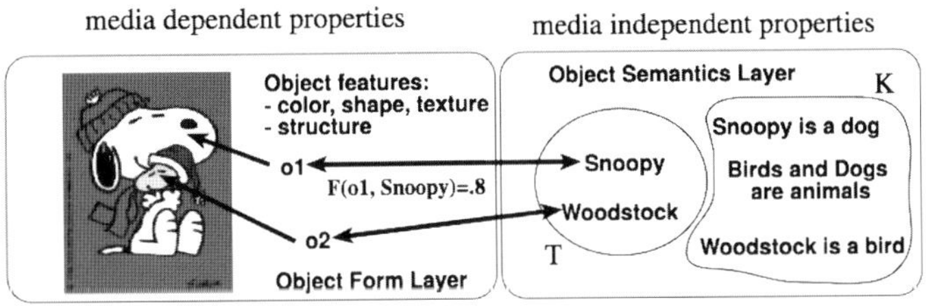

Fig. 1. LMIR model layers and objects.

o1 is about Snoopy") is called *semantic annotation*. According to the fuzzy information retrieval model [7,29,33,47], semantic annotation can be formalized as a membership function $F : \mathbb{O} \times \mathbb{T} \to [0, 1]$ describing the *correlation* between multimedia objects and semantic index terms. The value $F(o, t)$ indicates to which degree the multimedia object o deals with the semantic index term t. The meaning of a semantic index term t may in this context be represented as a fuzzy subset of multimedia objects in $\mathbb{O}$, $m(t)$, with the quantitative measure of aboutness being the values of function F for a given semantic index term t, i.e. $m(t) = \{\langle o, f_t(o)\rangle : o \in \mathbb{O}\}$, in which $f_t(o) = F(o, t)$. $m(t)$ is the meaning of term t. The function F acts as the membership function of $m(t)$. Depending on the context, the function F may be computed automatically (e.g., for text we may have [10], for images we may have an automated image annotation (classification) tool, as, e.g., [17]). Note that the function F will be a source for assertions of the form $\langle(\mathtt{o}, \mathtt{t}){:}\mathtt{F}, F(o, t)\rangle$ (see [42]). In practice, the scenario depicted in Figure 1 may roughly be represented with the following knowledge base K:

$$
\begin{aligned}
K \supseteq\ & \{\mathtt{Bird} \lessdot \mathtt{Animal}, \mathtt{Dog} \lessdot \mathtt{Animal}\} \cup \\
& \{\langle \mathtt{snoopy{:}Dog}, 1\rangle, \langle \mathtt{woodstock{:}Bird}, 1\rangle, \\
& \langle(\mathtt{o1}, \mathtt{snoopy}){:}\mathtt{F}, 0.8\rangle, \langle(\mathtt{o2}, \mathtt{woodstock}){:}\mathtt{F}, 0.7\rangle\}.
\end{aligned}
$$

Now, consider the query concept $Q = \mathtt{Object} \sqcap \exists \mathtt{F.Animal}$, i.e. retrieve all objects about animals. Then it is easily verified that we retrieve both objects, but with different *Retrieval Status Values* [3], which indicate their relatedness to the query ($K \models \langle \mathtt{snoopy}{:}Q, 0.8\rangle$, $K \models \langle \mathtt{woodstock}{:}Q, 0.7\rangle$).

3.1. *Decision algorithms in $\mathcal{L}$-$\mathcal{ALC}$*

Deciding satisfiability of a KB requires a calculus. Without loss of generality we consider purely assertional KBs only. We develop a calculus in the style of the constraint propagation method, as this method is usually proposed in the context of DLs [2]. Essentially, it generalizes the calculus presented in [58] for $\mathcal{L}_{[0,1]}$, to any certainty lattice $\mathcal{L}$. We first address the entailment problem and then the BCVB problem.

To guarantee soundness and completeness of the calculus, we make the following restrictions. We assume that in the certainty lattice $\mathcal{L} = \langle \mathcal{T}, \preceq\rangle$ the set of certainty values $\mathcal{T}$ is *finite*. From a practical point of view this is a limitation we can live with, especially taking into account that computers have finite resources, and thus, only a finite set of certainty values can be represented. In particular, this includes also the case of the rational numbers

Table 1
Conjugated pairs.

	$\langle\alpha \not\geq c'\rangle$	$\langle\alpha \preceq c'\rangle$
$\langle\alpha \succeq c\rangle$	$\neg(\exists c''.c'' \succeq c \wedge c'' \not\succeq c')$	$c \not\preceq c'$
$\langle\alpha \not\preceq c\rangle$	$\neg(\exists c''.c'' \not\preceq c \wedge c'' \not\succeq c')$	$\neg(\exists c''.c'' \not\preceq c \wedge c'' \preceq c')$

in $[0,1] \cap \mathbb{Q}$ under a given fixed precision p a computer can work with. We also point out an error in our early version [62], in which we do not rely on this assumption. Without this assumption, the results in [62] do not hold.

At first, we generalize $\mathcal{L}$-assertions to the form $\langle\alpha \succeq c\rangle$ and $\langle\alpha \preceq c\rangle$ with obvious semantics. Of course, $\langle\alpha, c\rangle$ is a syntactic variant of $\langle\alpha \succeq c\rangle$. We also assume that an assertion is always in *Negation Normal Form* where a negation appears in the head of an atom only. Furthermore, an expression $\langle a{:}\neg A \succeq c\rangle$ is always replaced with the expression $\langle a{:}A \preceq \neg c\rangle$. Similarly, an expression $\langle a{:}\neg A \preceq c\rangle$ is replaced with the expression $\langle a{:}A \succeq \neg c\rangle$.

An $\mathcal{L}$-*constraint* (denoted ψ) is inductively defined as follows: (i) an $\mathcal{L}$-assertion is an $\mathcal{L}$-constraint; (ii) if ψ and ψ' are $\mathcal{L}$-constraints, then so are $\neg\psi$, $\psi \wedge \psi'$ and $\psi \vee \psi'$ (e.g., $\langle\alpha_1 \preceq c_1\rangle \vee \neg\langle\alpha_2 \succeq c_2\rangle$). A *literal* is a primitive $\mathcal{L}$-assertion or its Boolean negation. Without loss of generality, we assume that $\mathcal{L}$-constraints are always in *Negation Normal Form* (NNF), where a Boolean negation appears in the head of an $\mathcal{L}$-assertion only. The definition of satisfiability (of a set) of $\mathcal{L}$-constraints is easy: e.g., I satisfies $\neg\psi$ iff I does not satisfy ψ, while I satisfies $\psi \vee \psi'$ iff I does satisfy either ψ or ψ'. It follows that

$$K \models_{\mathcal{L}} \langle\alpha \succeq c\rangle \quad \text{iff} \quad K \cup \{\neg\langle\alpha \succeq c\rangle\} \text{ not satisfiable.}$$

Therefore, it is sufficient to develop a calculus for the satisfiability problem.

Note that in case $\mathcal{L}$ is a total order, then we have the equivalences between $\neg\langle\alpha \succeq c\rangle$ and $\langle\alpha \prec c\rangle$, and between $\neg\langle\alpha \preceq c\rangle$ and $\langle\alpha \succ c\rangle$. For instance, this property is used in the calculus developed for fuzzy $\mathcal{ALC}$ [58]. In general, these equivalences do not hold (e.g., in $\mathcal{L}_B$, $\neg\langle\alpha \succeq u\rangle$ is equivalent to $\langle\alpha \preceq i\rangle$ and not to $\langle\alpha \prec u\rangle$). For ease, sometimes we write $\langle\alpha \not\succeq c\rangle$ in place of $\neg\langle\alpha \succeq c\rangle$ and $\langle\alpha \not\preceq c\rangle$ in place of $\neg\langle\alpha \preceq c\rangle$.

Our calculus, determining whether a finite set S of $\mathcal{L}$-constraints is satisfiable or not, is based on a set of constraint propagation rules transforming a set S into "simpler" satisfiability preserving sets S_i until either all S_i contain an inconsistency, called *clash* (indicating that from all the S_i no model of S can be build), or some S_i is completed and clash-free, that is, no rule can be further be applied to S_i and S_i contains no clash (indicating that from S_i a model of S can be build). A set of $\mathcal{L}$-constraints S contains a *clash* (inconsistency) iff it contains a set of literals $\{\langle\alpha\ r_j\ c_j\rangle\}_{j\in J}$ (with $r_j \in \{\succeq, \preceq, \not\succeq, \not\preceq\}$) such that the set of constraints $\{\langle x\ r_j\ c_j\rangle\}_{j\in J}$ has no solution for the variable x in the lattice $\mathcal{L}$.[3] For instance, S contains a clash if it contains either $\langle a{:}\bot \succeq c\rangle$ (where $c \succ f$), or $\langle a{:}\top \preceq c\rangle$ (where $c \prec t$), or $\langle a{:}\bot \not\preceq c\rangle$, or $\langle a{:}\top \not\succeq c\rangle$, or $\langle\alpha \not\succeq f\rangle$, or $\langle\alpha \not\preceq t\rangle$, or S contains a conjugated pair of $\mathcal{L}$-constraints. Each entry in Table 1 says us under which condition the row-column pair of $\mathcal{L}$-constraints is a *conjugated pair*.

[3] A solution for x is a certainty value c such that all constraints in $\{\langle c\ r_j\ c_j\rangle\}_{j\in J}$ are satisfied, i.e. for all $j \in J$, $c\ r_j\ c_j$ holds in $\mathcal{L}$.

For instance, in $\mathcal{L}_B$, $\langle a{:}A \succeq i\rangle$ and $\langle a{:}A \preceq u\rangle$ is a conjugated pair as $i \not\preceq u$. While in total ordered lattices checking inconsistency is easy, this may not be the case for arbitrary lattices. Given an $\mathcal{L}$-constraint ψ, with ψ^c we indicate a conjugate of ψ (if there exists one). A conjugate of an $\mathcal{L}$-constraint may be not unique, as there could be infinitely many. For instance, in $\mathcal{L}_{[0,1]}$-$\mathcal{ALC}$, any $\langle a{:}A \preceq c\rangle$ with $c \prec c'$ is a conjugate of $\langle a{:}A \succeq c'\rangle$.

There are obvious tableaux rules for the Boolean connectives $\wedge$ and $\vee$. For each connective $\sqcap, \sqcup, \neg, \forall, \exists$ there is a rule for each relation $\succeq, \preceq, \not\succeq$ and $\not\preceq$. In what follows, for any $c \in \mathcal{T}$, with $D_{\mathcal{L}}(c)$ we indicate the set $D_{\mathcal{L}}(c)$ as the set of all minimal elements of the set $\{(c_1, c_2) \in \mathcal{T} \times \mathcal{T}\colon c_1 \oplus c_2 \succeq c\}$ where the order over pairs is $(c_1, c_2) \preceq (c_3, c_4)$ iff $c_1 \preceq c_3$ and $c_2 \preceq c_4$. The purpose of $D_{\mathcal{L}}(c)$ is to identify meaningful candidates c_1 and c_2, making, e.g., a disjunction of the form $\langle a{:}A \sqcup B \succeq c\rangle$ true, whenever we reduce $\langle a{:}A \sqcup B \succeq c\rangle$ to $\{\langle a{:}A \succeq c_1\rangle \wedge \langle a{:}B \succeq c_2\rangle\}$. The choice is non-deterministic and there could $|D_{\mathcal{L}}(c)|$ many choices, and $|D_{\mathcal{L}}(c)|$ depends on c and $\mathcal{L}$. Note that $(c_1, c_2) \in D_{\mathcal{L}}(c)$ iff $(c_2, c_1) \in D_{\mathcal{L}}(c)$ and $|D_{\mathcal{L}}(c)| \succeq 2$ (except the case where $c = f$, in which case $|D_{\mathcal{L}}(c)| = 1$) as $\{(c, f), (f, c)\} \subseteq D_{\mathcal{L}}(c)$, e.g., in $\mathcal{L}_{\{0,1\}}$, $D_{\mathcal{L}_{\{0,1\}}}(1) = \{(1, 0), (0, 1)\}$ indicating that, e.g., $\langle a{:}A \sqcup B \succeq c\rangle$ can be branched into either $\{\langle a{:}A \succeq 1\rangle\}$ or $\{\langle a{:}B \succeq 1\rangle\}$. Note also that in $\mathcal{L}_B$, we have $D_{\mathcal{L}_B}(t) = \{(t, f), (f, t), (u, i), (i, u)\}$ and, thus, e.g., $\langle a{:}A \sqcup B \succeq c\rangle$ can be branched into four alternatives. As $\mathcal{L}$ has finite $\mathcal{T}$, then for any $c \in \mathcal{T}$, $D_{\mathcal{L}}(c)$ is finite as well. If $\mathcal{L}$ is a total order, like $\mathcal{L}_4$ or $\mathcal{L}_{[0,1]}$, then $D_{\mathcal{L}}(c) = \{(c, f), (f, c)\}$, for any $c \in \mathcal{T}$.

The limitation to the finiteness of $\mathcal{L}$ will guarantee both the termination of our decision procedure and its correctness and completeness.

We present rules for $\sqcap, \sqcup, \neg, \forall, \exists$, and $\succeq$ and $\not\succeq$ only. The rules for $\preceq, \not\preceq$ can be derived from $\succeq$ and $\not\succeq$, respectively. Indeed, the rules for $\langle a{:}C \preceq c\rangle$, $\langle a{:}C \sqcup D \preceq c\rangle$, $\langle a{:}C \sqcap D \preceq c\rangle$, $\langle a{:}\exists R.C \preceq c\rangle$ and $\langle a{:}\forall R.C \preceq c\rangle$ can be derived from the equivalent expressions $\langle a{:}\neg C \succeq \neg c\rangle$, $\langle a{:}\neg C \sqcap \neg D \succeq \neg c\rangle$, $\langle a{:}\neg C \sqcup \neg D \succeq \neg c\rangle$, $\langle a{:}\forall R.\neg C \succeq \neg c\rangle$ and $\langle a{:}\exists R.\neg C \succeq \neg c\rangle$, respectively. Similarly for $\not\preceq$ (e.g., we have the following equivalence: $\langle a{:}C \not\preceq c\rangle \equiv \neg\langle a{:}C \preceq c\rangle \equiv \neg\langle a{:}\neg C \succeq \neg c\rangle \equiv \langle a{:}\neg C \not\succeq \neg c\rangle$).

The rules below rely on the following properties over a lattice $\mathcal{L} = \langle \mathcal{T}, \preceq\rangle$: for any $c, c_1, c_2 \in \mathcal{T}$, (i) $c_1 \otimes c_2 \succeq c$ iff $c_1 \succeq c$ and $c_2 \succeq c$; (ii) $c_1 \otimes c_2 \not\succeq c$ iff $c_1 \not\succeq c$ or $c_2 \not\succeq c$; (iii) $c_1 \oplus c_2 \succeq c$ iff $c_1 \succeq c'$ and $c_2 \succeq c''$, for $(c', c'') \in D_{\mathcal{L}}(c)$; and (iv) $c_1 \oplus c_2 \not\succeq c$ iff not $(c_1 \oplus c_2 \succeq c)$ iff for all $(c', c'') \in D_{\mathcal{L}}(c)$, either $c_1 \not\succeq c'$ or $c_2 \not\succeq c''$.

In the following rules we assume that a new constraint is added to a constraint set S if it is not subsumed in S (a constraint $\langle \alpha \succeq c\rangle$ is *subsumed* by a constraint $\langle \alpha \succeq c'\rangle$ iff $c' \succeq c$ – similarly for the other relations $\not\succeq, \preceq$, and $\not\preceq$). We also avoid adding constraints of the form $\langle \alpha \succeq f\rangle$ and $\langle \alpha \preceq t\rangle$.

1. $S \rightarrow_{\sqcap_{\succeq}} S \cup \{\langle a{:}C \succeq c\rangle \wedge \langle a{:}D \succeq c\rangle\}$
   if $\psi = \langle a{:}C \sqcap D \succeq c\rangle \in S$ and the $(\sqcap_{\succeq})$ rule has not yet been applied to $\psi \in S$.
2. $S \rightarrow_{\sqcap_{\not\succeq}} S \cup \{\langle a{:}C \not\succeq c\rangle \vee \langle a{:}D \not\succeq c\rangle\}$
   if $\psi = \langle a{:}C \sqcap D \not\succeq c\rangle \in S$ and the $(\sqcap_{\not\succeq})$ rule has not yet been applied to $\psi \in S$.
3. $S \rightarrow_{\sqcup_{\succeq}} S \cup \{\psi'\}$
   if $\psi = \langle a{:}C_1 \sqcup C_2 \succeq c\rangle \in S$, the $(\sqcup_{\succeq})$ rule has not yet been applied to $\psi \in S$ and $\psi' = \bigvee_{(c_1,c_2)\in D_{\mathcal{L}}(c)} \langle a{:}C_1 \succeq c_1\rangle \wedge \langle a{:}C_2 \succeq c_2\rangle$.
4. $S \rightarrow_{\sqcup_{\not\succeq}} S \cup \{\psi'\}$
   if $\psi = \langle a{:}C_1 \sqcup C_2 \not\succeq c\rangle \in S$, the $(\sqcup_{\not\succeq})$ rule has not yet been applied to $\psi \in S$ and $\psi' = \bigwedge_{(c_1,c_2)\in D_{\mathcal{L}}(c)} (\langle a{:}C_1 \not\succeq c_1\rangle \vee \langle a{:}C_2 \not\succeq c_2\rangle)$.

5. $S \rightarrow_{\forall_{\succeq}} S \cup \{\psi'\}$
   if $\{\langle a{:}\forall R.C \succeq c\rangle, \psi^c\} \subseteq S$, $\psi = \langle (a,b){:}R \preceq \neg c\rangle$, the $(\forall_{\succeq})$ rule has not yet been applied in S to the pair $(\langle a{:}\forall R.C \succeq c\rangle, \psi)$ and ψ' is the expression $\bigvee_{(c_1,c_2)\in D_{\mathcal{L}}(c)\setminus\{(c,f)\}} \langle (a,b){:}R \preceq \neg c_1\rangle \wedge \langle b{:}C \succeq c_2\rangle$.
6. $S \rightarrow_{\forall_{\not\succeq}} S \cup \{\psi'\}$
   if $\psi = \langle a{:}\forall R.C \not\succeq c\rangle, \in S$, the $(\forall_{\not\succeq})$ rule has not yet been applied to $\psi \in S$ and $\psi' = \bigwedge_{(c_1,c_2)\in D_{\mathcal{L}}(c)} \langle (a,b){:}R \not\preceq \neg c_1\rangle \vee \langle b{:}C \not\succeq c_2\rangle$, for a new constant b.
7. $S \rightarrow_{\exists_{\succeq}} S \cup \{\langle (a,b){:}R \succeq c\rangle \wedge \langle b{:}C \succeq c\rangle\}$
   if $\psi = \langle a{:}\exists R.C \succeq c\rangle \in S$, b new constant and the $(\exists_{\succeq})$ rule has not yet been applied to $\psi \in S$.
8. $S \rightarrow_{\exists_{\not\succeq}} S \cup \{\psi'\}$
   if $\{\langle a{:}\exists R.C \succeq c\rangle, \psi^c\} \subseteq S$, $\psi = \langle (a,b){:}R \not\succeq c\rangle$, the $(\exists_{\not\succeq})$ rule has not yet been applied in S to the pair $(\langle a{:}\exists R.C \succeq c\rangle, \psi)$ and $\psi' = \langle b{:}C \not\succeq c_2\rangle$.

Some of the above rules deserve some explanation. The $(\sqcup_{\succeq})$ rule is a generalization of the classical disjunction rule. The main difference is that, depending on the lattice $\mathcal{L}$, there may be more than the usual two branches (likely as many as $|D_{\mathcal{L}}(c)|$). For instance, in $\mathcal{L}_B$, a constraint $\langle a{:}C_1 \sqcup C_2 \succeq t\rangle \in S$ may give rise to four branches, each containing $\langle a{:}C_1 \succeq t\rangle$, $\langle a{:}C_2 \succeq t\rangle$, $\langle a{:}C_1 \succeq i\rangle \wedge \langle a{:}C_2 \succeq u\rangle$ and $\langle a{:}C_1 \succeq u\rangle \wedge \langle w{:}C_2 \succeq i\rangle$, respectively. Note that if $\mathcal{L}$ is a total order then there are at most two branches. The $(\sqcup_{\not\succeq})$ rule is derived from the $(\sqcup_{\succeq})$ rule. The $\forall_{\succeq}$ is a specialization of the $(\sqcup_{\succeq})$ rule as well and is a generalization of the classical rule for $\forall$. Indeed, according to the semantics of $\mathcal{L}$-$\mathcal{ALC}$, $\langle a{:}\forall R.C \succeq c\rangle$ may be viewed as a disjunction and with decomposition rule

- $S \rightarrow_{\forall'_{\succeq}} S \cup \{\langle (a,b){:}R \preceq \neg c_1\rangle, \langle b{:}C \succeq c_2\rangle\}$
  if $\langle a{:}\forall R.C \succeq c\rangle \in S$, b occurs in S and $(c_1, c_2) \in D_{\mathcal{L}}(c)$.

But, observing that $(c, f) \in D_{\mathcal{L}}(c)$, we obtain a constraint set S containing $\psi = \langle (a,b){:}R \preceq \neg c\rangle$ (e.g., in the classical $\mathcal{L}_{\{t,f\}}$, with $c = t$, we have $\langle (a,b){:}R \preceq f\rangle \in S$). This constraint can only be clashed if S contains a conjugate to ψ (e.g., in the classical case $\langle (a,b){:}R \succeq t\rangle$ must be in S). This motivates the $\forall_{\succeq}$ rule as a refinement of the above $\forall'_{\succeq}$ rule. Note that the $(\forall_{\not\succeq})$ rule can be worked out by a similar argumentation, by relying on the $(\sqcup_{\not\succeq})$ rule, and is left to the reader.

A constraint set S is *complete* if no rule is applicable to it. A complete set S_2 obtained from a set S_1 by applying the above rules is called a *completion* of S_1. Note that more than one completion can be obtained. It can be verified that for finite certainty lattices, the above calculus has the *termination property*, i.e. any completion of a finite set of $\mathcal{L}$-constraints S can be obtained after a finite number of rule applications.

Moreover, in a similar way as in [58, Proposition 3], it is easily verified, by case analysis, that the above rules are sound, i.e. if S_1 is satisfiable then there is a satisfiable completion S_2 of S_1. Here we use the fact that for finite lattices, if, e.g.,

$$(\exists R.C)^{\mathcal{I}}(d) = \bigoplus_{d'\in\Delta^{\mathcal{I}}} \{R^{\mathcal{I}}(d,d') \otimes C^{\mathcal{I}}(d')\} \succeq c$$

then there is $d' \in \Delta^{\mathcal{I}}$ such that $R^{\mathcal{I}}(d,d') \succeq c$ and $C^{\mathcal{I}}(d') \succeq c$, which motivates rule 7. The case for $\forall R.C$ is similar.

PROPOSITION 2. *For a finite certainty lattice $\mathcal{L}$, a finite set of $\mathcal{L}$-constraints S is satisfiable iff there exists a clash free completion of S.*

From a computational complexity point of view, the satisfiability problem is in the same complexity class (PSPACE-complete [53]) as for $\mathcal{ALC}$. Indeed, let us indicate with $|\mathcal{L}|$ the finite dimension of representing $\mathcal{L}$ and with $|S|$ the dimension of a constraint set S. With *combined complexity* we intend the complexity w.r.t. $|\mathcal{L}| + |S|$.

In a similar way as in [58, Proposition 4]:

PROPOSITION 3. *The satisfiability problem is PSPACE-complete w.r.t. combined complexity for finite certainty lattices.*

The above result says us that no additional computational cost has to be paid for the major expressive power (if $\mathcal{L}$ is finite, of course).

Concerning the BCVB problem, for any finite certainty lattice $\mathcal{L} = \langle \mathcal{T}, \preceq \rangle$ we may apply a simple iterative approximation algorithm for determining the *glb*:

1. let $\mathcal{T}_0 = \mathcal{T}$ and $\bar{c} := f$. For $j = 0, 1, 2, \ldots$
2. take a value $c \in \mathcal{T}_j$; if there is no such value then set $glb(K, \alpha) := \bar{c}$ and exit;
3. if $K \models_{\mathcal{L}} \langle \alpha \succeq c \rangle$ then set $\bar{c} := c$ and $\mathcal{T}_{j+1} = \mathcal{T}_j \setminus (\{c' : c' \not\succeq \bar{c}\} \cup \{c\})$, otherwise $\mathcal{T}_{j+1} = \mathcal{T}_j \setminus \{c' : c' \succeq c\}$;
4. go to step 2.

The procedure converges after at most $|\mathcal{T}|$ steps. Of course, the speed of convergence depends on the 'goodness' of the chosen value c in step 2.

4. Uncertainty in description logic programs

We are going now to define an extension of DLPs towards the management of uncertainty. Classically, n-ary predicates may be seen as functions from their domain into $\{0, 1\}$, where 0 stands for *false*, while 1 stands for *true*. We extend this notion by mapping n-ary predicates into functions from their domain into a certainty lattice $\mathcal{L} = \langle \mathcal{T}, \preceq \rangle$. The associated value $c \in \mathcal{T}$ indicates to which *certainty* a predicate is *true*.

Syntax. From a syntax point of view, as the description logic component, we consider $\mathcal{L}$-$\mathcal{ALC}$. We then extend the DL component with rules, where the management of uncertainty is based on so-called annotation terms. Annotation terms and the form of the rules are like those in the Generalized Annotated Logic Programming framework of Kifer and Subrahmanian [31], except that now DL atoms and roles may appear in a rule body.

Let us define an *annotation function* of arity n to be a total, *monotone* and computable function[4] $f : (\mathcal{T})^n \to \mathcal{T}$. Assume a new alphabet of *annotation variables*, which will denote a value in $\mathcal{T}$ and can only appear in so-called *annotation terms*. An *annotation term*, λ, is defined inductively: (i) as a constant $c \in \mathcal{T}$, or as an annotation variable (ν), or (ii) is of the form $f(\lambda_1, \ldots, \lambda_n)$, where f is an n-ary annotation function and all λ_i are annotation terms. Annotation terms are supposed to denote elements of $\mathcal{T}$. Now, let A be an atom and λ an annotation term. An *a-atom*, denoted $\overline{A}$, is of the form $A{:}\lambda$. The intended meaning is that "the certainty of A is greater or equal to λ". A *a-logic program* (or, simply *a*-program), $\overline{\mathcal{P}}$, is a finite set of *a-rules* of the form

$$H{:}\lambda \leftarrow B_1{:}\lambda_1, \ldots, B_n{:}\lambda_n,$$

[4] The result of f is computable in a finite amount of time.

where $H{:}\lambda$ and all $B_1{:}\lambda_1$ are a-atoms. Finally, an *Annotated Description Logic Program* (ADLP), denoted $\overline{\mathcal{DP}}$, is a pair $\langle K, \overline{\mathcal{P}}\rangle$, where K is a $\mathcal{L}$-$\mathcal{ALC}$ knowledge base and $\overline{\mathcal{P}}$ is a a-program.

Semantics. In order to avoid straightforward repetition, if not stated otherwise, definitions related to a-logic programs, parallels those for classical logic programs. Furthermore, in grounding an a-atom $A{:}\lambda$, we assume that the annotation term λ is grounded as well, i.e. annotation variables and constants are replaced with values in $\mathcal{T}$ and annotation terms of the form $f(\lambda_1, \ldots, \lambda_n)$ are replaced with the result of the computation of $f(\lambda_1, \ldots, \lambda_n)$. A is grounded with the (individual) constants appearing in $\overline{\mathcal{DP}}$. Note that a grounded a-program $\overline{\mathcal{P}}$ may contain an infinite number of rules due to the grounding of annotation terms. For a grounded a-program $\overline{\mathcal{P}}$, the Herbrand base $\mathcal{B}_{\overline{\mathcal{P}}}$ is the set of ground non-DL atoms A that can be constructed using the predicate symbols in $\overline{\mathcal{P}}$ (annotations terms are not considered in the grounding of A). With $\mathtt{ground}(\overline{\mathcal{P}})$ we denote the set of all ground rules of $\overline{\mathcal{P}}$.

An *a-interpretation* I of a grounded a-program $\overline{\mathcal{P}}$ is a function $I : \mathcal{B}_{\overline{\mathcal{P}}} \to \mathcal{T}$. We say I is a *model* of a ground a-atom $A{:}\lambda$ w.r.t. K, denoted $I \models_K A{:}\lambda$, iff

- $I(A) \succeq \lambda$, if $A \in \mathcal{B}_{\mathcal{P}}$ (A is a non-DL atom);
- $K \models_{\mathcal{L}} \langle A, \lambda\rangle$ if A is a DL atom.

The notion of model is then straightforwardly extended to a-rules, a-programs $\overline{\mathcal{P}}$ and ADLPs $\overline{\mathcal{DP}}$. The notion of entailment, $\overline{\mathcal{DP}} \models A{:}\lambda$, is as usual.

We say that $I \preceq I'$ iff for all atoms $A \in \mathcal{B}_{\overline{\mathcal{P}}}$, $I(A) \preceq I'(A)$. Like for the crisp case (see Proposition 1) we have:

PROPOSITION 4. *Let $\overline{\mathcal{DP}} = \langle K, \overline{\mathcal{P}}\rangle$ be a ADLP. If $\overline{\mathcal{DP}}$ is satisfiable, then there exists a unique model $M_{\overline{\mathcal{DP}}}$ of $\overline{\mathcal{DP}}$ such that $M_{\overline{\mathcal{DP}}} \preceq I$ for all models I of $\overline{\mathcal{DP}}$. Furthermore, for any ground a-atom $\bar{A}$, $\overline{\mathcal{DP}} \models \bar{A}$ iff $M_{\overline{\mathcal{DP}}} \models_K \bar{A}$.*

Like in [31], the minimal model coincides with the least fixed-point of the following monotone operator. Let $\overline{\mathcal{DP}} = \langle K, \overline{\mathcal{P}}\rangle$ be a ADLP. Define the operator $T_{\overline{\mathcal{DP}}}$ on interpretations as follows: for every interpretation I, for all $A \in \mathcal{B}_{\overline{\mathcal{P}}}$ let

$$T_{\overline{\mathcal{DP}}}(I)(A) = \bigoplus\{\lambda \mid A{:}\lambda \leftarrow B_1{:}\lambda_1, \ldots, B_n{:}\lambda_n \in \overline{\mathcal{P}}, I \models_K B_i{:}\lambda_i, \text{ for all } i\}.$$

Note however, that $T_{\overline{\mathcal{DP}}}$ is not continuous in general. In fact the grounded a-program (over $\mathcal{L}_{[0,1]}$), containing all ground instances of the rules

$$\begin{aligned} &A{:}0.\\ A{:}&\frac{\nu+1}{2} \leftarrow A{:}\nu\\ &B{:}1 \leftarrow A{:}1 \end{aligned}$$

has unique minimal model $I(A) = I(B) = 1$, which is obtained after $\omega+1$ $T_{\overline{\mathcal{DP}}}$ iterations starting from $I_\bot$ (the interpretation assigning 0 to all atoms), where ω is the first limit ordinal. However, under the assumption that the certainty lattice $\mathcal{L} = \langle\mathcal{T}, \preceq\rangle$ is finite, the minimal model can be computed after at most $\mathcal{O}(|\overline{\mathcal{DP}}||\mathcal{T}|)$ steps (any rule can be applied at most $\mathcal{O}(|\mathcal{T}|)$ times). In particular, in the example above, in case we consider two digit precision of [0, 1] numbers, the minimal model is computed after at most 300 (precisely, 10) applications of the $T_{\overline{\mathcal{DP}}}$ iterations starting from $I_\perp$.

Like for the crisp case, from an implementation point of view, given $\mathcal{DP} = \langle K, \mathcal{P}\rangle$, we can compute the set of entailment assertions $F = \{\langle A, \lambda\rangle : K \models_{\mathcal{L}} \langle A, \lambda\rangle,$ where A is a DL atom$\}$ first and then add them as facts to $\overline{\mathcal{P}}$. Then, we can rely on theorem proving for $\overline{\mathcal{P}}$ either using a usual bottom-up computation or a top-down computation like [9,64].

EXAMPLE 2. Consider an insurance company, which has information about its customers used to determine the risk coefficient of each customer. The company has: (i) data grouped into a set of facts

$$\{\texttt{Experience(john)}{:}0.7, \texttt{Risk(john)}{:}0.5, \texttt{Sport\_car(john)}{:}0.8\};$$

and (ii) a set of rules (for ease we omit the terminological axioms about the involved concepts).

$$\begin{aligned}
&\texttt{Good_driver(x)}{:}\min(\nu, \nu'/2) &&\leftarrow \texttt{Experience(x)}{:}\nu, \texttt{Risk(x)}{:}\nu' \\
&\texttt{Risk(x)}{:}\nu \cdot 0.8 &&\leftarrow \texttt{Young(x)}{:}\nu \\
&\texttt{Risk(x)}{:}\nu \cdot 0.8 &&\leftarrow \texttt{Sport_car(x)}{:}\nu \\
&\texttt{Risk(x)}{:}\min(\nu, \nu'/2) &&\leftarrow \texttt{Experience(x)}{:}\nu, \texttt{Good_driver(x)}{:}\nu'
\end{aligned}$$

It turns out that

$$\begin{aligned}
&\texttt{Risk(john)}{:}0.64 \\
&\texttt{Sport_car(john)}{:}0.8 \\
&\texttt{Young(john)}{:}0 \\
&\texttt{Good_driver(john)}{:}0.32 \\
&\texttt{Experience(john)}{:}0.7
\end{aligned}$$

are entailed from the above facts and the rules, which establish that `john`'s degree of `Risk` is greater or equal than 0.64.

5. Related work

While the combination of DLs with LPs is not new, to best of our knowledge, the generality of DLPs and integration of the management of uncertainty as well, has not been investigated yet.

Concerning ADLPs, the DL component mainly relies on the work of Straccia [62]. Concerning the rule component, as anticipated, the Generalized Annotated Logic Programming framework of Kifer and Subrahmanian [31] inspires it.

We recall that concerning the combination of DLs with LPs, they do not consider function symbols and, the works can roughly be divided into (i) hybrid approaches, which use description logics to specify structural constraints in the bodies of logic programs rules; and (ii) approaches that reduce description logic inference to logic programming. The basic idea behind (i) is to combine the semantic and computational strengths of the two systems, while the main rationale of (ii) is to use powerful logic programming technology for inference in description logics.

In the former case fall approaches like [12,25,35]. In particular, [12] combines plain Datalog (no disjunction and negation) with the description logic $\mathcal{ALC}$, where the integration lies in using concepts from the terminological component as constraints in rule bodies. It also presents a technique for answering conjunctive queries (existentially quantified conjunctions of atoms) with such constraints, where SLD resolution is integrated with an inference method for $\mathcal{ALC}$. More related to our approach is [35], which combines Horn rules with the description logic $\mathcal{ALCNR}$, where concepts and roles may appear in the body of Horn rules. Like in [12], [35] devices an SLD resolution integrated with an inference method for $\mathcal{ALCNR}$. Finally, [25] is similar to [35], except that the atoms of a rule refer to OWL classes and properties rather than to $\mathcal{ALCNR}$.

In the latter case (reducing description logic inference to function-free logic programming) fall approaches like [1,19,21,43,27,45,44,65,68]. We remark that (i) [68] reduces knowledge bases in the DL $\mathcal{ALCN}$ ($\mathcal{ALC}$ with number restrictions) to open logic programs; (ii) [1,65] reduce reasoning in the DL $\mathcal{ALCQI}$ ($\mathcal{ALC}$ with qualified number restrictions and inverse roles) to query answering from answer sets of normal logic programs; (iii) [19] shows how a subset of the DL $\mathcal{SHOIQ}$ can be reduced to a subset of Horn programs (positive normal programs); (iv) [21] reduces the DL $\mathcal{SHIF}$ with transitive role closure to disjunctive logic programs (in [61] we use a similar reduction); and [43, 27,45,44] reduces reasoning in the DL $\mathcal{SHIQ}^-$ to reasoning in disjunctive Datalog programs by translating DL expressions into FOL clauses, saturates the clauses by resolution and translates the saturated clauses to disjunctive Datalog programs.

Finally, not in the above classification fall approaches like [14,15,70]. Roughly, [14,15] combines the DL $\mathcal{SHOIN}(D)$, which is $\mathcal{SHOIN}$ with concrete domains (e.g., integers, reals), with normal logic programs. The main characteristics of [14] lies in the use of a weaker semantics to be used to get decidable reasoning. Indeed, concept and role predicates appearing in rules can be managed as system calls to a DL reasoner. Our approach is inspired on this work, but considered uncertainty components. Finally, [70] is based on *F-logic* [30].

6. Conclusion

We have presented a DL framework for the management of uncertain information. Our main feature is that a sentence is not just true or false like in classical DLs, but *certain* to some degree, where the certainty value is taken from a certainty lattice. Syntax, semantics and a sound and complete tableaux algorithm for reasoning in it has been presented. We also extended it with a rule component based on annotated logic programs, forming the class of annotated description logic programs. We also discussed how to reason in it.

Acknowledgements

I'm in debt with the anonymous reviewers who provided useful comments to correct and improve an early version of this paper.

References

[1] G. Alsaç, C. Baral, *Reasoning in description logics using declarative logic programming*, Technical Report, Department of Computer Science and Engineering, Arizona State University, USA, 1997.

[2] F. Baader, D. Calvanese, D. McGuinness, D. Nardi, P.F. Patel-Schneider (Eds.), The Description Logic Handbook: Theory, Implementation, and Applications, Cambridge University Press, 2003.

[3] R.A. Baeza-Yates, B. Ribeiro-Neto, *Modern Information Retrieval*, Addison-Wesley Longman Publishing Co., Inc., 1999.

[4] N.D. Belnap, *A useful four-valued logic*, in: G. Epstein, J.M. Dunn (Eds.), Modern Uses of Multiple-Valued Logic, Reidel, Dordrecht, 1977, pp. 5–37.

[5] H. Boley, S. Tabet, G. Wagner, *Design rationale of RuleML: a markup language for semantic web rules*, in: Proceedings of the 1st Semantic Web Working Symposium (SWWS-01), Stanford, 2001, pp. 105–113.

[6] P. Bonatti, A. Tettamanzi, *Some complexity results on fuzzy description logics*, in: A. Petrosino, V. Di Gesù, F. Masulli (Eds.), WILF 2003 International Workshop on Fuzzy Logic and Applications, Lecture Notes in Comput. Sci., vol. 2955, Springer-Verlag, Berlin, 2004.

[7] G. Bordogna, P. Carrara, G. Pasi, *Query term weights as constraints in fuzzy information retrieval*, Inform. Process. Management **27** (1) (1991), 15–26.

[8] C.V. Damásio, J. Medina, M.O. Aciego, *Sorted multi-adjoint logic programs: termination results and applications*, in: Proceedings of the 9th European Conference on Logics in Artificial Intelligence (JELIA-04), Lecture Notes in Comput. Sci., vol. 3229, Springer-Verlag, 2004, pp. 252–265.

[9] C.V. Damásio, J. Medina, M.O. Aciego, *A tabulation proof procedure for residuated logic programming*, in: Proceedings of the 6th European Conference on Artificial Intelligence (ECAI-04), 2004.

[10] S. Dill, N. Eiron, D. Gibson, D. Gruhl, R. Guha, A. Jhingran, T. Kanungo, S. Rajagopalan, A. Tomkins, J.A. Tomlin, J.Y. Zien, *SemTag: and Seeker: bootstrapping the semantic web via automated semantic annotation*, in: The Twelfth International World Wide Web Conference (WWW-03), Budapest, Hungary, 2003.

[11] F.M. Donini, M. Lenzerini, D. Nardi, W. Nutt, A. Schaerf, *An epistemic operator for description logics*, Artificial Intelligence **100** (1–2) (1998), 225–274.

[12] F.M. Donini, M. Lenzerini, D. Nardi, A. Schaerf, *AL-log: integrating datalog and description logics*, J. Intelligent Inform. Systems **10** (3) (1998), 227–252.

[13] D. Dubois, J. Lang, H. Prade, *Towards possibilistic logic programming*, in: Proc. of the 8th Int. Conf. on Logic Programming (ICLP-91), MIT Press, 1991, pp. 581–595.

[14] T. Eiter, T. Lukasiewicz, R. Schindlauer, H. Tompits, *Combining answer set programming with description logics for the semantic web*, in: Proceedings of the 9th International Conference on Principles of Knowledge Representation and Reasoning (KR-04), AAAI Press, 2004.

[15] T. Eiter, T. Lukasiewicz, R. Schindlauer, H. Tompits, *Well-founded semantics for description logic programs in the semantic web*, in: Proceedings RuleML 2004 Workshop, International Semantic Web Conference, Lecture Notes in Comput. Sci., vol. 3323, Springer-Verlag, 2004, pp. 81–97.

[16] E. Franconi, S. Tessaris, *Rules and queries with ontologies: a unified logical framework*, in: Workshop on Principles and Practice of Semantic Web Reasoning (PPSWR-04), 2004.

[17] Th. Gevers, A.W.M. Smeulders, *Content-based image retrieval: an overview*, in: G. Medioni, S.B. Kang (Eds.), Emerging Topics in Computer Vision, Prentice-Hall, 2004.

[18] T.R. Giugno, T. Lukasiewicz, *P-SHOQ(D): a probabilistic extension of SHOQ(D) for probabilistic ontologies in the semantic web*, in: Proceedings of the 8th European Conference on Logics in Artificial Intelligence (JELIA'02), Lecture Notes in Artificial Intelligence, vol. 2424, Springer-Verlag, 2002.

[19] B.N. Grosof, I. Horrocks, R. Volz, S. Decker, *Description logic programs: combining logic programs with description logic*, in: Proceedings of the Twelfth International Conference on World Wide Web, ACM Press, 2003, pp. 48–57.

[20] J. Heinsohn, *Probabilistic description logics*, in: R. Lopez de Mantara, D. Pool (Eds.), Proceedings of the 10th Conference on Uncertainty in Artificial Intelligence, 1994, pp. 311–318.

[21] S. Heymans, D. Vermeir, *Integrating description logics and answer set programming*, in: Principles and Practice of Semantic Web Reasoning (PPSWR-03), Mumbai, India, Lecture Notes in Comput. Sci., vol. 2901, Springer-Verlag, 2003, pp. 146–159.

[22] S. Hölldobler, H.-P. Störr, T.D. Khang, *The subsumption problem of the fuzzy description logic ALC_{FH}*, in: Proceedings of the 10th International Conference on Information Processing and Management of Uncertainty in Knowledge-Based Systems (IPMU-04), 2004.

[23] B. Hollunder, *An alternative proof method for possibilistic logic and its application to terminological logics*, in: 10th Annual Conference on Uncertainty in Artificial Intelligence, Seattle, Washington, Morgan Kaufmann, 1994, pp. 327–335.

[24] I. Horrocks, P.F. Patel-Schneider, *Three theses of representation in the semantic web*, in: Proceedings of the 12th International Conference on World Wide Web, ACM Press, 2003, pp. 39–47.

[25] I. Horrocks, P.F. Patel-Schneider, *A proposal for an OWL rules language*, in: Proc. of the Thirteenth International World Wide Web Conference (WWW-04), ACM, 2004.

[26] I. Horrocks, P.F. Patel-Schneider, F. van Harmelen, *From SHIQ and RDF to OWL: the making of a web ontology language*, J. Web Semantics **1** (1) (2003), 7–26.

[27] U. Hustadt, B. Motik, U. Sattler, *Reasoning in description logics with a concrete domain in the framework of resolution*, in: Proceedings of the 6th European Conference on Artificial Intelligence (ECAI-04), 2004.

[28] M. Jäger, *Probabilistic reasoning in terminological logics*, in: Proceedings of KR-94, 5-th International Conference on Principles of Knowledge Representation and Reasoning, Bonn, FRG, 1994, pp. 305–316.

[29] E.E. Kerre, R.B. Zenner, R.M. De Caluwe, *The use of fuzzy set theory in information retrieval and databases: a survey*, J. Amer. Soc. Inform. Sci. **37** (5) (1986), 341–345.

[30] M. Kifer, G. Lausen, J. Wu, *Logical foundations of object-oriented and frame-based languages*, J. ACM **42** (4) (1995), 741–843.

[31] M. Kifer, V.S. Subrahmanian, *Theory of generalized annotated logic programming and its applications*, J. Logic Programming **12** (1992), 335–367.

[32] D. Koller, A. Levy, A. Pfeffer, *P-CLASSIC: a tractable probabilistic description logic*, in: Proc. of the 14th Nat. Conf. on Artificial Intelligence (AAAI-97), 1997, pp. 390–397.

[33] D.H. Kraft, D. Buel, *Fuzzy sets and generalised Boolean retrieval systems*, Int. J. Man–Machine Studies **19** (1983), 45–56.

[34] L.V.S. Lakshmanan, N. Shiri, *A parametric approach to deductive databases with uncertainty*, IEEE Trans. on Knowledge and Data Engineering **13** (4) (2001), 554–570.

[35] A.Y. Levy, M.-C. Rousset, *Combining horn rules and description logics in CARIN*, Artificial Intelligence **104** (1998), 165–209.

[36] J.W. Lloyd, *Foundations of Logic Programming*, Springer, Heidelberg, 1987.

[37] A. Lopatenko, E. Franconi, G. Kuper, L. Serafini, *A robust logical and computational characterisation of peer-to-peer database systems*, in: Proceedings of the VLDB International Workshop on Databases, Information Systems and Peer-to-Peer Computing (DBISP2P-03), 2004.

[38] Y. Loyer, U. Straccia, *The approximate well-founded semantics for logic programs with uncertainty*, in: 28th International Symposium on Mathematical Foundations of Computer Science (MFCS-2003), Bratislava, Slovak Republic, Lecture Notes in Comput. Sci., vol. 2747, Springer-Verlag, 2003, pp. 541–550.

[39] T. Lukasiewicz *Probabilistic logic programming*, in: Proc. of the 13th European Conf. on Artificial Intelligence (ECAI-98), Brighton (England), August 1998, pp. 388–392.

[40] C. Lutz, F. Wolter, M. Zakharyaschev, *A tableau algorithm for reasoning about concepts and similarity*, in: Proceedings of the Twelfth International Conference on Automated Reasoning with Analytic Tableaux and Related Methods TABLEAUX 2003, Rome, Italy, Lecture Notes in Artificial Intelligence, vol. 2796, Springer, 2003.

[41] C. Mateis, *Quantitative disjunctive logic programming: semantics and computation*, AI Communications **13** (2000), 225–248.

[42] C. Meghini, F. Sebastiani, U. Straccia, *A model of multimedia information retrieval*, J. ACM **48** (5) (2001), 909–970.

[43] B. Motik, U. Sattler, U. Hustadt, *Reducing $\mathcal{SHIQ}^-$ description logic to disjunctive datalog programs*, in: Proceedings of the 9th International Conference on Principles of Knowledge Representation and Reasoning (KR-04), AAAI Press, 2004.

[44] B. Motik, U. Sattler, R. Studer, *Query answering for OWL-DL with rules*, in: Proceedings of the International Semantic Web Conference, Lecture Notes in Comput. Sci., vol. 3298, Springer-Verlag, 2004, pp. 549–563.

[45] B. Motik, R. Volz, A. Maedche, *Optimizing query answering in description logics using disjunctive deductive databases*, in: 10th International Workshop on Knowledge Representation meets Databases (KRDB-03), 2003, pp. 39–50.

[46] B. Nebel, *Reasoning and Revision in Hybrid Representation Systems*, Springer, Heidelberg, 1990.

[47] C.V. Negoita, P. Flondor, *On fuzziness in information retrieval*, Int. J. Man–Machine Studies **8** (1976), 711–716.

[48] R. Ng, V.S. Subrahmanian, *Probabilistic logic programming*, Inform. and Comput. **101** (2) (1993), 150–201.

[49] J.Z. Pan, et al., Specification of coordination of rule and ontology languages, Technical Report, Knowledgeweb Network of Excellence, EU-IST-2004-507482, 2004. Deliverable D2.5.1.
[50] P.F. Patel-Schneider, *A four-valued semantics for terminological logics*, Artificial Intelligence **38** (1989), 319–351.
[51] D. Roth, *On the hardness of approximate reasoning*, Artificial Intelligence J. **82** (1996), 273–302.
[52] D. Sánchez, G.B. Tettamanzi, *Generalizing quantification in fuzzy description logics*, in: Proceedings 8th Fuzzy Days in Dortmund, 2004.
[53] M. Schmidt Schauß, G. Smolka, *Attributive concept descriptions with complements*, Artificial Intelligence **48** (1991), 1–26.
[54] F. Sebastiani, *A probabilistic terminological logic for modelling information retrieval*, in: Proceedings of SIGIR-94, 17th ACM International Conference on Research and Development in Information Retrieval, Dublin, IRL, Springer-Verlag, Heidelberg, 1994, pp. 122–130.
[55] U. Straccia, *A sequent calculus for reasoning in four-valued description logics*, in: Proc. of the Int. Conf. on Analytic Tableaux and Related Methods (TABLEAUX-97), Pont-à-Mousson, France, Lecture Notes in Artificial Intelligence, vol. 1227, 1997, pp. 343–357.
[56] U. Straccia, *A fuzzy description logic*, in: Proc. of the 15th Nat. Conf. on Artificial Intelligence (AAAI-98), Madison, USA, 1998, pp. 594–599.
[57] U. Straccia, *A framework for the retrieval of multimedia objects based on four-valued fuzzy description logics*, in: F. Crestani, G. Pasi (Eds.), Soft Computing in Information Retrieval: Techniques and Applications, Physica-Verlag (Springer-Verlag), Heidelberg, 2000, pp. 332–357.
[58] U. Straccia, *Reasoning within fuzzy description logics*, J. Artificial Intelligence Research **14** (2001), 137–166.
[59] U. Straccia, *Towards a fuzzy description logic for the semantic web*, Technical Report 2004-TR-64, Istituto di Scienza e Tecnologie dell'Informazione, Consiglio Nazionale delle Ricerche, Pisa, Italy, 2004.
[60] U. Straccia, *Transforming fuzzy description logics into classical description logics*, in: Proceedings of the 9th European Conference on Logics in Artificial Intelligence (JELIA-04), Lisbon, Portugal, Lecture Notes in Comput. Sci., vol. 3229, Springer-Verlag, 2004, pp. 385–399.
[61] U. Straccia, *Uncertainty in description logic programs*, Technical Report ISTI-2004-TR-01, Istituto di Scienza e Tecnologie dell'Informazione, Consiglio Nazionale delle Ricerche, Pisa, Italy, 2004.
[62] U. Straccia, *Uncertainty in description logics: a lattice-based approach*, in: Proceedings of the 10th International Conference on Information Processing and Managment of Uncertainty in Knowledge-Based Systems (IPMU-04), 2004, pp. 251–258.
[63] U. Straccia, *Fuzzy description logics with concrete domains*, Technical Report 2005-TR-03, Istituto di Scienza e Tecnologie dell'Informazione, Consiglio Nazionale delle Ricerche, Pisa, Italy, 2005.
[64] U. Straccia, *Uncertainty management in logic programming: simple and effective top-down query answering*, in: 9th International Conference on Knowledge-Based & Intelligent Information & Engineering Systems (KES-05), Melbourne, Australia, Lecture Notes in Comput. Sci., Springer-Verlag, 2005.
[65] T. Swift, *Deduction in ontologies via ASP*, in: Proceedings of the 7th International Conference in Logic Programming and Nonmonotonic Reasoning (LPNMR-04), Fort Lauderdale, FL, USA, Lecture Notes in Artificial Intelligence, vol. 2923, Springer-Verlag, 2004, pp. 275–288.
[66] A. Tarski, *A lattice-theoretical fixpoint theorem and its applications*, Pacific J. Math. (5) (1955), 285–309.
[67] C. Tresp, R. Molitor, *A description logic for vague knowledge*, in: Proc. of the 13th European Conf. on Artificial Intelligence (ECAI-98), Brighton (England), August 1998.
[68] K. van Belleghem, M. Denecker, D. de Schreye, *A strong correspondence between description logics and open logic programs*, in: Proc. of the 13th Int. Conf. on Logic Programming (ICLP-97), MIT Press, 1997, pp. 346–360.
[69] P. Vojtáš, *Fuzzy logic programming*, Fuzzy Sets and Systems **124** (2004), 361–370.
[70] G. Yang, M. Kifer, C. Zhao, *Flora-2: a rule-based knowledge representation and inference infrastructure for the semantic web*, in: Second International Conference on Ontologies, Databases and Applications of Semantics (ODBASE-03), Catania, Sicily, Italy, Lecture Notes in Comput. Sci., vol. 2888, Springer-Verlag, 2003, pp. 671–688.
[71] J. Yen, *Generalizing term subsumption languages to fuzzy logic*, in: Proceedings of the 12th International Joint Conference on Artificial Intelligence (IJCAI-91), Sydney, Australia, 1991, pp. 472–477.

CHAPTER 8

Fuzzy Quantification in Fuzzy Description Logics

Daniel Sánchez

Department of Computer Science and A.I., University of Granada, Periodista Daniel Saucedo Aranda s/n, 18071 Granada, Spain
E-mail: daniel@decsai.ugr.es

Andrea G.B. Tettamanzi

Department of Information Technologies, University of Milan, Via Bramante 65, I-26013 Crema (CR), Italy
E-mail: andrea.tettamanzi@unimi.it

Abstract

This chapter introduces reasoning procedures for $\mathcal{ALCQ}_F^+(\mathrm{D})$, a fuzzy description logic with extended qualified quantification [D. Sánchez, A.G.B. Tettamanzi, *Generalizing quantification in fuzzy description logics*, in: Proceedings 8th Dortmund Fuzzy Days, Dortmund, Germany, 2004]. The language allows for the definition of fuzzy quantifiers of the absolute and relative kind by means of piecewise linear functions on $\mathbb{N}$ and $\mathbb{Q} \cap [0, 1]$ respectively. In order to reason about instances, the semantics of quantified expressions is defined by using method GD [M. Delgado, D. Sánchez, M. Vila, *Fuzzy cardinality based evaluation of quantified sentences*, Int. J. Approximate Reasoning **23** (2000), 23–66], which is based on recently developed measures of the cardinality of fuzzy sets. The main contribution of this chapter is a procedure to calculate the fuzzy satisfiability of a fuzzy concept, which is a very important reasoning task. The procedure considers several different cases and provides direct solutions for the most frequent types of fuzzy concepts. In order to distinguish between these cases, a novel idea of concept independence is also introduced.

Keywords

fuzzy description logics, fuzzy quantification, fuzzy cardinality, fuzzy satisfiability

FUZZY LOGIC AND THE SEMANTIC WEB
Edited by Elie Sanchez
© 2006 Elsevier B.V. All rights reserved

1. Introduction

Description logics (DLs) [1] are well-suited for the representation of and reasoning about terminological knowledge, configurations, ontologies, database schemata, etc. The need of expressing and reasoning with imprecise knowledge and the difficulties arising in classifying individuals with respect to an existing terminology motivates research on non-classical DL semantics, suited for these purposes.

To cope with this problem, fuzzy DLs have been proposed that allow for imprecise concept description by using fuzzy sets and fuzzy relations. For instance, a fuzzy version of the description logic $\mathcal{SHOIN}(\mathrm{D})$ is described in Chapter 4 of this book [11]. Work is under way to provide a sound and complete calculus for it as an extension of the one developed for the fuzzy counterpart of the description logic $\mathcal{ALC}$ [10], with complete algorithms for solving the entailment problem, the subsumption problem, as well as the best truth-value bound problem.

However, little attention so far has been paid to the quantification issue. For instance, the fuzzy counterpart of $\mathcal{SHOIN}(\mathrm{D})$ extends the semantics of $\exists$, $\forall$ and *crisp* number restrictions. However, not only concepts, but also quantifiers can be imprecise (e.g., "around two", "most") in many situations typical of real-world applications.

For example, suppose you are the marketing director of a supermarket chain. You are about to launch a new line of low-calorie products. In order to set up your budget, you need to project the sales of this new line of products. This can be done either by means of an expensive market research, or by means of some kind of inference based on your knowledge of customer habits. For instance, you could expect prospective buyers of this new line of products to be essentially faithful customers who mostly buy foods with low energy value. We have here all the ingredients of imprecise knowledge: a "faithful customer" is a fuzzy concept; "low" energy value is a linguistic value, which might be modeled as a fuzzy number; to "mostly" buy a given kind of product is equivalent to a quantified statement of the form "most of the bought products are of this kind", where "most" is an imprecise quantifier.

In [9] we introduced $\mathcal{ALCQ}_F^+$, a fuzzy description logic with extended qualified quantification[1] that allows for the definition of fuzzy quantifiers of the absolute and relative kind by means of piecewise linear functions on $\mathbb{N}$ and $\mathbb{Q} \cap [0, 1]$ respectively. These quantifiers extends the usual (qualified) $\exists$, $\forall$ and number restrictions.

Zadeh [13] showed that imprecise quantifiers can be defined by using fuzzy sets, and by incorporating them into the language and providing the tools to define their semantics we can provide a very powerful knowledge representation tool, with greater expressive power, and closer to the humans' way of thinking.

---

[1] In keeping with DL naming conventions, the superscript plus is to suggest that, in addition to qualified number restrictions available in the description logic $\mathcal{ALCQ}$ introduced by De Giacomo and Lenzerini [2], we provide also more general fuzzy linguistic quantifiers. The subscript F means that the language deals with infinitely many truth-values, as in the language $\mathcal{ALC}_{F_M}$ of Tresp and Molitor [12].

2. Fuzzy description logics

The fundamental definitions of fuzzy DLs are given in Chapter 4 of this book [11]. Here, we will just point out and comment on some slight differences arising in a few definitions that are largely matter of subjective preferences.

From a notational point of view, we prefer to use the symbols n, m, etc., to stand for natural numbers and α, β, etc., for degrees of truth; we will reserve upper-case Greek letters like Φ and Ψ for fuzzy assertions.

2.1. *Fuzzy assertions and truth-value bounds*

For us, a *fuzzy assertion* has exactly the same syntax of an assertion in a crisp DL: what makes it a *fuzzy* assertion is the fact that it is an assertion about a fuzzy concept or role, and it can thus have a truth value in [0, 1]. A fuzzy assertion Ψ thus defined, has to be distinguished from what Straccia calls a *fuzzy concept or role assertion axiom*, of the form $\langle\Psi \leqslant \alpha\rangle$ or $\langle\Psi \geqslant \alpha\rangle$: these could as well be called *truth-value bounds* (TVBs). One important thing to keep in mind is that a TVB is either true or false in a given interpretation; therefore, whereas it is a proposition *about* a fuzzy assertion, a TVB is itself a *crisp* logical proposition.

2.2. *Fuzzy terminological axioms*

Axioms and queries can be of two kinds: specializations and definitions.

A fuzzy concept specialization is a statement of the form $C \sqsubseteq D$, where C and D are concepts. A fuzzy interpretation $\mathcal{I}$ satisfies a fuzzy concept specialization $C \sqsubseteq D$ if, for all $a \in \Delta^{\mathcal{I}}$, $C^{\mathcal{I}}(a) \leqslant D^{\mathcal{I}}(a)$.

A fuzzy concept definition is a statement of the form $C \equiv D$, which can be understood as an abbreviation of the pair of statements $\{C \sqsubseteq D, D \sqsubseteq C\}$.

If it were not for the fuzzy generalized quantifiers that we are going to introduce, the TBox of a fuzzy knowledge base would have formally nothing fuzzy with it, for the syntax of the terminological part of the fuzzy $\mathcal{ALC}$ language as defined in [10] is identical to its crisp counterpart; only the ABox can contain fuzzy assertions.

Admitting TVBs to be part of the fuzzy TBox, as is the case with $\mathcal{SHOIN}(\mathrm{D})$, is equivalent to allowing the formulation of *uncertain* axioms, i.e. axioms whose truth is not absolute. This is a type of uncertainty in a sense *orthogonal* to the uncertainty modeled by means of fuzzy generalized quantification.

3. Cardinality and fuzzy quantification

Crisp quantification is strongly linked to crisp cardinality since a crisp quantifier Q represents a crisp subset of absolute (values in $\mathbb{N}$) or relative (values in $\mathbb{Q} \cap [0, 1]$) cardinalities, we call $S(Q)$. By the same token, fuzzy quantification is strongly linked to fuzzy cardinality.

We may consider two different kinds of cardinalities: *absolute* cardinality measures the number of elements in a set, while *relative* cardinality measures the percentage of elements of one set that also belong in another set (called *referential*).

3.1. *Absolute cardinality*

The most widely used definition of fuzzy set cardinality introduced in [7] is the following: given a fuzzy set F defined on a countable universe set Δ,

$$|F| = \sum_{x \in \Delta} F(x), \tag{1}$$

which is, in general, a real number, and not an integer as it is the case with classical set cardinality. This definition lends itself to many objections and leads to paradoxes. A more satisfactory cardinality measure is *ED*, introduced in [4].

DEFINITION 3.1 [4]. The fuzzy cardinality of a set G is the set $ED(G)$ defined for each $0 \leqslant k \leqslant |\mathrm{supp}(G)|$ as

$$ED(G)(k) = \begin{cases} \alpha_i - \alpha_{i+1}, & \alpha_i \in \Lambda(G) \text{ and } |G_{\alpha_i}| = k, \\ 0, & \text{otherwise} \end{cases} \tag{2}$$

with G_α being the α-cut of G and $\Lambda(G)$ the level set of G, i.e.

$$\Lambda(G) = \big\{G(x) \mid x \in \Delta\big\} \cup \{1\}$$

assuming $\Lambda(G) = \{\alpha_1, \ldots, \alpha_p\}$ and $\alpha_i > \alpha_{i+1}$ for every $i \in \{1, \ldots, p\}$, and $\alpha_{p+1} = 0$.

3.2. *Relative cardinality*

An extension to the *ED* measure to cope with relative cardinality is *ER* introduced in [4].

DEFINITION 3.2 [4]. The fuzzy relative cardinality of a set G with respect to a set F is the set $ER(G/F)$ defined for each $0 \leqslant q \leqslant 1$ as

$$ER(G/F)(q) = \sum_{\alpha_i | C(G/F, \alpha_i) = q} (\alpha_i - \alpha_{i+1}) \tag{3}$$

with

$$C(G/F, \alpha_i) = \frac{|G \cap F|}{|F|}$$

with $\Lambda(F) \cup \Lambda(G \cap F) = \{\alpha_1, \ldots, \alpha_p\}$ and $\alpha_i > \alpha_{i+1}$ for every $i \in \{1, \ldots, p\}$, and $\alpha_{p+1} = 0$.

ER is an extension of *ED* since, if F is crisp and $|F| = n$, then $ER(G/F)(k/n) = ED(G)(k)$ [4].

3.3. *Fuzzy quantifiers*

The concept of fuzzy linguistic quantifier is due to L.A. Zadeh [13]. Fuzzy quantifiers are linguistic labels representing imprecise quantities or percentages. It is usual to distinguish absolute and relative quantifiers, based respectively on absolute and relative cardinality.

Absolute quantifiers represent imprecise natural quantities and are modeled as fuzzy subsets of the naturals, whilst relative quantifiers represent imprecise percentages and are modeled as fuzzy subsets of the rational interval [0, 1].

3.4. *Evaluation of quantified sentences*

Quantified sentences are natural language sentences involving fuzzy linguistic quantifiers, and therefore they express claims about the (fuzzy) quantity or percentage of elements of a (possibly fuzzy) set that verify a certain imprecise property.

According to Zadeh [13], there are two main types of quantified sentences, whose general structure is the following:

Type I sentences: Q of X are G

Type II sentences: Q of F are G

where Q is a linguistic quantifier, X is a crisp finite set, and F and G are two fuzzy subsets of X that represent imprecise properties. Examples are

Type I: *Around* 30 *students are young*

Type II: *Most of the proficient students are young*

Type I sentences are suitable for both absolute and relative quantifiers, whilst type II sentences only make sense for relative quantifiers (i.e. a sentence like "Around 2 F are G" is in fact a type I sentence "Around 2 X are $F \cap G$").

The evaluation of a quantified sentence is the process of calculating its fuzzy accomplishment degree. There are several methods available in the literature (some methods are discussed in [3], recent developments are [6,5]).

In order to extend fuzzy DLs with absolute and relative quantifiers, we shall employ method *GD* [3]. This method calculates the accomplishment degree of a quantified sentence "Q of F are G" as the compatibility degree between the fuzzy relative cardinality measure $ER(G/F)$ [4] and Q (if Q is relative) or between its absolute counterpart $ED(G \cap F)$ [4] and Q (if Q is absolute). A convenient formulation of *GD* is the following [3]:

DEFINITION 3.3. The method *GD* obtains the evaluation of a quantified sentence "Q of F are G" as

$$GD_Q(G/F) = \sum_{\alpha_i \in \Lambda(G/F)} (\alpha_i - \alpha_{i+1}) Q\left(\frac{|(G \cap F)_{\alpha_i}|}{|F_{\alpha_i}|}\right) \tag{4}$$

for relative quantifiers, and

$$GD_Q(G/F) = \sum_{\alpha_i \in \Lambda(G/F)} (\alpha_i - \alpha_{i+1}) Q\left(\left|(F \cap G)_{\alpha_i}\right|\right) \tag{5}$$

for absolute ones, where $F \cap G$ is computed using the minimum, and $\Lambda(G/F) = \Lambda(G \cap F) \cup \Lambda(F)$. We label these values as $\Lambda(G/F) = \{\alpha_1, \ldots, \alpha_p\}$ with $\alpha_i > \alpha_{i+1}$ for every $i \in \{1, \ldots, p\}$ and $\alpha_{p+1} = 0$.

As discussed in [3], $GD_Q(G/F)$ is undefined for type II sentences when F is not normalized since $F_\alpha = \emptyset$ for at least one $\alpha \in [0, 1]$ (in particular for $\alpha = 1$). Possible solutions to this problem are to normalize F (applying the same factor to $F \cap G$ after) or to define the value of any relative quantifier when the relative cardinality is undefined (we note this value as $Q(u)$, by considering $|\emptyset|/|\emptyset| = u$). In this work we shall employ this last option. In particular, $\exists(u) = 0$ and $\forall(u) = 1$.

4. Extending FDLs with fuzzy quantification

Quantified expressions in DLs are expressions of the form $QR.C$, where Q is a quantifier, R is a role and C is a concept, called sometimes a *qualifier*. These expressions are useful for defining new concepts, like $D \equiv QR.C$, meaning that an individual a is an instance of concept D if Q of the individuals which fill its role R are instances of C.

In crisp DLs, quantification is limited to the classical (relative) quantifiers $\exists$ and $\forall$, as well as to the so-called number restrictions, $\leqslant n$, $\geqslant n$, and $= n$ (crisp absolute quantifiers). In this section we propose a fuzzy description logic capable of representing and reasoning with general absolute and relative quantifiers. The goal is to be able to express such fuzzy definitions as, e.g., $D \equiv QR.C$, where Q is now a fuzzy quantifier, meaning that an individual a is an instance of concept D *to the extent that* Q of the individuals which fill its role R are instances of C.

4.1. *The $\mathcal{ALCQ}^+_F(\mathrm{D})$ language*

The extended language we introduce, $\mathcal{ALCQ}^+_F(\mathrm{D})$, where the symbol D is to suggest the fact that it uses concrete domains, has the following syntax, where A is an atomic concept, R a role, C a concept, Q_{abs} is an absolute fuzzy quantifier, and Q_{rel} a relative fuzzy quantifier:

$$\begin{aligned} C \longrightarrow\ & \bot \mid \top \mid A \mid C_1 \sqcap C_2 \mid C_1 \sqcup C_2 \mid \neg C \mid \\ & \forall R.C \mid \exists R.C \mid Q_{\text{abs}} R.C \mid Q_{\text{rel}} R.C. \end{aligned}$$

In this extension, the semantics of fuzzy absolute and relative quantifiers is given by using Straccia's notion of *fuzzy concrete predicate* [11]. For relative quantifiers, a membership degree must be defined for the "undefined" case, which arises when the referential set with respect to which the relative cardinality is calculated is empty. This value could depend on the subjective view of the user or on the application at hand, although it can be conventionally agreed upon for some quantifiers like $\exists$ and $\forall$.

4.2. *Fuzzy knowledge base*

A $\mathcal{ALCQ}^+_F(\mathrm{D})$ fuzzy knowledge base comprises two components: a TBox and an ABox.

A TBox $\mathcal{T}$ consists of a finite set of fuzzy terminological axioms of the form $C \sqsubseteq D$ or $C \equiv D$. We require that the TBox be acyclic, i.e. that all concept symbols on the left-hand side of inclusion axioms appear only once in the TBox.

An ABox $\mathcal{A}$ consists of a finite set of fuzzy assertions of the form $C(a)$ or $R(a,b)$.

4.3. *Fuzzy interpretation*

A fuzzy interpretation $\mathcal{I}$ with respect to a concrete domain D consists of a non-empty domain $\Delta^{\mathcal{I}}$, disjoint from Δ_{D}, and a fuzzy interpretation function $\cdot^{\mathcal{I}}$, which maps every atomic concept A onto a fuzzy subset $A^{\mathcal{I}}$ of $\Delta^{\mathcal{I}}$, every atomic role R onto a fuzzy binary relation $R^{\mathcal{I}} \subseteq \Delta^{\mathcal{I}} \times \Delta^{\mathcal{I}}$, and every individual name a onto an element $a^{\mathcal{I}} \in \Delta^{\mathcal{I}}$.

The mapping $\cdot^{\mathcal{I}}$ is extended to all constructs of the language as follows: for all $a \in \Delta^{\mathcal{I}}$,

$$\begin{aligned}
\bot^{\mathcal{I}}(a) &= 0;\\
\top^{\mathcal{I}}(a) &= 1;\\
(C \sqcap D)^{\mathcal{I}}(a) &= \min\{C^{\mathcal{I}}(a), D^{\mathcal{I}}(a)\};\\
(C \sqcup D)^{\mathcal{I}}(a) &= \max\{C^{\mathcal{I}}(a), D^{\mathcal{I}}(a)\};\\
(\neg C)^{\mathcal{I}}(a) &= 1 - C^{\mathcal{I}}(a).
\end{aligned}$$

For general quantification, the idea is to directly translate a general expression like $QR.C$ into the usual notation of quantified sentences introduced in the previous section as follows: if Q is an absolute quantifier, the degree to which an individual a verifies concept D is the result of the evaluation of the quantified sentence

$$Q \text{ of } \Delta^{\mathcal{I}} \text{ are } R_a^{\mathcal{I}} \cap C^{\mathcal{I}},$$

whilst for relative quantifiers, it is the result of the evaluation of the sentence

$$Q \text{ of } R_a^{\mathcal{I}} \text{ are } C^{\mathcal{I}},$$

where $R_a^{\mathcal{I}}$ is the projection of fuzzy relation $R^{\mathcal{I}}$ on individual a: for all $b \in \Delta^{\mathcal{I}}$, $R_a^{\mathcal{I}}(b) = R^{\mathcal{I}}(a,b)$.

However, the semantics should be well-defined for infinite domains; this means that

(1) the level set Λ might be infinite;

(2) for relative quantifiers, the referential set might be infinite, making the argument of Q undefined.

The first issue can be settled by expressing the truth value of a quantification by means of an integral over the [0, 1] interval. The second issue requires the use of limits.

As usual, we consider two cases depending on whether the quantifier is absolute:

$$(Q_{\text{abs}} R.C)^{\mathcal{I}}(a) = \int_0^1 Q_{\text{abs}}\big(\big|\{b\colon\ R^{\mathcal{I}}(a,b) \geqslant \tau \wedge C^{\mathcal{I}}(b) \geqslant \tau\}\big|\big)\,\mathrm{d}\tau, \tag{6}$$

or relative:

$$(Q_{\text{rel}} R.C)^{\mathcal{I}}(a) = \int_0^1 Q_{\text{rel}}\left(\lim_{n\to\infty} \frac{|\{b\colon\ R_n^{\mathcal{I}}(a,b) \geqslant \tau \wedge C^{\mathcal{I}}(b) \geqslant \tau\}|}{|\{b\colon\ R_n^{\mathcal{I}}(a,b) \geqslant \tau\}|}\right)\mathrm{d}\tau, \tag{7}$$

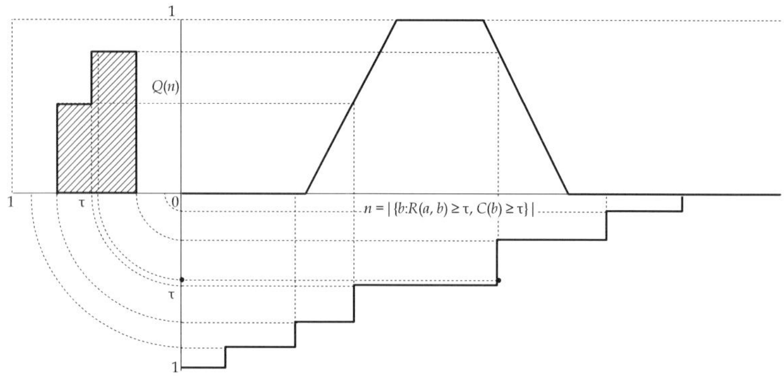

Fig. 1. A graphical depiction of the interpretation of fuzzy concept $QR.C$ with Q an absolute fuzzy quantifier whose membership function is shown in the upper-left part of the figure; the bottom-left part of the figure graphs the cardinality of the τ-cuts of $R^{\mathcal{I}}(a,b) \cap C^{\mathcal{I}}(b)$ for $0 \leqslant \tau \leqslant 1$; the upper-right part of the figure shows the polygon resulting from applying the membership function of Q to the former cardinalities, whose area represents the degree of truth of $QR.C$.

where $\{R_n^{\mathcal{I}}\}_{n=1,2,\ldots}$ is a sequence of fuzzy relations such that

(1) for all $n \geqslant 1$, $R_n^{\mathcal{I}} \subseteq R_{n+1}^{\mathcal{I}}$;

(2) $\lim_{n\to\infty} R_n^{\mathcal{I}} = R^{\mathcal{I}}$;

(3) for all $n \geqslant 1$, $\operatorname{supp}(R_n^{\mathcal{I}})$ is a finite set.

Of course, given any finite interpretation, the above integrals become discrete summations because the arguments of the $Q_{\text{abs}}(\cdot)$ and $Q_{\text{rel}}(\cdot)$ membership functions turn out to be step functions with a finite set of levels, yielding their finite versions

$$(Q_{\text{abs}} R.C)^{\mathcal{I}}(a) = GD_{Q_{\text{abs}}}\big((R_a^{\mathcal{I}} \cap C^{\mathcal{I}})/\Delta^{\mathcal{I}}\big), \tag{8}$$

$$(Q_{\text{rel}} R.C)^{\mathcal{I}}(a) = GD_{Q_{\text{rel}}}\big(C^{\mathcal{I}}/R_a^{\mathcal{I}}\big) \tag{9}$$

where GD is the evaluation method introduced in [3]. Figure 1 depicts an example of how the integral is calculated for the absolute quantifier case.

In particular, the semantics for the quantifiers $\exists$ and $\forall$ is:

$$(\exists R.C)^{\mathcal{I}}(a) = \sum_{(R_a^{\mathcal{I}})_{\alpha_i} \cap (C^{\mathcal{I}})_{\alpha_i} \neq \emptyset} (\alpha_i - \alpha_{i+1}), \tag{10}$$

$$(\forall R.C)^{\mathcal{I}}(a) = \sum_{(R_a^{\mathcal{I}})_{\alpha_i} \subseteq (C^{\mathcal{I}})_{\alpha_i}} (\alpha_i - \alpha_{i+1}) \tag{11}$$

with $\alpha_i \in \Lambda(G/F)$. According to the properties of GD [3], the semantics of the existential quantification turns out to have its usual, intuitive definition, i.e. Equation (10) is equivalent to

$$(\exists R.C)^{\mathcal{I}}(a) = \sup_{b \in \Delta^{\mathcal{I}}} \min\big\{R_a^{\mathcal{I}}(b), C^{\mathcal{I}}(b)\big\}.$$

As a particular case, when the referential F is crisp, GD verifies De Morgan's laws, i.e. $1 - GD_{\forall}(G/F) = GD_{\exists}(\neg G/F)$ and $1 - GD_{\exists}(G/F) = GD_{\forall}(\neg G/F)$. Hence, if $R_a^{\mathcal{I}}$ is crisp then Equation (11) is equivalent to

$$(\forall R.C)^{\mathcal{I}}(a) = \inf_{b \in \Delta^{\mathcal{I}}} \max\big\{1 - R_a^{\mathcal{I}}(b), C^{\mathcal{I}}(b)\big\}.$$

However, this is not true in general: let's consider the fuzzy sets whose membership functions are $R_a^{\mathcal{I}}$ and $C^{\mathcal{I}}$ respectively (we shall use the same notation for the functions and the fuzzy subsets); De Morgan's laws hold only if

$$R_a^{\mathcal{I}} \subseteq C^{\mathcal{I}} \Leftrightarrow R_a^{\mathcal{I}} \cap (\neg C)^{\mathcal{I}} = \emptyset,$$

i.e. only when $R_a^{\mathcal{I}}$ (or $C^{\mathcal{I}}$) is crisp. Otherwise we could have a situation where both sets are fuzzy and $R_a^{\mathcal{I}} \subseteq C^{\mathcal{I}}$ (and hence the evaluation using $\forall$ is expected to yield 1), but $R_a^{\mathcal{I}} \cap (\neg C)^{\mathcal{I}} \neq \emptyset$ (and hence the evaluation using $\exists$ is expected to yield a value greater than 0). For example, suppose that

$$R_a^{\mathcal{I}} = 1/b_1 + 0.6/b_2 + 0.4/b_3,$$
$$C^{\mathcal{I}} = 1/b_1 + 0.9/b_2 + 0.5/b_3.$$

Then $R_a^{\mathcal{I}} \subseteq C^{\mathcal{I}}$ and $R_a^{\mathcal{I}} \cap (\neg C)^{\mathcal{I}} = 0.1/b_2 + 0.4/b_3 \neq \emptyset$, consequently $GD_\forall(C^{\mathcal{I}}/R_a^{\mathcal{I}}) = 1$, but $GD_\exists((\neg C)^{\mathcal{I}}/R_a^{\mathcal{I}}) = 0.4 \neq 0$.

4.4. *An example*

Let us go back to the example of the marketing director of a supermarket chain about to launch a line of low-calorie products.

The knowledge base describing the business of running a supermarket chain could contain, among others, the following terminological axioms:

$$\mathsf{FaithfulCustomer} \sqsubseteq \mathsf{Customer} \sqsubseteq \top$$
$$\mathsf{FoodProduct} \sqsubseteq \mathsf{Product} \sqsubseteq \top$$
$$\mathsf{LowCalorie} \sqsubseteq \mathsf{EnergyMeasure} \sqsubseteq \top$$
$$\mathsf{LowCalorieFood} \equiv \mathsf{FoodProduct} \sqcap \forall \mathsf{energyValue}.\mathsf{LowCalorie}$$

The ABox describing facts about your supermarket chain might contain TVBs which we might summarize as follows:

- given an individual customer c and a product p, buys(c, p) might be interpreted as

$$\mathsf{buys}(c, p) = f\big(\text{weeklyrevenue}(c, p)\big),$$

  where $f : \mathbb{R} \to [0, 1]$ is non-decreasing, and weeklyrevenue(c, p) : $\mathsf{Customer}^{\mathcal{I}} \times \mathsf{Product}^{\mathcal{I}} \to \mathbb{R}$ returns the result of a database query which calculates the average revenue generated by product p on customer c in all the stores operated by the chain;
- given an individual customer c, FaithfulCustomer(c) might be interpreted as

$$\mathsf{FaithfulCustomer}(c) = g\big(\text{weeklyrevenue}(c)\big),$$

  where $g : \mathbb{R} \to [0, 1]$ is non-decreasing, and weeklyrevenue(c) : $\mathsf{Customer}^{\mathcal{I}} \to \mathbb{R}$ returns the result of a database query which calculates the average revenue generated by customer c in all the stores operated by the chain;

Table 1
The energy value, membership in the LowCalorieFood, and the degree to which customer CARD0400009324198 buys them for a small sample of products.

Product	Energy [kJ/hg]	LowCalorieFood(·)	buys(CARD ..., ·)
GTIN8001350010239	1680	0.320	0.510
GTIN8007290330987	1475	0.525	0.050
GTIN8076809518581	1975	0.025	0.572
GTIN8000113004003	1523	0.477	0.210
GTIN8002330006969	498	1.000	1.000
GTIN8005410002110	199	1.000	1.000
GTIN017600081636	1967	0.033	0.184

- finally, LowCalorie(x), where x is an average energy value per 100 g of product measured in kJ, could be interpreted as

$$\mathsf{LowCalorie}(x) = \begin{cases} 1, & x < 1000, \\ \dfrac{2000 - x}{1000}, & 1000 \leqslant x \leqslant 2000, \\ 0, & x > 2000. \end{cases}$$

By using the $\mathcal{ALCQ}_F^+(\mathrm{D})$ language, it is now possible to express the notion of a faithful customer who mostly buys food with low energy value as

$$C \equiv \mathsf{FaithfulCustomer} \sqcap (\mathsf{Most})\mathsf{buys.LowCalorieFood},$$

where $(\mathsf{Most}) \equiv (0/u + 0/0.5 + 1/0.75)$.

A useful deduction this new axiom allows you to make is, for instance, calculating the extent to which a given individual customer or, more precisely, a fidelity card, say CARD0400009324198, is a C. For instance, you could know that

$$\mathsf{FaithfulCustomer(CARD0400009324198)} = 0.8,$$

and, by querying the sales database, you might get all the degrees to which that customer buys each product. For sake of example, we give a small subset of those degrees of truth in Table 1, along with the energy values of the relevant products.

According to the semantics of $\mathcal{ALCQ}_F^+(\mathrm{D})$,

$$C(\mathsf{CARD0400009324198}) \approx 0.742,$$

i.e. the degree to which most of the items purchased by this customer are low-calorie is around 0.742. This seems to be in accordance with the data in Table 1, where we can see that four products (those products p in rows 2, 4, 5, and 6) verify

$$\mathsf{buys(CARD0400009324198}, p) \leqslant \mathsf{LowCalorieFood}(p)$$

while for the products in rows 1 and 7 the difference between being purchased and being low-calorie food is not so high. Only the item in row 3 seems to be a clear case of item purchased but not low-calorie.

As another justification of why this result appears in agreement with the data, in Table 2 we show the percentage of purchased items that are low-calorie at α-cuts of the same level. At any other level, the percentage obtained is one of those shown in Table 2.

Table 2
Percentage of purchased items that are low-calorie at significant levels.

Level	Percentage
1.000	$1.000 = 2/2$
0.572	$0.667 = 2/3$
0.510	$0.500 - 2/4$
0.320	$0.750 = 3/4$
0.210	$0.800 = 4/5$
0.184	$0.667 = 4/6$
0.050	$0.714 = 5/7$
0.033	$0.857 = 6/7$

At many levels the percentage is above 0.75, therefore fitting the concept of Most as we have defined it. At level 0.050 the percentage is almost 0.75. The only level that clearly doesn't fit Most is 0.510, but at the next level (0.320) we have again 0.75 and $\mathsf{Most}(0.75) = 1$.

5. Fuzzy satisfiability

One particularly important reasoning task is to calculate the fuzzy satisfiability of a fuzzy concept D, i.e. the tightest interval of values $\mathcal{S}(D) = [\beta_D, \tau_D]$ such that for any interpretation $\mathcal{I}$ and any object $a \in \mathcal{I}$, the degree of truth of an assertion of the form $D(a)$ under $\mathcal{I}$ verifies $D^{\mathcal{I}}(a) \in [\beta_D, \tau_D]$ (i.e. the maximum interval $[\beta, \tau]$ such that D is $[\beta, \tau]$-satisfiable in the sense of Navara's definition [8]).

One might wonder whether it is really *satisfiability* we are dealing with here, or rather *validity*. The fact is, these two categories, which can be clearly distinguished in the realm of classical logic, become just the two extremes of a continuum in the case of fuzzy logic: as a matter of fact, a fuzzy concept D could be considered *valid* in the classical sense if $\mathcal{S}(D) = \{1\}$, and *satisfiable* (but not valid) in the classical sense if $\mathcal{S}(D) = [0, 1]$. It is all a matter of conventions: one might say, e.g., that, if $\mathcal{S}(D) = [\frac{3}{4}, 1]$, D is more valid than just satisfiable, whereas D might be considered almost unsatisfiable if $\mathcal{S}(D) = [0, \frac{1}{4}]$. Or we might define β_D as the degree of validity of D, and τ_D as the degree of satisfiability of D. However, for the sake of convenience, here we will use the term *fuzzy satisfiability* to cover the whole gamut of such possibilities.

We now illustrate a polynomial-space algorithm to calculate the fuzzy satisfiability of a fuzzy concept. Though infinite interpretations are taken into account in the definition of the semantics of the language $\mathcal{ALCQ}_F^+(\mathrm{D})$, in practice and due to the physical limitations of computers, we are going to deal with a finite number of individuals. The same limitations put a bound on the number of different membership degrees we can deal with. Therefore we shall calculate the fuzzy satisfiability of a fuzzy concept up to a certain precision degree, given as a number of decimals p.

5.1. *Concept independence*

In addition to the general algorithm, we have obtained some results showing that we can calculate the fuzzy satisfiability of a fuzzy concept directly in some (the most common) cases. Some of these results are based on a previous result about *independence* of fuzzy concepts. We introduce the following definition:

DEFINITION 5.1. Two concepts A and B are *independent* of each other if, for all $\alpha \in \mathcal{S}(A)$ and $\beta \in \mathcal{S}(B)$, there exists an interpretation $\mathcal{I}$ containing an individual $d \in \Delta_{\mathcal{I}}$ such that

$$A^{\mathcal{I}}(d) = \alpha \quad \text{and} \quad B^{\mathcal{I}}(d) = \beta.$$

In other words, A and B are independent if the degree of truth of A does not affect the degree of truth of B and *vice versa*.

The following results help to determine whether two concepts are independent.

PROPOSITION 5.1. *Two concepts A and B, not containing quantifiers, which are neither tautologies nor contradictions, are independent if and only if the following four concepts are satisfiable in the crisp sense*: $A \sqcap B$; $A \sqcap \neg B$; $\neg A \sqcap B$; $\neg A \sqcap \neg B$.

PROOF. We first show that independence of A and B implies the satisfiability of the four conjuncts: since A and B are neither tautologies nor contradictions, $\mathcal{S}(A) = \mathcal{S}(B) = [0, 1]$; therefore, it is possible to construct an interpretation with four individuals d_1, d_2, d_3, d_4 such that

$$\begin{aligned} A(d_1) &= 1, & B(d_1) &= 1,\\ A(d_2) &= 1, & B(d_2) &= 0,\\ A(d_3) &= 0, & B(d_3) &= 1,\\ A(d_4) &= 0, & B(d_4) &= 0; \end{aligned}$$

now, d_1 satisfies concept 1, d_2 satisfies concept 2, d_3 satisfies concept 3, and d_4 satisfies concept 4: all four conjuncts are satisfiable in the crisp sense.

Next we show that if the four conjuncts are satisfiable in the crisp sense, then A and B are independent. The four conjuncts are satisfiable in the crisp sense if there exists an interpretation $\mathcal{I}$ containing four individuals d_1, d_2, d_3, d_4 such that

$$\begin{aligned} A(d_1) &= 1, & B(d_1) &= 1,\\ A(d_2) &= 1, & B(d_2) &= 0,\\ A(d_3) &= 0, & B(d_3) &= 1,\\ A(d_4) &= 0, & B(d_4) &= 0; \end{aligned}$$

now, $\mathcal{S}(A)$ and $\mathcal{S}(B)$ are convex sets; therefore, if $0 \in \mathcal{S}(A)$ and $1 \in \mathcal{S}(A)$, it must be $\mathcal{S}(A) = [0, 1]$; for the same reason, it must be $\mathcal{S}(B) = [0, 1]$. First of all, neither A nor B can be a tautology or a contradiction; furthermore, when $A(d) = 0$, $B(d)$ can be 0 or 1, whence $B(d) \in [0, 1]$; the same holds when $A(d) = 1$. The possible constraints which could bind the truth values $\alpha = A(d)$ and $\beta = B(d)$, for all d, are only three:

- $\alpha \leqslant \beta$, excluded by individual d_2;
- $\beta \leqslant \alpha$, excluded by individual d_3;
- $\alpha = 1 - \beta$, excluded by individuals d_1 and d_4.

We must therefore conclude that for all $\alpha \in [0, 1]$ and $\beta \in [0, 1]$, it is possible to construct an interpretation containing an individual d such that $A(d) = \alpha$ and $B(d) = \beta$. □

PROPOSITION 5.2. *Let D_1 and D_2 be two independent concepts. Then, no atomic concept and no role appears in the expansion of both.*

PROOF. Trivial. □

PROPOSITION 5.3. *Let D_1 be a concept that doesn't contain quantifiers in its expansion, and let $D_2 \equiv QR.C$. Then, D_1 and D_2 are independent, regardless of whether D_1 and C are independent or not.*

PROOF. The truth of $D_2(a)$ in any interpretation depends on Q, R and C, where Q and R don't appear in D_1, and C applies to individuals $b_1, \ldots, b_n$ with $b_i \neq a$. Hence, $\forall \alpha_1 \in \mathcal{S}(D_1)$ and $\alpha_2 \in \mathcal{S}(D_2)$ we can build an interpretation $\mathcal{I}$ such that $D_1^{\mathcal{I}}(a) = \alpha_1$ and $D_2^{\mathcal{I}}(a) = \alpha_2$. □

PROPOSITION 5.4. *Let $D_1 \equiv Q_1 R_1.C_1$ and let $D_2 \equiv Q_2 R_2.C_2$. If $R_1 \neq R_2$ or C_1 and C_2 are independent, then D_1 and D_2 are independent.*

PROOF. We consider both conditions:

- Let $R_1 \neq R_2$. Then we can build an interpretation $\mathcal{I}$ such that $R_1 \cap R_2 = \emptyset$, i.e. the set of individuals b_i^1 related to a by role R_1 is different from the set of individuals b_j^2 related to a by role R_2. Since only degrees of truth of $C_1(b_i^1)$ and $C_2(b_j^2)$ are relevant in order to calculate $D_1(a)$ and $D_2(a)$, and these are independent, both concepts are independent.
- Let $R_1 = R_2$ and let C_1 and C_2 be independent. Then, though the set of individuals b_i related to a in each case is the same, there is no dependence between the degrees we can assign to $C_1(b_i)$ and $C_2(b_i)$ for every i. Hence, D_1 and D_2 are independent. □

A general procedure to determine whether two concepts are independent can be obtained from Proposition 5.1 using the crisp procedure to check unsatisfiability. However, Propositions 5.2, 5.3, and 5.4 can make things easier in some cases.

5.2. *Calculating fuzzy satisfiability*

We shall consider different cases related to the syntactic rules of our language. For each case we want to provide the minimum number of individuals needed and, when it is possible to derive it directly, the fuzzy satisfiability.

Since we are dealing with a fixed task, we shall denote the minimum number of individuals needed to calculate the fuzzy satisfiability of a fuzzy concept D as n_D. In general, $n_D = \min \mathbb{N}_D$, where $\mathbb{N}_D$ is the set of possible cardinalities of $\Delta_{\mathcal{I}}$ that allow us to calculate the fuzzy satisfiability.

Before one starts the process the TBox must be eliminated, which is always possible in $\mathcal{ALCQ}^+_F(\mathrm{D})$, since the TBox is acyclic.

5.3. *Atomic concepts*

Let A be an atomic concept, then $\mathcal{S}(A) = [0, 1]$ (if A is fuzzy) and $\mathcal{S}(A) = \{0, 1\}$ (if A is crisp[2]). In both cases, the number of individuals required in an interpretation is in the set $\mathbb{N}_A = \{n \in \mathbb{N}\colon n \geqslant 1\}$, hence $n_A = 1$.

5.4. *Negation*

We consider several cases:

(1) Let A be an atomic concept. Then $\mathcal{S}(\neg A) = \mathcal{S}(A)$, $\mathbb{N}_A = \mathbb{N}_{\neg A}$, and $n_{\neg A} = n_A$.
(2) $\mathcal{S}(\neg\neg C) = \mathcal{S}(C)$, $\mathbb{N}_C = \mathbb{N}_{\neg\neg C}$ and $n_{\neg\neg C} = n_C$.
(3) Let $C = C_1 \sqcap \cdots \sqcap C_r$. Then $\neg C = \neg C_1 \sqcup \cdots \sqcup \neg C_r$ and $\mathcal{S}(\neg C)$, $\mathbb{N}_{\neg C}$, and $n_{\neg C}$ are calculated by following the procedure we provide for disjunction.
(4) In the same way, let $C = C_1 \sqcup \cdots \sqcup C_r$. Then $\neg C = \neg C_1 \sqcap \cdots \sqcap \neg C_r$ and $\mathcal{S}(\neg C)$, $\mathbb{N}_{\neg C}$, and $n_{\neg C}$ are calculated by following the procedure we provide for conjunction.
(5) Let $D \equiv QR.C$. Then,

$$\mathcal{S}(\neg D) = \mathcal{S}\bigl(\neg(QR.C)\bigr) = \mathcal{S}\bigl((\neg Q)R.C\bigr), \tag{12}$$

since, by the properties of method GD

$$1 - GD_Q(G/F) = GD_{\neg Q}(G/F).$$

In addition, $\mathbb{N}_{\neg D} = \mathbb{N}_{(\neg Q)R.C}$ and $n_{\neg D} = n_{[(\neg Q)R.C]}$.

5.5. *Absolute quantifiers*

Satisfiability of $D \equiv QR.C$ depends on Q and the satisfiability of C. The set $\mathbb{N}_D$ (and hence the value n_D) depends on Q and $\mathbb{N}_C$.

We assume that $\mathrm{core}(Q) \neq \emptyset$ and $\mathbb{N}\backslash\mathrm{supp}(Q) \neq \emptyset$. This means it is always possible to find $n_0^Q, n_1^Q \in \mathbb{N}$ such that $Q(n_0^Q) = 0$ and $Q(n_1^Q) = 1$ and such that $n < n_0^Q$ implies $Q(n) > 0$ and $n < q_1$ implies $Q(n) < 1$.

PROPOSITION 5.5. *Let $D \equiv QR.C$ and let $\mathcal{S}(C) = [\beta_C, \tau_C]$. Then $\mathcal{S}(D) = \mathcal{S}(QR.C) = [\beta_D, \tau_D]$ with*

$$\beta_D = (1 - \tau_C)Q(0), \tag{13}$$

$$\tau_D = \max\bigl\{\bigl(\tau_C + (1 - \tau_C)Q(0)\bigr), Q(0)\bigr\}. \tag{14}$$

In addition,

$$\mathbb{N}_{QR.C} = \bigl\{n + n'\colon \bigl(n \geqslant \max\bigl(n_0^Q, n_1^Q\bigr)\bigr) \wedge (n' \in \mathbb{N}_C)\bigr\}.$$

[2] A concept C is *crisp* if an axiom of the form $C \sqcap \neg C \sqsubseteq \bot$ is found in the TBox.

PROOF. Let a be an individual. We know that $\mathcal{S}(R_a) = [0, 1]$, hence $\mathcal{S}(R_a \cap C) = [0, \tau_C]$. The extreme values for $D(a)$ can be reached for an interpretation such that the possible cardinalities of the set $R_a \cap C$ match the quantifier with degree 0 or 1, and the simplest such interpretation is one where all the individuals b_i verify $(R_a \cap C)(b_i) = (R_a \cap C)(b_j)$ and such that the number of b's is a certain q with $Q(q) = 0$ or $Q(q) = 1$. We distinguish three types of interpretations:

- We introduce n_0^Q individuals $b_1, \ldots, b_{n_0^Q}$ such that $(R_a \cap C)(b_i) = (R_a \cap C)(b_j) = \alpha$ $\forall i, j$. Then $ER(R_a \cap C) = (1-\alpha)/0 + \alpha/n_0^Q$, $Q(n_0^Q) = 0$, and $D(a) = (1-\alpha)Q(0)$. Since $\alpha \in [0, \tau_C]$ we have $D(a) \in [(1 - \tau_C)Q(0), Q(0)]$. In this interpretation there are at least $n_0^Q + n_C$ individuals.
- Let $q_1 = 0$. We define $R_a = \emptyset$ and hence $ER(R_a \cap C) = 1/0$ and $D(a) = 1 = Q(0) = \max\{(\tau_C + (1 - \tau_C)Q(0)), Q(0)\}$. Here we need just one individual (a).
- Let $q_1 > 0$. We introduce q_1 individuals $b_1, \ldots, b_{n_1^Q}$ such that $(R_a \cap C)(b_i) = (R_a \cap C)(b_j) = \alpha$ $\forall i, j$. Then $ER(R_a \cap C) = (1 - \alpha)/0 + \alpha/q_1$ and $D(a) = \alpha + (1 - \alpha)Q(0)$. Since $\alpha \in [0, \tau_C]$ and $D(a)$ is non-decreasing in α, we have $D(a) \in [Q(0), \tau_C + (1 - \tau_C)Q(0)]$. In this interpretation there are at least $n_1^Q + n_C$ individuals.

Hence,

$$\beta_D = \min\{(1 - \tau_C)Q(0), Q(0)\} = (1 - \tau_C)Q(0),$$
$$\tau_D = \max\{(\tau_C + (1 - \tau_C)Q(0)), Q(0)\}$$

and we need at least $\max(n_0^Q, n_1^Q) + n_C$ individuals. □

COROLLARY 5.1. *If Q is a quantifier with $Q(0) = 0$, or $\tau_C = 1$, then*

$$\mathcal{S}(QR.C) = [0, \tau_C].$$

5.6. *Relative quantifiers*

Again, satisfiability of $D \equiv QR.C$ depends on Q and the satisfiability of C, and the set $\mathbb{N}_D$ (and hence the value n_D) depends on Q and $\mathbb{N}_C$.

We also assume that $\text{core}(Q) \neq \emptyset$ and $[0, 1] \cap \mathbb{Q} \backslash \text{supp}(Q) \neq \emptyset$. This means it is always possible to find $q_0^Q, q_1^Q \in [0, 1] \cap \mathbb{Q}$ such that $Q(q_0^Q) = 0$ and $Q(q_1^Q) = 1$ and such that $q < q_0^Q$ implies $Q(q) > 0$ and $q < q_1$ implies $Q(q) < 1$.

Since $q_0^Q, q_1^Q \in [0, 1] \cap \mathbb{Q}$ it is always possible to find $n_0^Q, d_0^Q, n_1^Q, d_1^Q \in \mathbb{N}$ such that $q_0^Q = \frac{n_0^Q}{d_0^Q}$ and $q_1^Q = \frac{n_1^Q}{d_1^Q}$.[3]

As in the case of absolute quantifiers, in order to calculate the whole range of truth values that form $\mathcal{S}(D)$ we will consider those interpretations such that the number of degrees of truth is the minimum possible, and verifying that all the important cardinalities $(0, 1, q_0^Q, q_1^Q$ and the undefined case $0/0)$ are considered. Hence, we shall consider interpretations of the following kind:

---

[3] If $q_0^Q = 0$ (resp. $q_1^Q = 0$) we shall assume $d_0^Q = d_1^Q$ and $n_0^Q = 0$ (resp. $n_1^Q = 0$).

- We consider an individual a in order to evaluate $D(a)$.
- The number of individuals related to a by role R, namely d^Q, will be a multiple of both d_0^Q and d_1^Q, in particular $d^Q = mcm(d_0^Q, d_1^Q)$ is the minimum such value. We shall denote such individuals as $b_1, \ldots, b_{d^Q}$. Notice that by considering a fixed precision given by a number of decimals, we can bound this value by 10^p, where p is the number of decimals taken into account, since every possible rational number described with a precision of p decimals can be expressed as $n/10^p$, so in the worst case $mod(10^p, d_0^Q) = mod(10^p, d_1^Q) = 0$ and hence 10^p is a multiple of both.[4]
- $R(a, b_i) = r \in [0, 1]\ \forall 1 \leqslant i \leqslant d^Q$.
- $C(b_i) = \alpha_1\ \forall 1 \leqslant i \leqslant d$ and $C(b_j) = \alpha_2\ \forall (d+1) \leqslant j \leqslant d^Q$ with $d \in \{d_0^Q, d_1^Q\}$, $\alpha_1, \alpha_2 \in \mathcal{S}(C)$, and $\alpha_1 \geqslant \alpha_2$.

With this kind of interpretations, the following cases are possible:

- $1 \geqslant r \geqslant \alpha_1$: the relative cardinality of C with respect to R_a is

  $$ER(C/R_a) = (1-r)/(0/0) + (r-\alpha_1)/0 + (\alpha_1 - \alpha_2)/q + \alpha_2/1,$$

  with $q \in \{q_0^Q, q_1^Q\}$. Hence

  $$D(a) = (1-r)u^Q + (r-\alpha_1)Q(0) + (\alpha_1 - \alpha_2)Q(q) + \alpha_2 Q(1),$$

  where u^Q is the value assigned to the quantifier Q in the case "undefined", i.e. $Q(0/0) = u^Q$.
- $\alpha_1 \geqslant r \geqslant \alpha_2$: the relative cardinality of C with respect to R_a is

  $$ER(C/R_a) = (1-r)/(0/0) + (r-\alpha_2)/0 + \alpha_2/1.$$

  Hence

  $$D(a) = (1-r)u^Q + (r-\alpha_2)Q(0) + \alpha_2 Q(1).$$

- $\alpha_2 \geqslant r \geqslant 0$: the relative cardinality of C with respect to R_a is

  $$ER(C/R_a) = (1-r)/(0/0) + r/1.$$

  Hence

  $$D(a) = (1-r)u^Q + rQ(1).$$

5.6.1. *Number of individuals*

From the kind of interpretations we consider, it is obvious that the number of individuals required to calculate $\mathcal{S}(D)$ is any value in the set

$$\mathbb{N}_D = \{n + n' \colon (mod(n, d_0^Q) = mod(n, d_1^Q) = 0) \wedge (n' \in \mathbb{N}_C)\}.$$

[4] For the sake of simplicity we could consider $d_0^Q = d_1^Q = d^Q = 10^p$, though this is not the lower possible value of d^Q in general.

5.6.2. *Calculating $\mathcal{S}(D)$*

From the expressions above it is clear that $\mathcal{S}(D)$ depends on Q (fixed), r, α_1 and α_2 with the restrictions $r \in [0,1]$ and $\alpha_1, \alpha_2 \in \mathcal{S}(C)$ with $\alpha_1 \geqslant \alpha_2$. Let $\mathcal{S}(C) = [\beta_C, \tau_C]$. By considering only values of a certain precision of p decimals, $\mathcal{S}(D)$ can be computed by Algorithm 1, where the round operator rounds the result to the precision of p decimals employed. The complexity of this algorithm is O(1) if we consider that the precision p is fixed. The algorithm performs $O(10^{3p})$ operations.

5.6.3. *Some particular cases*

Most of the relative quantifiers usually employed verify $Q(0), Q(1) \in \{0,1\}$. Hence, providing analytical results for these cases can help to improve the average cost of reasoning algorithms.

ALGORITHM 1. Algorithm to compute $S(D)$.

(1) Calculate $\mathcal{S}(C)$.
(2) $\tau_D = 0$, $\beta_D = 1$.
(3) For every $r \in [0,1]$ and every pair $\alpha_1, \alpha_2 \in \mathcal{S}(C)$ with $\alpha_1 \leqslant \alpha_2$, all three with precision of p decimals
 (a) If $1 \geqslant r \geqslant \alpha_1$, then for every $Q(q) \in \{0,1\}$
  (i) $v \leftarrow round((1-r)u^Q + (r-\alpha_1)Q(0) + (\alpha_1 - \alpha_2)Q(q) + \alpha_2 Q(1))$.
  (ii) If $v < \beta_D$ then $\beta_D \leftarrow v$.
  (iii) If $v > \tau_D$ then $\tau_D \leftarrow v$.
 (b) If $\alpha_1 \geqslant r \geqslant \alpha_2$,
  (i) $v \leftarrow round((1-r)u^Q + (r-\alpha_2)Q(0) + \alpha_2 Q(1))$.
  (ii) If $v < \beta_D$ then $\beta_D \leftarrow v$.
  (iii) If $v > \tau_D$ then $\tau_D \leftarrow v$.
 (c) If $\alpha_2 \geqslant r \geqslant 0$,
  (i) $v \leftarrow round((1-r)u^Q + rQ(1))$.
  (ii) If $v < \beta_D$ then $\beta_D \leftarrow v$.
  (iii) If $v > \tau_D$ then $\tau_D \leftarrow v$.

PROPOSITION 5.6. *Let $Q(0) = 0$, $Q(1) = 1$. Then*

$$\mathcal{S}(D) = \left[\min\left\{u^Q, \beta_C\right\}, \tau_C + (1-\tau_C)u^Q\right].$$

PROOF. We have the following cases:

- Let $1 \geqslant r \geqslant \alpha_1$. Then

$$D(a) = \begin{cases} (1-r)u^Q + \alpha_2, & Q(q) = 0, \\ (1-r)u^Q + \alpha_1, & Q(q) = 1, \end{cases}$$

 that is $D(a) = (1-r)u^Q + \alpha$ with $1 \geqslant r \geqslant \alpha$ and $\alpha \in \mathcal{S}(C)$ with $\mathcal{S}(C) = [\beta_C, \tau_C] \cup V_C$. Then for $\alpha \in [\beta_C, \tau_C]$ we have $r \in [\beta_C, 1]$ and we obtain $[\beta_C, \tau_C + (1-\tau_C)u^Q] \subseteq \mathcal{S}(D)$.
- Let $\alpha_1 \geqslant r \geqslant \alpha_2$. Then

$$D(a) = (1-r)u^Q + \alpha_2.$$

Hence $\beta_C \leqslant \alpha_2 \leqslant r \leqslant \tau_C$ and we obtain $[(1-\tau_C)u^Q+\beta_C, (1-\tau_C)u^Q+\tau_C] \subseteq \mathcal{S}(D)$.

- Let $\alpha_2 \geqslant r \geqslant 0$. Then

$$D(a) = (1-r)u^Q + r.$$

Hence $0 \leqslant r \leqslant \alpha_2 \leqslant \tau_C$ and we obtain $[u^Q, (1-\tau_C)u^Q + \tau_C] \subseteq \mathcal{S}(D)$.

In summary, from the set of values $\mathcal{S}(C) = [\beta_C, \tau_C]$ we obtain

$$\begin{aligned}
&\left[\beta_C, \tau_C + (1-\tau_C)u^Q\right] \subseteq \mathcal{S}(D),\\
&\left[(1-\tau_C)u^Q + \beta_C, (1-\tau_C)u^Q + \tau_C\right] \subseteq \mathcal{S}(D),\\
&\left[u^Q, (1-\tau_C)u^Q + \tau_C\right] \subseteq \mathcal{S}(D).
\end{aligned}$$

Hence

$$\left[\min\left\{u^Q, \beta_C\right\}, \tau_C + (1-\tau_C)u^Q\right] \subseteq \mathcal{S}(D).$$

Finally, since we have taken into account all the possible cardinalities with the maximum possible degrees, inclusion can be changed by equality in the final results. □

COROLLARY 5.2. *Quantifiers* $\exists$ *and* $\forall$ *verify*

$$\begin{aligned}
&\mathcal{S}(\exists R.C) = [0, \tau_C],\\
&\mathcal{S}(\forall R.C) = [\beta_C, 1].
\end{aligned}$$

COROLLARY 5.3. *Quantifiers* $\exists$ *and* $\forall$ *verify*

$$\begin{aligned}
&\mathcal{S}(\exists R.C) = \mathcal{S}(\neg\forall R.\neg C),\\
&\mathcal{S}(\forall R.C) = \mathcal{S}(\neg\exists R.\neg C).
\end{aligned}$$

COROLLARY 5.4. *Let* $Q(0) = 1$, $Q(1) = 0$. *Then*

$$\mathcal{S}(QR.C) = \left[1 - \left(\tau_C + (1-\tau_C)u^Q\right), 1 - \min\left\{u^Q, \beta_C\right\}\right].$$

PROOF. Immediate by the properties of GD, $\neg(QR.C) \equiv (\neg Q)R.C$ and taking into account that $(\neg Q)(0) = 0$ and $(\neg Q)(1) = 1$. □

PROPOSITION 5.7. *Let* $Q(0) = 0$, $Q(1) = 0$. *Then*

$$\mathcal{S}(D) = \left[0, \max\left\{u^Q, (1-\tau_C)u^Q + (\tau_C - \beta_C)\right\}\right].$$

PROOF. We have the following cases:

- Let $1 \geqslant r \geqslant \alpha_1$. Then

$$D(a) = \begin{cases} (1-r)u^Q, & Q(q) = 0,\\ (1-r)u^Q + (\alpha_1 - \alpha_2), & Q(q) = 1. \end{cases}$$

Hence for those values $\alpha \in [\beta_C, \tau_C]$ we have $r \in [\beta_C, 1]$ and we obtain $[0, (1-\beta_C)u^Q] \subseteq \mathcal{S}(D)$ (first expression) and $[0, (1-\tau_C)u^Q + (\tau_C - \beta_C)] \subseteq \mathcal{S}(D)$ (second expression, 0 obtained when $r = 1$ and $\alpha_1 = \alpha_2$), with $(1-\tau_C)u^Q + (\tau_C - \beta_C) \geqslant (1-\beta_C)u^Q$.

- Let $\alpha_1 \geqslant r \geqslant \alpha_2$ or $\alpha_2 \geqslant r \geqslant 0$. Then

$$D(a) = (1 - r)u^Q.$$

Hence $r \in [0, \tau_C]$ and we obtain $[(1 - \tau_C)u^Q, u^Q] \subseteq \mathcal{S}(D)$.

In summary, from the set of values $\mathcal{S}(C) = [\beta_C, \tau_C]$ we obtain

$$\big[0, (1 - \tau_C)u^Q + (\tau_C - \beta_C)\big] \subseteq \mathcal{S}(D),$$
$$\big[(1 - \tau_C)u^Q, u^Q\big] \subseteq \mathcal{S}(D).$$

Hence

$$\big[0, \max\big\{u^Q, (1 - \tau_C)u^Q + (\tau_C - \beta_C)\big\}\big] \subseteq \mathcal{S}(D).$$

Finally, since we have taken into account all the possible cardinalities with the maximum possible degrees, inclusion can be changed by equality in the final results. □

COROLLARY 5.5. *Let* $Q(0) = 1$, $Q(1) = 1$. *Then*

$$\mathcal{S}(QR.C) = \big[1 - \max\big\{u^Q, (1 - \tau_C)u^Q + (\tau_C - \beta_C)\big\}, 1\big].$$

PROOF. Immediate by the properties of *GD*, $\neg(QR.C) \equiv (\neg Q)R.C$ and taking into account that $(\neg Q)(0) = 0$ and $(\neg Q)(1) = 0$. □

5.7. *Disjunction*

In this section we deal with the task of calculating the satisfiability of a fuzzy concept D of the form

$$D \equiv D_1 \sqcup D_2 \sqcup \cdots \sqcup D_s.$$

It is trivial to show that if $\mathcal{S}(D_i) = [\beta_{D_i}, \tau_{D_i}]$, in general, $\mathcal{S}(D) = [\beta_D, \tau_D]$ with

$$\beta_D \geqslant \max_{i \in \{1,\dots,s\}} \{\beta_{D_i}\},$$
$$\tau_D = \max_{i \in \{1,\dots,s\}} \{\tau_{D_i}\}.$$

The value of β_D can only be bounded in general because it can be the case that the different D_i are not *independent*. When two concepts C and C' are not independent, it can happen that there is no interpretation $\mathcal{I}$ such that $a \in \Delta_{\mathcal{I}}$ with $C^{\mathcal{I}}(a) = \beta_C$ and $C'^{\mathcal{I}}(a) = \beta_{C'}$, and hence $\max\{\beta_C, \beta_{C'}\} \notin \mathcal{S}(C \sqcup C')$, so as a consequence $\beta_{C \sqcup C'} \geqslant \max\{\beta_C, \beta_{C'}\}$.

As a simple counterexample, consider the concepts A and $\neg A$. Obviously, $\mathcal{S}(A) = \mathcal{S}(\neg A) = [0, 1]$, but $\mathcal{S}(A \sqcup \neg A) = [0.5, 1]$, i.e. $\beta_{A \sqcup \neg A} = 0.5 > 0 = \max\{\beta_A, \beta_{\neg A}\}$.

5.7.1. *Calculating the lower satisfiability bound in the absence of independence*

The previous results allow us to calculate the lower satisfiability bound directly when the concepts involved in the disjunction are pairwise-independent. Now, we must provide a procedure to calculate the lower bound when independence does not hold.

A natural way to proceed is to divide the set of concepts D_i in groups of independent concepts, calculate the satisfiability of the disjunction of the concepts in each group (for which we shall provide an algorithm later) and then to combine the results.

Let $Set(D) = \{D_1 \dots D_s\}$ be the set of concepts whose disjunction defines D. Then, the groups of dependent concepts in $Set(D)$ are the elements of the quotient set $Set(D)/\bowtie\, = \{G_1^D, \dots, G_r^D\}$ with $r \leqslant s$. By definition, this set is a partition of the elements in $Set(D)$.

We shall denote $\sqcup_{G_i^D}$ the concept defined by the disjunction of the concepts in group G_i^D, i.e.

$$\sqcup_{G_i^D} = \sqcup_{D_j \in G_i^D} D_j.$$

With these definitions, the procedure can be formalized as follows:

(1) Calculate $Set(D)/\bowtie$, the groups of dependent concepts in $Set(D)$ (using Propositions 5.2, 5.3, and 5.4).

(2) For each $G_i^D \in Set(D)/\bowtie$, calculate the satisfiability of $\sqcup_{G_i^D}$. Let it be $\mathcal{S}(\sqcup_{G_i^D}) = [\beta_i, \tau_i]$.

(3) $\mathcal{S}(D) = [\max_i \beta_i, \max_i \tau_i]$.

The third step is immediate since it is obvious that

$$D = \sqcup_{G_1^D} \sqcup \cdots \sqcup_{G_r^D}$$

and $\sqcup_{G_i^D} \not\bowtie \sqcup_{G_j^D}$ $\forall 1 \leqslant i < j \leqslant r$. However, as we mentioned before, we must provide a way to calculate the satisfiability of each $\sqcup_{G_i^D}$, since the concepts involved in it are not independent.

Finally, let us remark a very important consequence from Proposition 5.3: all the concepts appearing in a group G_i^D are either quantified or not quantified. In the two next sections we propose algorithms to deal with each case.

5.7.2. *Satisfiability of the union of dependent and non-quantified concepts*

It is well known that the satisfiability of concepts involving just atomic concepts connected by means of $\neg$, $\sqcup$, and $\sqcap$ can only be [0, 1], [0, 0.5] (for crisp contradictions), or [0.5, 1] (for crisp tautologies).

Though in this section we are interested in calculating the satisfiability of the union of dependent and not-quantified concepts, we can provide a general set of rules that are sufficient in order to calculate the satisfiability of any not-quantified concept, i.e. any concept whose expansion involves just atomic concepts connected by $\neg$, $\sqcup$, and $\sqcap$.

Some of these rules have been described before, but it is worth to include them here again because it is possible to check for independence at the same time the satisfiability is being calculated. This means the procedure in the previous section can be simplified by creating a single group G_1^D containing the non-quantified concepts, and checking independence just between quantified concepts in order to form the rest of the groups. Then, the satisfiability of $\sqcup_{G_1^D}$ can be obtained by recursively using the following rules:

(1) If A is an atomic concept then $\mathcal{S}(A) = [0, 1]$.

(2) Let $\mathcal{S}(C) = [\beta_C, \tau_C]$. Then $\mathcal{S}(\neg C) = [1 - \tau_C, 1 - \beta_C]$.

(3) The satisfiability of $C \sqcap D$ can be obtained from $\mathcal{S}(C)$ and $\mathcal{S}(D)$ as it is shown in Table 3.

Table 3
Satisfiability of the conjunction of non-quantified concepts.

$\mathcal{S}(C \sqcap D)$		$\mathcal{S}(D)$		
		[0, 1]	[0, 0.5]	[0.5, 1]
	[0, 1]	?	[0, 0.5]	[0, 1]
$\mathcal{S}(C)$	[0, 0.5]	[0, 0.5]	[0, 0.5]	[0, 0.5]
	[0.5, 1]	[0, 1]	[0, 0.5]	[0.5, 1]

Table 4
Satisfiability of the disjunction of non-quantified concepts.

$\mathcal{S}(C \sqcup D)$		$\mathcal{S}(D)$		
		[0, 1]	[0, 0.5]	[0.5, 1]
	[0, 1]	?	[0, 1]	[0.5, 1]
$\mathcal{S}(C)$	[0, 0.5]	[0, 1]	[0, 0.5]	[0.5, 1]
	[0.5, 1]	[0.5, 1]	[0.5, 1]	[0.5, 1]

The case $\mathcal{S}(C) = \mathcal{S}(D) = [0, 1]$ can be easily solved as follows: if $C \sqcap D$ is crisp satisfiable then $\mathcal{S}(C \sqcap D) = [0, 1]$, otherwise $C \sqcap D$ is a crisp contradiction and hence $\mathcal{S}(C \sqcap D) = [0, 0.5]$.

(4) The satisfiability of $C \sqcup D$ can be obtained from $\mathcal{S}(C)$ and $\mathcal{S}(D)$ as it is shown in Table 4.

Again, the case $\mathcal{S}(C) = \mathcal{S}(D) = [0, 1]$ can be easily solved as follows: if $\neg(C \sqcup D)$ is crisp satisfiable then $\mathcal{S}(C \sqcup D) = [0, 1]$, otherwise $C \sqcup D$ is a crisp tautology and hence $\mathcal{S}(C \sqcup D) = [0.5, 1]$.

5.7.3. *Satisfiability of the union of dependent and quantified concepts*

Let $D \equiv \sqcup_{G_k^D} \equiv Q_1 R.C_1 \sqcup Q_2 R.C_2 \sqcup \cdots \sqcup Q_t R.C_t$, where $Q_1, \ldots, Q_t$ are quantifiers and $\{C_1, \ldots, C_t\}$ is a group of dependent concepts, i.e. $C_i \bowtie C_j\ \forall 1 \leqslant i < j \leqslant t$. Let $D_i = Q_i R.C_i$. We consider now the problem of deciding the minimum extent β_D to which D is satisfiable. We already know the maximum extent τ_D to which D is satisfiable, and also that $\beta_D \geqslant \bar{\beta}_D$ with $\bar{\beta}_D = \max_{i \in \{1,\ldots,t\}}\{\beta_{D_i}\}$.

By definition, given an individual $a \in \Delta$,

$$\begin{aligned}
\beta_D &= \inf_{\mathcal{I}} \max_i \{ D_i^{\mathcal{I}}(a) \} \\
&= \inf_{\mathcal{I}} \max_i \left\{ \int_0^1 Q_i\left(\left| \{ b \in \Delta^{\mathcal{I}} \colon\ R(a,b) \geqslant \tau \wedge C_i(b) \geqslant \tau \} \right| \right) \mathrm{d}\tau \right\} \\
&= \inf_{\mathcal{I}} \max_i \left\{ \sum_j (\alpha_j - \alpha_{j+1}) Q_i\left(\left| (R_a \cap C_i)_{\alpha_j} \right| \right) \right\}.
\end{aligned}$$

We solve this problem by constructing an interpretation $\mathcal{I}^*$ which minimizes $\max_i\{D_i^{\mathcal{I}^*}(a)\}$. For this to be true, all the following inequalities must hold:

$$\begin{cases} D_1^{\mathcal{I}}(a) \leqslant \sigma, \\ \vdots \\ D_t^{\mathcal{I}}(a) \leqslant \sigma, \end{cases}$$

where $\sigma = \beta_D$, as well as all the TVBs in the expanded ABox, i.e. the ABox obtained after eliminating the TBox.

We are able to construct an interpretation $\mathcal{I}$ which satisfies all the above inequalities as well as all the TVBs in the expanded ABox for a given σ, provided there exists one. This could be done by means of Algorithm 2. If the algorithm has exhausted all guesses at step 2, then it must be concluded that the conjunction is not satisfiable to the degree σ.

However, we do not know the value of σ in advance, except that it must lie above the maximum of the β_{D_i}; therefore, we shall perform a dichotomic search on the values of σ by using Algorithm 3.

ALGORITHM 2. Building an interpretation that satisfies the inequalities and all the TVBs in the expanded ABox for a given σ.

(1) Determine the number n_D of individuals that must be included in the interpretation as fillers of role R, applying that

$$\mathbb{N}_{D_i \sqcup D_j} = \mathbb{N}_{D_i} \cap \mathbb{N}_{D_j}.$$

Let us remark that $\mathbb{N}_{D_i \sqcup D_j} \neq \emptyset$ since in the restrictions on the number of individuals we know that at least 10^p is a common value for all relative quantifiers (and any multiple of 10^p will be also in this set), and restrictions for absolute quantifiers take always the form "at least n individuals".

(2) "Guess" truth values for $R(a, b_k)$ and $C_i(b_k)$: these latter are actually calculated from the definition of C_i, after guessing the value of $A(b_k)$ for all primitive concepts A occurring in the definition of C_i; or recursively if C_i contains quantified concepts.

(3) Check whether any TVB in the ABox or any of the t inequalities is violated. If so, backtrack to the previous step; otherwise $\mathcal{I}$ is an interpretation satisfying the conjunction to the degree σ.

ALGORITHM 3. Performing a dichotomic search on the values of σ.

(1) $\sigma \leftarrow \bar{\beta}_D$, $\beta_D \leftarrow \tau_D$;

(2) By using Algorithm 2 try to construct an interpretation $\mathcal{I}$ that satisfies all the inequalities as well as all the TVBs in the ABox; if that is possible,

$$\beta_D \leftarrow \sigma;$$
$$\sigma \leftarrow \frac{\sigma + \bar{\beta}_D}{2};$$

otherwise,

$$\bar{\beta}_D \leftarrow \sigma;$$
$$\sigma \leftarrow \frac{\sigma + \tau_D}{2};$$

(3) if $\bar{\beta}_D - \sigma < 10^{-p}$, stop;
(4) go back to step 2.

It is easy to verify that this algorithm always stops with the correct value of β_D, up to a precision 10^{-p}, which represents the precision to which the results are needed. Furthermore, it always stops after at most $\log_2 10^p$ iterations. Although we do not provide here a formal proof for that, it can be proven that Algorithm 3 runs in polynomial space.

5.8. *Conjunction*

The satisfiability of a conjunction can be solved easily since a conjunction can be turn into the negation of a disjunction by using De Morgan's laws, and we have provided a general procedure for disjunction in the previous section.

5.9. *Summary*

We can summarize these results as follows:

- If A is an atomic concept, then $\mathcal{S}(A) = [0, 1]$.
- (Negation) If $\mathcal{S}(D) = [\beta_D, \tau_D]$, then $\mathcal{S}(\neg D) = [1 - \tau_D, 1 - \beta_D]$. This verifies $\mathcal{S}(\neg\neg D) = \mathcal{S}(D)$ and:
  - if A is atomic, $\mathcal{S}(\neg A) = \mathcal{S}(A) = [0, 1]$;
  - if $C = C_1 \sqcap \cdots \sqcap C_r$, then $\mathcal{S}(\neg C) = \mathcal{S}(\neg C_1 \sqcup \cdots \sqcup \neg C_r)$;
  - if $C = C_1 \sqcup \cdots \sqcup C_r$, then $\mathcal{S}(\neg C) = \mathcal{S}(\neg C_1 \sqcap \cdots \sqcap \neg C_r)$;
  - if $D \equiv QR.C$,

$$\mathcal{S}(\neg D) = \mathcal{S}\big(\neg(QR.C)\big) = \mathcal{S}\big((\neg Q)R.C\big). \tag{15}$$

- Let Q be an absolute quantifier such that $\mathrm{core}(Q) \neq \emptyset$ and $\mathbb{N}\backslash\mathrm{supp}(Q) \neq \emptyset$. If $D \equiv QR.C$ and $\mathcal{S}(C) = [\beta_C, \tau_C]$, $\mathcal{S}(D) = \mathcal{S}(QR.C) = [\beta_D, \tau_D]$, with

$$\beta_D = (1 - \tau_C)Q(0), \tag{16}$$

$$\tau_D = \max\big\{\big(\tau_C + (1 - \tau_C)Q(0)\big), Q(0)\big\}. \tag{17}$$

- Let Q be a relative quantifier such that $\mathrm{core}(Q) \neq \emptyset$ and $[0, 1] \cap \mathbb{Q}\backslash\mathrm{supp}(Q) \neq \emptyset$ with $u^Q = Q(x/0)$ (i.e. u^Q is the value returned by the quantifier when the relative cardinality is undefined). If $D \equiv QR.C$ and $\mathcal{S}(C) = [\beta_C, \tau_C]$,
  - If $Q(0) = 0$ and $Q(1) = 1$,

$$\mathcal{S}(D) = \big[\min\{u^Q, \beta_C\}, \tau_C + (1 - \tau_C)u^Q\big].$$

  In particular, for quantifiers $\exists$ and $\forall$ we have $u^\exists = 0$, $u^\forall = 1$, and

$$\mathcal{S}(\exists R.C) = [0, \tau_C],$$

$$\mathcal{S}(\forall R.C) = [\beta_C, 1].$$

  - If $Q(0) = 1$ and $Q(1) = 0$,

$$\mathcal{S}(QR.C) = \big[1 - \big(\tau_C + (1 - \tau_C)u^Q\big), 1 - \min\{u^Q, \beta_C\}\big].$$

– If $Q(0) = 0$ and $Q(1) = 0$,

$$\mathcal{S}(D) = \big[0, \max\big\{u^Q, (1 - \tau_C)u^Q + (\tau_C - \beta_C)\big\}\big].$$

– If $Q(0) = 1$ and $Q(1) = 1$,

$$\mathcal{S}(QR.C) = \big[1 - \max\big\{u^Q, (1 - \tau_C)u^Q + (\tau_C - \beta_C)\big\}, 1\big].$$

The remaining cases are solved by means of Algorithm 1 in O(1) with a fixed precision of p decimals.

- If $D \equiv D_1 \sqcup D_2 \sqcup \cdots \sqcup D_s$ with $\mathcal{S}(D_i) = [\beta_{D_i}, \tau_{D_i}]$, then $\mathcal{S}(D) = [\beta_D, \tau_D]$, with

$$\beta_D \geqslant \max_{i \in \{1,\dots,s\}} \{\beta_{D_i}\},$$
$$\tau_D = \max_{i \in \{1,\dots,s\}} \{\tau_{D_i}\}.$$

In particular, if the fuzzy concepts D_i are pairwise independent,

$$\beta_D = \max_{i \in \{1,\dots,s\}} \{\beta_{D_i}\},$$
$$\tau_D = \max_{i \in \{1,\dots,s\}} \{\tau_{D_i}\}.$$

The value of β_D with a precision degree of p decimals, when the D_i are not independent, is obtained by means of Algorithm 3 that performs a dichotomic search after guessing values of the atomic concepts and roles that appear in the D_i's.

- The satisfiability of conjunctions can be obtained by using De Morgan's laws and the latter result.

6. Conclusions

$\mathcal{ALCQ}_F^+(\mathrm{D})$ allows for concept description involving fuzzy linguistic quantifiers of the absolute and relative kind, and using qualifiers. In this chapter we have introduced algorithms to perform two important reasoning tasks with this logic: reasoning about instances, using the semantics of $\mathcal{ALCQ}_F^+(\mathrm{D})$, and calculating the fuzzy satisfiability of a fuzzy concept, for which a polynomial-space algorithm is provided. Though infinite interpretations are taken into account in the definition of the semantics of the language $\mathcal{ALCQ}_F^+(\mathrm{D})$, in practice and due to the physical limitations of computers, we have considered a finite number of individuals. Since the same limitations put a bound on the number of different membership degrees we can deal with, we calculate the fuzzy satisfiability of a fuzzy concept up to a certain precision degree, given as a number of decimals p. In fact, we believe a precision of one decimal digit is already enough for most applications and a precision of more than two decimal digits would be absolutely unnecessary in all applications.

In addition to the general algorithm, we have obtained some results showing that we can calculate the fuzzy satisfiability of a fuzzy concept directly in some (the most common) cases, hence improving the performance of the reasoning process. Some of these results are based on a result about *independence* of fuzzy concepts that we also introduce as a remarkable result.

References

[1] F. Baader, D. Calvanese, D. McGuinness, D. Nardi, P. Patel-Schneider (Eds.), The Description Logic Handbook: Theory, Implementation and Applications, Cambridge University Press, Cambridge, 2003.

[2] G. De Giacomo, M. Lenzerini, *A uniform framework for concept definitions in description logics*, J. Artificial Intelligence Research **6** (1997), 87–110.

[3] M. Delgado, D. Sánchez, M. Vila, *Fuzzy cardinality based evaluation of quantified sentences*, Int. J. Approximate Reasoning **23** (2000), 23–66.

[4] M. Delgado, M. Martín-Bautista, D. Sánchez, M. Vila, *A probabilistic definition of a nonconvex fuzzy cardinality*, Fuzzy Sets and Systems **126** (2) (2002), 41–54.

[5] F. Díaz-Hermida, A. Bugarín, P. Cariñena, S. Barro, *Voting model based evaluation of fuzzy quantified sentences: a general framework*, Fuzzy Sets and Systems, submitted for publication.

[6] I. Glöckner, *Fundamentals of fuzzy quantification: plausible models, constructive principles, and efficient implementation*, Technical Report TR2002-07, Technical Faculty, University Bielefeld, 33501 Bielefeld, Germany, 2002.

[7] A.D. Luca, S. Termini, A definition of a nonprobabilistic entropy in the setting of fuzzy sets theory, Inform. and Control **20** (1972), 301–312.

[8] M. Navara, *Satisfiability in fuzzy logics*, Neural Networks World **10** (5) (2000), 845–858.

[9] D. Sánchez, A.G.B. Tettamanzi, *Generalizing quantification in fuzzy description logics*, in: Proceedings 8th Dortmund Fuzzy Days, Dortmund, Germany, 2004.

[10] U. Straccia, *Reasoning within fuzzy description logics*, J. Artificial Intelligence Research **14** (2001), 137–166.

[11] U. Straccia, , *A fuzzy description logic for the semantic web*, in: E. Sanchez (Ed.), Fuzzy Logic in the Semantic Web, Elsevier, Amsterdam, 2006 , this issue.

[12] C.B. Tresp, R. Molitor, *A description logic for vague knowledge*, in: Proceedings of the 13th biennial European Conference on Artificial Intelligence (ECAI'98), Wiley, Brighton, UK, 1998, pp. 361–365.

[13] L.A. Zadeh, *A computational approach to fuzzy quantifiers in natural languages*, Comput. Math. Appl. **9** (1) (1983), 149–184.

Search and Protoforms

CHAPTER 9

From Search Engines to Question Answering Systems – The Problems of World Knowledge, Relevance, Deduction and Precisiation[1]

Lotfi A. Zadeh

Professor in the Graduate School and Director, Berkeley initiative in Soft Computing (BISC), Computer Science Division and the Electronics Research Laboratory, Department of EECS, University of California, Berkeley, CA 94720-1776, USA

E-mail: zadeh@eecs.berkeley.edu

Dedicated to Elie Sanchez

Abstract

Existing search engines, with Google at the top, have many truly remarkable capabilities. Furthermore, constant progress is being made in improving their performance. But what is not widely recognized is that there is a basic capability which existing search engines do not have: deduction capability – the capability to synthesize an answer to a query by drawing on bodies of information which reside in various parts of the knowledge base. By definition, a question-answering system, or Q/A system for short, is a system which has deduction capability. Can a search engine be upgraded to a question-answering system through the use of existing tools – tools which are based on bivalent logic and probability theory? A view which is articulated in the following is that the answer is: No.

The first obstacle is world knowledge – the knowledge which humans acquire through experience, communication and education. Simple examples are: "Icy roads are slippery," "Princeton usually means Princeton University," "Paris is the capital of France," and "There are no honest politicians." World knowledge plays a central role in search, assessment of relevance and deduction. The problem with world knowledge is that much of it is perception-based. Perceptions – and especially perceptions of probabilities – are intrinsically imprecise, reflecting the fact that human sensory organs, and ultimately the brain, have a bounded ability to resolve detail and store information. Imprecision of perceptions stands in the way of using conventional techniques – techniques which are based on bivalent logic and probability theory – to deal with perception-based information. A further complication is that much of world knowledge is negative knowledge in the sense that it

[1] This paper draws in part on *Toward a generalized theory of uncertainty (GTU) – an outline*, Inform. Sci. **172** (2005), 1–40.

FUZZY LOGIC AND THE SEMANTIC WEB
Edited by Elie Sanchez
© 2006 Elsevier B.V. All rights reserved

relates to what is impossible and/or non-existent. For example, "A person cannot have two fathers," and "Netherlands has no mountains."

The second obstacle centers on the concept of relevance. There is an extensive literature on relevance, and every search engine deals with relevance in its own way, some at a high level of sophistication. But what is quite obvious is that the problem of assessment of relevance is quite complex and far from solution.

There are two kinds of relevance: (a) question relevance and (b) topic relevance. Both are matters of degree. For example, on a very basic level, if the question is q: Number of cars in California? and the available information is p: Population of California is 37,000,000, then what is the degree of relevance of p to q? Another example: To what degree is a paper entitled "A New Approach to Natural Language Understanding" of relevance to the topic of machine translation.

Basically, there are two ways of approaching assessment of relevance: (a) semantic; and (b) statistical. To illustrate, in the number of cars example, relevance of p to q is a matter of semantics and world knowledge. In existing search engines, relevance is largely a matter of statistics, involving counts of links and words, with little if any consideration of semantics. Assessment of semantic relevance presents difficult problems whose solutions lie beyond the reach of bivalent logic and probability theory. What should be noted is that assessment of topic relevance is more amendable to the use of statistical techniques, which explains why existing search engines are much better at assessment of topic relevance then question relevance.

The third obstacle is deduction from perception-based information. As a basic example, assume that the question is q: What is the average height of Swedes?, and the available information is p: Most adult Swedes are tall. Another example is: Usually Robert returns from work at about 6 pm. What is the probability that Robert is home at about 6:15 pm? Neither bivalent logic nor probability theory provide effective tools for dealing with problems of this type. The difficulty is centered on deduction from premises which are both uncertain and imprecise.

Underlying the problems of world knowledge, relevance and deduction is a very basic problem – the problem of natural language understanding. Much of world knowledge and web knowledge is expressed in a natural language. A natural language is basically a system for describing perceptions. Since perceptions are intrinsically imprecise, so are natural languages, especially in the realm of semantics.

A prerequisite to mechanization of question-answering is mechanization of natural language understanding, and a prerequisite to mechanization of natural language understanding is precisiation of meaning of concepts and proposition drawn from a natural language. To deal effectively with world knowledge, relevance, deduction and precisiation, new tools are needed. The principal new tools are: Precisiated Natural Language (PNL); Protoform Theory (PFT); and the Generalized Theory of Uncertainty (GTU). These tools are drawn from fuzzy logic – a logic in which everything is, or is allowed to be, a matter of degree.

The centerpiece of new tools is the concept of a generalized constraint. The importance of the concept of a generalized constraint derives from the fact that in PNL and GTU it serves as a basis for generalizing the universally accepted view that information is statistical in nature. More specifically, the point of departure in PNL and GTU is the fundamental premise that, in general, information is representable as a system of generalized constraints, with statistical information constituting a special case. This, much more general, view of information is needed to deal effectively with world knowledge, relevance, deduction, precisiation and related problems.

In summary, the principal objectives of this paper are: (a) to make a case for the view that a quantum jump in search engine IQ cannot be achieved through the use of methods based on bivalent logic and probability theory; and (b) to introduce and outline a collection of non-standard concepts, ideas and tools which open the door to addition of deduction capability to search engines.

1. Introduction

If I were asked, "What is the most challenging problem in the realm of information science and technology?" my unequivocal answer would be; conception and design of question-answering systems. And if I were asked what is likely to be the most important application area for fuzzy logic in coming years, my answer would be (a) improvement of performance of search engines; and (b) upgrading search engines to question-answering systems, or Q/A systems, for short.

Existing search engines are not perfect. Nevertheless, they are extremely useful. Without access to a search engine, many of us would be paralyzed. It is evident that existing search engines, with Google at the top but faced with growing competition, have many truly remarkable capabilities. But there is a very important capability which they do not have – deduction capability – the capability to synthesize an answer to a question by drawing on bodies of information which reside in various parts of the knowledge base. What should be noted, however, is that there are many widely used special purpose Q/A systems which have limited deduction capability. Examples of such systems are driving direction systems, reservation systems, diagnostic systems and specialized expert systems, especially in the domain of medicine.

It is of historical interest to note that question-answering systems were an object of considerable attention in the early seventies. The literature abounded with papers dealing with them. A few examples: L.S. Coles [6], M. Nagao and J. Tsujii [18], J.R. McSkimin and J. Minker [16], A.R. Aronson, B.E. Jacobs and J. Minker [1], W.J.H.J. Bronnenberg et al. [5]. There was no discussion of search engines because the needed computer technology was not in existence. Interest in question-answering systems dwindled in the early eighties, when it became obvious that AI was not advanced enough to provide the needed tools and technology.

In recent years, significant progress toward enhancement of web intelligence has been achieved through the use of concepts and techniques related to the Semantic Web [3], OWL [23], CYC [14] and other approaches. But can such approaches, based as they are on bivalent logic and probability theory, be employed to upgrade search engines to question-answering systems? The answer, in my view, is: No. The reason, which is not widely recognized as yet, is that bivalent logic and bivalent-logic-based probability theory have intrinsic limitations. To circumvent these limitations what are needed are new tools based on fuzzy logic and fuzzy-logic-based probability theory. What distinguishes fuzzy logic from standard logical systems is that in fuzzy logic everything is, or is allowed to be graduated, that is, be a matter of degree. Furthermore, in fuzzy logic everything is allowed to be granulated, with a granule being a clump of values drawn together by indistinguishability, similarity or proximity. It is these fundamental features of fuzzy logic that give it a far greater power to deal with problems related to web intelligence than standard tools based on bivalent logic and probability theory. An analogy which will be elaborated at a later point is: In general, a valid model of a nonlinear system cannot be constructed through the use of linear components.

There are three major obstacles to upgrading a search engine to a question-answering system: (a) the problem of world knowledge; (b) the problem of relevance; and (c) the underlying problem of mechanization of natural language understanding and, in particular, the basic problem of precisiation of meaning. A very brief discussion of these problems follows. We will return to a discussion of these problems at a later point. Since the issues to be discussed are not restricted to web-related problems, our discussion will be general in nature.

2. The principal problems

2.1. *The problem of world knowledge*

World knowledge is the knowledge which humans acquire through experience, education and communication. Simple examples are:

- Few professors are rich
- There are no honest politicians
- It is not likely to rain in San Francisco in midsummer
- Most adult Swedes are tall
- There are no mountains in Holland
- Usually Princeton means Princeton University
- Paris is the capital of France
- In Europe, the child-bearing age ranges from about sixteen to about forty-two

The problem with world knowledge is that much of it is perception-based. Examples:

- Most adult Swedes are tall
- Most adult Swedes are much taller than most adult Italians
- Usually a large house costs more than a small house
- There are no honest politicians

Perception-based knowledge is intrinsically imprecise, reflecting the bounded ability of sensory organs, and ultimately the brain, to resolve detail and store information. More specifically, perception-based knowledge is f-granular in the sense that (a) the boundaries of perceived classes are unsharp (fuzzy); and (b) the values of perceived attributes are imprecise (fuzzy). Bivalent-logic-based approaches provide no methods for deduction from perception-based knowledge. For example, given the datum: Most adult Swedes are tall, existing bivalent-logic-based methods cannot be employed to come up with valid answers to the questions q_1: How many adult Swedes are short; and q_2: What is the average height of adult Swedes?

2.2. *The problem of relevance*

The importance of the concept of relevance is hard to exaggerate. Relevance is central to search. Indeed, the initial success of Google is due, in large measure, to its simple but ingenious page ranking algorithm for assessment of relevance [4].

Despite its importance, there are no satisfactory definitions of relevance in the literature, nor can the problem of assessment of relevance be considered to be on the way to solution. In fact, it may be argued that, as in the case of world knowledge, the concept of relevance is much too complex to lend itself to treatment within the limited conceptual framework of bivalent logic and bivalent-logic-based probability theory. An immediate problem is that relevance is not a bivalent concept. Relevance is a matter of degree, that is, it is a fuzzy concept. To define fuzzy concepts, what is needed is the conceptual structure of fuzzy logic. As was stated earlier, in fuzzy logic everything is, or is allowed to be, a matter of degree.

For concreteness, it is convenient to define a relevance function, $R(q/p)$, as a function in which the first argument, q, is a question or a topic; the second argument, p, is a proposition, topic, document, web page or a collection of such objects; and R is the degree to which p is relevant to q.

When q is a question, computation of $R(q/p)$ involves an assessment of the degree of relevance of p to q, with p playing the role of question-relevant information. For example, if q: What is the number of cars in California, and p: Population of California is 37 million, then p is question-relevant to q in the sense that p constrains, albeit imprecisely, the number of cars in California. The constraint is a function of world knowledge.

If q is a topic, e.g., q: Ontology, then a document entitled p: What is ontology?, is of obvious relevance to q, i.e., p is topic-relevant. The problem in both cases is that of assessment of degree of relevance. The problem is far from solution. Basically, what we need is a method of computing the degree of relevance based on the meaning of q and p, that is, we need semantic relevance. Existing search engines have a very limited capability to deal with semantic relevance. Instead, what they use is what may be called statistical relevance. In statistical relevance, what is used is, in the main, statistics of links and counts of words. Performance of statistical methods of assessment of relevance is unreliable, as can be seen from examples which will be considered later in this section.

A major source of difficulty in assessment of relevance relates to non-compositionality of the relevance function. More specifically, assume that we have a question, q, and two propositions p and r. Can the value of $R(q/p, r)$ be composed from the values of $R(q/p)$ and $R(q/r)$? The answer, in general, is: No. As a simple, not web-related, example, suppose that q: How old is Vera; p: Vera's age is the same as Irene's; r: Irene is 65. In this case, $R(q/p) = 0$; $R(q/r) = 0$ and yet $R(q/p, r) = 1$. What this implies is that, in general, relevance cannot be assessed in isolation. This suggests a need for differentiation between relevance and what may be called i-relevance, that is, relevance in isolation. In other words, a proposition, p, is i-relevant if it is relevant by itself, and it is i-irrelevant if it is not of relevance by itself but might be relevant in combination with other propositions. An example is considered in [40,41].

2.3. *The problem of precisiation of meaning – a prerequisite to mechanization of natural language understanding*

Much of world knowledge and web knowledge is expressed in a natural language. This is why issues relating to natural language understanding and natural language reasoning are of direct relevance to search and, even more so, to question-answering.

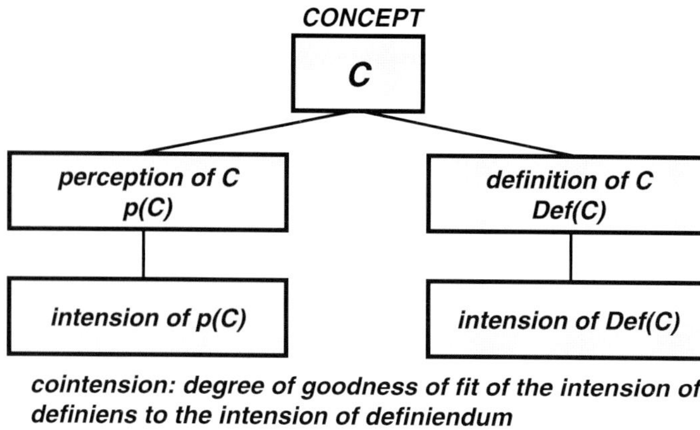

Fig. 1. Cointension of definition.

Humans have no difficulty in understanding natural language, but machines have many. One basic problem is that of imprecision of meaning. A human can understand an instruction such as "Take a few steps," but a machine cannot. To execute this instruction, a machine needs a precisiation of "few."

Precisiation of propositions drawn from a natural language is the province of PNL (Precisiated Natural Language) [40,41]. A forerunner of PNL is PRUF [34]. In PNL, precisiation is interpreted as meaning precisiation rather than value precisiation. A proposition is precisiated through translation into the Generalized Constraint Language (GCL), as will be dismissed in greater detail in Section 3. An element of GCL which precisiates p is referred to as a precisiand of p, $\mathrm{GC}(p)$, with $\mathrm{GC}(p)$ representing a generalized constraint. A precisiand may be viewed as a model of meaning.

A concept which plays a key role in precisiation is cointension, with intension used in its usual logical sense as attribute-based meaning. Thus, p and q are cointensive if the meaning of p is a close approximation to that of q. In this sense, a precisiand, $\mathrm{GC}(p)$, is valid if $\mathrm{GC}(p)$ is cointensive with p.

The concept of cointensive precisiation has an important implication for validity of definitions of concepts. More specifically, if C is a concept and $\mathrm{Def}(C)$ is its definition, then for $\mathrm{Def}(C)$ to be a valid definition, $\mathrm{Def}(C)$ must be cointensive with C (Figure 1).

The concept of cointensive definition leads to an important conclusion: In general, a cointensive definition of a fuzzy concept cannot be formulated within the conceptual structure of bivalent logic and bivalent-logic-based probability theory. This is why cointensive definitions of such concepts as causality, relevance, summary and mountain cannot be found in the literature. The concept of cointensive precisiation will be discussed in greater detail in Section 4.

3. Test queries (Google)

There is no question that existing search engines, with Google at the top, have many remarkable capabilities and are extremely useful. Search engines are extremely useful because the user of a search engine employs human intelligence to compensate for its limitations, especially in relation to world knowledge, assessment of relevance and natural language understanding. A few examples of use of Google are intended to illustrate its strengths and limitations.

Topic 1: Precisiation
Top-ranking results
[UAI] The concept of cointensive **precisiation**
... from data expressed in a natural language is **precisiation** of meaning. ...
In this perspective, the problem of **precisiation** is that of picking a ...
[PPT] www.cs.berkeley.edu/~nikraves/bisc/Present/Fall0/l ...
... **precisiation** of c, i, and q can be reduced to precisiation of p ... the issues of definability, **precisiation**, recognition, organization and search loom ...
AI Magazine: Precisiated natural language
... Basically, what PNL provides is a system for **precisiation** of ... provide a computational framework for **precisiation** of meaning rather than to serve as a ...
Precisiation of Meaning via Translation into PRUF, Lotfi A. Zadeh, In Vaina, L. and Hintikka, J., Eds., *Cognitive Constraints on Communication*, D. Reidel, 1984: 373–401.

COMMENT. In this test, Google does a remarkably good job. Precisiation is a neologism. I used it for the first time in a paper published in 1984 – a paper which Google finds. The top-ranking result relates to a presentation which I made just a few weeks ago.

Rating of performance: Excellent

Topic 2: Precisiand
Top-ranking results
[UAI] The concept of cointensive precisiation
... the meaning of p is referred to as a **precisiand** of p, denoted as GC(p). ...
A cointensive **precisiand** would be GC(p): If speed is less than about 70 mph, ...
[PDF] Microsoft PowerPoint – Zadeh_presentation
... In general, **precisiand** of p is not unique. If GC.1.(p), ... , .GC.n.(p) are possible precisiands of p, then a basic. question which arises is: which of ...

COMMENT. Precisiand is a neologism which I introduced in a recent presentation. In this test, Google does a remarkably good job.

Rating of performance: Excellent

Topic 3: Tikhonov regularization
Top-ranking results
Tikhonov Regularization and Total Least Squares
... stable solutions to these systems it is necessary to apply **regularization** methods. We show how **Tikhonov's regularization** method, which in its original ...
Tikhonov Regularization for Finitely and Infinitely Smoothing ...
... The main goal of this paper is to obtain a unified theory of **Tikhonov regularization**, ... **Tikhonov regularization**, Hilbert scales, spectral measure ...
Finite Dimensional Approximation and **Tikhonov Regularization** Method
... of the **regularization** functional (**Tikhonov** & Arsenin, 1977, Morozov, 1984, ... and then applying the **Tikhonov regularization** method in the frequency space. ...

COMMENT. Tikhonov regularization is a specialized topic. Google does extremely well.

Rating of performance: Excellent

Topic 4: Knowledge representation
Top-ranking results
Book results for **knowledge representation** Knowledge Representation – by Arthur B. Markman – 200 pages
What is a **Knowledge Representation**?
An introductory critical paper by R. Davis, H. Shrobe, and P. Szolovits in AI Magazine – 1993.
Knowledge Representation Book
The cover art for John Sowa's book is Ramon Lull's tree of nature and logic. It is reproduced here, along with Sowa's explanation.

COMMENT. This is a much broader topic, and yet Google does extremely well.

Rating of performance: Excellent

Question 1: What is the capital of New York?
Top-ranking results
Web definitions for capital of new york Albany: state capital of New York; located in eastern New York State on the west bank of the Hudson river
News results for what is the capital of New York – View today's top stories After the twin-tower nightmare, New York is back on form, says ... – Economist –
The New Raiders – BusinessWeek –
Brascan acquires New York-based Hyperion Capital for $50M US

COMMENT. There is some ambiguity in the meaning of New York: Does it refer to the city of New York or the state of New York? Google resolves the ambiguity and comes up with correct answer. The results which follow the correct answer are of little, if any, relevance.

Rating of performance: Very good

Question 2: What is the population of the capital of New York?
Top-ranking results
News results for population of New York – View today's top stories After the twin-tower nightmare, New York is back on form, says ...
UN: World's population is aging rapidly –
New, deadly threat from AIDS virus

COMMENT. Google fails to come up with the correct answer. However, Google comes up with a correct answer to the query: What is the population of Albany? What this example shows is that Google lacks the ability to synthesize an answer to the question by combining the answers to: What is the capital of New York?, and What is the population of Albany?

Rating of performance: Poor

Question 3: What is the distance between the largest city in Spain and the largest city in Portugal?
Top-ranking results
Porto – Oporto – Portugal Travel Planner
Munich Germany Travel Planner – Hotels Restaurants Languange . . .

COMMENT. This example is a slightly more complicated version of the previous question. Similar problem with deduction incapability.

Rating of performance: Poor

Topic 5: Chirac
Top-ranking results
News results for **Chirac** – View today's top stories **Chirac**, Schroeder meet on EU constitution, cooperation
en.wikipedia.org/wiki/Jacques_Chirac
BBC NEWS | World | Europe | **Chirac** makes case for EU treaty
The French president goes on TV in a bid to boost the faltering Yes campaign on the EU constitution.
Biography – Website of the Office of the French President
. . . Jacques **Chirac** Born on 29 November 1932 in the 5th arrondissement of Paris.
Son of François **Chirac**, a company director, and Marie-Louise, née Valette. . . .

COMMENT. Google correctly assumes that Chirac is the president of France. Relevance of many results is low.

Rating of performance: Fairly good

Topic 6: Chirac's son
Top-ranking results
Guardian Unlimited | Special reports | **Chirac's** nemesis
. . . Two events destroyed his position as **Chirac's** spiritual **son**. The first, a little talked about extra-marital romance with the President's daughter Claude . . .
lgf: Moore: a BIG Hit in Britain
. . . 'intellectual' circles, like Mitterand's hidden daughter, or, today, **Chirac's** hidden **son** in Japan. . . . I was not talking of **Chirac's son**, nor mazarine. . . .
International PoliticsandElections.com
. . . **son** of late French president Francois Mitterand. . . . **Chirac's** candidacy appears to be weakend substantially by allegations and investigations of . . .

COMMENT. Most of the results are irrelevant.

Rating of performance: Poor

Question 4: Age of Chirac's son
Top-ranking results
Guardian Unlimited | Special reports | **Chirac's** nemesis
. . . Their **son** plunged himself into politics at a precocious **age**, and was active

in the ... Two events destroyed his position as **Chirac's** spiritual **son**. ...
Jacques Chirac
... An only child, Chirac is the **son** of a bank clerk and later an executive for
... However, **Chirac's** extensive family acquaintances had him ranked back at ...
Politics and Elections
... Jean-Christophe Mitterand, **son** of former President Francois Mitterand, ...
Chirac's candidacy appears to be weakend substantially by allegations and ...

COMMENT. No relevant results. No deduction capability.

Rating of performance: Very poor

Question 5: What is the number of cars in California?
Top-ranking results
California Lemon Law – Lemon Check CARFAX Report, used **car** history ...
California Lemon Law provides FREE CARFAX Report, Kelley Blue Book, VIN **number**, used **car** history, auto insurance, auto warranty, lemon check, ...
CBS News | States Mull Taxing Drivers By Mile | February 16, 2005 ...
... Officials in **car**-clogged **California** are so worried they may be considering a ... So Too Does **Number** Of Fakes, • Softer **Cars** To Save Pedestrians ...
Ev Archive for January 2001
... of a decade-long bureaucratic and political fight over electric **cars** in **California**. ... should be a greater **number** of **cars** that produce no pollution. ...

COMMENT. No relevant results.

Rating of performance: Poor

Question 6: How many Ph.D. degrees in mathematics were granted by European universities in 1986?
Top-ranking results
A History of the **University of** Podlasie
... to grant master's **degrees** in the following programmes: **mathematics**, ... its inception in **1986**, the Faculty was **granted** rights to confer **Ph.D**. degree in ...
1999 Gairdner Foundation Winners!
... SB **degrees** in **Mathematics** and Economics from MIT in 1968 and his **Ph.D**. ... of Science **degrees** from the **University of** Manitoba and McGill University. ...
Enabling technology for distributed multimedia applications ...
... **Ph.D**. degree in computer science from the **University of** Waterloo in **1986**. ... degree in **mathematics** from the **University of** Regina, and a **Ph.D**. degree in ...

COMMENT. No relevant results. No deduction capability.

Rating of performance: Very poor

What these tests show is that Google has some remarkable capabilities and serious incapabilities, especially in relation to assessment of semantic relevance and deduction. Google is extremely useful because a skilled human user can exploit Google's strengths and compensate for Google's limitations.

4. The new tools

The principal thesis of the proceeding discussion is that upgrading a search engine to a question-answering system is beyond the reach of methods based on bivalent logic and bivalent-logic-based probability theory. The principal obstacles are the problems of world knowledge, relevance and precisiation.

To deal with these and related problems it is necessary to move from bivalent logic to fuzzy logic – a logic that is far more general than bivalent logic. With fuzzy logic as the base, a complex of new tools can be constructed. The structure of this complex is shown in Figure 2.

In what follows, we discuss the basics of some of the new tools. The centerpiece of the new tools is the concept of a generalized constraint [35].

4.1. *The concept of a generalized constraint*

Constraints are ubiquitous. A typical constraint is an expression of the form $X \in C$, where X is the constrained variable and C is the set of values which X is allowed to take. A typical constraint is hard (inelastic) in the sense that if u is a value of X then u satisfies the constraint if and only if $u \in C$.

The problem with hard constraints is that most real-world constraints are not hard, meaning that most real-world constraints have some degree of elasticity. For example, the constraints "check-out time is 1 pm," and "speed limit is 100 km/h," are, in reality, not hard. How can such constraints be defined? The concept of a generalized constraint is motivated by questions of this kind.

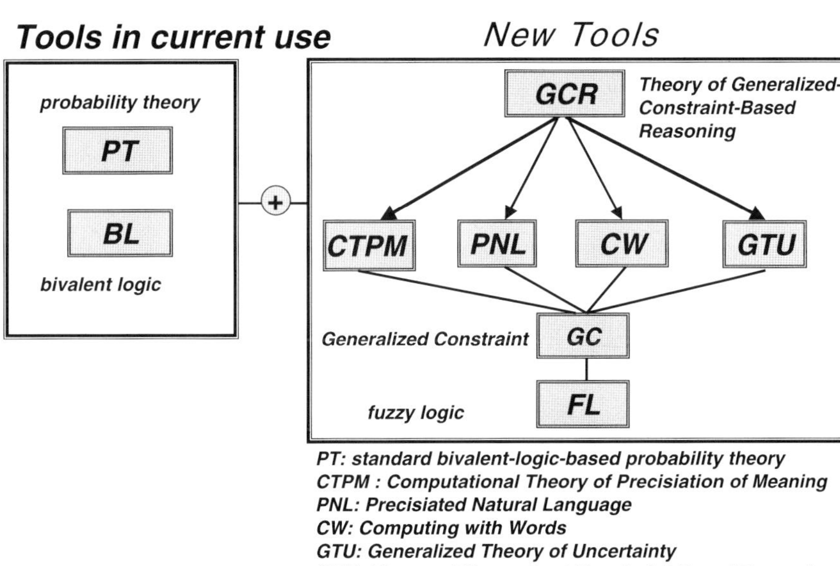

Fig. 2. Structure of new tools.

Real-world constraints may assume a variety of forms. They may be simple in appearance and yet have a complex structure. Reflecting this reality, a generalized constraint, GC, is defined as an expression of the form.

$$\text{GC: } X \text{ is} r\ R,$$

where X is the constrained variable; R is a constraining relation which, in general, is non-bivalent; and r is an indexing variable which identifies the modality of the constraint, that is, its semantics. R will be referred to as a granular value of X.

The constrained variable, X, may assume a variety of forms. In particular,

- X is an n-ary variable, $X = (X_1, \ldots, X_n)$
- X is a proposition, e.g., X = Leslie is tall
- X is a function
- X is a function of another variable, $X = f(Y)$
- X is conditioned on another variable, X/Y
- X has a structure, e.g., X = Location(Residence(Carol))
- X is a group variable. In this case, there is a group, $G[A]$; with each member of the group, Name_i, $i = 1, \ldots, n$, associated with an attribute-value, A_i. A_i may be vector-valued. Symbolically

  $$G[A]\text{: } \text{Name}_1/A_1 + \cdots + \text{Name}_n/A_n.$$

  Basically, $G[A]$ is a relation.
- X is a generalized constraint, $X = Y$ is$r$ R.

A generalized constraint, GC, is associated with a test-score function, $ts(u)$ [31] which associates with each object, u, to which the constraint is applicable, the degree to which u satisfies the constraint. Usually, $ts(u)$ is a point in the unit interval. However, if necessary, the test-score may be a vector, an element of a semi-ring [21], an element of a lattice [11] or, more generally, an element of a partially ordered set, or a bimodal distribution – a constraint which will be described later in this section. The test-score function defines the semantics of the constraint with which it is associated.

The constraining relation, R, is, or is allowed to be, non-bivalent (fuzzy). The principal modalities of generalized constraints are summarized in the following.

4.1.1. *Principal modalities of generalized constraints*

(a) *Possibilistic* (r = blank)

$$X \text{ is } R$$

with R playing the role of the possibility distribution of X. For example:

$$X \text{ is } [a, b]$$

means that $[a, b]$ is the set of possible values of X. Another example:

$$X \text{ is small.}$$

In this case, the fuzzy set labeled small is the possibility distribution of X. If μ_{small} is the membership function of small, then the semantics of "X is small" is defined by

$$\text{Poss}\{X = u\} = \mu_{\text{small}}(u)$$

where u is a generic value of X.

(b) *Probabilistic* ($r = \text{p}$)

$$X \text{ isp } R,$$

with R playing the role of the probability distribution of X. For example.

$$X \text{ isp } N\left(m, \sigma^2\right)$$

means that X is a normally distributed random variable with mean m and variance σ^2.

If X is a random variable which takes values in a finite set $\{u_1, \ldots, u_n\}$ with respective probabilities $p_1, \ldots, p_n$, then X may be expressed symbolically as

$$X \text{ isp } (p_1 \backslash u_1 + \cdots + p_n \backslash u_n),$$

with the semantics

$$\text{Prob}(X = u_i) = p_i, \quad i = 1, \ldots, n.$$

What is important to note is that in the Generalized Theory of Uncertainty (GTU) [42] a probabilistic constraint is viewed as an instance of a generalized constraint.

When X is a generalized constraint, the expression

$$X \text{ isp } R$$

is interpreted as a probability qualification of X, with R being the probability of X [29]. For example.

$$(X \text{ is small}) \text{ isp likely},$$

where small is a fuzzy subset of the real line, means that the probability of the fuzzy event $\{X \text{ is small}\}$ is likely. More specifically, if X takes values in the interval $[a, b]$ and g is the probability density function of X, then the probability of the fuzzy even "X is small" may be expressed as [25]

$$\text{Prob}(X \text{ is small}) = \int_a^b \mu_{\text{small}}(u) g(u)\, \text{d}u.$$

Hence

$$ts(g) = \mu_{\text{likely}}\left(\int_b^a g(u)\mu_{\text{small}}(u)\, \text{d}u\right).$$

This expression for the test-score function defines the semantics of probability qualification of a possibilistic constraint.

(c) *Veristic* ($r = \text{v}$)

$$X \text{ isv } R,$$

where R plays the role of a verity (truth) distribution of X. In particular, if X takes values in a finite set $\{u_1, \ldots, u_n\}$ with respective verity (truth) values $t_1, \ldots, t_n$, then X may be expressed as

$$X \text{ isv } (t_1|u_1 + \cdots + t_n|u_n),$$

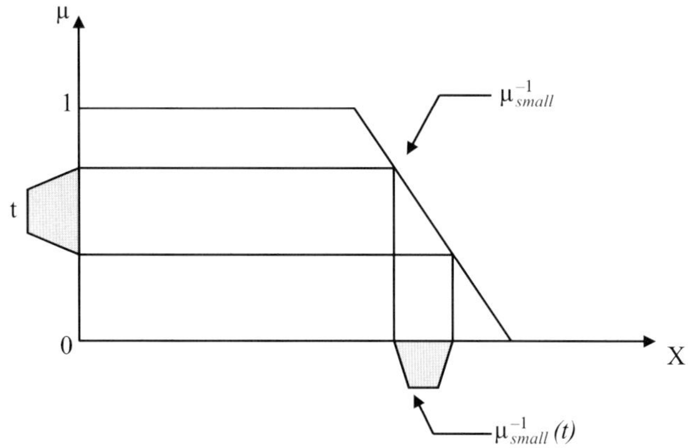

Fig. 3. Truth-qualification: (X is small) is t.

meaning that $\mathrm{Ver}(X = u_i) = t_i$, $i = 1, \ldots, n$.

For example, if Robert is half German, quarter French and quarter Italian, then

$$\text{Ethnicity(Robert) isv } (0.5|\text{German} + 0.25|\text{French} + 0.25|\text{Italian}).$$

When X is a generalized constraint, the expression

$$X \text{ isv } R$$

is interpreted as verity (truth) qualification of X. For example,

$$(X \text{ is small}) \text{ isv very.true},$$

should be interpreted as "It is very true that X is small." The semantics of truth qualification is defined in [29,30]

$$\mathrm{Ver}(X \text{ is } R) \text{ is } t \to X \text{ is } \mu_R^{-1}(t),$$

where μ_R^{-1} is inverse of the membership function of R and t is a fuzzy truth value which is a subset of [0, 1], Figure 3.

NOTE. There are two classes of fuzzy sets: (a) possibilistic, and (b) veristic. In the case of a possibilistic fuzzy set, the grade of membership is the degree of possibility. In the case of a veristic fuzzy set, the grade of membership is the degree of verity (truth). Unless stated to the contrary, a fuzzy set is assumed to be possibilistic.

(d) *Usuality* ($r = \mathrm{u}$)

$$X \text{ isu } R.$$

The usuality constraint presupposes that X is a random variable, and that probability of the event $\{X \text{ isu } R\}$ is usually, where usually plays the role of a fuzzy probability which is a fuzzy number [13]. For example.

$$X \text{ isu small}$$

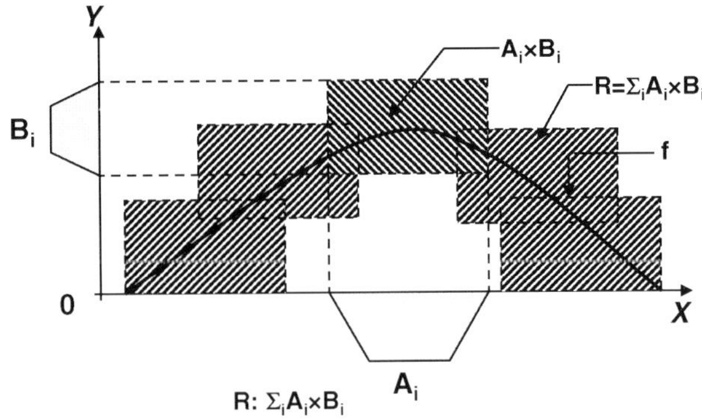

Fig. 4. Fuzzy graph.

means that "usually X is small" or, equivalently,

$$\text{Prob}\{X \text{ is small}\} \text{ is usually.}$$

In this expression, small may be interpreted as the usual value of X. The concept of a usual value has the potential of playing a significant role in decision analysis, since it is more informative than the concept of expected value.

(e) *Random-set* ($r = \text{rs}$)
In

$$X \text{ isrs } R,$$

X is a fuzzy-set-valued random variable and R is a fuzzy random set.

(f) *Fuzzy-graph* ($r = \text{fg}$)
In

$$X \text{ isfg } R,$$

X is a function, f, and R is a fuzzy graph [27] which constrains f (Figure 4). A fuzzy graph is a disjunction of Cartesian granules expressed as

$$R = A_1 \times B_1 + \cdots + A_n \times B_n,$$

where the A_i and B_i, $i = 1, \ldots, n$, are fuzzy subsets of the real line, and $\times$ is the Cartesian product. A fuzzy graph is frequently described as a collection of fuzzy if–then rules [26, 36,20,2].

$$R\text{: if } X \text{ is } A_1 \text{ then } Y \text{ is } B_1, \quad i = 1, \ldots, n.$$

The concept of a fuzzy-graph constraint plays an important role in applications of fuzzy logic [2,10,12].

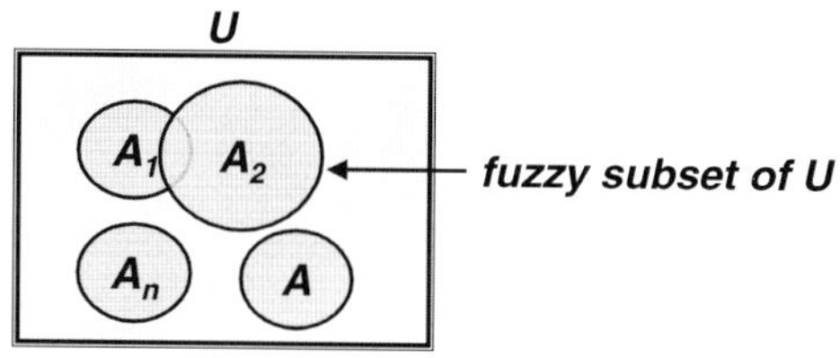

Fig. 5. Bimodal distribution. The A_i are fuzzy subsets of U.

(g) *Bimodal* (r = bm)

In the bimodal constraint,

$$X \text{ isbm } R,$$

R is a bimodal distribution of the form

$$R\colon \sum_i P_i \backslash A_i, \quad i = 1, \ldots, n,$$

which means that Prob(X is A_i) is P_i [39].

To clarify the meaning of a bimodal distribution it is expedient to start with an example. I am considering buying Ford stock. I ask my stockbroker, "What is your perception of the near-term prospects for Ford stock?" He tells me, "A moderate decline is very likely; a steep decline is unlikely; and a moderate gain is not likely." My question is: What is the probability of a large gain?

Information provided by my stock broker may be represented as a collection of ordered pairs:

- Price: ((unlikely, steep.decline), (very.likely, moderate.decline), (not.likely, moderate.gain)).

In this collection, the second element of an ordered pair is a fuzzy event or, generally, a possibility distribution, and the first element is a fuzzy probability. The expression for Price is an example of a bimodal distribution.

The importance of the concept of a bimodal distribution derives from the fact that in the context of human-centric systems, most probability distributions are bimodal. Bimodal distributions can assume a variety of forms. The principal types are Type 1, Type 2 and Type 3 [29,30]. Type 1, 2 and 3 bimodal distributions have a common framework but differ in important detail (Figure 5). A bimodal distribution may be viewed as an important generalization of standard probability distribution. For this reason, bimodal distributions of Type 1, 2, 3 are discussed in greater detail in the following.

- Type 1 (default): X is a random variable taking values in U
  $A_1, \ldots, A_n, A$ are events (fuzzy sets)
  $p_i = \text{Prob}(X \text{ is } A_i)$, Prob($X$ is A_i) is P_i, $i = 1, \ldots, n$,
  $\sum_i p_i$ is unconstrained

  BD: bimodal distribution: $((P_1, A_1), \ldots, (P_n, A_n))$
  or, equivalently,

$$X \text{ isbm } (P_1 \backslash A_1 + \cdots + P_n \backslash A_n)$$

PROBLEM. What is the probability, p, of A? In general, this probability is fuzzy-set-valued.

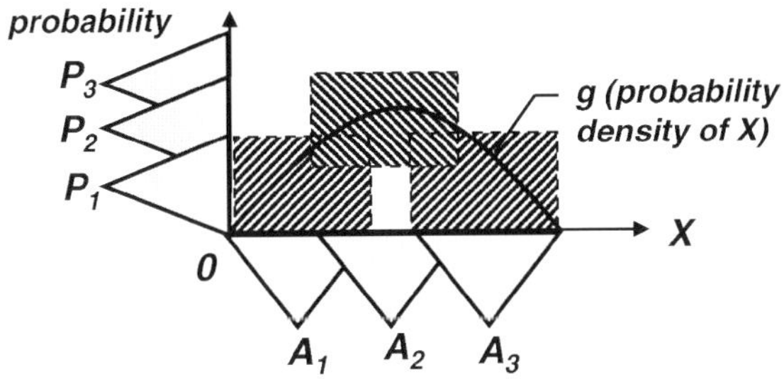

Fig. 6. Basic bimodal distribution.

A special case of bimodal distribution of Type 1 is the basic bimodal distribution (BBD). In BBD, X is a real-valued random variable, and X and P are granular (Figure 6).

- Type 2 (fuzzy random set): X is a fuzzy-set-valued random variable with values $A_1, \ldots, A_n$ (fuzzy sets)
  $p_i = \text{Prob}(X = A_i)$, $\text{Prob}(X \text{ is } A_i)$ is P_i, $i = 1, \ldots, n$

  BD: X isrs $(P_1 \backslash A_1 + \cdots + P_n \backslash A_n)$

  $$\sum_i P_i = 1,$$

  where the P_i are granular probabilities.

PROBLEM. What is the probability, P, of A? P is not definable. What are definable are (a) the expected value of the conditional possibility of A given BD, and (b) the expected value of the conditional necessity of A given BD.

- Type 3 (Dempster–Shafer) [22]: X is a random variable taking values $X_1, \ldots, X_n$ with probabilities $p_1, \ldots, p_n$.

  $$X_i \text{ is a random variable taking values in } A_i, \quad i = 1, \ldots, n$$

  Probability distribution of X_i in A_i, $i = 1, \ldots, n$, is not specified.

PROBLEM. What is the probability, p, that X is in A? Because probability distributions of the X_i in the A_i are not specified, p is interval-valued. What is important to note is that the concepts of upper and lower probabilities break down when the A_i are fuzzy sets.

NOTE. In applying Dempster–Shafer theory, it is important to check on whether the data fit Type 3 model. In many cases, the correct model is Type 1 rather than Type 3.

The importance of bimodal distributions derives from the fact that in many realistic settings a bimodal distribution is the best approximation to our state of knowledge. An example is assessment of degree of relevance, since relevance is generally not well defined. If I am asked to assess the degree of relevance of a book on knowledge representation to summarization, my state of knowledge about the book may not be sufficient to justify an answer such as 0.7. A better approximation to my state of knowledge may be "likely to be high." Such an answer is an instance of a bimodal distribution.

What is the expected value of a bimodal distribution? This question is considered in Section 4.

(h) *Group* ($r = g$)

In

$$X \text{ isg } R,$$

X is a group variable, $G[A]$, and R is a group constraint on $G[A]$. More specifically, if X is a group variable of the form

$$G[A]\colon \text{Name}_1/A_1 + \cdots + \text{Name}_n/A_n$$

or

$$G[A]\colon \sum_i \text{Name}_i/A_i, \quad \text{for short}, \; i = 1, \ldots, n,$$

then R is a constraint on the A_i. To illustrate, if we have a group of n Swedes, with Name_i being the name of ith Swede, and A_i being the height of Name_i, then the proposition "most Swedes are tall," is a constraint on the A_i which may be expressed as [40]

$$\frac{1}{n} \sum \text{Count}(\textit{tall.Swedes}) \text{ is most}$$

or, more explicitly,

$$\frac{1}{n}\big(\mu_{\text{tall}}(A_1) + \cdots + \mu_{\text{tall}}(A_n)\big) \text{ is most},$$

where most is a fuzzy quantifier which is interpreted as a fuzzy number.

4.1.2. *Operations on generalized constraints*

There are many ways in which generalized constraints may be operated on. The basic operations – expressed in symbolic form – are the following.

(a) *Conjunction*

$$\frac{\begin{array}{l} X \text{ isr } R \\ Y \text{ iss } S \end{array}}{(X, Y) \text{ ist } T}.$$

EXAMPLE (possibilistic constraints).

$$\frac{\begin{array}{l} X \text{ is } R \\ Y \text{ is } S \end{array}}{(X, Y) \text{ is } R \times S},$$

where $\times$ is the Cartesian product.

EXAMPLE (probabilistic/possibilistic).

$$\frac{\begin{array}{l} X \text{ isp } R \\ (X, Y) \text{ is } S \end{array}}{(X, Y) \text{ isrs } T}.$$

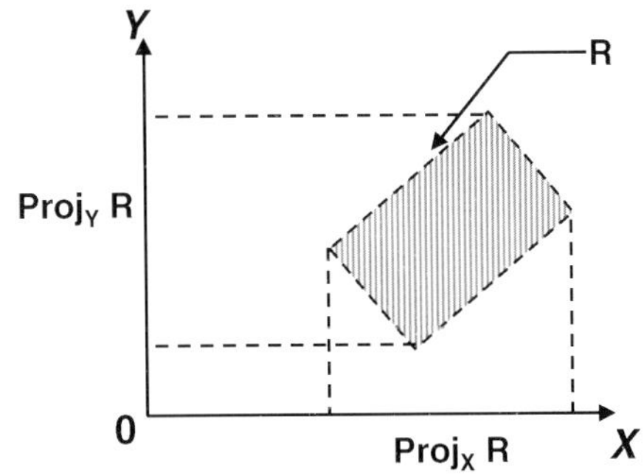

Fig. 7. Projection.

In this example, if S is a fuzzy relation then T is a fuzzy random set. What is involved in this example is a conjunction of a probabilistic constraint and a possibilistic constraint. This type of probabilistic/possibilistic constraint plays a key role in the Dempster–Shafer theory of evidence [22], and in its extension to fuzzy sets and fuzzy probabilities [29,30].

EXAMPLE (possibilistic/probabilistic).

$$\frac{\begin{array}{l} X \text{ is } R \\ (X, Y) \text{ isp } S \end{array}}{Y/X \text{ isp } T}.$$

This example, which is a dual of the proceeding example, is an instance of conditioning.

(b) *Projection (possibilistic) (Figure 7)*

$$\frac{(X, Y) \text{ is } R}{X \text{ is } S},$$

where X takes values in $U = \{u\}$; Y takes values in $V = \{v\}$; and the projection

$$S = \mathrm{Proj}_X R,$$

is defined as

$$\mu_S(u) = \mu_{\mathrm{Proj}_X R}(u) = \max_v \mu_R(u, v),$$

where μ_R and μ_S are the membership functions of R and S, respectively.

(c) *Projection (probabilistic)*

$$\frac{(X, Y) \text{ isp } R}{X \text{ isp } S},$$

where X and Y are real-valued random variables, and R and S are the probability distributions of (X, Y) and X, respectively. The probability density function of S, p_S, is related to that of R, p_R, by the familiar equation

$$p_S(u) = \int p_R(u, v)\, \mathrm{d}v$$

with the integral taken over the real line.

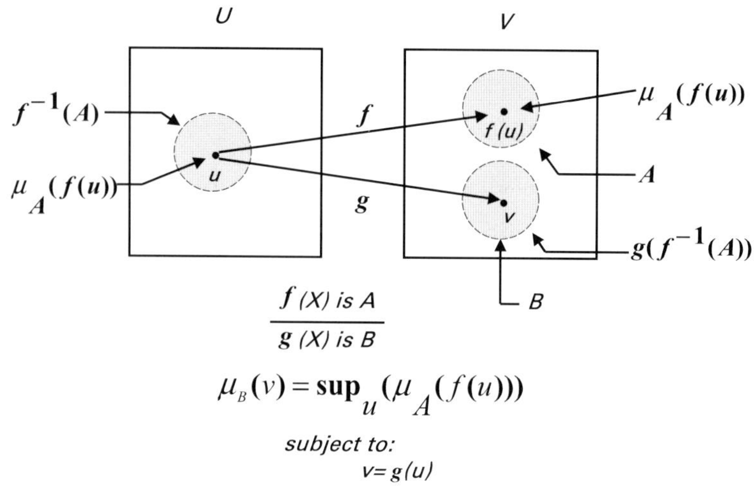

Fig. 8. Extension principle.

(d) *Propagation*

$$\frac{f(X) \text{ isr } R}{g(X) \text{ iss } S},$$

where f and g are functions or functionals.

EXAMPLE (possibilistic constraints) (Figure 8).

$$\frac{f(X) \text{ is } R}{g(X) \text{ is } S},$$

where R and S are fuzzy sets. In terms of the membership function of R, the membership function of S is given by the solution of the variational problem

$$\mu_S(v) = \sup_u \bigl(\mu_R\bigl(f(u)\bigr)\bigr)$$

subject to

$$v = g(u).$$

NOTE. The constraint propagation rule described in this example is the well-known extension principle of fuzzy logic [24,28]. Basically, this principle provides a way of computing the possibilistic constraint on $g(X)$ given a possibilistic constraint on $f(X)$.

4.1.3. *Primary constraints, composite constraints and standard constraints*

Among the principal generalized constraints there are three that play the role of primary generalized constraints. They are:

Possibilistic constraint: X is R

Probabilistic constraint: X isp R

and

Veristic constraint: X isv R

A special case of primary constraints is what may be called standard constraints: bivalent possibilistic, probabilistic and bivalent veristic. Standard constraints form the basis for the conceptual framework of bivalent logic and probability theory.

A generalized constraint, GC, is composite if it can be generated from other generalized constraints through conjunction, and/or projection, and/or constraint propagation, and/or qualification and/or possibly other operations. For example, a random-set constraint may be viewed as a conjunction of a probabilistic constraint and either a possibilistic or veristic constraint. The Dempster–Shafer theory of evidence is, in effect, a theory of possibilistic random-set constraints. The derivation graph of a composite constraint defines how it can be derived from primary constraints.

The three primary constraints – possibilistic, probabilistic and veristic – are closely related to a concept which has a position of centrality in human cognition – the concept of partiality. In the sense used here, partial means: a matter of degree or, more or less equivalently, fuzzy. In this sense, almost all human concepts are partial (fuzzy). Familiar examples of fuzzy concepts are: knowledge, understanding, friendship, love, beauty, intelligence, belief, causality, relevance, honesty, mountain and, most important, truth, likelihood and possibility. Is a specified concept, C, fuzzy? A simple test is: If C can be hedged, then it is fuzzy. For example, in the case of relevance, we can say: very relevant, quite relevant, slightly relevant, etc. Consequently, relevance is a fuzzy concept.

The three primary constraints may be likened to the three primary colors: red, blue and green. In terms of this analogy, existing theories of uncertainty may be viewed as theories of different mixtures of primary constraints. For example, the Dempster–Shafer theory of evidence is a theory of a mixture of probabilistic and possibilistic constraints. The Generalized Theory of Uncertainty (GTU) [42] embraces all possible mixtures. In this sense the conceptual structure of GTU accommodates most, and perhaps all, of the existing theories of uncertainty.

4.1.4. *The generalized constraint language and standard constraint language*

A concept which has a position of centrality in PNL is that of Generalized Constraint Language (GCL). Informally, GCL is the set of all generalized constraints together with the rules governing syntax, semantics and generation. Simple examples of elements of GCL are:

$$\big((X, Y) \text{ isp } A\big) \wedge (X \text{ is } B)$$
$$(X \text{ isp } A) \wedge \big((X, Y) \text{ isv } B\big)$$
$$\mathrm{Proj}_Y\big((X \text{ is } A) \wedge \big((X, Y) \text{ isp } B\big)\big),$$

where $\wedge$ is conjunction.

A very simple example of a semantic rule is:

$$(X \text{ is } A) \wedge (Y \text{ is } B) \longrightarrow \mathrm{Poss}(X = u, Y = v) = \mu_A(u) \wedge \mu_B(v),$$

where u and v are generic values of X, Y, and μ_A and μ_B are the membership functions of A and B, respectively.

In principle, GCL is an infinite set. However, in most applications only a small subset of GCL is likely to be needed.

In PNL, the set of all standard constraints together with the rules governing syntax, semantics and generation constitute the Standard Constraint Language (SCL). SCL is a subset of GCL.

4.2. *The concept of cointensive precisiation*

As was pointed out already, much of world knowledge and web knowledge is expressed in a natural language. For this reason, mechanization of natural language understanding is of direct relevance to enhancement of web intelligence.

In recent years, considerable progress has been made in areas of computational linguistics which relate to mechanization of natural language understanding. But what is widely unrecognized is that there is a fundamental limitation to what can be achieved through the use of commonly-employed methods of meaning representation. The aim of what follows is, first, to highlight this limitation and, second, to suggest ways of removing it.

To understand the nature of the limitation, two facts have to be considered. First, as was pointed out earlier, a natural language, NL, is basically a system for describing perceptions; and second, perceptions are intrinsically imprecise, reflecting the bounded ability of human sensory organs, and ultimately the brain, to resolve detail and store information.

A direct consequence of imprecision of perceptions is semantic imprecision of natural languages. Semantic imprecision of natural languages is not a problem for humans, but is a major problem for machines.

To clarify the issue, let p be a proposition, concept, question or command. For p to be understood by a machine, it must be precisiated, that is, expressed in a mathematically well-defined language. A precisiated form of p, $\text{Pre}(p)$, will be referred to as a precisiand of p and will be denoted as p^*. The object of precisiation, p, will be referred to us precisiend.

To precisiate p we can employ a number of meaning-representation languages, e.g., Prolog, predicate logic, semantic networks, conceptual graphs, LISP, SQL, etc. The commonly-used meaning-representation languages are bivalent, i.e., are based on bivalent logic. Are we moving in the right direction when we employ such languages for mechanization of natural language understanding? The answer is: No. The reason relates to an important issue which we have not addressed: cointension of p^*, with intension used in its logical sense as attribute-based meaning. More specifically, cointension is a measure of the goodness of fit of the intension of a precisiand, p^*, to the intended intension of precisiend, p. Thus, cointension is a desideratum of precisiation. What this implies is that mechanization of natural language understanding requires more than precisiation – it requires cointensive precisiation. Note that definition is a form of precisiation. In plain words, a definition is cointensive if its meaning is a good fit to the intended meaning of the definiendum.

Here is where the fundamental limitation which was alluded to earlier comes into view. In a natural language, NL, most p's are fuzzy, that is, are in one way or another, a matter of degree. Simple examples: propositions "most Swedes are tall" and "overeating causes obesity;" concepts "mountain" and "honest;" question "is Albert honest?" and command "take a few steps."

Employment of commonly-used meaning-representation languages to precisiate a fuzzy p leads to a bivalent (crisp) precisiand p^*. The problem is that, in general, a bi-

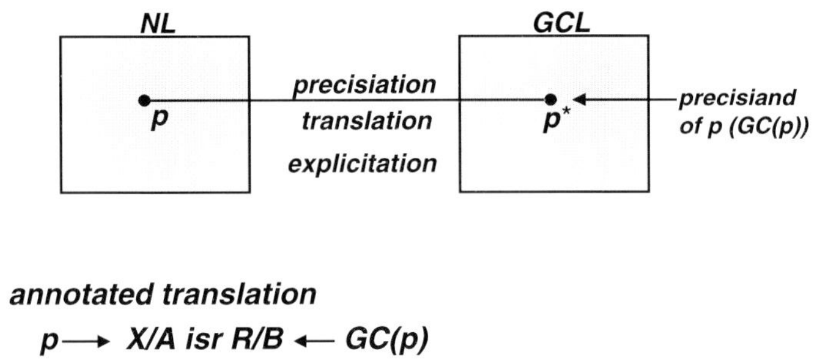

Fig. 9. Precisiation = translation into GCL.

valent p^* is not cointensive. As a simple illustration, consider the concept of recession. The standard definition of recession is: A period of general economic decline; specifically, a decline in GDP for two or more consecutive quarters. Similarly, a definition of bear market is: We classify a bear market as a 30 percent decline after 50 days, or a 13 percent decline after 145 days. (Robert Shuster, Ned Davis Research.) Clearly, neither definition is cointensive. Another example is the classical definition of stability. Consider a ball of diameter D which is placed on an open bottle whose mouth is of diameter d. If D is somewhat larger than d, the configuration is stable: Obviously, as D increases, the configuration becomes less and less stable. But, according to Lyapounov's bivalent definition of stability, the configuration is stable for all values of D greater than d. This contradiction is characteristic of crisp definitions of fuzzy concepts – a well-known example of which is the Greek sorites (heap) paradox. The magnitude of the problem becomes apparent when we consider that many concepts in scientific theories are fuzzy, but are defined and treated as if they are crisp. This is particularly true in fields in which the concepts which are defined are descriptions of perceptions.

To remove the fundamental limitation, bivalence must be abandoned. Furthermore, new concepts, ideas and tools must be developed and deployed to deal with the issues of cointensive precisiation, definability and deduction. The principal tools are Precisiated Natural Language (PNL); Protoform Theory (PFT); and the Generalized Theory of Uncertainty (GTU). These tools form the core of what may be called the Computational Theory of Precisiation of Meaning (CTPM). The centerpiece of CTPM is the concept of a generalized constraint [35].

The concept of a generalized constraint plays a key role in CTPM by providing a basis for precisiation of meaning. More specifically, if p is a proposition or a concept, its precisiand, Pre(p), is represented as a generalized constraint, GC. Thus, Pre(p) $=$ GC. In this sense, the concept of a generalized constraint may be viewed as a bridge from natural languages to mathematics (Figure 9).

Representing precisiands of p as elements of GCL is the pivotal idea in CTPM. Each precisiand is associated with the degree to which it is cointensive with p. Given p, the problem is that of finding those precisiands which are cointensive, that is, have a high degree of cointension. If p is a fuzzy proposition or concept, then in general there are no cointensive precisiands in SCL (Figure 10).

In CTPM, a refinement of the concept of precisiation is needed. First, a differentiation is made between v-precision (precision in value) and m-precision (precision in meaning) (Figure 11). For example, proposition p: X is 5, is both v-precise and m-precise; p: X

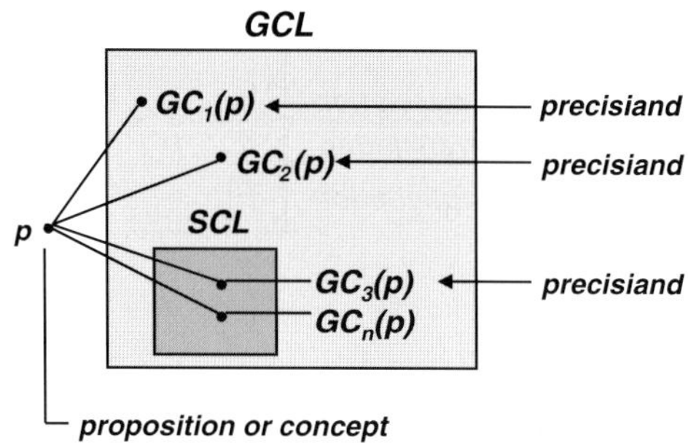

Fig. 10. Precisiands of p.

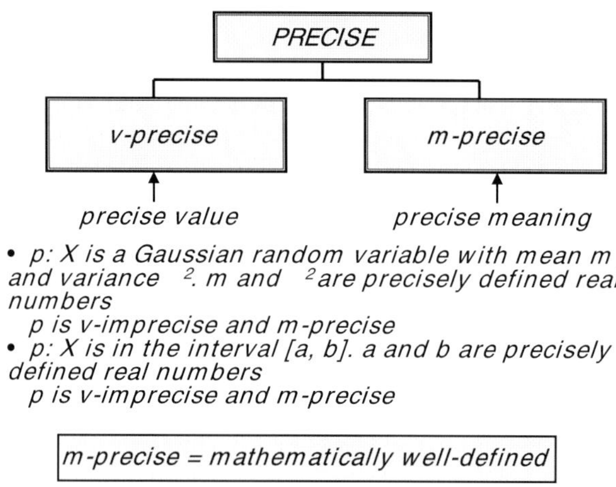

Fig. 11. Two meanings of precise.

is between 5 and 7, is v-imprecise and m-precise; and p: X is small, is both v-imprecise and m-imprecise; however, p can be m-precisiated by defining small as a fuzzy set or a probability distribution. A perception is v-imprecise and its description is m-imprecise. PNL makes it possible to m-precisiate descriptions of perceptions.

Granulation of a variable, e.g., representing the values of age as young, middle-aged and old, may be viewed as a form of v-imprecisiation. Granulation plays an important role in human cognition by serving as a means of (a) exploiting a tolerance for imprecision through omission of irrelevant information; (b) lowering precision and thereby lowering cost; and (c) facilitating understanding and articulation. In fuzzy logic, granulation is m-precisiated through the use of the concept of a linguistic variable.

Further refinement of the concept of precisiation relates to two modalities of m-precisiation: (a) human-oriented, denoted as mh-precisiation; and (b) machine-oriented, denoted as mm-precisiation (Figure 12). Unless stated to the contrary, in CTPM, precisiation should be understood as mm-precisiation.

In a bimodal dictionary or lexicon, the first entry, p, is a concept or proposition; the second entry, p^*, is mh-precisiand of p; and the third entry is mm-precisiand of p. To illustrate,

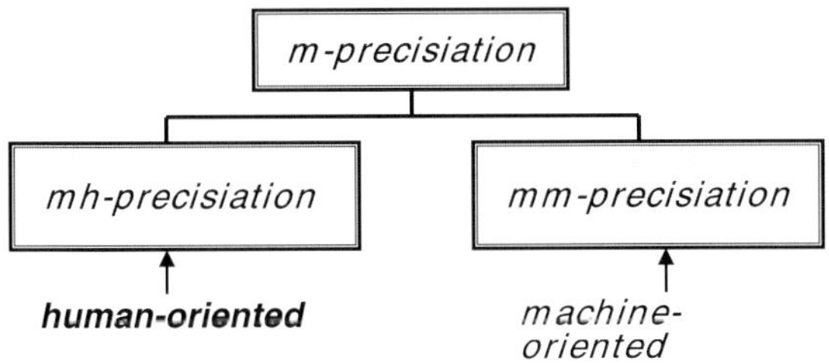

Fig. 12. Modalities of m-precisiation.

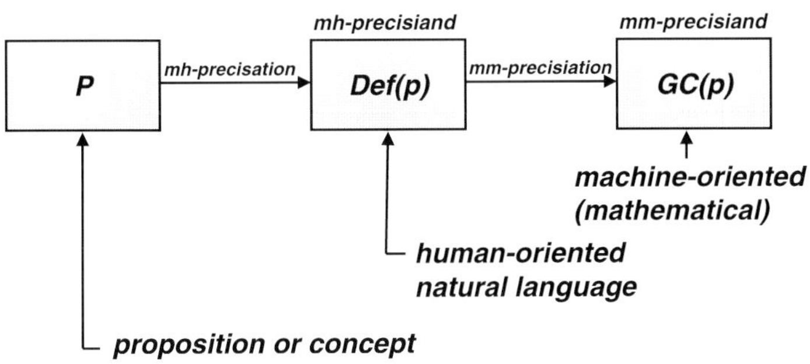

Fig. 13. Bimodal lexicon (PNL).

the entries for recession might read: *mh*-precisiand – a period of general economic decline; and *mm*-precisiand – a decline in GDP for two or more consecutive quarters (Figure 13).

There is a simple analogy which helps to understand the meaning of cointensive precisiation. Specifically, a proposition, p, is analogous to a system, S; precisiation is analogous to modelization; a precisiand, expressed as a generalized constraint, GC(p), is analogous to a model, $M(S)$, of S; test-score function is analogous to input–output relation; cointensive precisiand is analogous to well-fitting model; GCL is analogous to the class of all fuzzy-logic-based systems; and SCL is analogous to the subclass of all bivalent-logic-based systems (Figure 14). To say that, in general, a cointensive definition of a fuzzy concept cannot be formulated within the conceptual structure of bivalent logic and probability theory, is similar to saying that, in general, a linear system cannot be a well-fitting model of a nonlinear system.

Ramifications of the concept of cointensive precisiation extend well beyond mechanization of natural language understanding. A broader basic issue is validity of definitions in scientific theories, especially in the realms of human-oriented fields such as law, economics, medicine, psychology and linguistics. More specifically, the concept of cointensive precisiation calls into question the validity of many of the existing definitions of basic concepts – among them the concepts of causality, relevance, independence, stability, complexity and optimality.

Translation of p into GCL is made more transparent though annotation. To illustrate,

(a) p: Monika is young $\longrightarrow$ X/Age(Monika) is R/young
(b) p: It is likely that Monika is young $\longrightarrow$ Prob(X/Age(Monika) is R/young) is S/likely

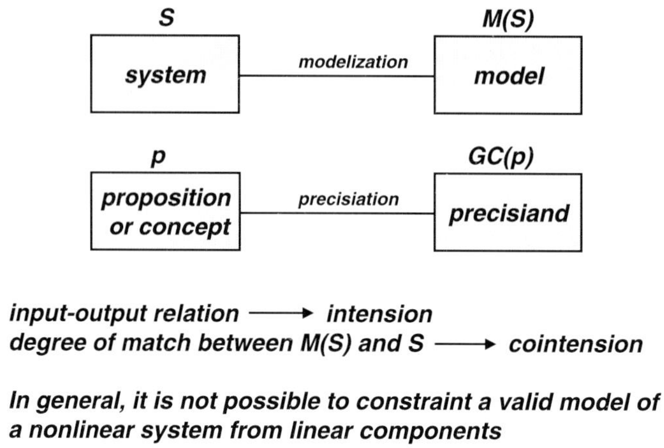

Fig. 14. Analogy between precisiation and modelization.

NOTE. Example (b) is an instance of probability qualification.

More concretely, let $g(u)$ be the probability density function of the random variable, Age(Monika). Then, with reference to our earlier discussion of probability qualification, we have

$$\text{Prob}\big(\text{Age(Monika) is young}\big) \text{ is likely} \longrightarrow \int_0^{100} g(u)\mu_{\text{young}}(u)\,\mathrm{d}u \text{ is likely}$$

or, in annotated form,

$$\mathrm{GC}(g) = X \Big/ \int_0^{100} g(u)\mu_{\text{young}}(u)\,\mathrm{d}u \text{ is } R/\text{likely}.$$

The test-score of this constraint on g is given by

$$ts(g) = \mu_{\text{likely}}\left(\int_0^{100} g(u)\mu_{\text{young}}(u)\,\mathrm{d}u\right).$$

(c) p: Most Swedes are tall

Following (b), let $h(u)$ be the count density function of Swedes, meaning that

$h(u)\,\mathrm{d}u =$ fraction of Swedes whose height lies in the interval $[u, u + \mathrm{d}u]$.

Assume that height of Swedes lies in the interval $[a, b]$. Then,

$$\text{fraction of tall Swedes:} \quad \int_a^b h(u)\mu_{\text{tall}}(u)\,\mathrm{d}u \text{ is most.}$$

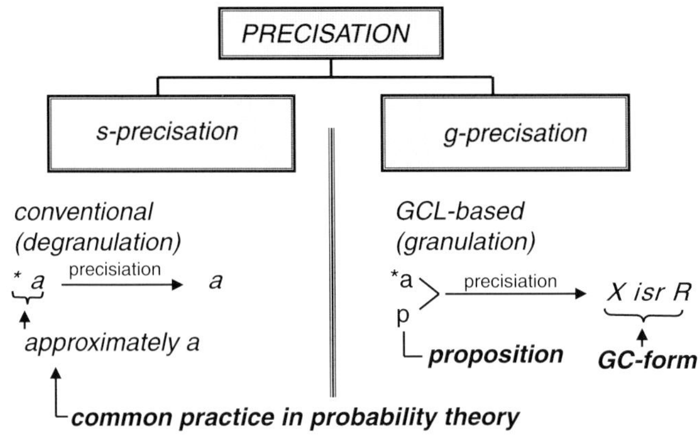

Fig. 15. *s*-precisiation and *g*-precisiation.

Interpreting this relation as a generalized constraint on h, the test-score may be expressed as

$$ts(h) = \mu_{\text{most}}\left(\int_0^h h(u)\mu_{\text{tall}}(u)\,\mathrm{d}u\right).$$

In summary, precisiation of "Most Swedes are tall" may be expressed as the generalized constraint.

$$\text{Most Swedes are tall} \longrightarrow \text{GC}(h) = \mu_{\text{most}}\left(\int_a^b h(u)\mu_{\text{tall}}(u)\,\mathrm{d}u\right).$$

An important application of the concept of precisiation relates to precisiation of propositions of the form "X is approximately a," where a is a real number. How can "approximately a," or $^*a$ for short, be precisiated? In other words, how can the uncertainty associated with the value of X which is described as $^*a$, be defined precisely?

There is a hierarchy of ways in which this can be done. The simplest is to define $^*a$ as a. This mode of precisiation will be referred to as singular precisiation, or s-precisiation, for short (Figure 15) s-precisiation is employed very widely, especially in probabilistic computations in which an imprecise probability, $^*a$, is computed with as if it were an exact number, a.

The other ways (Figures 15, 16) will be referred to as granular precisiation, or g-precisiation, for short. In g-precisiation, $^*a$ is treated as a granule. What we see is that various modes of precisiating $^*a$ are instances of the generalized constraint.

The concept of precisiation has an inverse – the concept of imprecisiation, which involves replacing a with $^*a$, with the understanding that $^*a$ is not unique.

Imprecisiation has a negative connotation. In fact, imprecisiation serves an important purposes. More specifically, consider a proposition p of the form

$$p\colon\ X \text{ is } V,$$

where X is a variable and V is its value. X may assume a variety of forms. In particular, X may be a real-valued variable, an n-ary variable, a function or a relation. The value, V, is

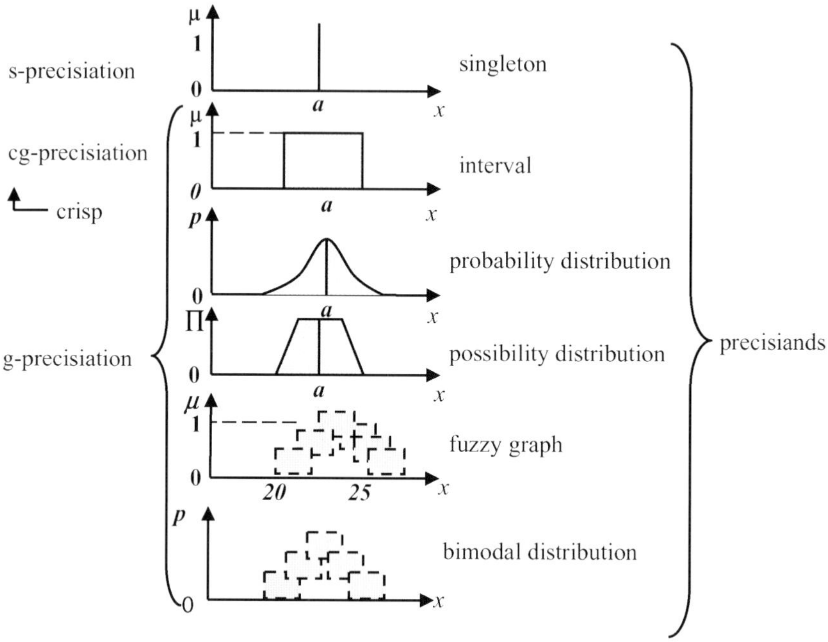

Fig. 16. Precisiands of "X is approximately a."

v-precise if it is singular, that is, V is a singleton. V is v-imprecise if it is granular. In this framework, v-imprecisiation may be interpreted as a transition from singular to granular value of V.

v-imprecisiation is forced (necessary) when the value of V is not known precisely. v-imprecisiation is deliberate (optional) if there is no need for V to be known precisely. In this case, what may be called v-imprecisiation principle comes into play.

v-imprecisiation principle: Precision carries a cost. If there is a tolerance for imprecision, exploit it by employing v-imprecisiation to achieve lower cost, robustness, tractability, decision-relevance and higher level of confidence.

A word about confidence. If V is uncertain, the confidence in p, Con(p), may be defined as the probability that p is true. Generally, v-imprecisiation of V serves to increase Con(p). For example, Con(Carol is young) $>$ Con(Carol is 23). Thus, as a rule, confidence increases when specificity decreases.

An important example is granulation. In fuzzy logic, granulation may be interpreted as v-imprecisiation followed by *mm*-precisiation. In this perspective, the concept of granulation – in combination with the associated concept of a linguistic variable – may be viewed as one of the major contributions of fuzzy logic.

A basic problem which relates to imprecisiation is the following. Assume for simplicity that we have two linear equations involving real-valued coefficients and real-valued variables:

$$a_{11}X + a_{12}Y = b_1,$$
$$a_{21}X + a_{22}Y = b_2.$$

Solutions of these equations read,

$$X = \frac{a_{22}b_1 - a_{12}b_2}{a_{11}a_{22} - a_{12}a_{21}},$$
$$Y = \frac{a_{11}b_2 - a_{21}b_1}{a_{11}a_{22} - a_{12}a_{21}}.$$

Now suppose that we imprecisiate the coefficients, replacing, a_{ij} with ${}^*a_{ij}$, $i, j = 1, 2$, and replacing b_i with ${}^*b_i$, $i = 1, 2$. How can we solve these equations when imprecisiated coefficients are defined as generalized constraints?

There is no general answer to this question. Assuming that all coefficients are defined in the same way, the method of solution will depend on the modality of the constraint. For example, if the coefficients are interval-valued, the problem falls within the province of interval analysis [17]. If the coefficients are fuzzy-interval-valued, the problem falls within the province of the theory of relational equations [8,7]. And if the coefficients are real-valued random variables, we are dealing with the problem of solution of stochastic equations. In general, solution of a system of equations with imprecisiated coefficients may present complex problems.

One complication is the following. If (a) we solve the original equations, as we have done above; (b) imprecisiate the coefficients in the solution; and (c) employ the extension principle to complete X and Y, will we obtain solutions of imprecisiated equations? The answer, in general, is: No.

Nevertheless, when we are faced with a problem which we do not know how to solve correctly, we proceed as if the answer is: Yes. This common practice may be described as Precisiation/Imprecisiation Principle which is defined in the following.

4.3. *Precisiation/imprecisiation principle (P/I principle)*

Informally, let f be a function or a functional. $Y = f(X)$, where X and Y are assumed to be imprecise, $Pr(X)$ and $Pr(Y)$ are precisiations of X and Y, and ${}^*Pr(X)$ and ${}^*Pr(Y)$ are imprecisiations of $Pr(X)$ and $Pr(Y)$, respectively. In symbolic form, the P/I principle may be expressed as

$$f(X) \;{}^*\!= {}^*f\big(Pr(X)\big),$$

where ${}^*\!=$ denotes "approximately equal," and ${}^*f$ is imprecisiation of f. In words, to compute $f(X)$ when X is imprecise, (a) precisiate X, (b) compute $f(Pr(X))$; and (c) imprecisiate $f(Pr(X))$. Then, usually, ${}^*f(Pr(X))$ will be approximately equal to $f(X)$. An underlying assumption is that approximations are commensurate in the sense that the closer $Pr(X)$ is to X, the closer $f(Pr(X))$ is to $f(X)$. This assumption is related to the concept of gradual rules of Dubois and Prade [9].

As an illustration, suppose that X is a real-valued function; f is the operation of differentiation, and ${}^*X$ is the fuzzy graph of X. Then, using the P/I principle, ${}^*f(X)$ will have the form shown in Figure 17. It should be underscored that imprecisiation is an imprecise concept.

Use of the P/I principle underlies many computations in science, engineering, economics and other fields. In particular, as was alluded to earlier, this applies to many computations in probability theory which involve imprecise probabilities. It should be emphasized that the

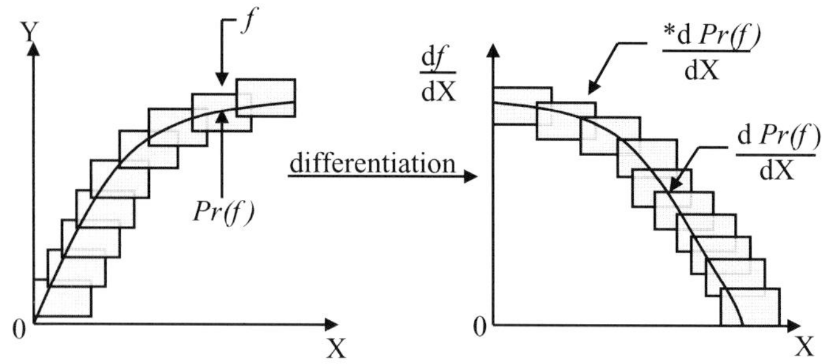

Fig. 17. Illustration of P/I principle.

P/I principle is neither normative (prescriptive) nor precise; it merely describes imprecisely what is common practice – without suggesting that common practice is correct.

4.3.1. *Precisiation of propositions*

In preceding discussion, we focused our attention on precisiation of propositions of the special form "X is $^*a$." In the following, we shall consider precisiation in a more general setting. In this setting, the concept of precisiation in PNL opens the door to a wide-ranging enlargement of the role of natural languages in scientific theories, especially in fields such as economics, law and decision analysis. Our discussion will be brief; details may be found in [40,41].

Within CTPM, precisiation of propositions – and the related issues of precisiation of questions, commands and concepts – falls within the province of PNL. As was stated earlier, the point of departure in PNL is representation of a precisiand of a proposition, p, as a generalized constraint.

$$p \longrightarrow X \text{ isr } R.$$

To illustrate precisiation of propositions and questions, it will be convenient to consider the examples which were discussed earlier in Section 4.

(a) *The Robert example*

p: Usually Robert returns from work at about 6 pm.

Q: What is the probability that Robert is home at about 6:15 pm?

Precisiation of p may be expressed as

p: Prob(Time(Return(Robert)) is *6:00 pm) is usually

where "usually" is a fuzzy probability.

Assuming that Robert stays home after returning from work, precisiation of q may be expressed as

$$q\text{: Prob}\big(\text{Time}\big(\text{Return(Robert)}\big) \text{ is } \leqslant \circ\, 6\text{:}15 \text{ pm}\big) \text{ is } A?$$

where $\circ$ is the operation of composition, and A is a fuzzy probability

(b) *The balls-in-box problem*

p_1: A box contains about 20 black and white balls

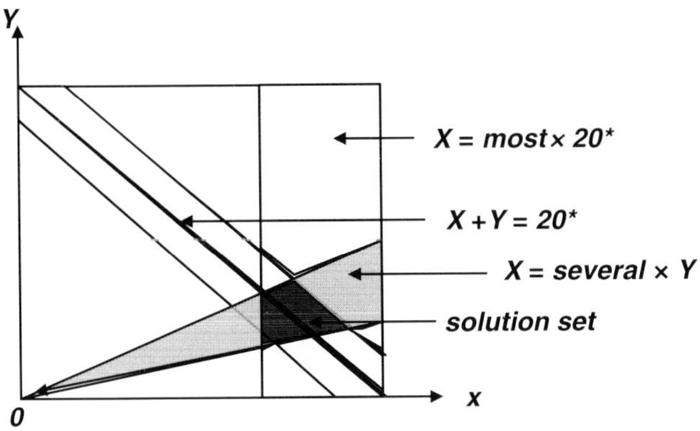

Fig. 18. Fuzzy integer programming.

p_2: Most are black
p_3: There are several times as many black balls as white balls
q_1: What is the number of white balls?
q_2: What is the probability that a ball drawn at random is white?

Let X be the number of black balls and let Y be the number of white balls. Then, in precisiated form, the statement of the problem may be expressed as:

$$\left.\begin{array}{l} p_1\colon (X+Y) \text{ is } {}^*20 \\ p_2\colon X \text{ is most } \times {}^*20 \\ p_3\colon X \text{ is several } \times Y \end{array}\right\} \text{data}$$

$$\left.\begin{array}{l} q_1\colon Y \text{ is } ?A \\ q_2\colon \dfrac{Y}{{}^*20} \text{ is } ?B \end{array}\right\} \text{questions}$$

where $Y/{}^*20$ is the granular probability that a ball drawn at random is white.

Solution of these equations reduces to an application of fuzzy integer programming (Figure 18).

(c) *The tall Swedes problem*

p: Most Swedes are tall.
Q: What is the average height of Swedes?
Q: How many Swedes are short?

As was shown earlier,

$$p\colon \text{Most Swedes are tall} \longrightarrow \int_a^b h(u)\mu_{\text{tall}}(u)\,du \text{ is most},$$

where h is the count density function.

Precisiations of q_1, and q_2 may be expressed as

$$q_1\colon \int_a^b uh(u)\,du \text{ is } ?A,$$

where A is a fuzzy number which represents the average height of Swedes, and

$$q_2: \int_a^b h(u)\mu_{\text{short}}(u)\,\mathrm{d}u \text{ is } ?B,$$

where μ_{short} is the membership function of short, and B is the fraction of short Swedes.

(d) *The partial existence problem*

X is a real number. I am uncertain about the value of X. What I know about X is:

p_1: X is much larger than approximately a,

p_2: X is much smaller than approximately b,

where a and b are real numbers, with $a < b$.

What is the value of X?

In this case, precisiations of data may be expressed as

p_1: X is much.larger $\circ\, {}^*a$

p_2: X is much smaller $\circ\, {}^*b$,

where $\circ$ is the operation of composition. Precisiation of the question is:

$$q: X \text{ is } ?A,$$

where A is a fuzzy number. The solution is immediate:

$$X \text{ is } (\text{much.larger} \circ {}^*a \wedge \text{much.smaller} \circ {}^*b),$$

when $\wedge$ is min or a t-norm. In this instance, depending on a and b, X may exist to a degree.

These examples point to an important aspect of precisiation. Specifically, to precisiate p we have to precisiate or, equivalently, calibrate its lexical constituents. For example, in the case of "Most Swedes are tall," we have to calibrate "most" and "tall." Likewise, in the case of the Robert example, we have to calibrate "about 6:00 pm," "about 6:15 pm" and "usually." In effect, we are composing the meaning of p from the meaning of its constituents. This process is in the spirit of Frege's principle of compositionality [32,33], Montague grammar [19] and the semantics of programming languages.

An important aspect of precisiation which will not be discussed here relates to precisiation of concepts. It is a deep-seated tradition in science to base definition of concepts on bivalent logic. In probability theory, for example, independence of events is a bivalent concept. But, in reality, independence is a matter of degree, i.e., is a fuzzy concept. PNL, used as a definition language, makes it possible, more realistically, to define independence and other bivalent concepts in probability theory as fuzzy concepts. For this purpose, when PNL is used as a definition language, a concept is first defined in a natural language and then its definition is precisiated through the use of PNL.

4.4. *The concept of a protoform*

Viewed in a broader perspective, what should be noted is that precisiation of meaning is not the ultimate goal – it is an intermediate goal. Once precisiation of meaning is achieved, the next goal is that of deduction from decision-relevant information. The ultimate goal is decision.

In CTPM, a concept which plays a key role in deduction is that of a protoform – an abbreviation for prototypical form. Informally, a protoform of an object is its abstracted summary. More specifically, a protoform is a symbolic expression which defines the deep semantic structure of an object such as a proposition, question, command, concept, scenario, or a system of such objects. In the following, our attention will be focused on protoforms of propositions, with PF(p) denoting a protoform of p (Figure 19). Abstraction has levels, just as summarization does. For this reason, an object may have a multiplicity of protoforms (Figure 20). Conversely, many objects may have the same protoform. Such objects are said to be protoform-equivalent, or PF-equivalent, for short. The set of protoforms of all precisiable propositions in NL, together with rules which govern propagation of generalized constraints, constitute what is called the Protoform Language (PFL).

EXAMPLES.

- Monika is young ⟶

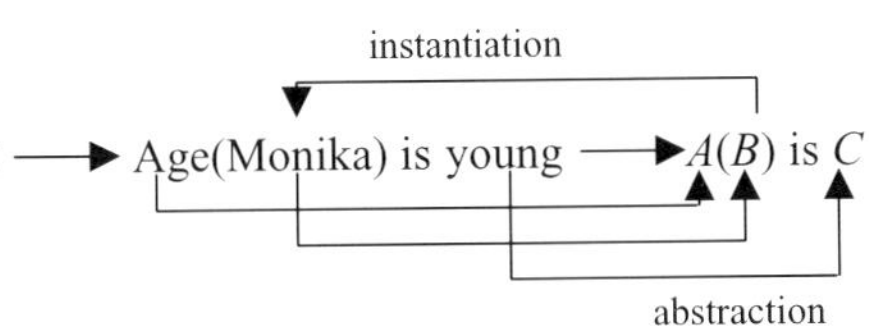

- Monika is much younger than Pat ⟶

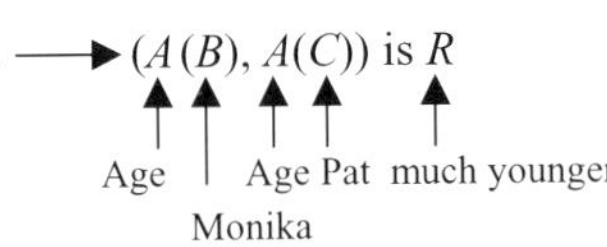

- distance between New York and Boston is about 200 mi ⟶ $A(B, C)$ is R
- usually Robert returns from work at about 6 pm ⟶

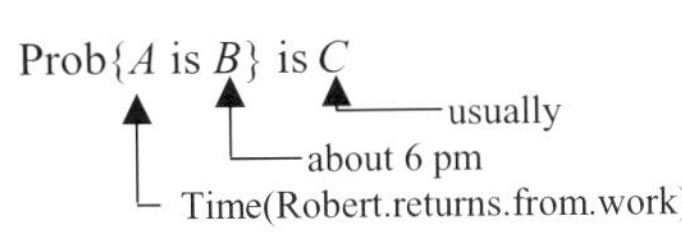

- Carol lives in a small city near San Francisco ⟶

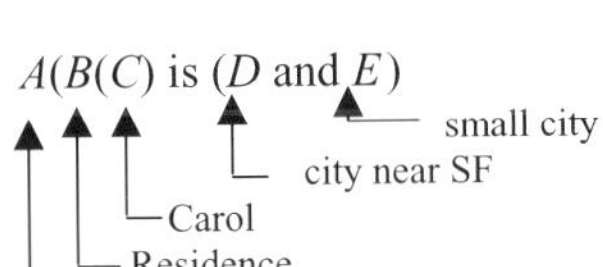

- 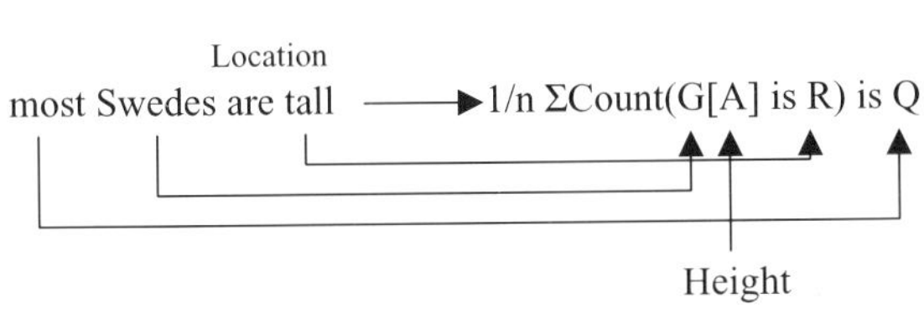

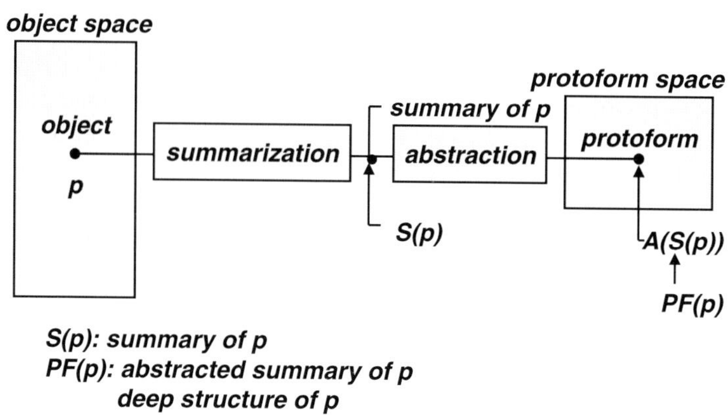

Fig. 19. Definition of protoform of p.

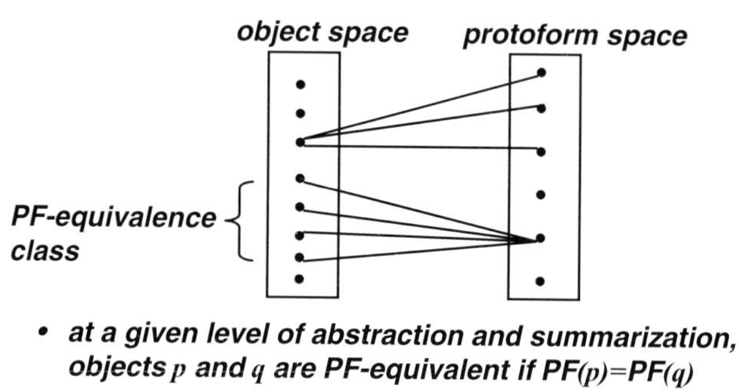

Fig. 20. Protoforms and PF-equivalence.

Alan has severe back pain. He goes to see a doctor. The doctor tells him that there are two options: (1) do nothing; and (2) do surgery. In the case of surgery, there are two possibilities: (a) surgery is successful, in which case Alan will be pain free; and (b) surgery is not successful, in which case Alan will be paralyzed from the neck down.

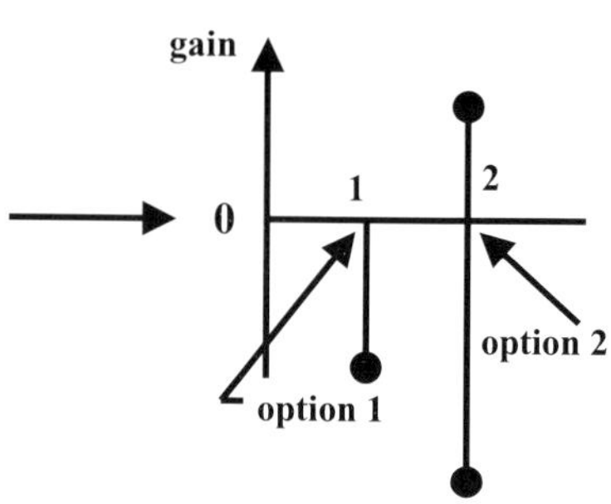

4.4.1. *Protoformal deduction*

The rules of deduction in CTPM are, basically, the rules which govern constraint propagation. In CTPM, such rules reside in the Deduction Database (DDB), Figure 21. The Deduction Database comprises a collection of agent-controlled modules and submodules, each of which contains rules drawn from various fields and various modalities of generalized constraints (Figure 22). A typical rule has a symbolic part, which is expressed in

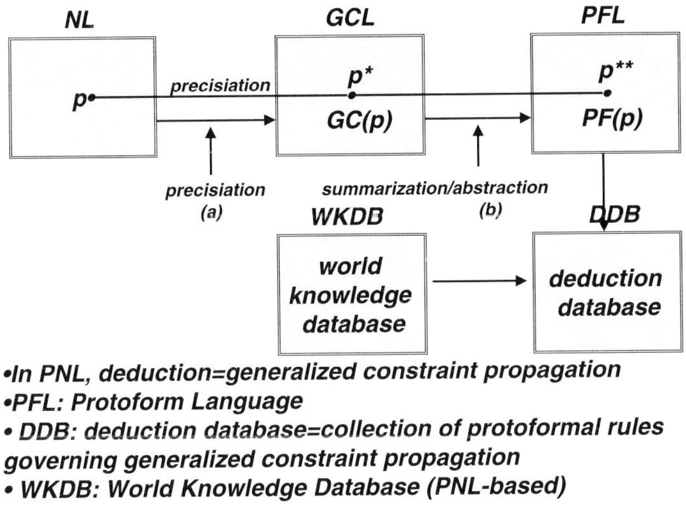

Fig. 21. Basic structure of PNL.

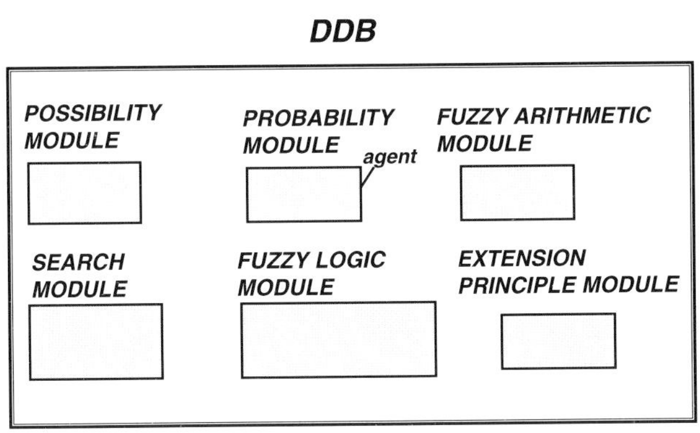

Fig. 22. Structure of deduction database.

terms of protoforms; and a computational part which defines the computation that has to be carried out to arrive at a conclusion. In what follows, we describe briefly some of the basic rules, and list a number of other rules without describing their computational parts. The motivation for doing so is to point to the necessity of developing a set of rules which is much more complete than the few rules which are used as examples in this section.

(a) *Computational rule of inference* [37]

Symbolic part	Computational part
X is A	
(X, Y) is B	$\mu_C(v) = \max_u\big(\mu_A(u) \wedge \mu_B(u, v)\big)$
Y is C	

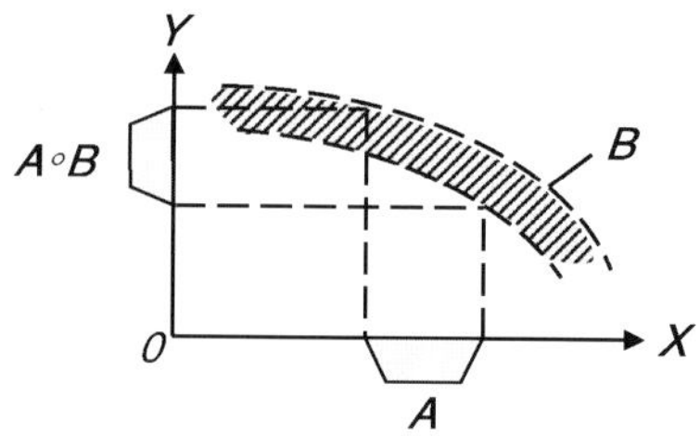

Fig. 23. Compositional rule of inference.

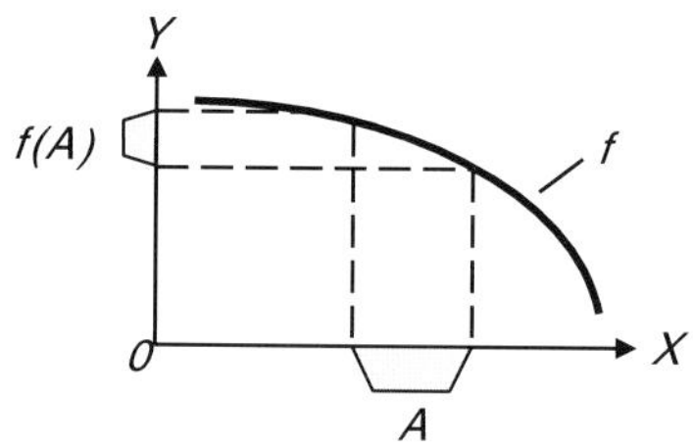

Fig. 24. Basic extension principle.

A, B and C are fuzzy sets with respective membership functions μ_A, μ_B, μ_C. $\wedge$ is min or t-norm (Figure 23).

(b) *Intersection/product syllogism* [32,33]

Symbolic part	Computational part
$Q_1 A$'s are B's $Q_2(A\&B)$'s are C's ——— $Q_3 A$'s are $(B\&C)$'s	$Q_3 = Q_1 * Q_2$

Q_1 and Q_2 are fuzzy quantifiers; A, B, C are fuzzy sets; $*$ is product in fuzzy arithmetic [13].

(c) *Basic extension principle* [24]

Symbolic part	Computational part
X is A ——— $f(X)$ is B	$\mu_B(v) = \sup_u(\mu_A(u))$ subject to $v = f(u)$

g is a given function or functional; A and B are fuzzy sets (Figure 24).

(d) *Extension principle* [38]

This is the principal rule governing possibilistic constraint propagation (Figure 8)

Symbolic part	Computational part
$\dfrac{f(X) \text{ is } A}{g(X) \text{ is } B}$	$\mu_B(v) = \sup\limits_u(\mu_B(f(u)))$ subject to $v = g(u)$

NOTE. The extension principle is a primary deduction rule in the sense that many other deduction rules are derivable from the extension principle. An example is the following rule.

(e) *Basic probability rule*

Symbolic part	Computational part
$\dfrac{\text{Prob}(X \text{ is } A) \text{ is } B}{\text{Prob}(X \text{ is C}) \text{ is } D}$	$\mu_D(v) = \sup\limits_r\left(\mu_B\left(\int_U \mu_A(u)r(u)\,du\right)\right)$

subject to

$$v = \int_U \mu_C(u)r(u)\,du$$

$$\int_U r(u)\,du = 1.$$

X is a real-valued random variable; A, B, C and D are fuzzy sets: r is the probability density of X; and $U = \{u\}$. To derive this rule, we note that

$$\text{Prob}(X \text{ is } A) \text{ is } B \longrightarrow \int_U r(u)\mu_A(u)\,du \text{ is } B$$

$$\text{Prob}(X \text{ is } C) \text{ is } D \longrightarrow \int_U r(u)\mu_C(u)\,du \text{ is } D$$

which are generalized constraints of the form

$$f(r) \text{ is } B$$

$$g(r) \text{ is } D.$$

Applying the extension principle to these expressions, we obtain the expression for D which appears in the basic probability rule.

(f) *Bimodal interpolation rule*

The bimodal interpolation rule is a rule which resides in the Probability module of DDB. With reference to Figure 25, the symbolic and computational parts of this rule are:

Symbolic

$$\frac{\text{Prob}(X \text{ is } A_i) \text{ is } P_i}{\text{Prob}(X \text{ is } A) \text{ is } Q}, \quad i = 1, \ldots, n$$

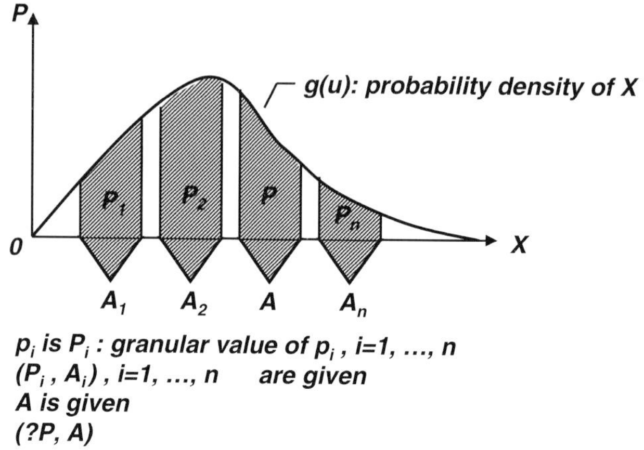

Fig. 25. Interpolation of a bimodal distribution.

Computational

$$\mu_Q(v) = \sup_r \left(\mu_{P_1} \left(\int_U \mu_{A_1}(u) r(u)\, \mathrm{d}u \right) \wedge \cdots \wedge \mu_{P_n} \left(\int_U \mu_{A_n}(u) r(u)\, \mathrm{d}u \right) \right)$$

subject to

$$v = \int_U \mu_A(u) r(u)\, \mathrm{d}u$$

$$\int_U r(u)\, \mathrm{d}u = 1$$

In this rule, X is a real-valued random variable; r is the probability density of X; and U is the domain of X.

NOTE. The probability rule is a special case of the bimodal interpolation rule.

What is the expected value, $E(X)$, of a bimodal distribution? The answer follows through application of the extension principle:

$$\mu_{E(X)}(v) = \sup_r \left(\mu_{P_1} \left(\int_U \mu_{A_1}(u) r(u)\, \mathrm{d}u \right) \wedge \cdots \wedge \mu_{P_n} \left(\int_U \mu_{A_n}(u) r(u)\, \mathrm{d}u \right) \right)$$

subject to

$$v = \int_U u r(u)\, \mathrm{d}u$$

$$\int_U r(u)\, \mathrm{d}u = 1$$

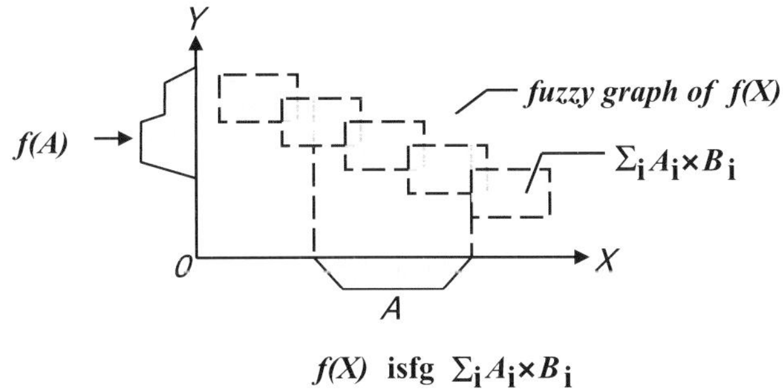

Fig. 26. Fuzzy-graph interpolation.

NOTE. $E(X)$ is a fuzzy subset of U.

(g) *Fuzzy-graph interpolation rule*

This rule is the most widely used rule in applications of fuzzy logic [36]. We have a function, $Y = f(X)$, which is represented as a fuzzy graph (Figure 26). The question is: What is the value of Y when X is A? The A_i, B_i and A are fuzzy sets.

Symbolic part

$$\begin{array}{l} X \text{ is } A \\ Y = f(X) \\ f(X) \text{ isfg } \sum_i A_i \times B_i \\ \hline Y \text{ is } C \end{array}$$

Computational part

$$C = \sum_i m_i \wedge B_i,$$

where m_i is the degree to which A matches A_i

$$m_i = \sup_u \big(\mu_A(u) \wedge \mu_{A_i}(u)\big), \quad i = 1, \ldots, n.$$

When A is a singleton, this rule reduces to

$$X = a$$

$$Y = f(X)$$

$$f(X) \text{ isfg } \sum_i A_i \times B_i, \quad i = 1, \ldots, n.$$

$$Y = \sum_i \mu_{A_i}(a) \wedge B.$$

In this form, the fuzzy-graph interpolation rule coincides with the Mamdani rule – a rule which is widely used in control and related applications [15] (Figure 27).

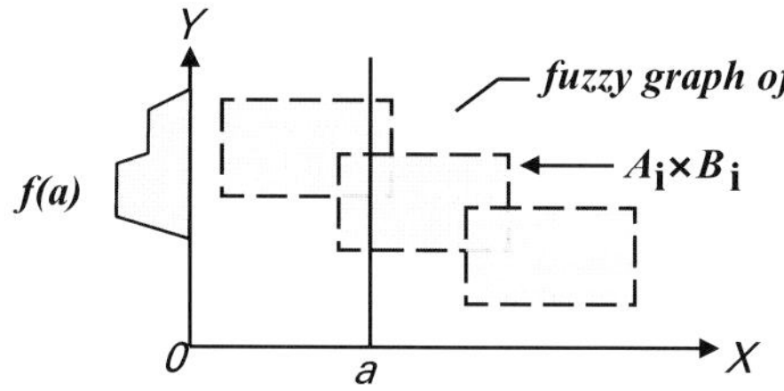

Fig. 27. Mamdani interpolation.

In the foregoing, we have summarized some of the basic rules in DDB which govern generalized constraint propagation. Many more rules will have to be developed and added to DDB. A few examples of such rules are the following.

(a) *Probabilistic extension principle*

$$\frac{f(X) \text{ isp } A}{g(X) \text{ isr } ?B}$$

(b) *Usuality-qualified extension principle*

$$\frac{f(X) \text{ isu } A}{g(X) \text{ isr } ?B}$$

(c) *Usuality-qualified fuzzy-graph interpolation rule*

$$\frac{\begin{array}{l} X \text{ is } A \\ Y = f(X) \\ f(X) \text{ isfg } \sum_i \text{ if } X \text{ is } A_i \text{ then } Y \text{ isu } B_i \end{array}}{Y \text{ isr } ?B}$$

(d) *Bimodal extension principle*

$$\frac{\begin{array}{l} X \text{ isbm } \sum_i P_i \backslash A_i \\ Y = f(X) \end{array}}{Y \text{ isr } ?B}$$

(e) *Bimodal, binary extension principle*

$$\frac{\begin{array}{l} X \text{ isr } R \\ Y \text{ iss } S \\ Z = f(X, Y) \end{array}}{Z \text{ ist } T}$$

In the instance, bimodality means that X and Y have different modalities, and binary means that f is a function of two variables. An interesting special case is one in which X is R and Y isp S.

The deduction rules which were briefly described in the foregoing are intended to serve as examples. How can these rules be applied to reasoning under uncertainty? To illustrate, it will be convenient to return to the examples given in Section 1.

(a) *The Robert example*

p: Usually Robert returns from work at about 6:00 pm. What is the probability that Robert is home at about 6:15 pm?

First, we find the protoforms of the data and the query.

Usually Robert returns from work at about 6:00 pm $\longrightarrow$

$$\longrightarrow \text{Prob}\big(\text{Time(Return(Robert))} \text{ is } {}^*6:00 \text{ pm}\big) \text{ is usually}$$

which in annotated form reads

$$\longrightarrow \text{Prob}\big(X/\text{Time(Return(Robert))} \text{ is } A/{}^*\text{6:00 pm}\big) \text{ is } B/\text{usually}$$

Likewise, for the query, we have

$$\text{Prob}\big(\text{Time(Return(Robert))} \text{ is } \leqslant \circ\, {}^*\text{6:15 pm}\big) \text{ is } ?D$$

which in annotated form reads

$$\longrightarrow \text{Prob}\big(X/\text{Time(Return(Robert))} \text{ is } C/\leqslant \circ\, {}^*\text{6:15 pm}\big) \text{ is } D/\text{usually}$$

Searching the Deduction Database, we find that the basic probability rule matches the protoforms of the data and the query

$$\frac{\text{Prob}(X \text{ is } A) \text{ is } B}{\text{Prob}(X \text{ is } C) \text{ is } D},$$

where

$$\mu_D(v) = \sup_g \left(\mu_B\left(\int_U \mu_A(u) g(u)\, \mathrm{d}u\right)\right)$$

subject to

$$v = \int_U \mu_C(u) g(u)\, \mathrm{d}u$$

$$\int_U g(u)\, \mathrm{d}u = 1$$

Instantiating A, B, C and D, we obtain the answer to the query:

Probability that Robert is home at about 6:15 pm is D, where

$$\mu_D(v) = \sup_g \left(\mu_{\text{usually}}\left(\int_U \mu_{*\text{6:00 pm}}(u) g(u)\, \mathrm{d}u\right)\right)$$

subject to

$$v = \int_U \mu_{\leqslant\circ *\text{6:15 pm}}(u) g(u)\, \mathrm{d}u$$

and

$$\int_U g(u)\, \mathrm{d}u = 1$$

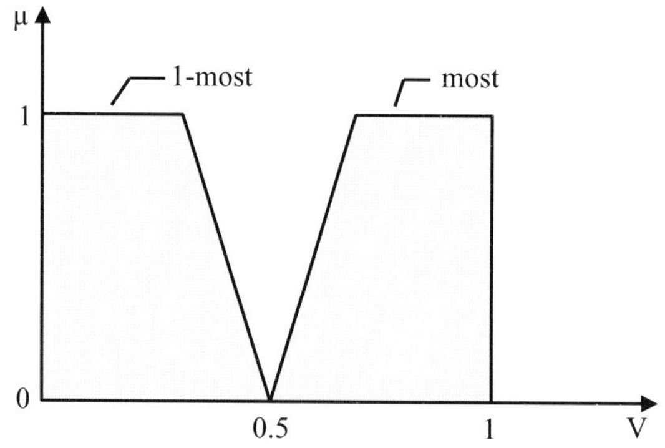

Fig. 28. "Most" and antonym of "most."

(b) *The tall Swedes problem*

We start with the data

p: Most Swedes are tall.

Assume that the queries are:

q_1: How many Swedes are not tall

q_2: How many are short

q_3: What is the average height of Swedes

In our earlier discussion of this example, we found that p translates into a generalized constraint on the count density function, h.

Thus

$$p \longrightarrow \int_a^b h(u)\mu_{\text{tall}}(u)\,\mathrm{d}u \text{ is most}$$

Precisiations of q_1, q_2 and q_3 may be expressed as

$$q_1\colon \longrightarrow \int_a^b h(u)\mu_{\text{not.tall}}(u)\,\mathrm{d}u$$

$$q_2\colon \longrightarrow \int_a^b h(u)\mu_{\text{short}}(u)\,\mathrm{d}u$$

$$q_3\colon \longrightarrow \int_a^b uh(u)\,\mathrm{d}u.$$

Considering q_1, we note that

$$\mu_{\text{not.tall}}(u) = 1 - \mu_{\text{tall}}(u).$$

Consequently

$$q_1 \longrightarrow 1 - \int_a^b h(u)\mu_{\text{tall}}(u)\,\mathrm{d}u$$

which may be rewritten as

$$q_2 \longrightarrow \text{1-most}$$

where 1-most plays the role of the antonym of most (Figure 28).
Considering q_2, we have to compute

$$A\colon \int_a^b h(u)\mu_{\text{short}}(u)\,\mathrm{d}u$$

given that $\int_a^b h(u)\mu_{\text{tall}}(u)\,\mathrm{d}u$ is most.

Applying the extension principle, we arrive at the desired answer to the query:

$$\mu_A(v) = \sup\left(\mu_{\text{most}}\left(\int_a^b h(u)\mu_{\text{tall}}(u)\,\mathrm{d}u\right)\right)$$

subject to

$$v = \int_a^b h(u)\mu_{\text{short}}(u)\,\mathrm{d}u$$

and

$$\int_a^b h(u)\,\mathrm{d}u = 1.$$

Likewise, for q_3 we have as the answer

$$\mu_A(v) = \sup_u\left(\mu_{\text{most}}\left(\int_a^b h(u)\mu_{\text{tall}}(u)\,\mathrm{d}u\right)\right)$$

subject to

$$v = \int_a^b uh(u)\,\mathrm{d}u$$

and

$$\int_a^b h(u)\,\mathrm{d}u = 1.$$

As an illustration of application of protoformal deduction to an instance of this example, consider

p: Most Swedes are tall
q: How many Swedes are short?

We start with the protoforms of p and q (see earlier example):
Most Swedes are tall $\rightarrow 1/n \sum \text{Count}(G[A \text{ is } R]) \text{ is } Q$
?T Swedes are short $\longrightarrow 1/n \sum \text{Count}(G[A \text{ is } S]) \text{ is } T$,
where

$$G[A] = \sum_i \text{Name}_i / A_i, \quad i = 1, \ldots, n.$$

An applicable deduction rule in symbolic form is:

$$\frac{1/n \sum \text{Count}(G[A \text{ is } R]) \text{ is } Q}{1/n \sum \text{Count}(G[A \text{ is } S]) \text{ is } T}$$

The computational part of the rule is expressed as

$$\frac{1/n \sum_i \mu_R(A_i) \text{ is } Q}{1/n \sum_i \mu_S(A_i) \text{ is } T}$$

where

$$\mu_T(v) = \sup_{A_i, \ldots, A_n} \mu_Q\left(\sum_i \mu_R(A_i)\right)$$

subject to

$$v = \sum_i \mu_S(A_i).$$

What we see is that computation of the answer to the query, q, reduces to the solution of a variational problem, as it does in the earlier discussion of this example in which protoformal deduction was not employed.

The foregoing examples are merely elementary instances of reasoning through the use of generalized constraint propagation. What should be noted is that the chains of reasoning in these examples are very short. More generally, what is important to recognize is that shortness of chains of reasoning is an intrinsic characteristic of reasoning processes which take place in an environment of substantive imprecision and uncertainty. What this implies is that, in such environments, a conclusion arrived at the end of a long chain of reasoning is likely to be vacuous or of questionable validity.

4.5. *Deduction (extension) principle*

Underlying almost all examples involving computation of an answer to a question, is a basic principle which may be referred to as the Deduction Principle. This principle is closely related to the extension principle of fuzzy logic [24,28].

Assume that we have a database, D, and database variables $X_1, \ldots, X_n$, with u_i being a generic value of X_i, $i = 1, \ldots, n$.

Suppose that q is a given question and that the answer to q, Ans(q), is a function of the u_i.

$$\text{Ans}(q) = g(u_1, \ldots, u_n), \quad u = (u_1, \ldots, u_n).$$

I do not know the exact values of the u_i. My information about the u_i, $I(u_1, \dots, u_n)$, is a generalized constraint on the u_i. The constraint is defined by its test-score function $ts(u) = f(u_1, \dots, u_n)$.

At this point, the problem is that of constraint propagation from $ts(u)$ to $g(u)$. Employing the extension principle, we are led to the membership function of the answer to q. More specifically,

$$\mu_{\mathrm{Ans}(q)}(v) = \sup_u \big(ts(u)\big)$$

subject to

$$v = g(u)$$

This, in brief, is the substance of the Deduction Principle.

As a simple illustration, let us consider an example that was discussed earlier. Suppose that q: What is the average height of Swedes. Assume that D consists of information about the heights of a population of Swedes, $\mathrm{Swede}_1, \dots, \mathrm{Swede}_n$, with height of ith Swede being h_i, $i = 1, \dots, n$. Thus, average height may be expressed as

$$\mathrm{Ave}(h) = 1/n(h_1 + \cdots + h_n).$$

Now, I do not know the h_i. What I am given is the datum d: Most Swedes are tall. This datum constrains the h_i. The test-score of this constraint is

$$ts(h) = \mu_{\mathrm{most}}\left(\frac{1}{n}\sum \mu_{\mathrm{tall}}(h_i)\right), \qquad h = (h_1, \dots, h_n).$$

The generalized constraint on the h_i induces a generalized constraint on Ave(h). Thus

$$\mu_{\mathrm{Ave}(h)}(v) = \sup\left(\mu_{\mathrm{most}}\left(\frac{1}{n}\sum_i \mu_{\mathrm{tall}}(h_i)\right)\right), \qquad h = (h_1, \dots, h_n)$$

subject to

$$v = \frac{1}{n}\sum_i h_i.$$

5. Concluding remark

Existing search engines, with Google at the top, have many truly remarkable capabilities. But there is a basic limitation – search engines do not have deduction capability – a capability which a question-answering system is expected to have. Nevertheless, search engines are extremely useful because a skilled human user can get around search engine limitations. In this perspective, a search engine may be viewed as a semi-mechanized question-answering system.

Upgrading a search engine such as Google to a question-answering system is a task whose complexity is hard to exaggerate. To achieve success, new concepts, ideas and tools are needed to address difficult problems which arise when knowledge has to be dealt with in an environment of imprecision, uncertainty and partial truth.

The present paper sketches a conceptual framework which may be viewed as a step in this direction. What lies ahead is the challenging task of applying this framework to web-specific problems.

Acknowledgement

Research supported in part by ONR N00014-02-1-0294, BT Grant CT1080028046, Omron Grant, Chevron Texaco, Tekes Grant and the BISC Program of UC Berkeley.

References

[1] A.R. Aronson, B.E. Jacobs, J. Minker, *A note on fuzzy deduction*, J. ACM **27** (4) (1980), 599–603.
[2] A. Bardossy, L. Duckstein, *Fuzzy Rule-based Modelling with Application to Geophysical, Biological and Engineering Systems*, CRC Press, 1995.
[3] T. Berners-Lee, J. Hendler, O. Lassila, *The semantic web*, Scientific American **284** (5) (2001), 34–43.
[4] S. Brin, L. Page, *The anatomy of a large-scale hypertextual web search engine*, Computer Networks **30** (1–7) (1998), 107–117.
[5] W.J.H.J. Bronnenberg, M.C. Bunt, S.P.J. Lendsbergen, R.H.J. Scha, W.J. Schoenmakers, E.P.C. van Utteren, *The question answering system PHLIQA1*, in: L. Bola (Ed.), Natural Language Question Answering Systems, Macmillan, 1980.
[6] L.S. Coles, *Techniques for information retrieval using an inferential question-answering system with natural language input*, SRI Report, 1972.
[7] A. Di Nola, S. Sessa, W. Pedrycz, W. Pei-Zhuang, *Fuzzy relation equation under a class of triangular norms: a survey and new results*, in: Fuzzy Sets for Intelligent Systems, Morgan Kaufmann Publishers, San Mateo, CA, 1993, pp. 166–189.
[8] A. Di Nola, S. Sessa, W. Pedrycz, E. Sanchez, *Fuzzy Relation Equations and their Applications to Knowledge Engineering*, Kluwer Academic Publishers, Dordrecht, 1989.
[9] D. Dubois, H. Prade, *Gradual inference rules in approximate reasoning*, Inform. Sci. **61** (1–2) (1992), 103–122.
[10] D. Filev, R.R. Yager, *Essentials of Fuzzy Modeling and Control*, Wiley-Interscience, 1994.
[11] J.A. Goguen, *The logic of inexact concepts*, Synthese **19** (1969), 325–373.
[12] M. Jamshidi, A. Titli, L.A. Zadeh, S. Boverie (Eds.), Applications of Fuzzy Logic – Towards High Machine Intelligence Quotient Systems, Environmental and Intelligent Manufacturing Systems Series, vol. 9, Prentice-Hall, Upper Saddle River, NJ, 1997.
[13] A. Kaufmann, M.M. Gupta, *Introduction to Fuzzy Arithmetic: Theory and Applications*, Van Nostrand, New York, 1985.
[14] D.B. Lenat, *CYC: a large-scale investment in knowledge infrastructure*, Comm. ACM **38** (11) (1995), 32–38.
[15] E.H. Mamdani, S. Assilian, *An experiment in linguistic synthesis with a fuzzy logic controller*, Int. J. Man–Machine Studies **7** (1975), 1–13.
[16] J.R. McSkimin, J. Minker, *The use of a semantic network in a deductive question-answering system*, in: IJCAI, 1977, pp. 50–58.
[17] R.E. Moore, *Interval Analysis*, SIAM Studies in Applied Mathematics, vol. 2, Philadelphia, PA, 1979.
[18] M. Nagao, J. Tsujii, *Mechanism of deduction in a question-answering system with natural language input*, in: ICJAI, 1973, pp. 285–290.
[19] B.H. Partee (Ed.), Montague Grammar, Academic Press, New York, 1976.
[20] W. Pedrycz, F. Gomide, *Introduction to Fuzzy Sets*, MIT Press, Cambridge, MA, 1998.
[21] F. Rossi, P. Codognet (Eds.), Soft Constraints, Special issue on Constraints, vol. 8, N. 1, Kluwer Academic Publishers, 2003.
[22] G. Shafer, *A Mathematical Theory of Evidence*, Princeton University Press, Princeton, NJ, 1976.
[23] M.K. Smith, C. Welty, D. McGuinness (Eds.), OWL Web Ontology Language Guide, W3C Working Draft 31, 2003.
[24] L.A. Zadeh, *Fuzzy sets*, Inform. and Control **8** (1965), 338–353.
[25] L.A. Zadeh, *Probability measures of fuzzy events*, J. Math. Anal. Appl. **23** (1968), 421–427.
[26] L.A. Zadeh, *Outline of a new approach to the analysis of complex systems and decision processes*, IEEE Trans. on Systems Man Cybernet. **3** (1973), 28–44.
[27] L.A. Zadeh, *On the analysis of large scale systems*, in: H. Gottinger (Ed.), Systems Approaches and Environment Problems, Vandenhoeck and Ruprecht, Gottingen, 1974, pp. 23–37.

[28] L.A. Zadeh, *The concept of a linguistic variable and its application to approximate reasoning, Part I*, Inform. Sci. **8** (1975), 199–249; *Part II*, Inform. Sci. **8** (1975), 301–357; *Part III*, Inform. Sci. **9** (1975), 43–80.
[29] L.A. Zadeh, *Fuzzy sets and information granularity*, in: M. Gupta, R. Ragade, R. Yager (Eds.), Advances in Fuzzy Set Theory and Applications, North-Holland Publishing Co, Amsterdam, 1979, pp. 3–18.
[30] L.A. Zadeh, *A theory of approximate reasoning*, in: J. Hayes, D. Michie, L.I. Mikulich (Eds.), Machine Intelligence, vol. 9, Halstead Press, New York, 1979, pp. 149–194.
[31] L.A. Zadeh, *Test-score semantics for natural languages and meaning representation via PRUF*, in: B. Rieger (Ed.), Empirical Semantics, Brockmeyer, Bochum, W. Germany, 1982, pp. 281–349. Also Technical Memorandum 246, AI Center, SRI International, Menlo Park, CA, 1981.
[32] L.A. Zadeh, *A computational approach to fuzzy quantifiers in natural languages*, Computers and Mathematics **9** (1983), 149–184.
[33] L.A. Zadeh, *A fuzzy-set-theoretic approach to the compositionality of meaning: propositions, dispositions and canonical forms*, J. Semantics **3** (1983), 253–272.
[34] L.A. Zadeh, *Precisiation of meaning via translation into PRUF*, in: L. Vaina, J. Hintikka (Eds.), Cognitive Constraints on Communication, Reidel, Dordrecht, 1984, pp. 373–402.
[35] L.A. Zadeh, *Outline of a computational approach to meaning and knowledge representation based on a concept of a generalized assignment statement*, in: M. Thoma, A. Wyner (Eds.), Proceedings of the International Seminar on Artificial Intelligence and Man–Machine Systems, Springer-Verlag, Heidelberg, 1986, pp. 198–211.
[36] L.A. Zadeh, *Fuzzy logic and the calculi of fuzzy rules and fuzzy graphs*, Multiple-Valued Logic **1** (1996), 1–38.
[37] L.A. Zadeh, *Toward a theory of fuzzy information granulation and its centrality in human reasoning and fuzzy logic*, Fuzzy Sets and Systems **90** (1997), 111–127.
[38] L.A. Zadeh, *From computing with numbers to computing with words – from manipulation of measurements to manipulation of perceptions*, IEEE Trans. on Circuits and Systems **45** (1) (1999), 105–119.
[39] L.A. Zadeh, *Toward a perception-based theory of probabilistic reasoning with imprecise probabilities*, J. Statist. Plann. Inference **105** (2002), 233–264.
[40] L.A. Zadeh, *Precisiated natural language (PNL)*, AI Magazine **25** (3) (2004), 74–91.
[41] L.A. Zadeh, *A note on web intelligence, world knowledge and fuzzy logic*, Data and Knowledge Engineering **50** (2004), 291–304.
[42] L.A. Zadeh, *Toward a generalized theory of uncertainty (GTU) – an outline*, Inform. Sci. **172** (2005), 1–40.

Further reading

[43] J. Arjona, R. Corchuelo, J. Pena, D. Ruiz, *Coping with web knowledge*, in: Advances in Web Intelligence, Springer-Verlag, Berlin, 2003, pp. 165–178.
[44] A. Bargiela, W. Pedrycz, *Granular Computing – An Introduction*, Kluwer Academic Publishers, Boston, 2003.
[45] Z. Bubnicki, *Analysis and Decision Making in Uncertain Systems*, Springer-Verlag, 2004.
[46] P.P. Chen, *Entity-relationship Approach to Information Modeling and Analysis*, North-Holland, 1983.
[47] M. Craven, D. DiPasquo, D. Freitag, A. McCallum, T. Mitchell, K. Nigam, S. Slattery, *Learning to construct knowledge bases from the world wide web*, Artificial Intelligence **118** (1–2) (2000), 69–113.
[48] M.J. Cresswell, *Logic and Languages*, Methuen, London, UK, 1973.
[49] D. Dubois, H. Prade, *On the use of aggregation operations in information fusion processes*, Fuzzy Sets and Systems **142** (1) (2004), 143–161.
[50] T.F. Gamat, *Language, Logic and Linguistics*, University of Chicago Press, 1996.
[51] M. Mares, *Computation over Fuzzy Quantities*, CRC, Boca Raton, FL, 1994.
[52] V. Novak, I. Perfilieva, J. Mockor, *Mathematical Principles of Fuzzy Logic*, Kluwer Academic Publishers, Boston, 1999.
[53] V. Novak, I. Perfilieva (Eds.), Discovering the World with Fuzzy Logic, Studies in Fuzziness and Soft Computing, Physica-Verlag, Heidelberg, 2000.
[54] Z. Pawlak, *Rough Sets: Theoretical Aspects of Reasoning about Data*, Kluwer Academic Publishers, Dordrecht, 1991.
[55] M.K. Smith, C. Welty, *What is ontology? Ontology: towards a new synthesis*, in: Proceedings of the Second International Conference on Formal Ontology in Information Systems, 2002.

[56] J.F. Sowa, in: L. Albertazzi (Ed.), Ontological Categories. Shapes of Forms: From Gestalt Psychology and Phenomenology to Ontology and Mathematics, Kluwer Academic Publishers, Dordrecht, 1999, pp. 307–340.
[57] L.A. Zadeh, *A fuzzy-algorithmic approach to the definition of complex or imprecise concepts*, Int. J. Man–Machine Studies **8** (1976), 249–291.
[58] L.A. Zadeh, *Fuzzy logic, neural networks, and soft computing*, Comm. ACM – AI **37** (1994), 77–84.
[59] L.A. Zadeh, *A new direction in AI – toward a computational theory of perceptions*, AI Magazine **22** (1) (2001), 73–84.

CHAPTER 10

A Perception-based Web Search with Fuzzy Semantic[1]

Chris Tseng* and Toan Vu

CS department, San Jose State University, One Washington Square, San Jose, CA 95192, USA
E-mail: tseng@cs.sjsu.edu; toantai@gmail.com

Abstract

We propose a perception-based search methodology established on fuzzy semantic for improving the web search. Unlike most existing relevant work that focuses on filtering the returned search result, we study the search query semantic and expand the search to include information related to the linguistic and numerical fuzzy counterparts of the search query. The top 20 search results returned from this search methodology showed impressive improvement over the conventional one. Fifty percent average gain in search relevancy is obtained when our search methodology is applied to websites matching the chosen fuzzy semantic theme. We demonstrate the effectiveness of our methodology on the search domain of health and food.

Keywords

semantic search, fuzzy logic, search engine, GCL, semantic web

1. Introduction

One of the crucial steps in an efficient web search is to understand what users meant in their web search query. The search query is usually in the form of natural language. The current popular search engines like Yahoo and Google literally take the search input without much semantic interpretation and attempt to find web pages that may contain all or some of the keywords in the input query. This sometimes leads to the inclusion of web pages not

[1] The experimental search engine used in this paper is available at: http://130.65.86.103:8080/fss/domain_search/domain_search.jsp

* Corresponding author.

FUZZY LOGIC AND THE SEMANTIC WEB
Edited by Elie Sanchez
© 2006 Elsevier B.V. All rights reserved

relevant to the input query in the returned search result [2,21]. Some innovative schemes have been proposed by researchers to improve the search results by either studying the user's usage patterns or preferences [13]. Data mining techniques have been applied to improve search efficiency [5,14]. Others explored and examined the meaning of relevancy through user's feedback and learning systems [4,9,11]. Varying techniques of inference [10], rough fuzzy sets [17,18], and fuzzy query [7] have also been proposed to attempt to help improve Internet search results.

Most ideas in literatures aimed to improve search engine focused on filtering search results to extract more relevant ones [6,15] without much effort on studying the input search query. With the vast amount of data available on almost every Internet search query nowadays, users typically only look at the 10 or 20 items contained in the first page of the returned result and will not bother to go after the second page or thereafter. It is therefore of high interest to see if new search methodologies can yield more relevant search results in the top 10 or 20 return items. We will follow this need and focus on how to improve search in this direction.

One of the main reasons that fuzzy logic applications excel in numerous areas [12] is due to the recognition that the input to the system may be fuzzy and should not be treated as a crisp input. Garcia-Serrano [8] introduced the notion of extending search query to synonyms of the search keywords but ignore the fact that some synonyms may be closer to the chosen keyword than others. Fuzzy logic and fuzzy membership functions obviously can provide this important need and help to precisiate the query semantic for a more reliable web search.

We begin by presenting an overview of how GCL/PFL and RDF can be used to perform search deduction when the search query is being precisiated and abstracted. The fuzzy semantic search engine architecture is also described in Section 2. In Section 3, we study how the concept of Fuzzy Linguistic Semantic search can be used to improve web search with some experimental data. Similar experiments using our proposed Fuzzy Numeric Semantic search are being conducted and the improvement results are presented in Section 4. Future research direction is included in Section 5 of the conclusion.

2. Internet search with fuzzy semantic deduction

It is well known that many web search engines provide fuzzy search options. Most of these fuzzy search options make elementary use of fuzzy logic [23–25]. Zadeh proposed the advanced concept and technique to improve Internet search with the notion of Protoform deduction [28]. Application to web search can be interpreted as shown in Figure 1. A web search query q in the form of natural language (NL) is input to the web interface. The

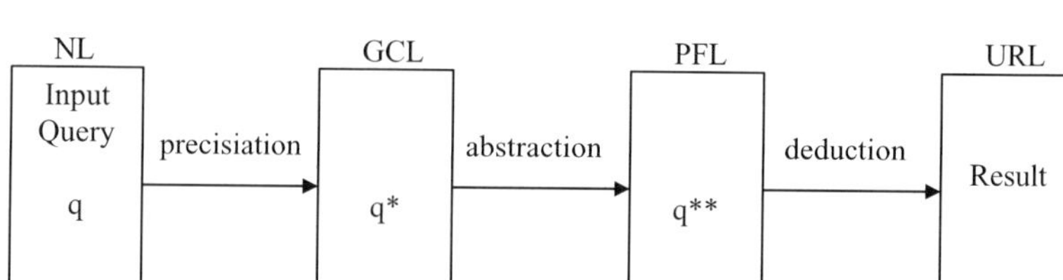

Fig. 1. Protoform deduction.

query is then precisiated by rewriting q in an appropriate semantic form and subsequently converting to q^* in the context of generalized constrained language (GCL). To facilitate the subsequent logical deduction purpose for retrieving matching web information, q^* is represented in protoform language (PFL) to perform appropriate web information mining of matching universal resource links (URLs).

If P is a proposition in natural language, then the precisiation of P results in a representation of the meaning of P in the form of what is referred to as a GCL [24]. GCL takes on a general form as follows:

$$X \text{ is} r \ R, \tag{2.1}$$

where X is the constrained variable, R is the constraining relation and r is the indexing variable whose value defines the way in which R constrains X.

Typical types of constraints and the associated values of r are the following:

- Equality constraint (r = "="); $X = R$.
- Inequality constraint (r = "$\leqslant$"); $X \leqslant R$.
- Subsethood constraint (r = "$\subset$"); $X \subset R$.
- Possibilistic constraint (r = blank); R is the possibility distribution of X.
- Veristic constraint (r = v); R is the verity distribution of X.
- Probabilistic constraint (r = p); R is the probability distribution of X.
- Usuality constraint (r = u); usually X is R.
- Random set constraint (r = rs); R is the Fuzzy-set-valued probability distribution of X.
- Fuzzy graph constraint (r = fg); X is a function and R is its Fuzzy graph.
- Pawlak set constraint (r = ps); X isps $(\overline{X}, \underline{X})$ – means that X is a set and $\overline{X}$ and $\underline{X}$ are the lower and upper approximation to X.

The set of all the composite generalized constraints and their associated rules of generation and interpretation, i.e., combination, modification and qualification, constitute the GCL elements. GCL serves to represent the meaning of perception-based information in perception-based probability theory.

Protoform is an abbreviation of prototypical form. This concept was introduced by Zadeh and one of its purposes is to facilitate the use of complex web query with fuzzy semantic [28]. This is considered to be one of the crucial tools to enable question-answering type of web query for the next-generation web search. The availability of searching the web with a valid question-answering format will allow web query to be in forms beyond just keywords. A stateful web query engine that can remember and analyze is consequently possible with this notion.

A protoform, A, of an object, B, is an abstracted summary of B and can be written as

$$A = \mathrm{PF}(B). \tag{2.2}$$

A can be a symbolic expression or a complex object. B is a lexical entity such as question, scenario, decision problem, etc. In a broader sense, B may be a relation, system, or an object of arbitrary complexity [27,28].

The Semantic Web, introduced by Tim Berners-Lee, uses Resource Description Framework (RDF) to add structure and meaning to Web applications [3]. The Semantic Web uses XML, Resource Description Framework (RDF) and ontologies to represent and analyze web information. Typically, a given information is transformed into the RDF format and subsequent deduction can be processed with suitable class hierarchy taxonomies, relations, or functions defined for the given data. RDF plays a similar role in abstraction of data by representing data in the following triple form

$$(\text{subject, predicate, object}). \tag{2.3}$$

A statement, or triple, consists of three individual components, namely, the subject, the predicate, and the object. The triple describes a relationship between a subject and an object by a property described in the predicate. The object of a statement can be another resource or a literal. A literal is a simple string or a data type defined by an XML Schema.

Existing RDF languages deal with hard semantics that are suitable for describing and processing crisp data. RDF cannot deal with soft semantics like perception descriptions inherent in natural languages.

The next example will illustrate how a natural language statement can be precisiated and abstracted by GCL/PFL and RDF, respectively.

EXAMPLE 2.1. Consider the statement:

$$\text{“Carol is 25”} \tag{2.4}$$

We can follow the steps below to represent it in RDF and PFL forms.

Step 1: Represent the given statement in a semantic form suitable for precisiation:

$$\text{Age (Carol) is 25.} \tag{2.5}$$

Step 2: Represent (2.5) in GCL format. Statement (2.4) can be represented in GCL (2.1) with the equality constraint, i.e., $r :=$, as

$$X = R, \tag{2.6}$$

where X is Age (Carol) and R is 25.

Step 3.P: Abstract (2.6) to its protoform format as follows.

$$A(B) \text{ is } R, \tag{2.7}$$

where A is Age, B is Carol, and R is 25.

Step 3.R: (2.5) can also be alternatively represented in an RDF format.

The corresponding RDF triple components for (2.4) can be as follows:

Subject: *Carol*
Predicate: *Age* (2.8)
Object: 25

A set of RDF triples, or statements, can be represented as a directed labeled graph with nodes and arcs. These nodes, typically show in ovals, represent subjects and objects whereas arcs are used to represent named predicates. Nodes that represent string literals are usually drawn as rectangles. The direction of the arc does matters. The arc always starts from the subject and points to the object of the statement. Statement (2.8) can be equivalently described as shown in Figure 2.

Note the notion of GCL/PFL allows us to represent soft semantic that is not available with RDF. In this example, if the age of 25 was to be a fuzzy description RDF will have trouble representing it.

In this paper we shall focus on the deduction aspect of web queries in keyword pairs that are already in precisiated and abstraction form.

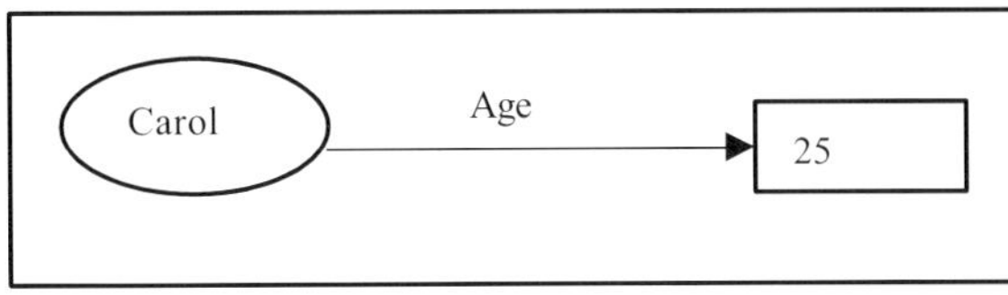

Fig. 2. RDF graph representation

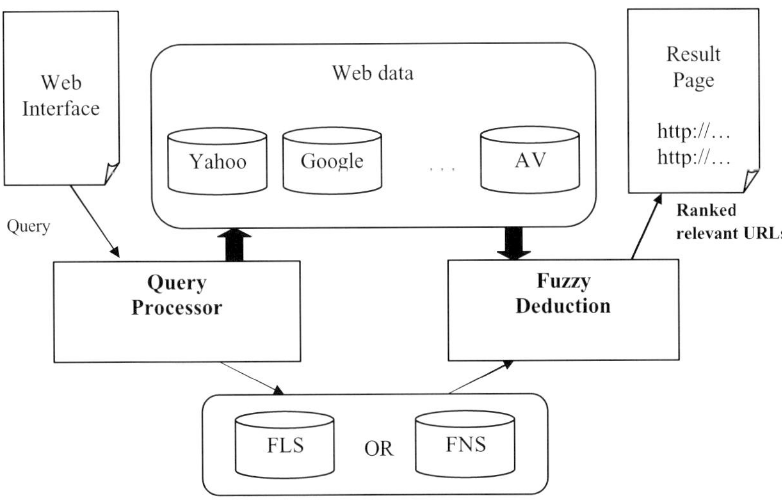

Fig. 3. Perception-based search engine with fuzzy semantic.

2.1. *Fuzzy semantic search architecture*

Our proposed perception-based search engine has the architecture as shown in Figure 3.

The web interface allows users to enter their query and select if they would like to use the Fuzzy Linguistic Semantic (FLS) search or Fuzzy Numeric Semantic (FNS) search. The Fuzzy Linguistic Semantic (FLS) module will generate semantically related keywords that are related to the fuzzy sets of the search query. Both the semantically related keywords and the original search query string are considered relevant to the search and will be used in the search deduction. Similarly, web data matching the fuzzy numeric interpretations of the search query will be considered if Fuzzy Numeric Semantic module is chosen. The returned search results from the search engines are subsequently processed to remove duplicate entries and ranked in accordance to the semantic relevance with fuzzy logic. We examine the top 20 URLs in the returned list for possible relevancy for the original search query.

Unlike most web search methodologies that usually attempts to filter returned URLs based on keywords only [6,19], our proposed search engine expands the search to other semantically related queries for more relevant results. As we will see in Section 3, this approach actually yield more relevant search results in the top URLs list.

The core of our perception-based fuzzy semantic search is described in Figure 4. We shall focus on the search queries that consist of two keywords, (X, Y), where X constrains Y in the search query. X typically is an adjective, adverb, or verb whereas Y can be a noun

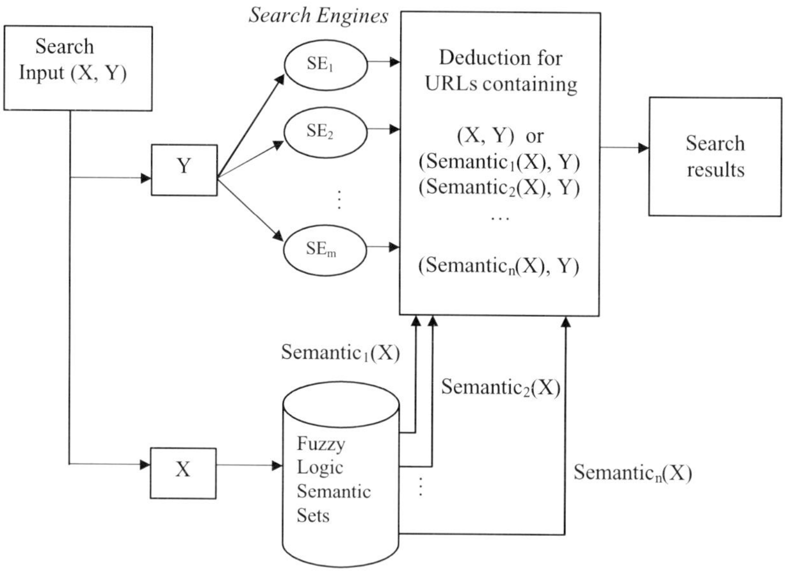

Fig. 4. Deduction of fuzzy semantic search engine.

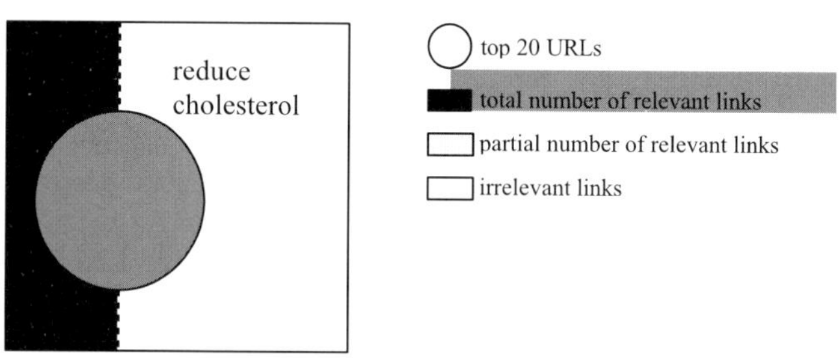

Fig. 5. Conventional search result.

or a noun clause in the natural language sense. Choi [7] studied these types of queries and suggested improvement using fuzzy logic without experimental statistics on its efficiency. The constrained variable Y is firstly submitted to the search engines to retrieve relevant web pages set $\boldsymbol{U}$. The second pass involves validating possible existence of linguistic or numeric variables, Semantic$_i(X)$, $1 \leqslant i \leqslant n$, and now the variables relate to the fuzzy sets of the constraining variable X in the set $\boldsymbol{U}$.

The reason why the proposed methodology depicted in Figure 4 can produce more relevant URLs can be deduced as follows. When the search query (X, Y), e.g., *reduce cholesterol*, is submitted to the conventional web search engines, typically some URLs in the return result set will not match the search query. This can be illustrated as shown in the white region as shown in Figure 5. When the users look at the first 20 URLs in the first page of the returned result, it is likely that they may see even fewer relevant URLs like the grey region in Figure 5.

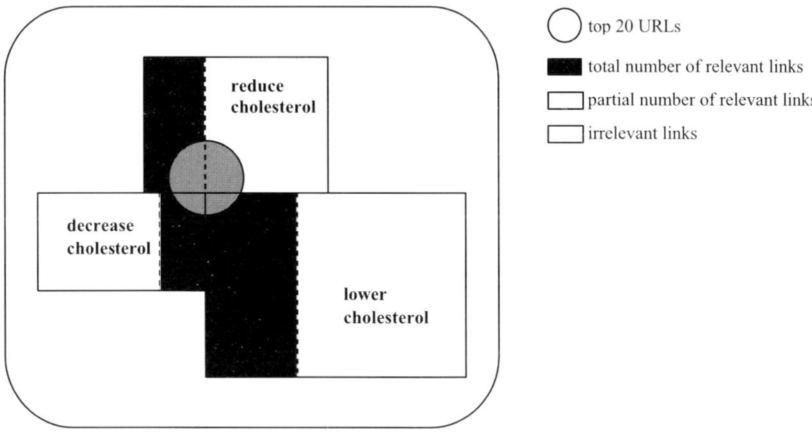

Fig. 6. Fuzzy Linguistic Semantic search result.

On the other hand, by expanding the search to additional queries that are related to the original one using fuzzy semantic we can get several more sets of relevant data in return. When one extracts the top 20 URLs out of these data sets, it is more likely to yield more relevant URLs than just one set of data. Our proposed search engine would expand the search of a query, say, *reduce cholesterol*, to several other queries, (Semantic$_i$(reduce), cholesterol), $i = 1, \ldots, n$, that are related to the original query in the fuzzy semantic sense. Examples of such related query include *lower cholesterol* and *decrease cholesterol* as both *lower* and *decrease* are related to the fuzzy set about *reduce*. This would enhance the likelihood of getting more relevant URLs from the first 20 URLs in the returned result based on all the relevant queries. This is illustrated with the proportional larger grey region in Figure 6. We will validate this by experimenting various queries in the next section.

3. Search improvement with fuzzy linguistic semantic

We shall focus on the search query in the form of

$$(X, Y), \tag{3.1}$$

where X is the constraining variable and Y is the constrained variable.

Many search queries occur naturally in manner. Take for example, in the search query *reduce cholesterol*, *reduce* (X) is considered as a variable constraining the variable *cholesterol* (Y). We will compare our proposed search methodology with the conventional web search by submitting the same query strings to both. In the case of conventional web search, the search query like (3.1) is submitted as the search string in the usual way. To take advantage of semantic structure in the search string, we first submit the constrained variable Y to the search engine to retrieve web pages that may contain Y. The system would then parse through the returned web pages for possible existence of the constraining variable X or its fuzzy linguistic semantic counterparts.

$$\text{FLSC} = \{\text{Semantic}_i(X),\ 1 \leqslant i \leqslant n\}. \tag{3.2}$$

The flow of fuzzy linguistic semantic search can be summarized as follow:

Fuzzy linguistic semantic search steps

LS1: Submit the keyword Y to the search engine and obtain the set, $\boldsymbol{U}$, of the search result web pages.

LS2: Parse through the returned web pages in $\boldsymbol{U}$ for possible existence of the constraining variable X or its fuzzy linguistic semantic counterparts (FLSC).

LS3: Calculate the membership function values, μ_X(FLSC), of the fuzzy linguistic semantic variable, FLSC, of the membership function with respect to the constraining keyword X.

LS4: Take into account how close X and Y are to each other. Calculate the distance ranking score, S, of Y and X or Y and its Fuzzy Linguistic Semantic counterparts (FLSC) for each web page in the search results $\boldsymbol{U}$ with the following formula.
Case 1: $S = 1$, if Y and X or Y and its FLSC are adjacent to each other in the returned web page.
Case 2: $S = 0.5$, if Y and X or Y and its FLSC are in the same sentence of the returned web page.
Case 3: $S = 0.2$, if Y and X or Y and its FLSC are in the same page of the returned web page.
Case 4: $S = 0$, if Y and X or Y and its FLSC do not both show up in the returned web page.

LS5: Calculate the overall relevancy index R of each of the returned web pages in set $\boldsymbol{U}$ by

$$R = S * \mu_X(\text{FLSC}). \tag{3.3}$$

EXAMPLE 3.1. An example of the above algorithm as applied to a search query like *prevent cancer* can be as follows. The constrained variable, *cancer*, is firstly submitted to the search engine to retrieve the set of all possible web pages that may contain the keyword *cancer*. A parser is then applied to seek for possible existence of the constraining variable *prevent* and its fuzzy linguistic counterparts in this returned URL set, $\boldsymbol{U}$. As shown in Figure 7, several words are related to the constraining variable, *prevent*. These FLSC for *cancer* include *avoid, prevent, inhibit* and *stop* as described by the fuzzy membership functions.

Each web page in the URL set $\boldsymbol{U}$ that contains the constraining variable or one of its linguistic counterparts, FLSC, will be noted with a respective membership function value, μ_{prevent}(FLSC) as described in Figure 7. Taking semantic into consideration in evaluating the relevancy of the search result, LS4 is performed and subsequently integrated in LS5 to form the overall relevancy index. The last step is to select the top 20 URLs and display them in descending order with respect to their overall relevancy index values. Part of the final result can be like that shown below in Figure 8.

Oyama and others showed how their search scheme in selective web domains could produce impressive results [16]. Most of the domains selected by various papers to illustrate the search engine improvement are typically in non-technical areas. Some of typical ones used include education [14,21], events [15,19], and health [5,18,29]. Non-technical subjects have replaced technical subjects in the top web search queries [13]. We will follow this convention and select the subject of health and food to illustrate the improvement of search efficiency with our proposed fuzzy semantic search engine.

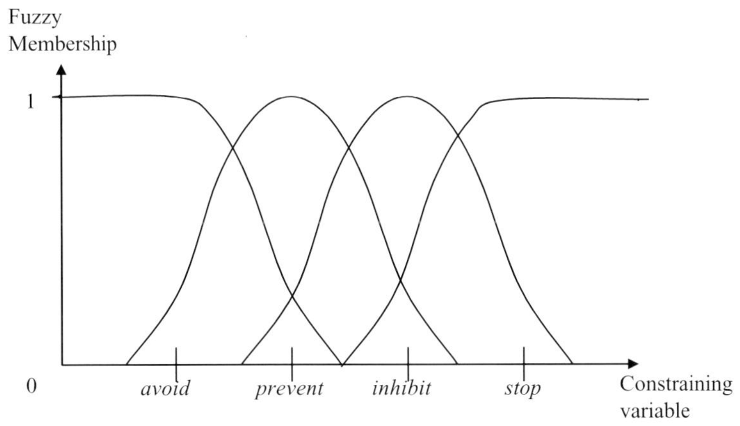

Fig. 7. Membership function of constraining variable *prevent* and its FLSC.

Fig. 8. Fuzzy Linguistic Semantic search result.

Two scenarios will be used to test the search performance.

(i) Domains with matching semantic theme: Several popular domains with matching theme of the chosen subject, i.e., health and food, will be used to test the search performance. In particular, the following websites are used.

(ii) World-wide web: In this case, the search is performed on world-wide-web by submitting the search queries Google, Yahoo, AltaVista, Alltheweb, and MSN-search.

3.1. *Case study*

In this section, we compare the conventional web search against our Fuzzy Linguistic Semantic search in terms of search relevancy in the top 20 URLs. We create 20 random fuzzy

- Whole Health MD http://www.wholehealthmd.com
- World's healthiest foods http://www.whfoods.com
- About Nutrition http://nutrition.about.com
- Mayo Clinic http://www.mayoclinic.com
- Foods 'n' Herbs http://www.foodsnherbs.com

List 1. Websites with health and food theme.

prevent: avoid, **prevent**, inhibit, stop
reduce: lose, lessen, decrease, lower, **reduce**, relieve, remove, eliminate, cure, treat
increase: promote, soothe, help, improve, enhance, **increase**
relieve: lose, lessen, decrease, lower, reduce, **relieve**, remove, eliminate, cure, treat
treat: lose, lessen, decrease, lower, reduce, relieve, remove, eliminate, cure, **treat**
beneficial: wholesome, aid, **beneficial**, useful, helpful, great, health, good, fresh, nutritious, fibrous, nourish.
significant: valuable, **significant**, great, rich, excellent, outstanding, exceptional, leading
rich: valuable, significant, great, **rich**, excellent, outstanding, exceptional, leading
improve: promote, soothe, help, **improve**, enhance, increase
cure: lose, lessen, decrease, lower, reduce, relieve, remove, eliminate, **cure**, treat
muscular: athletic, **muscular**, strong, healthy, fit, robust, invigorating

List 2. List of keywords and their fuzzy linguistic counterparts.

queries related to the health and food domain and perform a conventional and Fuzzy Linguistic Semantic searches. Two separate experiments are conducted. The first one focuses on domains with matching semantic theme of our chosen subject, i.e., health and food. The second one is submitted to the world-wide-web. For each query, we look at the top 20 URLs of the result set to determine the relevancy to the search query. The results are then visually inspected and the number of relevant results is determined. The percentage of relevance is determined by dividing the number of relevant URLs by the number of URLs examined, i.e., 20, in the result set. We then determine the average of the relevant percentage of the 20 queries submitted. Lastly, we calculate the improvement in terms of percentage to determine the effectiveness of the proposed Fuzzy Linguistic Semantic search method.

3.2. *Domains with matching semantic theme*

For the health and food domain we tested its search efficiency by first building a list of typical keywords and its possible *fuzzy linguistic counterparts*. The possible linguistic counterparts are obtained from *WordNet* [22] and appropriate membership functions are designed accordingly [23]. The lists of keywords and their fuzzy linguistic counterparts used in this experiment are as shown in List 1. The extended list of such in the health and

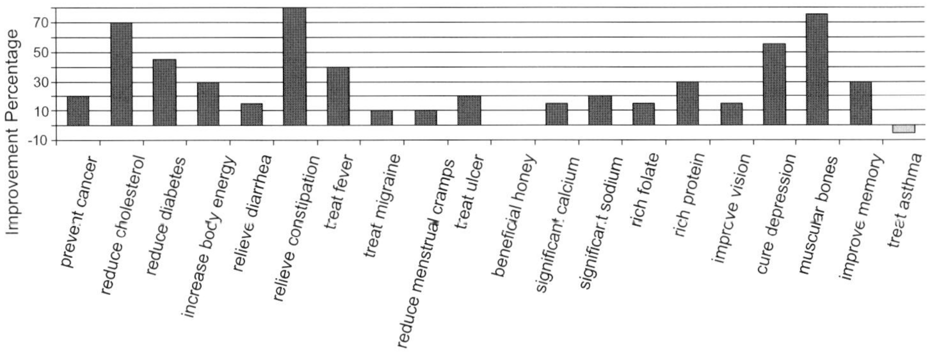

Figure 9. Improvement of search relevancy by using Fuzzy Linguistic Semantic search in domains with matching semantic theme.

Table 1
Relevancy result for conventional and Fuzzy Linguistic Semantic search methodologies in websites with matching semantic theme.

Keyword	Conventional search			Fuzzy Linguistic Semantic search			
	No. of URLs	Top 20 relevant URLs	Relevance %	No. of URLs	Top 20 relevant URLs	Relevance %	Search improvement %
prevent cancer	321	16/20	80	205	20/20	100	**20**
reduce cholesterol	208	6/20	30	433	20/20	100	**70**
reduce diabetes	289	6/20	30	162	15/20	75	**45**
increase body energy	397	8/20	40	343	14/20	70	**30**
relieve diarrhea	77	12/20	60	38	15/20	75	**15**
relieve constipation	94	5/20	20	59	20/20	100	**80**
treat fever	58	7/20	35	12	8/12	75	**40**
treat migraine	89	12/20	60	43	14/20	70	**10**
reduce menstrual cramps	215	13/20	65	25	15/20	75	**10**
treat ulcer	54	12/20	60	10	8/10	80	**20**
beneficial honey	183	20/20	100	82	20/20	100	**0**
significant calcium	326	17/20	85	262	20/20	100	**15**
significant sodium	392	16/20	80	297	20/20	100	**20**
rich folate	382	16/20	80	219	19/20	95	**15**
rich protein	217	14/20	70	424	20/20	100	**30**
improve eyesight	278	17/20	85	3	3/3	100	**15**
cure depression	88	6/20	30	34	17/20	85	**55**
muscular bones	54	4/20	20	43	19/20	95	**75**
improve brain memory	323	14/20	70	62	20/20	100	**30**
treat asthma	188	16/20	80	150	15/20	75	**−5**
Overall			**59**			**89**	**+30**

Table 2
Relevancy results for world-wide-web search using conventional and Fuzzy Linguistic Semantic search methodologies.

Keyword	Conventional search			Fuzzy Linguistic Semantic search			
	No. of URLs	Top 20 relevant URLs	Relevance %	No. of URLs	Top 20 relevant URLs	Relevance %	Search improvement %
prevent cancer	56	17/20	85	22	18/20	90	**5**
reduce cholesterol	61	19/20	95	31	20/20	100	**5**
reduce diabetes	45	16/20	80	16	14/16	92	**12**
increase body energy	55	19/20	95	32	18/20	90	**−5**
relieve diarrhea	46	17/20	85	28	18/20	90	**5**
relieve constipation	56	17/20	85	30	19/20	95	**10**
treat high fever	56	18/20	90	24	18/20	87	**−3**
treat migraine	55	19/20	95	28	20/20	100	**5**
reduce menstrual cramps	61	16/20	80	48	19/20	95	**15**
treat ulcer	42	18/20	90	44	20/20	100	**10**
beneficial honey	53	17/20	85	37	17/20	83	**−2**
significant calcium	46	16/20	80	47	19/20	94	**14**
significant sodium	40	11/20	55	12	11/12	92	**37**
rich folate	51	15/20	75	10	9/10	90	**15**
rich protein	44	18/20	90	7	7/7	100	**10**
improve eyesight	54	17/20	85	72	20/20	100	**15**
cure depression	55	17/20	85	12	12/12	100	**15**
muscular bones	51	17/20	85	8	8/8	100	**15**
improve brain memory	49	15/20	75	16	16/16	100	**25**
treat asthma	56	20/20	100	12	12/12	100	**0**
Overall			**85**			**95**	**+10**

food domain can be found in [1]. Note the order of words in each list is relevant in our work and such word ordering is ignored in the original WordNet synonym information. Take for example, *lower* will have a higher membership function values to the keyword *reduce* as compared with *decrease* in the fuzzy logic sense.

When 20 random queries in the form of (3.1) and containing the constraining variable from List 2 are composed and submitted for web search, the results returned are summarized in Table 1. Each query contains a pair of (X, Y) in the sense of (3.1). Fuzzy sets related to the domain of study, i.e., health and food are defined for possible X used. Table 1 clearly shows that, the Fuzzy Linguistic Semantic search method returned fewer and more relevant URLs. The proposed Fuzzy Linguistic Semantic method shows an average of 30% improvement in the search results. Furthermore, Figure 9 shows all, except one, queries did better by up to 80%.

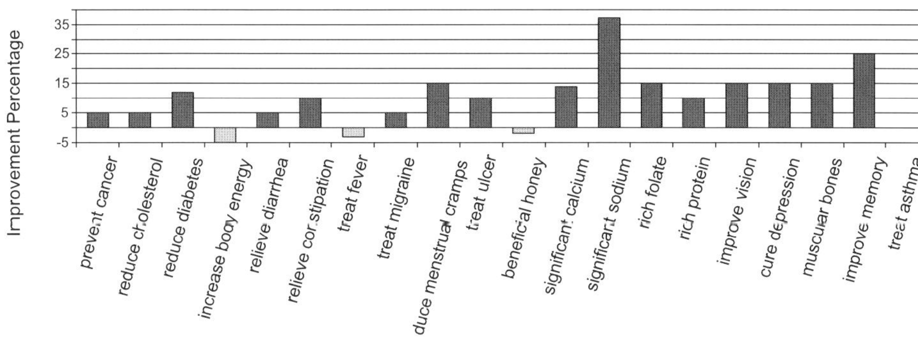

Figure 10. Improvement of search relevancy by using Fuzzy Linguistic Semantic search in world-wide-web search.

3.3. *World-wide-web search*

When the same 20 queries in Section 3.2 are submitted, the results returned by the conventional and our proposed ones are summarized in Table 2. Table 2 shows that the Fuzzy Linguistic Semantic search methodology return fewer and more relevant URLs among the top 20 items in the search result set. Our proposed Fuzzy Linguistic Semantic method shows 10% improvement over the conventional search in the search relevancy. Though the improvement is not as impressive as that applied to domains with matching themes, Figure 10 shows all, except three, queries gain search relevancy in the top 20 URLs returned.

4. Search improvement with fuzzy numeric semantic

One of powerful concept in fuzzy logic is its association of linguistic variables to numeric variable [25]. We shall take advantage of this concept to enhance the capability of web search engines in perceiving search queries [26].

We shall focus on the search query, (X, Y), in the same form as that in (3.1). The constraining variable X, i.e.,

$$\text{High, Medium, and Low,} \tag{4.1}$$

will be used to modify the selected constrained variables Y, with Y matches the chosen theme on health and food theme. Many search queries will not be able to return meaningful search results if they were submitted literally because the web page information may contain relevant data in a different form. Take for example, when one is looking for food with *medium Vitamin E* there may be lot of such relevant URLs that talks about food that contains Vitamin E with around 4 mg. However, one will probably not be able to find too many web pages that describe food as medium in Vitamin E. We will show in this section how this type of Fuzzy Numeric Semantic (FNS) search methodology can be used to improve search result relevancy.

The steps for FNS search are as follows:

Fuzzy Numeric Semantic search steps

NS1: Submit the constrained keyword Y to the search engine and obtain the set, $\boldsymbol{U}$, of the search result web pages.

NS2: Parse through the result set $\boldsymbol{U}$ for possible numbers z near the constrained keyword Y on web pages.

NS3: Calculate the membership function values, $\mu_{\text{FNS}}(z)$, of z with regards to the constraining keyword X with following Gaussian membership function formula:

$$\mu_{\text{FNS}}(z) = \mathrm{e}^{\frac{-(z-\bar{z})^2}{\sigma}}, \tag{4.2}$$

where $\bar{z}$ is the mean of the Gaussian membership function representing the fuzzy set for the keyword X and σ is a positive constant that determines the shape of the Gaussian membership function.

NS4: Take into account how close X and Y are to each other on the web page that contains them. Calculate the distance ranking score, i.e., S, of Y and X or Y and its Fuzzy Numeric Semantic counterparts (FNSC) for each webpage in the search results $\boldsymbol{U}$ with the following formula.
Case 1: $S = 1$, if Y and X or Y and its FNSC are adjacent to each other in the returned web page.
Case 2: $S = 0.5$, if Y and X or Y and its FNSC are in the same sentence of the returned web page.
Case 3: $S = 0.2$, if Y and X or Y and its FNSC are in the same page of the returned web page.
Case 4: $S = 0$, if Y and X or Y and its FNSC do not both show up in the returned web page.

NS5: Calculate the overall relevancy index R of each of the returned web pages in set $\boldsymbol{U}$ by

$$R = S * \mu_{\text{FNS}}(z). \tag{4.3}$$

EXAMPLE 4.1. An example of the above algorithm as applied to a search query like *Medium Vitamin E* can be as follows.

We start by submitting the constrained variable, *Vitamin E*, to the search engine to retrieve the set of all possible web pages that may contain the keyword *Vitamin E*. We then parse through this returned URL set, $\boldsymbol{U}$, for the presence of possible numerical counterparts of the constraining variable, i.e., Medium. The fuzzy membership functions of High, Medium, and Low for Vitamin E are as depicted in Figure 11.

Once the numbers near the constrained keyword *Vitamin E* is located on the web page, we compute its closeness to being Medium in the fuzzy sense by using (4.2). Take for example, if the number found near Vitamin E in the web page is 5.5 mg, the membership function will be

$$\mathrm{e}^{\frac{-(4-5.5)^2}{4}} = 0.78, \quad \text{with } s = 4. \tag{4.4}$$

Assuming the number 5.5 is found right next the constrained keyword *Vitamin E* (see Figure 12), the overall relevancy index, R, by using (4.3) for this URL will be

$$R = S * \mu_{\text{FNS}}(z) = 1 * 0.78 = 0.78.$$

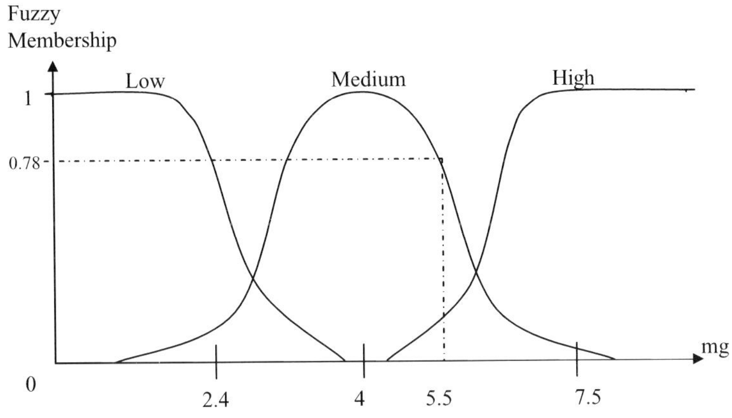

Figure 11. Fuzzy Numeric Semantic for Vitamin E.

Jicama
Numeric Value: **5.5** Score:0.78
Description: wholehealthmd. within in our healing centers you will discover a broad range of ways to manage over 100 health conditions and illnesses using integrative approaches in addition you will find personalized treatment advice using our step-by-step paths
Keywords: wholehealth. whole health. alternative medicine. complementary. integrative medicine. **vitamin**. herbs. supplements. acne. aging. alcoholism. allergies. alzheimer
Body: **vitamin** e (mg) **5.5** healing kitchen links jicama previous | next see recipes related to this food healing kitchen links > save to my wholehealth > send to a friend > printer-friendly version --> --> --> .disclaimer {font-family: verdana, arial, helvetica,

Figure 12. Matching URL with relevancy index as 0.78.

Table 3
Fuzzy Numeric Semantic for health and food ingredients.

Vitamin/Mineral	High	Medium	Low
calcium	250	150	80
calories	500	300	160
folate	100	60	32
iron	4.5	2.7	1.44
magnesium	875	525	280
potassium	875	525	280
saturated fat	5	3	1.6
sodium	600	360	192
protein	12.5	7.5	4
vitamin a	1250	750	400
vitamin c	15	9	4.8
vitamin e	7.5	4.5	2.4
vitamin k	20	12	6.4
carbohydrate	75	45	24
cholesterol	75	45	24
fat	16.25	9.75	5.2
saturated fat	5	3	1.6
fiber	6.25	3.75	2
biotin	75	45	24
vitamin d	100	60	32

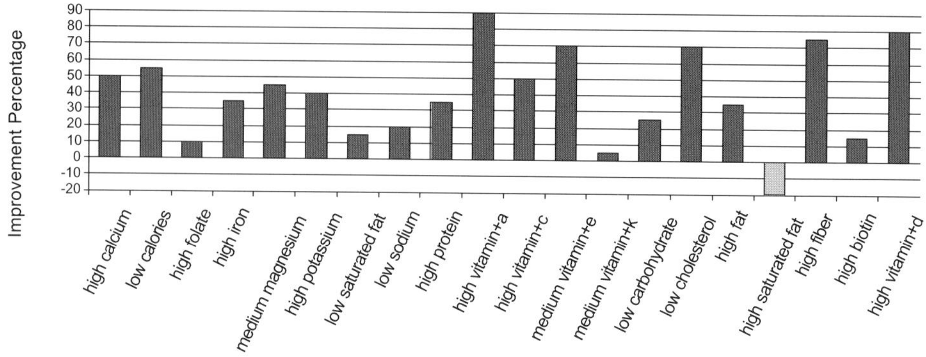

Figure 13. Improvement of search relevancy by using Fuzzy Numeric Semantic search in domains with matching semantic theme.

Table 4
Relevancy result for conventional and Fuzzy Numeric Semantic search methodologies in websites with matching semantic theme.

Keyword	Conventional search			Fuzzy Numeric Semantic search			
	No. of URLs	No. of relevant URLs	Relevance %	No. of URLs	No. of relevant URLs	Relevance %	Search improvement %
high calcium	270	10/20	50	159	20/20	100	**50**
low calories	329	9/20	45	165	20/20	100	**55**
high folate	299	16/20	80	187	18/20	90	**10**
high iron	253	12/20	60	187	19/20	95	**35**
medium magnesium	310	9/20	45	204	18/20	90	**45**
high potassium	251	12/20	60	193	20/20	100	**40**
low saturated fat	323	17/20	85	151	20/20	100	**15**
low sodium	316	15/20	75	285	19/20	95	**20**
high protein	610	10/20	50	339	17/20	85	**35**
high vitamin a	673	2/20	10	167	20/20	100	**90**
high vitamin c	673	8/20	40	231	18/20	90	**50**
medium vitamin e	90	4/20	20	150	18/20	90	**70**
medium vitamin k	90	1/20	10	13	2/13	15	**5**
low carbohydrate	316	15/20	75	279	20/20	100	**25**
low cholesterol	669	6/20	30	290	20/20	100	**70**
high fat	643	8/20	40	306	15/20	75	**35**
high saturated fat	323	16/20	80	153	12/20	60	**−20**
high fiber	611	5/20	25	366	20/20	100	**75**
high biotin	20	17/20	85	13	13/13	100	**15**
high vitamin d	673	4/20	20	23	20/20	100	**80**
Overall			**49**			**89**	**+50**

A snapshot of this particular URL matching such description is ranked 3rd in our FNS search and is shown in Figure 12.

Table 5
Relevancy results for world-wide-web search using conventional and Fuzzy Numeric Semantic search methodologies.

Keyword	Conventional search			Fuzzy Numeric Semantic search			
	No. of URLs	No. of relevant URLs	Relevance %	No. of URLs	No. of relevant URLs	Relevance %	Search improvement %
high calcium	53	18/20	90	7	6/7	83	**−7**
low calories	54	17/20	85	2	2/2	100	**15**
high folate	42	15/20	75	8	6/8	75	**0**
high iron	49	8/20	40	9	8/9	89	**49**
high manganese	52	16/20	80	6	6/6	100	**20**
low potassium	71	17/20	85	10	8/10	80	**−5**
low saturated fat	57	14/20	70	7	6/7	83	**13**
low sodium	47	17/20	85	7	6/7	83	**−2**
high protein	59	19/20	95	11	9/11	82	**−13**
high vitamin a	32	16/20	80	2	2/2	100	**20**
high vitamin c	68	18/20	90	9	8/9	89	**−1**
medium vitamin e	71	10/20	50	8	7/8	88	**38**
medium vitamin k	60	9/20	45	7	7/7	100	**55**
low carbohydrate	55	20/20	100	5	4/5	80	**−20**
low cholesterol	55	20/20	100	11	9/11	88	**−12**
low fat	53	17/20	90	4	3/4	75	**−15**
high saturated fat	48	16/20	80	10	7/10	70	**−10**
high fiber	41	17/20	85	7	6/7	83	**−2**
high biotin	48	16/20	80	4	4/4	100	**20**
high vitamin d	64	18/20	90	12	12/12	100	**10**
Overall			**80**			**87**	**+7**

In a similar manner we can compute the overall relevancy index for each of the URL in the returned set U. Subsequent ranking of the result can be done by sorting the URLs against their overall relevancy index values.

4.1. *Domains with matching semantic themes*

In this section, we compare our proposed Fuzzy Numeric Semantic search against the conventional search for possible search relevancy improvement in websites with theme on health and food. The health and food websites in List 1 will be used for this study. Vitamins, minerals, and other popular health ingredients listed in Table 4 are studied and assigned fuzzy membership functions for their High, Medium, and Low dosages. The numerical counterparts of High, Medium, and Low and their units in each ingredient are based on the published daily allowance according to USDA report [20].

Twenty random search queries related to the health and food subject are created in the form of (3.1), i.e., (X, Y), in which X is High, Medium, or Low, and Y is one of the keywords in the first column of Table 3. Searches are conducted with conventional search

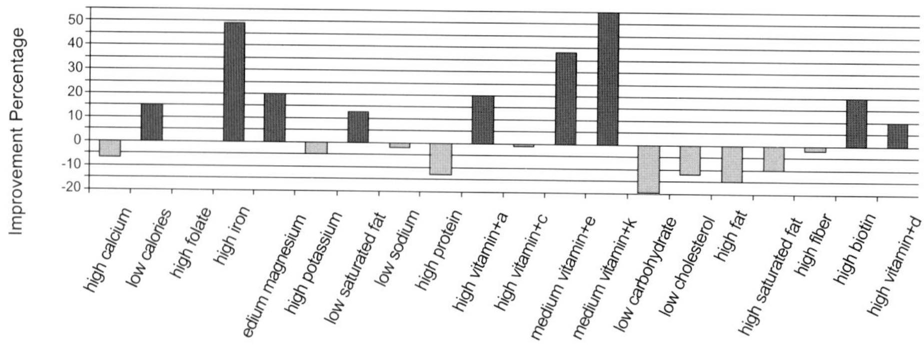

Figure 14. Improvement of search relevancy by using Fuzzy Numeric Semantic search in world-wide-web search.

and our proposed FNS search. Each of these 20 queries and the top 20 URLs for each query are checked for their relevancy under our two search methodologies. The results are tabulated in Table 4. The improvement in this case is even more impressive than that of FLS search in Section 3.2 with an average improvement over the conventional search as 50%. Figure 13 shows all, except one, queries with better and promising search relevancy in the top 20 URLs with our proposed FNS search methodology.

4.2. *World-wide-web search*

To test how semantic component in the search query may affect the search results, we submit the 20 queries in Table 4 to the world-wide-web. The conventional search engines like Google, Yahoo, AltaVista, Alltheweb, and MSN-search receive the submitted query. The returned results from these engines are combined to check on the relevancy of the top 20 URLs typically shown on the first page of the returned result set. Similarly, FNS search is conducted and the relevancy to the search query is checked and tabulated against the conventional search methodology. It is anticipated that the improvement in the world-wide-web search experiment may not be as impressive as the websites with matching theme. This is due to the fact that the world-wide-web data have varying semantic themes different from the one we designed. Table 5 still shows an average 7% improvement with our proposed FNS search over the conventional one. Figure 14 shows the gain of relevancy in the positive query test cases are also quite significant.

5. Conclusion

We present a perception-based search methodology to improve web search efficiency. The concept of fuzzy semantic for precisiating the search query proved to be very effective. Our experiments show a clear advantage of the proposed search methodology over the conventional and it is especially true when the websites match the fuzzy semantic in our design. This study shows the importance of semantic in improving web search. An important subject to be researched and validated experimentally is how complex queries with composite phrases can be abstracted for use in web search. GCL/PFL is an obvious choice for achieving this goal and will lead to a truly effective question-answering type search engine.

References

[1] H. Arcot, *Perception-based fuzzy information retrieval*, Master Thesis, San Jose State University, 2003.

[2] M.M. Beg, *A subjective measure of web search quality*, Inform. Sci. **169** (2005), 365–381.

[3] T. Berners-Lee, J. Hendler, O. Lassila, *The Semantic Web*, Scientific American **279** (5) (2001).

[4] F. Can, R. Nuray, A. Sevdik, *Automatic performance evaluation of Web search engines*, Inform. Process. Management **40** (2004), 495–514.

[5] M. Caramiaa, G. Felici, A. Pezzoli, *Improving search results with data mining in a thematic search engine*, Comput. Oper. Res. **31** (2004), 2387–2404.

[6] L. Chen, C. Luh, C. Jou, *Generating page clippings from web search results using a dynamically terminated genetic algorithm*, Inform. Systems **30** (2005), 299–316.

[7] D.Y. Choi, *Enhancing the power of Web search engines by means of fuzzy query*, Decision Support Systems **35** (2003), 31–44.

[8] A. Garcia-Serrano, P. Martinez, A. Ruiz, *Linguistic engineering approach to the enhancement of Web-searching*, in: Systems, Man, and Cybernetics, IEEE International Conference, 2001, pp. 447–452.

[9] E. Glover, G. Flake, S. Lawrence, W. Birmingham, A. Kruger, C. Giles, D. Pennock, *Improving category specific Web search by learning query modifications*, Applications and the Internet, 2001, pp. 23–32.

[10] Y. Horng, S. Chen, Y. Chang, C. Lee, *A new method for fuzzy information retrieval based on fuzzy hierarchical clustering and fuzzy inference techniques*, IEEE Trans. on Fuzzy Systems **13** (2) (2005), 216–228.

[11] B. Jansen, A. Spink, *An analysis of Web searching by European AlltheWeb.com users*, Inform. Process. Management **41** (2005), 361–381.

[12] G. Klir, B. Yuan, *Fuzzy Sets and Fuzzy Logic: Theory and Applications*, Pearson Publications, 1995.

[13] F. Liu, C. Yu, W. Meng, *Personalized Web search for improving retrieval effectiveness*, IEEE Trans. on Knowledge and Data Engineering **16** (1) (2004), 28–40.

[14] F. Menczer, *Complementing search engines with online web mining agents*, Decision Support Systems **35** (2003), 195–212.

[15] M. Okabea, S. Yamada, *Learning filtering rulesets for ranking refinement in relevance feedback*, Knowledge-Based Systems **18** (2005), 117–124.

[16] S. Oyama, T. Kokubo, T. Ishida, *Domain-specific Web search with keyword spices*, IEEE Trans. on Knowledge and Data Engineering **16** (1) (2004), 17–27.

[17] P. Rojanavasu, O. Pinngern, *Extended rough fuzzy sets for Web search agent*, in: Int. Conf. Information Technology Interfaces IT, 2003, pp. 403–407.

[18] S. Singh, L. Dey, *A rough-fuzzy document grading system for customized text information retrieval*, Inform. Process. Management **41** (2005), 195–216.

[19] W. Sunayama, M. Yachida, *Panoramic View System for extracting key sentences based on viewpoints and application to a search engine*, J. Network and Computer Applications **28** (2005), 115–127.

[20] USDA, *Tables: Dietary Reference Intakes: Vitamins, Food and Nutrition Board*, http://www.nal.usda.gov/fnic/etext/000105.html#q2, 2001.

[21] L. Vaughan, *New measurements for search engine evaluation proposed and tested*, Inform. Process. Management **40** (2004), 677–691.

[22] WordNet, Princeton, NJ, http://www.cogsci.princeton.edu/~wn/, 1998.

[23] L. Zadeh, *Knowledge representation in fuzzy logic*, IEEE Trans. on Knowledge and Data Engineering **1** (1) (1989), 89–100.

[24] L. Zadeh, *A new direction in AI, toward a computational theory of Perceptions*, AI Magazine **22** (1) (2001), 73–84.

[25] L. Zadeh, *From computing with numbers to computing with words. From manipulation of measurements to manipulation of perceptions*, Ann. NY Acad. Sci. **929** (2001), 221–252.

[26] L. Zadeh, *Toward a perception-based theory of probabilistic reasoning with imprecise probabilities*, J. Statist. Plann. Inference **105** (1) (2002), 233–264.

[27] L. Zadeh, *Web intelligence, world knowledge and fuzzy logic*, Powerpoint presentation slides, 2004.
[28] L. Zadeh, *Precisiated natural language*, AI Magazine **25** (3) (2004), 74–91.
[29] Q. Zeng, S. Kogan, R. Plovnick, J. Crowell, E. Lacroix, R. Greenes, *Positive attitudes and failed queries: an exploration of the conundrums of consumer health information retrieval*, Internat. J. Medical Informatics **73** (2004), 45–55.

CHAPTER 11

Using Knowledge Trees for Semantic Web Querying

Ronald R. Yager

Machine Intelligence Institute, Iona College, New Rochelle, NY 10801, USA
E-mail: yager@panix.com

Abstract

The capability for intelligent human like question answering is a fundamental goal of the semantic web. This capability requires a semantically enhanced, human like representation of various types of knowledge. It also requires a framework for providing a global plan for mobilizing the available knowledge to address the question posed. This should also direct us in seeking additional knowledge to help in the task. It also needs rules for making deductions from subsets of the knowledge base, performing local reasoning. We describe how the semantically rich fuzzy set based theory of approximate reasoning and the related paradigm of precisiated natural language can aid in the process of representing knowledge. The idea of a knowledge tree is introduced to provide a global framework for mobilizing the knowledge base in response to a query. We discuss how protoforms can be used to aid in deduction and local manipulation of knowledge. In addition to considering ordinary categorical knowledge we look at some types commonsense and default knowledge. These additional types of knowledge requires us to address the complexity of the non-monotonicity that these types of knowledge often display.

Keywords

semantic modeling, fuzzy sets, approximate reasoning, knowledge trees

1. Introduction

The Internet and more specifically the world wide web has generated considerable interest in the electronic retrieval and manipulation of information. The semantic web [1] will provide a quantum leap in this information processing capability. At a fundamental level it is an attempt to add the capability of human like information management and reasoning to the existing web capabilities. Human reasoning is based on a deep understanding

FUZZY LOGIC AND THE SEMANTIC WEB
Edited by Elie Sanchez
© 2006 Elsevier B.V. All rights reserved

of the knowledge. Computational machines can't understand, they only can manipulate and process. An important task for the success of the semantic web is the development of tools that enable the computational representation of knowledge that emulates human deep understanding. One tool that helps in this task is ontologies [5]. Among other things they enable an efficient modeling of the relationships between different objects so central to human cognition. While the use of ontologies provide an important tool for the computational modeling of human like semantics additional tools are needed. Among these is the need for a sophisticated set based framework for the representation of knowledge. The use of a set based framework allows the modeling of the granularity and imprecision so central to human language and conceptualization. Fuzzy sets, with its additional ability to allow the modeling of gradularity, will help further in capturing the robustness and sophistication of the human mind.

Various types of software applications have been developed to support the electronic retrieval of information. We refer to these as an Application Useful for Seeking INformation (AUSIN). One widely used class of AUSINs are search engines. These applications retrieve pointers to pages, files or documents based on their matching some key words specified by the user. Another widely used class of AUSINs are data base system. Another class of AUSIN are question answering systems [2,7]. An important dimension along which question answering systems differ from search engines and most databases is in their reasoning ability. A fundamental characteristic of question answering systems is its ability to reason over its information base. This facility leads to a fundamental difference in the nature of the response to query. A question answering system, because of its reasoning capacity, can construct new knowledge that is not resident in its knowledge base in response to a query. While currently less pervasive then either of the other information seeking systems, question answering systems with reasoning ability have the potential of being much more useful with regards to there responding to the real desires of the human user. With the development of the semantic web the tools will become more available to help support question answering systems. Our focus here is on use of fuzzy logic in the development of question answering systems.

Our interest in this work is on the development of tools to help in the representation of various types of knowledge as well as on the development of a framework and mechanism for reasoning and manipulating the knowledge in response to users questions.

Our approach will be in the spirit of Zadeh's paradigm of precisiated natural language and his related Generalized Constraint Language [20,21]. In providing a schema for representing and locally manipulating knowledge we shall rely heavily on the fuzzy set based theory of approximate reasoning [18] and particularly the idea of protoforms [20]. Considerable use will be made of knowledge trees [10,11,15] to provide a global framework for structuring and directing the process of answering a question posed to the system. As we shall subsequently see a knowledge tree takes the contents of the knowledge base and structures it in a manner to answer the question presented. These trees can be seen as a mobilization of the knowledge base to address a particular task. At a meta level the knowledge tree provides a global plan for solving the problem, the answering of the question. It is worth noting a knowledge tree can be used to help point directions for going outside our knowledge base for information to improve our answer. It will help us decide which information is relevant for answering the question being asked.

2. Basic concepts of approximate reasoning

In [21] Zadeh described his General Theory of Uncertainty (GTU) and the related Generalized Constraint Language (GCL). These provide a framework for the representation of various types of knowledge. Approximate Reasoning (AR) is a very important special case of this more general framework. In AR the constraints generally have of a possibilistic nature. The primary elements in an AR representation of a knowledge base are a collection of attribute variables, V_j for $j = 1$ to n, called the atomic variables. It is information about the value of these variables and relationships between them that constitutes the knowledge of interest. Associated with each variable V_j is a set X_j allowable values for the variable. A joint variable is any tuple of one or more distinct atomic variables; V_3, (V_2, V_5) and (V_1, V_2, V_6) are examples of joint variables. Associated with any joint variable is a domain set consisting of the Cartesian product of the domains of the individual variables making up the joint variable. In AR it is implicitly assumes that the variables of interest can assume only one value.

A categorical proposition is of the form V *is* A, here V is a joint variable and A is a fuzzy subset of the domain of V. A proposition which involves only one variable is called a canonical proposition while those involving two or more variables are called relational or joint propositions. A proposition in AR is viewed as imposing constraints on the possible values of the variable(s) involved. For example, if A is a crisp subset then the meaning of the proposition V_1 *is* A is to indicate that the value for V_1 is restricted to be in the set A. The use of fuzzy subsets provides for a grading of this idea.

When using the AR framework we are generally involved in one of three tasks. The first task, translation or knowledge representation is the process of taking knowledge normally expressed in natural language and converting into an appropriate representation within the framework of AR. The second task is extracting information from the knowledge base, it is called the inference process. It involves the manipulation of propositions to obtain other propositions. The third task is that of retranslation. Here we convert the propositions obtained via the inference process into statements of natural language. Our focus here is mainly on the second task, the manipulation of propositions to obtain other propositions and to a lesser extend on the representation issue. A comprehensive investigation of the question of retranslation can be found in [14].

The basic operations used for knowledge manipulation in AR are conjoin, containment and negation. The conjoin operation provides the system with the facility for combining or fusing information. In the translation process it allows us to represent knowledge involving multiple variables and allows us to construct joint variables from atomic variables. In the inference process it allows us to combine individual pieces of knowledge. The operation of conjoin is a generalization and unification of the operations of conjunction and Cartesian product. It is closely related to the join operation used in databases. We denote the conjoin of two propositions as V_a *is* $D \times V_b$ *is* E. In the special when the two variables being conjoined are the same the conjoin operation reduces to the usual intersection of fuzzy sets. The cylindrical extension [13,18] is a special case of the conjoin operation. Assume V_a and V_b are two joint variables such that V_b contains all the variables that are in V_a. The cylindrical extension operation allows us to extend the proposition V_a *is* F to be a proposition expressed using the extended variable, V_b *is* G. If X is the domain of V_a and $X \times Y$ the domain of V_b then $G(x, y) = F(x)$.

The second basic operation in approximate reasoning is containment [13]. If V_a *is* D and V_b *is* F are two propositions, we say that V_a *is* D contains V_b *is* F denoted V_b *is* $F \subseteq V_a$ *is* D if $F^\circ(z) \leqslant D^\circ(z)$ for all z where F° and D° are the cylindrical extensions of F and D to the base set of variable V, the union of the atomic variables in V_a and V_b.

The third basic operation in AR is negation. The negation of V *is* A, denoted not(V *is* A), is the proposition V *is* $\bar{A}$, $\bar{A}(x) = 1 - A(x)$ for each x.

As we indicated an important step in the construction of in question–answer systems is the translation of our knowledge from statements in a natural language into propositions in AR. A study of Zadeh's [6,17] work will provide a large collection of translation rules. Here we introduce some of these.

Assume P is a statement that has representation as V_a *is* A and Q is a statement with representation as V_b *is* B. Using these we obtain the following translation rules.

- not(P) is represented as V_a *is* $\bar{A}$ where $\bar{A}$ is the negation of A, $\bar{A}(x) = 1 - A(x)$.
- P and Q is represented as (V_a, V_b) *is* E where $E(x, y) = \text{Min}[A(x), B(y)]$.
- P or Q is represented as (V_a, V_b) *is* E where $E(x, y) = \text{Max}[A(x), B(y)]$.
- If P then Q is represented as (V_a, V_b) *is* E where $E(x, y) = \text{Max}[\bar{A}(x), B(y)]$.
- V *is* A *is* α certain is represented as V *is* B where $B(x) = \bar{\alpha} \vee A(x)$.

We now introduce the basic rules of inference available in AR [13,18]. In the following we use the notation $(P_1, \ldots, P_n) \vdash Q$ to indicate that we can infer Q from the collection $(P_1, \ldots, P_n)$. The first rule of inference is called the *entailment principle*.

IR-1: $P_1 \vdash P_2$ if $P_1 \subseteq P_2$.

The second rule of inference relates to the combination of multiple propositions.

IR-2: $(P_1, P_2, \ldots, P_n) \vdash P_1 \times P_2 \times \cdots \times P_n$.

We are now in a position to understand the process of obtaining inferences from knowledge base consisting of the collection $\{P_1, P_2, \ldots, P_n\}$. The set of propositions in our knowledge base induces a combined restriction which is the conjunction of all propositions in the knowledge base, KB $\equiv P_1 \times P_2 \times \cdots \times P_n$. Thus our knowledge base is also a proposition, V *is* A, which is the conjunction of all pieces of knowledge. The use of IR-2 along with IR-1 provides the bases of the reasoning mechanism

Basic Inference: Proposition P is inferable from a KB of proposition $P_1, \ldots, P_n$ denoted, $(P_1, \ldots, P_n) \vdash P$ if $P_1 \times P_2 \times \cdots \times P_n \subseteq P$.

It is important to emphasize that the inference requires that we must use the whole knowledge base. However, as we shall subsequently see, in some cases we need not use the whole knowledge to make inferences. This simplification results in a making the inference process tractable and useful. In cases in which we must use the whole knowledge base the process of reasoning becomes particularly onerous and often impractical.

We now introduce the important concept of monotonicity. We say that a proposition Q is monotonic if for any proposition $P \equiv V$ *is* A it is always the case $P \times Q \subseteq P$. If Q doesn't satisfy this condition we call it a non-monotonic proposition. All the propositions which we have introduced so far are monotonic. As we shall subsequently see non-monotonicity often arises in the representation of common sense knowledge.

In the following we shall assume all the propositions in the knowledge base are monotonic, we have a monotonic knowledge base. Monotonicity allows for a great simplification in the reasoning process. It is the basis of the type of deduction that is common in reasoning. The assumption of monotonicity implies that if $(P_1, \ldots, P_n) \vdash P$ then $(P_1, \ldots, P_n, P_{n+1}) \vdash P$. It means that adding any proposition doesn't cause us to have to withdraw any inference. This also implies that if we make an inference from some subset of the knowledge base we are sure it is valid for the whole KB. More formally under monotonicity if $\{P_1, \ldots, P_n\}$ is our knowledge base and if $(\widetilde{P}_1, \ldots, \widetilde{P}_q\}$ and $\{\widehat{P}_1, \ldots, \widehat{P}_r\}$ are two subsets of the KB then:

M1. If $(\widetilde{P}_1, \ldots, \widetilde{P}_q\} \vdash \widetilde{P}$ then $\{P_1, \ldots, P_n\} \vdash \widetilde{P}$.

M2. If $(\widetilde{P}_1, \ldots, \widetilde{P}_q\} \vdash \widetilde{P}, \{\widehat{P}_1, \ldots, \widehat{P}_r\} \vdash \widehat{P}$ and if $(\widetilde{P}, \widehat{P}) \vdash P$ then $(P_1, \ldots, P_n) \vdash P$.

This property allows us to take advantage of various types of local reasoning patterns, such as modus ponens, and proceed in a step by step fashion to obtain complex inferences. This situation allows localization which greatly simplifies the inference process. In the following we shall refer to a subset of our knowledge base as a local knowledge base. The observation of the preceding is that can make valid inferences using local knowledge bases.

3. Protoforms

Recently, Zadeh introduced the idea of protoforms [19–21]. Protoforms are local reasoning patterns based on the structure of the component propositions. Formally a protoform consists of a collection $P_1, \ldots, P_q$ of premises and a consequent Q and is of the nature $(P_1, \ldots, P_q) \vdash Q$. The basis of the protoform is the structure of the premises and consequent. For the most part the justification of these protoforms is based up application of conjunction (IR-1) and entailment (IR-2). The importance of these protoforms is that if we can find a local knowledge base (subset) of our knowledge base consisting of propositions whose structure matches the premise of a protoform then we can infer the consequent. Furthermore given the above properties M1 and M2 these protoforms can be independent parts of a complex deductive reasoning chain in monotonic knowledge bases.

In the following we shall identify some protoforms available in AR. We use the notation PF-K to indicate protoform #K. The premises are identified by P_j and the consequent is denoted Q.

PF-1 provides a basic protoform for inferring information about a component variable from a joint relation.

PF-1. *Projection protoform*

P_1: $(V_1, V_2, \ldots, V_q)$ *is* H,

Q: V_1 *is* D,

where $D = \text{Proj}_{X_1}((V_1, V_2, \ldots, V_q) \textit{ is } H)$. We recall $D(x_1) = \text{Max}_y H(x_1, y)$ where $y = (x_2, \ldots, x_q) \in Y$ and $Y = X_2 \times X_3 \times \cdots \times X_q$.

PF-2 is a fundamental protoform. It provides the basic protoform for making inferences from a joint relationship and companion propositions about its components. It essentially involves a conjunction of the premises followed by a projection onto the relevant variable.

PF-2. *Conjunction/projection protoform*

P_1: V_1 *is* B_1,

P_2: V_2 *is* B_2,

P_q: V_q *is* B_q,

P_{q+1}: $(V_1, V_2, \ldots, V_q)$ *is* H,

Q: V_j *is* D.

Here $D = \text{Proj}_{X_1}(V_1, V_2, \ldots, V_q)$ *is* E where E is obtained as

$$E(x_1, \ldots, x_q) = \text{Min}\big[B_1(x_1) \wedge B_2(x_2) \wedge \cdots \wedge B_q(x_n) \wedge H(x_1, \ldots, x_q)\big].$$

NOTE. If any of the $P_1, \ldots, P_q$ are missing these still holds where we can replace B_i by X_i.

The following three protoforms are important special cases of the preceding PF-2.

PF-3. *Modus ponens*

P_1: If V *is* A then U *is* F,
P_2: V *is* B,
Q: U *is* D.

Here $D(y) = \text{Poss}[\bar{A}|B] \vee F(y)$. We recall that $\text{Poss}[\bar{A}|B] = \text{Max}_x[\bar{A}(x) \wedge B(x)]$.

PF-4. *Weighted modus ponens*

P_1: If V *is* A then U *is* F is α certain,
P_2: V *is* B,
Q: U *is* D where $D(y) = \text{Poss}[\bar{A}|B] \vee \bar{\alpha} \vee F(y)$.

PF-5. *Modus tollens*

P_1: If V *is* A then U *is* F,
P_2: U *is* C,
Q: V *is* G where $G(x) = \bar{A}(x) \vee \text{Poss}[F|C]$.

PF-6. *Extension principle*

P_1: $U = f(V)$,
P_2: V *is* A,
Q: U *is* $f(A)$.

In this case $U = f(V)$ is translated as (V, U) *is* H where $H(x, y) = \bigcup_{x \in X}\{\frac{1}{(x, f(x))}\}$ and as in the preceding $G(x, y) = H(x, y) \wedge A(x)$. Here again $D(y) = \text{Max}_x[H(x, y) \wedge A(x)]$ however in this case $D(y) = \text{Max}_x[H(x, y) \wedge A(x)] = \text{Max}_{x \text{ s.t. } f(x)=y}[A(x)]$. Hence $D = f(A)$.

NOTE. More generally if $U = f(V_1, \ldots, V_q)$ and V_i is A_i for all i then we infer U *is* $f(A_1, \ldots, A_q)$ and $f(A_1, \ldots, A_q) = D$ where

$$D(y) = \underset{(x_1, \ldots, x_q) \text{ s.t. } f(x_1, \ldots, x_q)=y}{\text{Max}} \big[A_1(x_1) \wedge \cdots \wedge A_q(x_q)\big].$$

PF-6b. *Inverse extension*

P_1: $U = f(V)$,
P_2: U *is* B,
Q: V *is* $f^{-1}(B)$.

Again, $U = f(V)$ is translated as (V, U) *is* H where $H(x, y) = \bigcup_{x \in X}\{\frac{1}{(x, f(x))}\}$. However here $G(x, y) = H(x, y) \wedge B(y)$ and $D(x) = \text{Max}_y[H(x, y) \wedge B(y)]$. In this case

$$D(x) = \underset{y \text{ s.t. } f^{-1}(y)=x}{\text{Max}} \big[B(y)\big].$$

PF-7. *Fuzzy systems modeling*

P_1: Fuzzy Systems Model: (U, V) *is* $(A_1 \cap B_1) \cup (A_2 \cap B_2) \cup \cdots \cup (A_n \cap B_n)$,

P_2: U *is* E,
Q: V *is* G where $G(y) = \text{Max}_i[(\text{Poss}[A_i|E] \wedge B_i(y))]$.

4. Knowledge trees

In the following we shall describe an approach to constructing question answering systems using the concept of a knowledge tree [10,11]. As we shall see a knowledge tree is a structure generated from a knowledge base in response to a query. It can be seen as a mobilization of the knowledge base to address a particular task. At a meta level it provides a plan for answering the question. We shall denote our knowledge base as KB and initially we assume it consists of a collection of monotonic propositions. Associated with KB is a collection of atomic variables $V_1, \ldots, V_n$ and their associated domains, $X_1, \ldots, X_n$. We view the KB as consisting of two classes of propositions. The first class which we denote as $\mathcal{D}$ consists of all the atomic propositions, propositions of the form V_j *is* A_j where A_j is a fuzzy subset of the universe of V_j. Among the propositions in $\mathcal{D}$ are the atomic propositions V_i *is* X_i, these just specify the domain of V_i. The second class of those that involve joint variables, we denote the collection of these as $\mathcal{J}$, they specify relationships between variables. Considerable use will be made of protoforms as means of making local inferences in the knowledge tree.

We shall concern ourselves here specifically with questions that ask about the value of some atomic variable V_i. Our procedure for answering a question will be a two phase process. The first phase is a generation of a knowledge tree in response to the posing of a question. The second phase is the contraction of the tree to find the value the variable queried about. The first phase can be seen as generating the direction and plan for answering the question

Before proceeding let us clarify the tree terminology we shall use. A node n_a will be called an ancestor of the node n_b if n_a appears anywhere on the path from the initial node of the tree to n_b. In this case n_b is called a descendant of n_a. A node n_a is called the parent of node n_b if n_b emanated directly from n_a. In this case n_b is called the child of n_a. The terms immediate ancestor and immediate descendant are equivalent to parent and child.

The following is the basic procedure for the *construction* of the knowledge tree from the knowledge base in response to a query about an atomic variable V.

BASIC TREE GENERATION ALGORITHM.

1. The posing of a question in the form

   ? V *is*

   initiates the knowledge tree with a triangle of the form .
2. The appearance, anywhere in the knowledge tree, of a triangle labeled with a variable *not having itself as an ancestor* causes the emanation from the triangle of a family immediate successor nodes:
   a. One corresponding to each data proposition in the knowledge base subset $\mathcal{D}$ involving the label of the triangle. A successor node corresponding to an element from $\mathcal{D}$ will be manifested by a rectangle labeled by the name of the data proposition, i.e. $\boxed{d_i\text{: } V \text{ is } A}$.

b. One corresponding to each proposition in $\mathcal{J}$ having the label of the parent triangle as one of its component variables. A successor node corresponding to an element from $\mathcal{J}$ is manifested by a circle labeled by the name of the proposition, i.e. (j_i).
c. If there are no appropriate elements from either a or b this causes the emanation of a rectangle labeled by the domain of the variable, $\boxed{V\ is\ X}$.

3. The appearance anywhere in the knowledge tree of a triangle labeled with a variable *having itself as an ancestor* causes the emanation from that node of one successor node, a rectangle labeled by variables domain, $\boxed{V\ is\ X}$.
4. The appearance of a node corresponding to an element from $\mathcal{J}$, a labeled circle, causes the emanation from it of a family of successor triangle nodes each labeled by a variable appearing in the proposition denoted by the circle.
5. The appearance of a labeled rectangle terminates that branch of the tree.
6. The construction of the knowledge tree is completed when all branches terminate in rectangles.
7. The completion of the knowledge tree generation phase initiates the evaluation phase.

In Figure 1 we illustrate a typical knowledge tree.

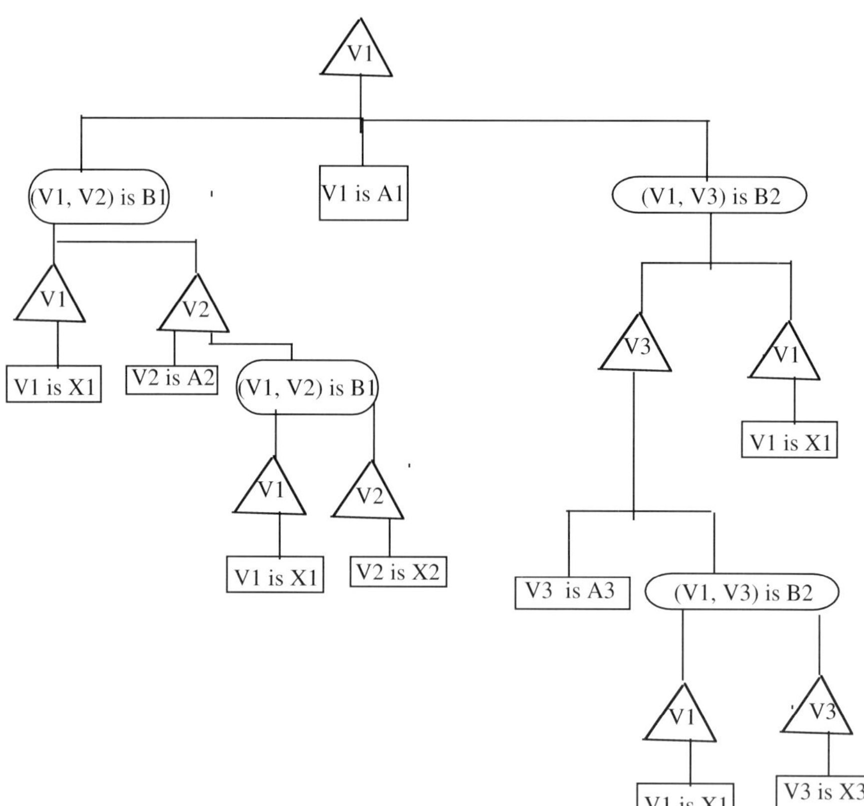

Fig. 1. Typical knowledge tree.

Once having completed the construction of the knowledge tree initiated by our query we must next apply an evaluation procedure to calculate the value of the desired variable. The evaluation algorithm is based upon the rules of approximate reasoning and makes considerable use of the protoforms introduced earlier to provide local inference tools.

EVALUATION ALGORITHM.

1. Any subtree consisting of a triangular node and a family of children which are all rectangular nodes can replaced by a rectangular node whose label is V *is* H, $\boxed{V \text{ } is \text{ } H}$. Here V is the label of the triangular node and H is the conjunction of the values of the children of this triangular node. (It is an application of the basic conjunction protoform.)
2. Any subtree consisting of a circular node, a relational proposition, and a family of children which are all rectangular nodes can be replaced by a rectangular node whose label is V_1 *is* G where V_1 is the label of the triangular node immediately preceding the circular node. Typically this is accomplished by application of PF-2, the conjunction/projection protoform. However more generally this may involves a two-step process. The first step is the application of the protoform associated with the circular node and its children. The second step is an application of a projection onto V_1 of the proposition resulting from Step 1.
3. The evaluation phase terminates when there exists one rectangular node whose label is the desired variable with its associated its value as the answer to the question.

5. Knowledge trees for directing knowledge search

The knowledge tree forms a useful framework for formulating the question answering process. As we see it takes a knowledge base and restructures the information in it to address a question being asked. We here briefly point out one feature of the knowledge tree approach which is a very useful possibility. In particular the knowledge tree can point us in a direction to seek additional information not in our KB that may help us getting a good answer. A simple illustration of this is the following. Assume we are interested in finding the value of V_1. Assume in our KB we have a relation (V_1, V_2) is G. This relation will appear as a decedent of our query to find V_1. Let us further assume we have no knowledge of V_2. Thus this branch will terminate with a rectangle $\boxed{V_2 \text{ } is \text{ } X_2}$. If as a result of our working on the whole tree we get an answer for V_1 is that is not sufficiently informative for the user we see that this branch provides a potential path for improving our answer. It is telling us if you are able to get better information about V_2 you can possibly improve you answer for V_1. A meta level this is essentially pointing in some direction for improving the quality of the answer.

A simple intuitive example of this is the following, Consider you are trying to obtain John's age. Assume you have in your knowledge base the relationship that people are about eighteen years old when they graduate from high school. This tells us that we can benefit by trying to find out when John graduated from high school. The fundamental issue here revolves around Zadeh's idea of relevance with respect to information retrieval. The concept of relevance plays an important role in determining which information we should seek on the web. Closely related to the concept of relevance is the idea of ease of extraction. Assume we are interested in some variable V and assume we have some relationship between another variable U and V. Thus implicit in this relationship is some information about V. In order to extract this information we need some information about the related variable U. The concept of ease of extraction is related how precise or knowledgeable we have to be about U in order to extract the knowledge about V contained in the relationship. Central to the formulation of the concept of relevance and the related idea of ease of extraction is the use of the measure of specificity [12] which enables a quantification of the amount of information in a possibility distribution.

6. Non-monotonic possibilistic propositions

In order to take full advantage of the wide array of different types of information available we extend our capability to allow the inclusion of the types of commonsense knowledge that play a substantial role in human reasoning. Examples of these are "generally it is difficult to find parking in New York City" and "rainy weather usually results in more traffic congestions" and "a cup of coffee costs about a dollar." This commonsense knowledge often is not the kind of categorical knowledge that can be simply represented by the type of propositions we have thus far considered. It is often tentative and context dependent. These types of propositions are often involved in inferences which must be withdrawn in face of gaining additional knowledge. This situation is formally referred to as non-monotonicity [4,9]. In the preceding we assumed all the knowledge to be categorical, of a monotonic type, this allowed us to use local reasoning techniques. Here we consider the handling of propositions that can display a non-monotonicity.

As we noted non-monotonic propositions often come from the inclusion of commonsense knowledge. One important class of non-monotonic propositions are of a default type, they provide values which we can associate with a variable in a situations in which our other available information about the variable doesn't provide a sufficiently precise value. An important aspect of non-monotonic propositions is that they are submissive to categorical information. For example assume that we are interested in John's age and we know that he is in high school. As commonsense knowledge we know that typically high school students are teenagers. Using this piece of commonsense knowledge our best guess is to conclude that John is a teenager. However assume now that we have the additional knowledge that John is in his thirties. In this situation we want the default based inference that John is a teenager to defer to the categorical knowledge, that John is in his thirties. Implicit in this situation is some idea of a prioritization of knowledge. In particular the default knowledge has a lower priority then categorical knowledge.

Another use of default propositions are as a simplifying tool. Here they are used to allow us to make inferences without requiring us to obtain some antecedent pieces of knowledge with complete certainty. In this case the possibility of some antecedent being true, it has not been falsified, is sufficient grounds for making an inference. An illustrative example of this is a rule about crossing streets: "if the light is red and the driver is sober then assume they will stop." Often when using this we can't realistically ascertain with complete certainty that a driver is sober, we usually settle for less certainty, he doesn't manifest any drunkenness. Here we can cross without having to test each driver.

The inclusion of non-monotonic propositions can add considerable complexity to the reasoning process. This complexity results from the fact that if P_1 is non-monotonic and $(P_1, P_2) \vdash Q$ it may be the case that using the whole knowledge base $(P_1, P_2, \ldots, P_n)$, doesn't justify the inferring of Q. Thus with inclusion of non-monotonic knowledge all *local* inferences may not be valid. A deeper understanding of the situation can benefit from the introduction of the concept of a "tail" associated with a non-monotonic proposition. Let $\mathcal{P} = (P_1, P_2, \ldots, P_n)$ be our knowledge base and let P^* be a proposition in $\mathcal{P}$. The tail T_{P^*} associated with P^* is a subset of $\mathcal{P}$ consisting of the minimal subset of propositions from $\mathcal{P}$ such that any proposition inferred from P^* along with its tail is also inferable from the whole knowledge base. The key point being that protoforms involving non-monotonic knowledge must include all members of its tail. This can lead to very complex protoforms which may involve many propositions and is context dependent upon what propositions exist in the KB.

We point out that monotonic, categorical, propositions have empty tails. In the monotonic world if we don't use all the available information about a subject we can end being less precise although not incorrect. In situations in which we use a non-monotonic piece of knowledge without considering its tail we can end up making an incorrect inference.

An important class of commonsense knowledge can be represented by propositions that have a possibilistic component. As we shall subsequently see this allows us to make inferences using the fact that some antecedent is possibly true rather the certainly true. This softening of the requirements brings with it the potential that we are wrong and hence introduces a non-monotonicity. In the following we shall look at some basic protoforms, associated with propositions having a possibilistic component.

In order to get an understanding of a paradigm that plays a fundamental role in many protoforms associated with possibilistic non-monotonic reasoning patterns we review the classical modus ponens protoform. In modus ponens we have $(P_1, P_2) \vdash Q$ where

P_1: If V *is* A then U *is* F,

P_2: V *is* B,

Q: U *is* D,

where $D(y) = (1 - \text{Cert}[A|B]) \vee F(y)$. Here we have used the fact that $\text{Cert}[A|B] = 1 - \text{Poss}[\overline{A}|B]$. We see that with this modus ponens protoform to infer U *is* F we require that we are *certain* that V *is* A.

We now consider a related non-monotonic protoform

NMPF-1: Basic Non-Monotonic Possibilistic Protoform: $(P^*, P_2) \vdash Q$ where[1]

P^*: If V *is* A *is possible* then U *is* F,

P_2: V *is* B,

Q: U *is* E,

where $E(y) = (1 - \text{Poss}[A|B]) \vee F(y)$. Here we see that we infer U *is* F if A is possible given we have B. Note in the usual modus ponens we require A is certain given we have B. Since $\text{Poss}[A|B] \geqslant \text{Cert}[A|B]$, then this is easier to satisfy.

A simple example illustrates this. Let V have domain $X = \{a, b, c, d\}$. Let $A = \{a\}$ and let $B = \{a, b\}$. In the case of modus ponens we get $D(y) = (1 - \text{Cert}[A|B]) \vee F(y)$. Since $\text{Cert}[A|B] = 1 - \text{Poss}[\overline{A}|B] = 1 - \text{Max}_x[\overline{A}(x) \wedge B(x)] = 0$ then $D(y) = (1 - \text{Cert}[A|B]) \vee F(y) = 1$ and we can't infer anything. In the case of P^* we get $E(y) = (1 - \text{Poss}[A|B]) \vee F(y)$. Since $\text{Poss}[A|B] = \text{Max}_x[A(x) \wedge B(x)] = 1$ then $E(y) = F(y)$. Thus here we infer U *is* F.

However we also can easily illustrate the non-monotonicity associated with the use of P^*. Assume that we get the additional piece of information P_3: V *is* G where $G =$

[1] We are implicitly assuming here that P_2 is the tail of P^*.

$\{b, c\}$. Here now the tail associated with P^* consists of both P_2 and P_3. In this case we must use (P^*, P_2, P_3) and our knowledge about V is that V *is* H where $H = G \cap B = \{b\}$. In this case $\text{Poss}[A|H] = 0$ and hence $E(y) = 1 - 0 \vee F(y) = 1$, $E = X$, and thus we can't infer anything.

A slightly more general version of the above protoform can be described. Let P^* be as in the preceding. If V *is* A *is possible* then U *is* F. Let T_{P^*} be the tail of P^*. Assume $T_{P^*} \vdash V$ *is* H then $(P^*, T_{P^*}) \vdash U$ *is* E where $E(y) = (1 - \text{Poss}[A|H]) \vee F(y)$. Let us formally denote this protoform.

NMPF-1b: *General* Non-Monotonic Possibilistic Protoform: $(P^*, T_{P^*}) \vdash Q$ where

P^*: If V *is* A *is possible* then U *is* F,

T_{p^*}: $\vdash V$ is H,

Q: U *is* E where $E(y) = (1 - \text{Poss}[A|H]) \vee F(y)$.

An issue here is the determination of the tail of P^*. In this case it is easily obtained by querying the knowledge base for the value of V. Thus in this case the appearance of P^* in a knowledge tree would generate a triangle labeled by V. This triangle would in turn generate a whole subtree which essentially constitutes T_{P^*}. One fundamental difference between this situation with P^* and that of the usual modus ponens with P_1 is the following. While the appearance of either of them in a knowledge tree will generate a trailing subtree initiated by a triangle labeled with V the modus ponens situation will allow us to prune this subtree and still obtain a resulting inference for U that is valid inference while the pruning of the subtree in the case of P^* can lead to invalid inferences.

An important class of non-monotonic possibilistic protoforms relates to reasoning with rules that supply default or typical values. In reasoning with these types of propositions we must often distinguish between what we know about a variable before we apply the proposition and value of the variable after we apply the protoform. Here our prior knowledge about the variable is important in that it effects how a protoform works. This knowledge is generally obtained from propositions with a higher priority, a stronger certainty, then the protoform being implemented. The following protoform relates to the reasoning with default or typical values

NMPF-2: *Typical Value Protoform*

P^*: Typically (V *is* A),

P_2: V *is* B (Prior Knowledge),

Q: V *is* D.

Here $D(x) = B(x) \wedge (A(x) \vee [1 - \text{Poss}(A|B)])$. In this case we represented the knowledge typically (V *is* A) as the possibilistic proposition.

If V *is* A is *possible* (with what we already know about V) then V *is* A.

Here what we already know is that V *is* B. We see that we are taking as D the conjunction of B and A is credible given B this gives us

$$D = B \cap (A \text{ is credible given } B).$$

A slightly more general version of the above protoform can be described. Let P^* be as in the preceding. Typically (V *is* A). Let T_{P^*} be the tail of P^*. Assume $T_{P^*} \vdash V$ is H then $(P^*, T_{P^*}) \vdash V$ *is* E where $E(x) = H(x) \wedge (A(x) \vee (1 - \text{Poss}(A|H)))$. We shall formally denote this protoform as NMPF-2b: *Typical Value Protoform*.

Another closely related protoform is

NMPF-3: *General Typical Value Protoform*

P^*: If V *is* C and U *is* A *is possible* then U *is* A,

P_2: V *is* G,
T_{p^*}: $\vdash U$ is B,
P_3: U^* *is* B,
Q: U *is* D where $D(y) = B(x) \wedge (A(x) \vee (1 - \mathrm{Cert}[C|G]) \vee (1 - \mathrm{Poss}[A|B]))$.

7. Non-monotonic propositions in knowledge trees

Here we are interested in the construction of knowledge trees from knowledge bases having non-monotonic propositions. However first we consider the formulation of knowledge bases having default or other types of non-monotonic knowledge. As we already noticed when working with non-monotonic propositions the idea of priority of proposition plays an important role. This prioritizing of propositions must be incorporated in the knowledge base. We note that the ordinary categorical type knowledge has the highest priority.

In constructing our knowledge base we shall add an additional class to $\mathcal{J}$ and $\mathcal{D}$ consisting of our possibilistic non-monotonic propositions. We shall denote this category as $\mathcal{N}$. Furthermore we shall assume that the propositions in $\mathcal{N}$ are ordered with respect to their priority, there is a hierarchy of these propositions. To avoid unnecessary complexities at this junction we assume this is a linear ordering. Thus $\mathcal{N}$ is a collection of non-monotonic propositions where

$$\mathcal{N} = N_q > N_{q-1} > \cdots > N_2 > N_1.$$

Here the bigger index the higher the priority. This hierarchy reflects the fact that the higher priority the sooner we want to consider the information in the reasoning process. Thus if N_i and N_j are such that $i > j$ this means that we want to use N_i before we use N_j. The categorical propositions in $\mathcal{D}$ and $\mathcal{J}$ are assumed of higher priority then any element in $\mathcal{N}$ and as such should be considered before any element in $\mathcal{N}$. We note prior consideration in the tree evaluation process means an appearance lower in the knowledge tree.

As is the case with the elements in $\mathcal{J}$ and $\mathcal{D}$ the propositions in $\mathcal{N}$ have associated variables and generally function by enforcing constraints on these variables.

The inclusion of the component $\mathcal{N}$ in our knowledge base effects the process of generating the knowledge tree described earlier. It particularly effects our action when we get a labeled triangle. In the following we shall use a labeled trapezoidal, N_k, to indicate the proposition N_k in $\mathcal{N}$.

We now provide a modified version of Step 2 in the tree generation process to account for the inclusion of non-monotonic elements in $\mathcal{N}$.

The appearance in the knowledge tree of a triangle labeled with a variable, V, not having itself as an ancestor causes the emanation from the triangle of a tail consisting of.

(i) A *serial* emanation of labeled trapezoids corresponding to all propositions in $\mathcal{N}$ having V as one of its components. This series begins with the *lowest* priority proposition and precedes in increasing order of priority. Furthermore emanating from each trapezoid is a special dashed triangle,[2] V, labeled by V as well as regular triangles labeled by each of the other variables associated with the preceding trapezoid.

[2] We use a dashed triangle so as to indicate that this is not a new appearance of the variable but part of the current tail.

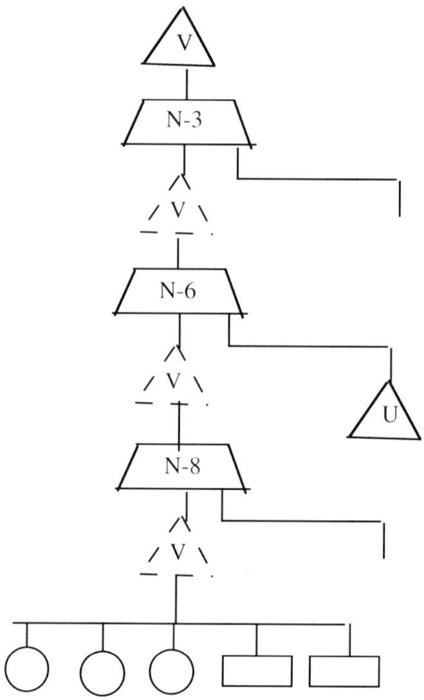

Fig. 2. Subtree from non-monotonic propositions.

(ii) The end of the chain of trapezoids causes the emanation from its trailing labeled dashed triangle of a parallel family of immediate successor nodes. (a) One corresponding to each proposition in the knowledge base subset $\mathcal{D}$ involving the label of the dashed triangle. These are labeled rectangles. (b) One corresponding to each proposition in $\mathcal{J}$ having the label of the parent dashed triangle as one of its component variables. These are labeled rectangles as described earlier. Finally if there are no appropriate elements from either (a) or (b) this causes the emanation of a rectangle labeled by the domain of the variable, $\boxed{V \ is \ X}$.

In Figure 2 we provide an illustration of a sub-branch of a knowledge tree generated using this revised method.

The evaluation step is only modified by an instruction for handling trapezoidal nodes. We add the following item.

3. Any subtree consisting of a trapezoidal node and a family of children which are all rectangular nodes can be replaced by a rectangular node whose label is determined by the preceding triangle and whose value is determined by the protoform associated with the proposition denoted by trapezoid.

8. Defuzzification

Let V be a variable taking its value in the domain X. In the preceding we considered the problem of finding the value of this variable from a knowledge base. In answering this

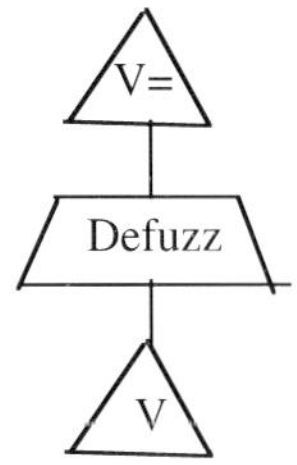

Fig. 3. Defuzzification subtree.

question we allowed our answer to be a fuzzy subset. More generally we allowed for an answer that contains some uncertainty or imprecision. In some cases we may be required to provide a more precise answer, a value that is an element from the set X. An example of this occurs in the widely used technology of fuzzy systems modeling [8]. If the available knowledge about V allows us only to infer that V *is* A where is a fuzzy subset we must provide a process to obtain a single element from X using this information. Here shall refer to this step as defuzzification, Defuzz. Other closely related terms for this process are choice, selection, decision or precisiation.

The process of defuzzification has been studied in the literature [16] and there exists a large number of methods for implementing this operation. In its most general sense the protoform associated defuzzification (choice) is an example of a *non-monotonic* operation. Let us look at this operation. In the following we provide a basic protoform of this operation

Protoform: *For* Defuzz(V *is* A)

P_1: V *is* A,

Q: $V = x_q$,

where $x_q = \text{Max}_{x \in X}[A(x)]$.

We note other protoforms of defuzz are possible. For example if X is the real line then Filev and Yager [3] defined the BADD defuzzification of V *is* A as $V = a$ where $a = \sum_{x_j} x_j \frac{A(x_j)^r}{\sum_{x_k} A(x_k)^r}$ with $r > 0$. We see if $r \to \infty$ we get the Max operator.

Let us see the non-monotonicity implicit in this operation. Assume we infer V *is* A from our current knowledge base. Using defuzzification we get DEFUZZ(V *is* A) $\Rightarrow V = a$. Assume we get more knowledge and subsequently obtain V *is* $\widetilde{A}$. In this case we get DEFUZZ(V *is* $\widetilde{A}$) $\Rightarrow V = \tilde{a}$. For the defuzzification process to be monotonic we must have $\tilde{a} = a$, a situation which cannot be guaranteed. At a meta level the reason for the non-monotonicity associated with the defuzzification operation is that in choosing a precise element we are assuming information, above and beyond what we have a right to do. We are essentially using an operation related to the principle of maximal entropy which is non-monotonic.

In using our knowledge tree approach we can include these types of exact questions in the following manner. We initiate our knowledge tree with a special triangle $V =$. The occurrence of this equal triangle causes the emanation of the defuzz protoform as seen in Figure 3. The appearance of the ordinary triangle induces a process of finding V.

9. Conclusion

The construction of systems to support intelligent question–answering is an important part of the agenda of the semantic web. In this work we described a number of technologies to help support the development of this capability. Our approach makes considerable use fuzzy sets and other related soft computing technologies and is in the spirit of Zadeh's paradigm of computing with words. Central to our approach is the use of knowledge trees to provide a framework for mobilizing the knowledge in the system in response to a query.

References

[1] G. Antoniou, F. van Harmelen, *A Semantic Web Primer*, MIT Press, Cambridge, MA, 2004.
[2] V. Chaudhri, R. Fikes, *Question Answering Systems*, AAAI Press, Menlo Park, 1999.
[3] D.P. Filev, R.R. Yager, *A generalized defuzzification method under BAD distributions*, Internat. J. Intelligent Systems **6** (1991), 687–697.
[4] M.L. Ginsberg, *Readings in Nonmonotonic Reasoning*, Morgan Kaufmann, Los Altos, CA, 1987.
[5] A. Gomez-Perez, M. Fernandez-Lopez, O. Corcho, *Ontological Engineering*, Springer, Heidelberg, 2004.
[6] G.J. Klir, B. Yuan (Eds.), Fuzzy Sets, Fuzzy Logic and Fuzzy Systems: Selected Papers by Lotfi A. Zadeh, World Scientific, Singapore, 1996.
[7] M. Maybury (Ed.), New Directions in Question Answering: Papers from the 2003 AAAI Spring Symposium, AAAI Press, Menlo Park, 2003.
[8] W. Pedrycz, F. Gomide, *An Introduction to Fuzzy Systems*, MIT Press, Cambridge, MA, 1998.
[9] R. Reiter, *Nonmonotonic reasoning*, Ann. Rev. Comput. Sci. **2** (1987), 147–186.
[10] R.R. Yager, *Querying knowledge base systems with linguistic information via knowledge trees*, Int. J. Man–Machine Studies **19** (1983), 73–95.
[11] R.R. Yager, *Knowledge trees in complex knowledge bases*, Fuzzy Sets and Systems **15** (1985), 45–64.
[12] R.R. Yager, *On measures of specificity*, in: O. Kaynak, L.A. Zadeh, B. Turksen, I.J. Rudas (Eds.), Computational Intelligence: Soft Computing and Fuzzy-Neuro Integration with Applications, Springer-Verlag, Berlin, 1998, pp. 94–113.
[13] R.R. Yager, *Approximate reasoning as a basis for computing with words*, in: L.A. Zadeh, J. Kacprzyk (Eds.), Computing with Words in Information/Intelligent Systems, vol. 1, Springer-Verlag, Heidelberg, 1999, pp. 50–77.
[14] R.R. Yager, *On the retranslation process in Zadeh's paradigm of computing with words*, IEEE Trans. on Systems Man Cybernet. B **34** (2004), 1184–1195.
[15] R.R. Yager, *Knowledge trees and protoforms in question answering systems*, J. Amer. Soc. Inform. Sci., submitted for publication.
[16] R.R. Yager, D.P. Filev, *On the issue of defuzzification and selection based on a fuzzy set*, Fuzzy Sets and Systems **55** (1993), 255–272.
[17] R.R. Yager, S. Ovchinnikov, R. Tong, H. Nguyen (Eds.), Fuzzy Sets and Applications: Selected Papers by L.A. Zadeh, Wiley, New York, 1987.
[18] L.A. Zadeh, *A theory of approximate reasoning*, in: J. Hayes, D. Michie, L.I. Mikulich (Eds.), Machine Intelligence, vol. 9, Halstead Press, New York, 1979, pp. 149–194.
[19] L.A. Zadeh, *Toward a perception-based theory of probabilistic reasoning with imprecise probabilities*, J. Statist. Plann. Inference **105** (2002), 233–264.
[20] L.A. Zadeh, *Precisiated natural language (PNL)*, AI Magazine **25** (3) (2004), 74–91.
[21] L.A. Zadeh, *Toward a generalized theory of uncertainty (GTU) – an outline*, Inform. Sci. **172** (2005), 1–40.

XML based Approaches, Building Ontologies, and (Conceptual) Graphs

CHAPTER 12

Fuzzy Data Mining for the Semantic Web: Building XML Mediator Schemas

A. Laurent and M. Teisseire

LIRMM–CNRS UMR 5506, 161, rue Ada, 34392 Montpellier Cedex 5, France
E-mail: {laurent,teisseire}@lirmm.fr

P. Poncelet

LGI2P, EMA–Site EERIE, Parc scientifique G. Besse, 30035 Nîmes Cedex 1, France
E-mail: pascal.poncelet@ema.fr

Abstract

As highlighted by the World Wide Web Consortium, XML has been proposed to deal with huge volumes of electronic documents and is playing an increasing important role in the exchange of a wide variety of data on the Web. However, when dealing with such large and heterogeneous data sources, it is necessary to have an idea on the way these data sources are structured. This information is indeed essential in order to build mediator schemas. These mediator schemas are required to query data in a uniform way. Moreover, this information is interesting since it provides users with a semantic structure of the data they can query. Recently schema mining approaches have been proposed to extract in an efficient way the commonly occurring schemas from a collection. Nevertheless, according to the semantic point of view, such approaches suffer from different drawbacks. In this work, we propose thus a fuzzy approach, showing why and how fuzziness is useful in order to extract frequent approximate schemas.

Keywords

fuzzy logic, semi-structured data, mediator schemas, XML, semantic structures, frequent patterns

FUZZY LOGIC AND THE SEMANTIC WEB
Edited by Elie Sanchez
© 2006 Elsevier B.V. All rights reserved

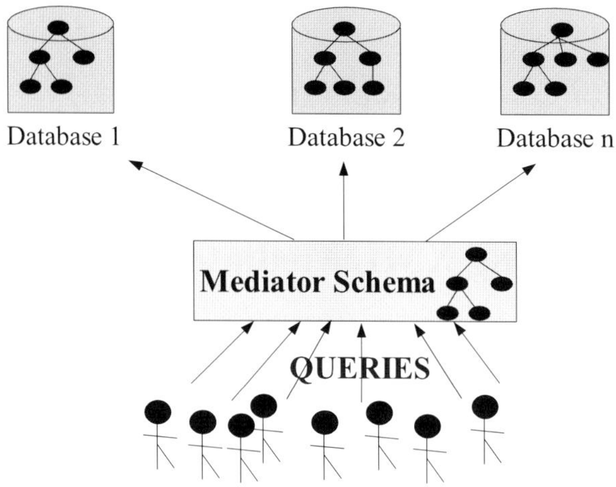

Fig. 1. Querying data distributed over the Web.

1. Introduction

Large amounts of data are available in XML format, and even if large volumes of legacy data are still marked up in HTML, efficient approaches have been proposed to transform HTML documents into XML documents [12]. These XML documents, stored in many sources distributed over the Web, contain useful information. For instance, many scientists have to manage and share large amounts of data (e.g., biologists have to deal with huge amounts of complex data). In this framework, XML is of great interest in order to represent this complicated knowledge.

Recently, sharing data has become very popular. For instance, Peer-based Data Management [23,14] has become a great challenge for scientific data sharing, as Peer-to-Peer (P2P) systems are becoming more and more popular. Nevertheless the information stored in heterogeneous collections of XML documents has to be integrated in order to be processed or to be queried in a uniform way. In all fields like biology and medicine, there will never be a centralized database, which would be impossible to build. Scientists need thus to be provided with tools in order to reach the data they require from multiple databases. Database concepts can help to solve these problems. One result of the research led in data integration and in database federations is mediator-based information systems [13]. For instance, Xyleme is a huge warehouse integrating XML data on the Web [30]. This integration can be performed by providing mediator schemas. A *mediator schema* can be considered as a shared structure through which queries can be defined [15], as shown on Figure 1. However, no complete automatic tool is available to extract semantic knowledge from these large amounts of distributed and heterogeneous data. It is indeed still hardly possible to automatically build mediator schemas, which are thus still hand-crafted.

In the past few years, data mining has been extensively studied in the framework of the semantic web. In this paper, we consider the problem of mining XML mediator schemas from a database perspective. By considering that a collection of XML documents could be modeled by labeled ordered trees [1,6] (e.g., Figure 2 illustrates such a transformation [19]), we focus on schema mining which provides tools to mine frequent sub-schemas from large databases.

```
<SigmodRecord>
<Issue>
   <Volume>19</Volume>
   <Number>7</Number>
   <Articles>
    <Article>
      <Title>Example XML</Title>
      <initPage>2</initPage>
      <endPage>3</endPage>
     <Authors>
        <Author>Peter</author>
     </Authors>
   </Article>
  </Articles>
 </Issue>
 </SigmodRecord>
```

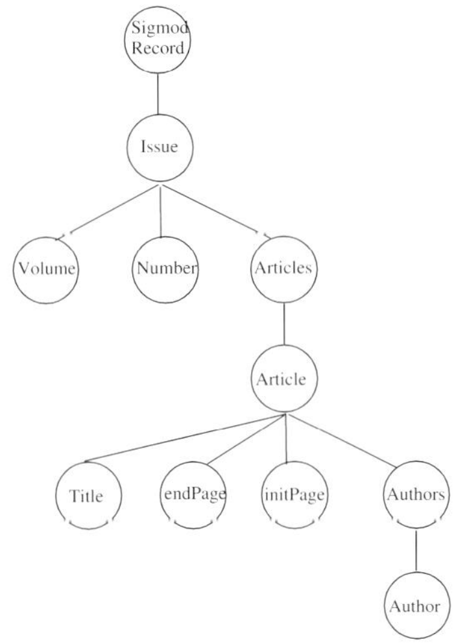

Fig. 2. An XML document modeled by a labeled ordered tree.

These frequent sub-schemas are then merged in order to build mediator schemas. We show that schema mining approaches have some drawbacks and that a fuzzy approach is very useful to mine approximate schemas. Moreover, frequent mediator schemas mined from our approach provide an interesting source of information about the data available on the web since these schemas can be seen as semantic structures describing the information. Our approach is based on the definition of the tree inclusion. We define several ways to introduce fuzzy logic in this problem. The main idea is to propose a definition of soft inclusion, meaning that a tree is no more *included or not* in another one, but *gradually included* within it. A degree of inclusion is defined, depending on the way the fuzzy inclusion is considered. Finally, we introduce fuzzy frequent patterns which aim at representing the strength of the links from a frequent tree.

The sequel of the paper is organized as follows. Section 2 goes deeper into presenting the problems of schema mining. In Section 3 we propose our motivations for introducing fuzzy logic in order to provide approximate mediator schemas. Section 4 addresses the Fuzzy Tree Inclusion while in Section 5 we propose to extend the knowledge on the mined semantic structures by providing fuzzy links within the frequent trees. Section 6 concludes the paper.

2. Schema mining principles

This section is devoted to stating the schema mining principles. As proposed approaches are mainly extensions of level-wise approaches for candidate and frequent generation, we first propose an overview of the association rules problem. Second we consider the schema mining problem. We broadly summarize the problem statement and give an overview of the core of this process.

2.1. *A brief overview of the association rules problem*

The problem of mining association rules, firstly introduced in 1993 in [2], has received a great deal of attention. In brief, the problem is the following.

Let $I = \{x_1, \ldots, x_n\}$ be a set of distinct literals called items. A set $X \subseteq I$ with $k = |X|$ is called a *k-itemset* or simply an *itemset*. Let a database D be a multi-set of subsets of I. Each $T \in D$ is called a transaction. We say that a transaction $T \in D$ supports an itemset $X \subseteq I$ if $X \subseteq T$ holds. An association rule is an expression $X \Rightarrow Y$, where X, Y are itemsets and $X \cap Y = \emptyset$ holds. The fraction of transactions T supporting an itemset X with respect to database D is called the *support* of X, $\text{supp}(X) = |\{t \in D \mid X \subseteq T\}|/|D|$. The support of a rule $X \Rightarrow Y$ is defined as $\text{supp}(X \Rightarrow Y) = \text{supp}(X \cup Y)$. The confidence of the rule is defined as $\text{conf}(X \Rightarrow Y) = \text{supp}(X \cup Y)/\text{supp}(X)$. As the number of rules grows exponentially with $|I|$, the rule sets are typically restricted by minimal thresholds for the measures of support and confidence, minSupp and minConf respectively.

According to this restriction, the problem can thus be decomposed into two subproblems:

- Find all itemsets having support greater than or equal to minSupp where the support for an itemset is defined as the number of transactions that support the itemset. Such itemsets are called frequent itemsets.
- Use the frequent itemsets to generate the set of rules according to minConf.

As solving the second subproblem is straightforward we will now have a look on how to obtain efficiently itemsets. In [3], the Apriori algorithm was defined to mine such associations. It makes multiple scans over the data (*level-wise* approach). In the first scan, the support of individual items is counted in order to determine which items are frequent. These frequent items are thus called frequent 1-itemsets. From them, new 2-candidates considered as potentially frequent itemsets are generated. A new scan is performed on the database in order to verify the support of each 2-candidate. The process goes on until no new frequent itemset is found.

2.2. *Schema mining: problem statement*

Firstly introduced by [27,28], schema mining has attracted a lot of attention [5,26,33, 37]. The problem is much more complicated than the classical association rule one since complex structures in the form of labeled hierarchical partially ordered data have to be considered.

The problem could be formulated as follows. A rooted labeled tree $T = (V, E)$ is a direct, acyclic, connected graph with $V = \{0, 1, \ldots, n\}$ as the set of vertices (nodes), $E = \{(x, y) \mid x, y \in V\}$ as the set of edges. We assume that there is a special vertex $r \in V$ designated as a root and for all $x \in V$, there is a unique path from r to x. Then if $x, y \in V$ and if there is a path from x to y then x is called an *ancestor* of y (i.e. y is a *descendant* of x). If the length of the path from two vertices x, y is reduced to one, then the ancestor relationship is considered as a *parent* relationship. For an internal node $x \in V$, we assume that its children $x_1, x_2, \ldots, x_n$ ($n \geqslant 0$) are ordered from left to right (i.e. there is a sibling relationship between children).

Let denote $\text{supp}(S)$ the number of occurrences of the subtree in a tree T. The support of a subtree in a database D is defined as the number of trees in D that contain at least one occurrence of S. A subtree is *frequent* if its support is greater than or equal to a user-specified minimum support (minSupp) value. The schema mining problem is thus reduced to mine all subtrees in D where the support value holds.

The core of the process for mining frequent subtrees is usually quite similar to the association rule approach and is briefly described in Algorithm 1. The *Gen_Cand* function builds candidates. If $k = 1$ then all nodes are considered as potential subtree candidates. Otherwise, candidates of size k (i.e. having k nodes) are built by considering frequent subtrees of size $k - 1$. At each step a scan is performed over the database in order to verify if the minSupp constraint holds. The process goes on until no new frequent subtree is found. Refer to [5,38] for details.

ALGORITHM 1. Mining Frequent SubTrees.

```
Data: Tree Database D
Result: Frequent Subtrees 𝓕
𝓕 ← ∅;
k ← 1;
repeat
    C_k ← Gen_Cand(k);
    foreach c ∈ C_k do
        c.cpt ← 0;
        foreach T ∈ D do
            if c is a subtree of T then
                c.cpt++;
        /* minSupp stands for a user-specified minimum support value */
        if c.cpt ⩾ minSupp then
            𝓕 ← 𝓕 ∪ {c};
    k++;
until 𝓕 does not grow any more;
```

3. Why considering fuzzy approaches

From the problem statement presented so far, one problem remains about the subtree inclusion. In [16], the author defines ten different tree inclusion relations and in [24], different definitions are proposed for strict, exact, ordered and weak tree inclusion. Let us have a closer look on these inclusion differences.

- The *strict* constraint addresses the labels of trees. When two nodes of different trees have similar labels then the inclusion is called strict.
- When the parent relationship is preserved, the inclusion is *exact* and the subtree is called *induced*. Otherwise, i.e. when the ancestor relationship is preserved, the inclusion is not exact and the subtree is called *embedded*.
- If the children order is preserved the inclusion is *ordered*.
- When two labels of a tree correspond to a unique label into the other tree, the inclusion is called *weak*.

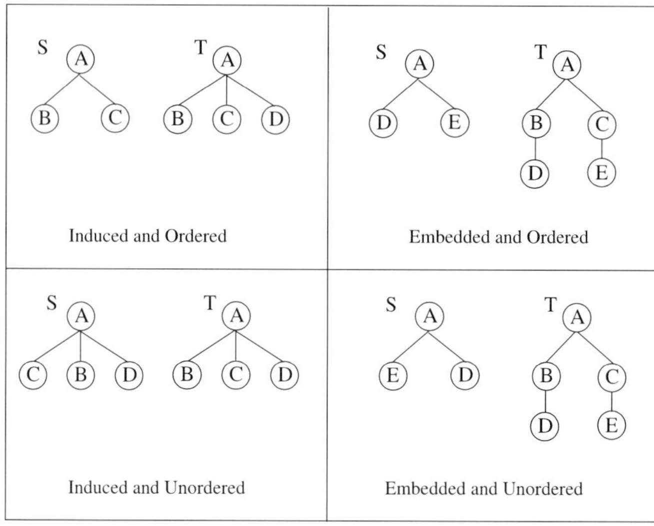

Fig. 3. Inclusion examples.

Figure 3 gives examples of a tree S included in a tree T.

According to the previous inclusion definitions, lot of algorithms were proposed for mining subtrees. Generally they consider strict and not weak inclusions. For example, in [27, 28], the authors propose a new approach for discovering frequently occurring subtrees in XML documents. This algorithm is an extension of the level-wise Apriori approach where induced subtrees are considered. The FREQT approach proposed in [5,7] also considers induced and ordered trees and an Apriori-like algorithm is proposed. In [36], TREEMINER was defined to mine labeled, embedded, and ordered subtrees. This work was extended in order to build a structural classifier for XML data [38]. In [37], the author introduces SLEUTH for mining frequent, unordered and embedded subtrees. The TREEFINDER algorithm [25] also considers embedded trees and its extension DRYADE [26] considers closed frequent subtrees, i.e. trees no more included in another one. Other approaches such as gSpan [32] and closeGraph [33] generalize the problem by mining subgraphs.

These works mainly focus on the database scan and on data representation in order to propose scalable algorithms. Furthermore all the existing approaches consider *crisp* inclusion when mining frequent subtrees, i.e. a subtree is included or not into another one, which is too restrictive from a semantic point of view. In order to avoid this crisp inclusion problem, we propose to fuzzify this definition in order to better describe the data available on the Web. It is indeed very interesting to mine approximate schemas in order to have a better idea of the semantic structures we are provided with.

In a general framework, some works have been proposed in order to integrate fuzzy mining methods with the Web [4,20]. In the more precise framework of tree and semi-structured data mining, some works have been proposed, especially to deal with approximate tree matching. [11] introduces a method to compare XML documents based on fuzzy bags. In [10,9], XML data are mined to build clusters of XML data, thus defining a distance measure between XML documents. [29] proposes an approximate method for graph schema extraction. Approximation is achieved by summarizing the semi-structured data graph using an incremental clustering method.

However as far as we know, no fuzzy method for mining frequent subtrees from huge databases has been proposed. We propose thus:

- to define the notion of fuzzy tree inclusion within this framework and
- to mine fuzzy frequent patterns.

4. Fuzzy tree inclusion

As shown on Algorithm 1, a tree candidate has to be fully included within a tree from the database to be considered when computing its support. In this work, we argue that a tree may be considered to a certain extent within this computation even if it not fully included. For this purpose, we define four ways to consider fuzzy inclusion of a tree within another one:

- The first way aims at considering the vertical paths of trees. Contrary to induced one, embedded inclusion allows us to soften the ancestor–descendant relationship. Nevertheless, no method exists to control this degree of relationship.
- In the second way, we address the horizontal paths of trees. While classical approaches consider that sibling nodes are ordered or not, there is no way to consider the proportion of sibling nodes included.
- The third way generalizes the previous one by considering the proportion of nodes.
- The last one considers similarities between nodes.

These definitions are then integrated within a new algorithm (Algorithm 2) which mines frequent subtrees in a fuzzy way. They aim at describing the extent to which a tree is included in another one. While classical approaches deal with crisp inclusion, meaning that a tree is or is not included within another one, we propose to use a degree, defined between 0 and 1. The four ways this degree can be obtained are described below. It corresponds to four ways the function *Fuzzy_Inclusion_Degree* from Algorithm 2 may be defined.

ALGORITHM 2. Mining Fuzzy Frequent SubTrees.

```
Data: Tree Database D
Result: Frequent Subtrees 𝓕
𝓕 ← ∅;
k ← 1;
repeat
    C_k ← Gen_Cand(k);
    foreach c ∈ C_k do
        c.cpt ← 0;
        foreach T ∈ D do
            c.cpt ← Compute_new_cpt(c.cpt, Fuzzy_Inclusion_Degree(c, T));
        /* minSupp stands for a user-specified minimum support value */
        if c.cpt ⩾ minSupp then
            𝓕 ← 𝓕 ∪ {c};
    k++;
until 𝓕 does not grow any more;
```

In Algorithm 2, the degree *cpt* of a candidate is updated after each tree scanning by the *Compute_new_cpt* function. Several possibilities are given when merging these degrees, depending on the way the counting is performed. These possibilities are described in Algorithm 3 by considering several counting methods: *sigma*-count, *thresholded*-count, *thresholded sigma*-count. This method is supposed to be known, as well as the associated threshold value ω if necessary.

ALGORITHM 3. Merging Fuzzy Degrees. *Compute_new_cpt* Function.

```
Data: od: old_degree, i: degree of fuzzy inclusion to be merged
Result: d: new degree
if counting_method = sigma-count then
    d ← od + i;

/* ω stands for a user-specified minimum degree */
if counting_method = thresholded count with minimum degree ω then
    if i ⩾ ω then
        d ← od + 1;
    else
        d ← od;

if counting_method = thresholded sigma-count with minimum degree ω then
    if i ⩾ ω then
        d ← od + i;
    else
        d ← od;

return d;
```

For all these four ways fuzzy inclusion may be defined. We consider a *node by node* computation as done in the classical approach. However, this computation does not aim at matching *exactly* one node from the candidate tree to another one from the database tree but rather to compute to which extend a node corresponds to another one. This computation node by node leads to a degree *deg* returned by a *Fuzzy_Match* function. This function takes into account the node n being considered, the tree S from which this node is originated, and the target tree T. The tree S has to be considered in order to keep the tree structure, and not only the nodes without any connections. We have thus to deal with a set of fuzzy degrees, ranging between 0 and 1 to be merged in order to compute the *Fuzzy_Inclusion_Degree*.

In order to provide the user with a large set of solutions, we consider the OWA operators (*Ordered Weighted Aggregator*) [31], as illustrated by Algorithm 4. An OWA operator of dimension n is a mapping

$$F\colon R^n \to R$$

that has an associated n vector $W = (w_1, w_2, \ldots, w_n)^{\mathrm{T}}$ such that $w_i \in [0, 1]$ and $\sum_{i=1}^{n} w_i = 1$. We have $F(a_1, a_2, \ldots, a_n) = \sum_{j=1}^{n} w_j \cdot b_j$ where b_j is the jth largest value of the a_i.

In this framework, it is thus possible to compute the final degree in multiple ways. In the most pessimistic way, the *minimum* is computed (logical *and*), representing the lower boundary. In the very optimistic way, the *maximum* is computed (logical *or*), representing the upper boundary. Between these two boundaries, all possibilities are offered by tuning the weights, defining several degrees of andness/orness. The measure of orness is defined as follows:

$$orness(W) = \frac{1}{n-1}\sum_{i=1}^{n}\bigl((n-i)\cdot w_i\bigr).$$

The measure of andness is defined as $andness(W) = 1 - orness(W)$.

Therefore, the user can choose whether he wants *all* nodes to be similar, *some of them* (using any desired degree of andness/orness), or *at least one* in a very flexible manner.

ALGORITHM 4. Computing the Fuzzy Inclusion Degree. *Fuzzy_Inclusion_Degree* Function.

Data: S candidate tree, T target tree
Result: Fuzzy Degree $d \in [0, 1]$
foreach *node* $n \in S$ **do**
  $deg_n \leftarrow Fuzzy\_Match(n, S, T)$;

$d \leftarrow OWA_{n=1}^{|S|} deg_n$;
return d;

Note that all the OWA operators are not associative. This is the reason why the computation of the degree d in Algorithm 4 can only be done after the whole scan of the database. In some cases, this degree is computed using an associative operator (e.g., minimum), which allows for computing it on the fly. This way, it may be possible to cut off the calculus. For instance, when computing the minimum, the process can be stopped as soon as the degree becomes lower than the minimum threshold, if one threshold is considered in Algorithm 3.

4.1. *Fuzzy vertical paths*

Let us consider fuzzy indirect links within trees. For instance, Figure 4 shows a tree S which is embedded within a tree T. When taken into account, embedded trees are counted within the final support whatever the ancestor–descendant relationship may be. There is no consideration of the number of nodes separating the ancestor node from the descendant one. Even if it is of great interest to consider two nodes as being related even if they are not directly linked, we argue that it becomes irrelevant to consider that two nodes are related if they are separated by a high number of nodes.

In our approach, we propose thus to give a scope to this ancestor–descendant relationship. This scope is defined considering the number of nodes between ancestor and descendant nodes. Since it does not make sense to consider crisp boundaries, we propose to consider a fuzzy scope for the ancestor–descendant relationship. We consider fuzzy membership functions describing the ancestor–descendant relationship depending on the number of nodes separating the two nodes being considered, as shown on Figure 5 when considering no more than five nodes in a fuzzy way. For this purpose, we use fuzzy quantifiers [34,35]. Fuzzy set theory enables to represent fuzzy quantifiers, such as *at least* 2,

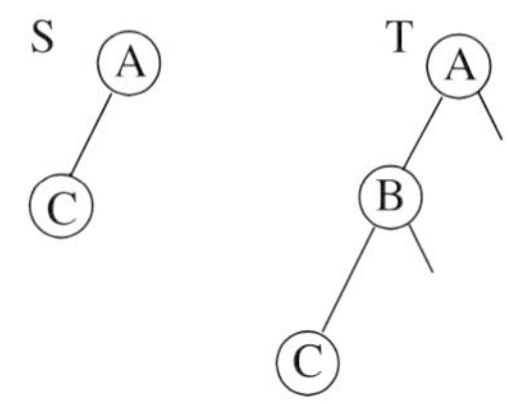

Fig. 4. Fuzzy vertical paths.

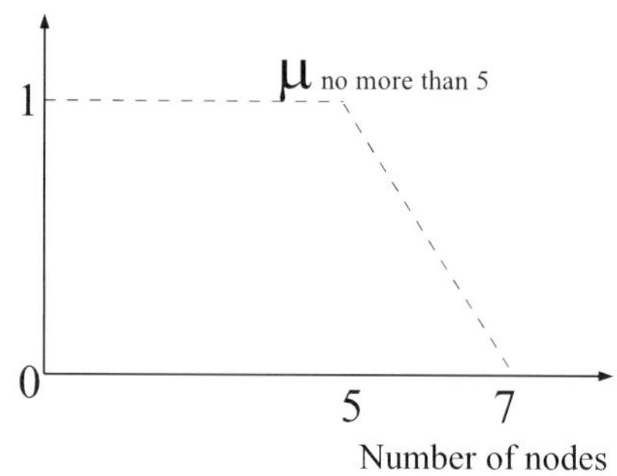

Fig. 5. Fuzzy ancestor–descendant degree. Absolute fuzzy quantifier.

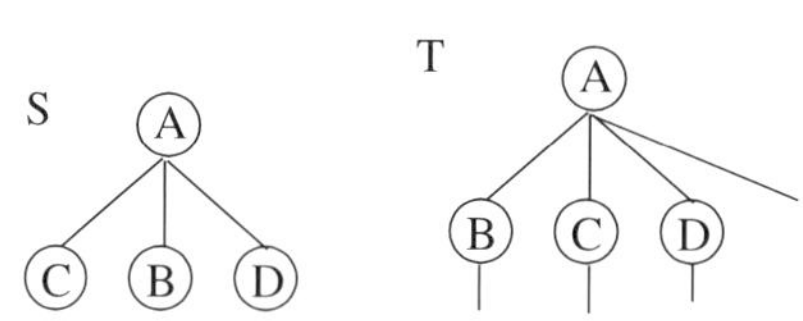

Fig. 6. Fuzzy horizontal paths.

most, by membership functions (as illustrated by Figure 5 and Figure 7). In this framework, *absolute* quantifiers like *at least* 2 are distinguished from *relative* ones (*most* for instance).

The *Fuzzy_Inclusion_Degree* from Algorithm 2 is then computed by merging the degrees to which nodes can be matched. This matching degree results from the extend to which a node may be linked to another one within a user-defined fuzzy ancestor–descendant relationship.

4.2. *Fuzzy horizontal paths*

In this section, we now consider fuzzy level inclusion. When considering ordered trees in the crisp approaches, a subtree S is included within a tree T only if all nodes of S can be mapped to nodes of T in the same order. In our approach, we propose to soften this definition by considering the proportion of nodes being included and well-ordered. For instance, Figure 6 shows an ordered tree S which is not embedded within an ordered tree T in the classical approaches since one of the node is misordered.

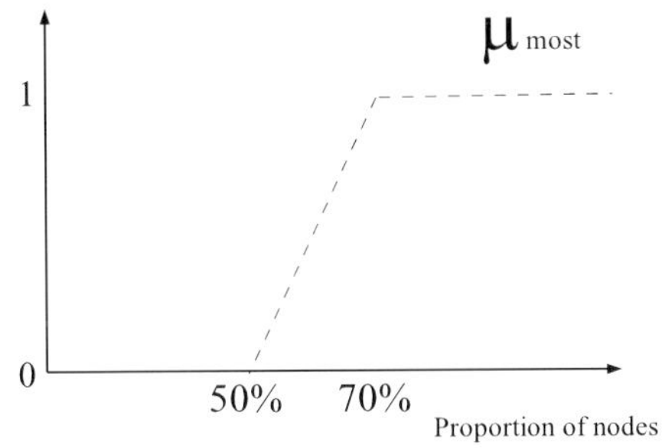

Fig. 7. Relative fuzzy quantifier: proportion of nodes.

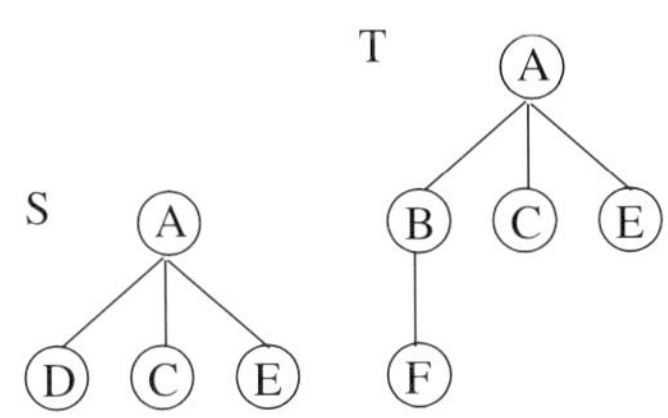

Fig. 8. Partial inclusion.

In our approach, we consider S as being included in T with a certain degree since the other nodes satisfy the inclusion. This degree is reported in the *Fuzzy_Inclusion_Degree* used in the algorithm. It depends on the proportion of nodes that are well-ordered within the nodes of the level. We consider thus a fuzzy membership function defining the degree (ranging from 0 to 1) to which *most* nodes are well-ordered, as illustrated by Figure 7.

4.3. *Partial inclusion*

In this section, we consider partial node inclusion. In a general way, *all* the nodes of a subtree S must be included in a tree T if S is included in T. However, we argue that this is too restrictive when mining data from the real world when imperfections are often present. For instance, Figure 8 shows a tree S having 75% of its nodes included in T.

In our approach, we propose thus to define partial inclusion by considering the proportion of nodes of S being present in T. This proportion is then related to a fuzzy quantifier which estimates to which extent the node has to be taken into account when computing the function *Fuzzy_Inclusion_Degree*.

Note that in the classical case, mining totally included trees allows to cut in the database scan since whenever a node cannot be matched, there is no need looking for the other ones. In our approach, outliers are accepted, which may be considered as a drawback considering scalability. However, it is still possible to cut off the search when the proportion has been overpassed.

4.4. *Similar trees*

In this section, we consider fuzzy similarities [8]. These similarities are related to some knowledge about the domain we are dealing with. For instance, when dealing with biological data, it is possible to know to which extent the node labeled by some bacteria is similar to the node labeled by another one. Let us now consider that we are provided with this kind of knowledge on the data. One of the drawbacks of schema mining approaches is that the inclusion detection is only performed on nodes having same labels (cf. Section 3 – strict inclusion). According to the semantic Web point of view, this restriction is usefulness since two different labels could describe similar concepts.

In our approach, we consider fuzzy approaches to overcome this drawback. For instance, Figure 8 shows a tree S that should be matched with T if we know that the concept D is close to B. For this purpose, we consider fuzzy ontologies as done in [21], and more generally fuzzy relations [17] describing to which extent two nodes are similar.

Let us recall that a fuzzy relation R between two reference sets X and Y is defined as a fuzzy subset from $X \times Y$. In particular, when X and Y are finite, the relation can be described by a matrix $M(R)$ including the values of its membership function.

It is possible to compute the transitive closure of a fuzzy relation. Let G be the graph associated to the fuzzy relation R and let f_R be its membership function. The *transitive closure* of R is the R_T such that $f_{R_T}(x, y) \neq 0$ iff there exists a path in G from the node x to the node y.

Let R_T be the *max–min* transitive closure of R and $R \circ R$ be the max–min composition of R and R. R_T is computed as follows:

(1) $R' = R$
(2) While R' changes
- $R = R'$
- $R' = R \cup (R \circ R)$

(3) Stop. $R_T = R'$

In our approach, we consider a fuzzy relation R defined on the universe of node labels N. We have thus $R : N \times N \to [0, 1]$ describing to which extent a node $n \in N$ is similar to a node $n' \in N$. By computing the transitive closure of R, all nodes can be compared to every other nodes. Symmetric relations are considered since we deal with similarities. This degree of similarity is taken into account in the *Fuzzy_Match* function when computing the *Fuzzy_Inclusion_Degree*, as illustrated by Algorithm 5. In this algorithm, the tree structure preservation is checked. This verification is required in order to prevent the tree structure from being destroyed. This verification must thus take into account the previous nodes that have been matched together.

Algorithm 5 proposes to compute the final degree as a merged value from all the similarity degrees with nodes from T. In the classical case, the algorithm is more efficient since there is no need to scan all nodes from T. Another possibility is thus to stop the scan of T as soon as a similar node $n' \in T$ is found, leading to Algorithm 6. In this case, we consider a user-defined γ threshold.

Fuzzy morphisms may also be considered in order to deal with this problem [22,18].

ALGORITHM 5. Computing the Fuzzy Node Matching. *Fuzzy_Match* Function.

Data: n node, S candidate tree, T target tree

Result: Fuzzy Degree $d \in [0, 1]$

foreach *node* $n' \in T$ **do**

  if n *matches* n' *and respects the* S *tree structure* **then**

    $deg_{n'} \leftarrow f_R(n, n')$;

$d \leftarrow OWA_{n'=1}^{|T|} deg_{n'}$;

return d;

ALGORITHM 6. Fuzzy Node Matching with Stop Criterion. *Fuzzy_Match* Function.

Data: n node, S candidate tree, T target tree

Result: Fuzzy Degree $d \in [0, 1]$

foreach *node* $n' \in T$ **do**

  /* γ *stands for a user-specified threshold* */

  if n *matches n' and respects the S tree structure and* $f_R(n, n') \geqslant \gamma$ **then**

    return $f_R(n, n')$;

5. Fuzzy frequent subtrees

Classical approaches mine frequent trees that do not provide much information about the occurrences within the database except the fact that these subtrees are embedded in a sufficient number of trees from the database (depending on the user-defined minimum support value). We propose thus to extend the knowledge on the mined semantic structures by providing fuzzy links within the frequent trees. These fuzzy links are more or less thick. They help knowing whether they are *very shared, middle shared* or *a little shared,* as illustrated on Figure 9.

In this framework, frequent subtrees are fuzzy acyclic graphs. Each link between two nodes is labeled by a degree ranging from 0 to 1. This degree is obtained by considering the proportion of trees from the database that contain the link. This proportion is then mapped to fuzzy quantifiers such as *very shared*, middle shared or little shared, as shown by Figure 10. It is graphically represented by taking into account the fuzzy quantifier having the greatest membership degree. Each fuzzy quantifier is then represented by a different thickness.

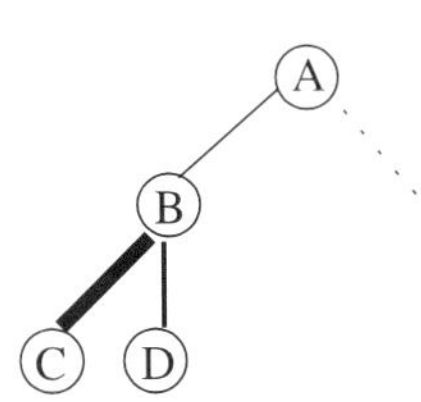

Fig. 9. Fuzzy tree.

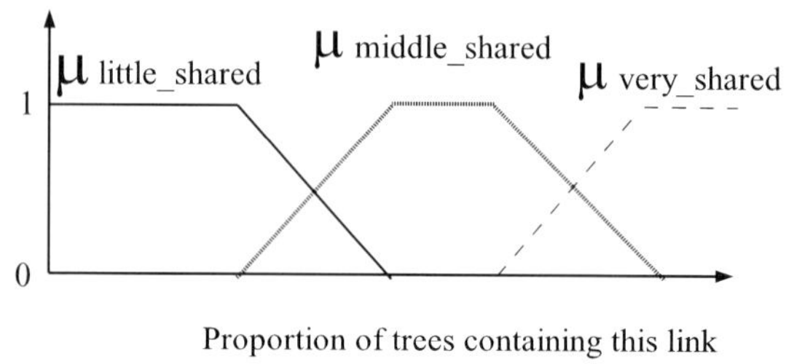

Fig. 10. Fuzzy quantifiers for fuzzy trees.

6. Conclusion

Recently mediator-based integration of heterogeneous data sources over the Web has shown that it is a very efficient way for dealing with different types of Web documents such as XML ones. One of the main advantages of such approaches is that an heterogeneous collection of XML documents could be integrated into mediator schemas in order to be processed or queried in a uniform way. However, we have seen that no complete automatic tool is available to extract semantic knowledge from these large amounts of distributed and heterogeneous data. It is indeed still hardly possible to automatically build mediator schemas, which are thus still hand-crafted. Even if schema mining approaches exist to extract in an efficient way the commonly occurring schemas from a collection, they suffer from different drawbacks. The main drawback is that they consider *crisp* inclusion when mining frequent subtrees, i.e. a subtree is included or not into another one. In the semantic Web context this constraint is too restrictive. In this paper we have thus proposed to soften such an inclusion in order to better describe the data available on the Web.

Data mining is of great interest for the semantic Web. In coming years, the Internet is likely to become one of the principal application areas of fuzzy logic. In this framework, mining fuzzy XML mediator schemas is crucial (i) in order to be automatically provided with a mediator schema to query data and (ii) in order to get semantic structures from the huge amounts of distributed and heterogeneous data. The approach we propose here can be used in order to mine frequent web structures from Web sites. Moreover, mediator schemas, fuzzy or not, can be used in the framework of fuzzy queries over Web data.

References

[1] S. Abiteboul, P. Buneman, D. Suciu, *Data on the Web*, Morgan Kaufmann, 2000.

[2] R. Agrawal, T. Imielinski, A. Swami, *Mining association rules between sets of items in large database*, in: Proceedings of the International Conference on Management of Data (ACM SIGMOD 93), 1993, pp. 207–216.

[3] R. Agrawal, R. Srikant, *Fast algorithms for mining association rules*, in: Proceedings of the 20th International Conference on Very Large Data Bases (VLDB 93), 1994, pp. 487–499, 12–15.

[4] D. Arotaritei, S. Mitra, *Web mining: a survey in the fuzzy framework*, Fuzzy Sets and Systems **148** (2004), 5–19.

[5] T. Asai, K. Abe, S. Kawasoe, H. Arimura, S. Arikawa, *Efficient substructure discovery from large semi-structured data*, in: Proceedings of the 2nd SIAM International Conference on Data Mining (SDM 02), Arlington, VA, USA, April 2002.

[6] T. Asai, H. Arimura, K. Abe, S. Kawasoe, S. Arikawa, *Online algorithms for mining semi structured data stream*, in: Proceedings of the IEEE International Conference on Data Mining (ICDM 02), Maebashi, Japan, December 2002, pp. 27–34.

[7] T. Asai, H. Arimura, T. Uno, S. Nakano, *Discovering frequent substructures in large unordered trees*, in: Proceedings of the 6th International Conference on Discovery Science, Sapporo, Japan, October 2003, pp. 47–61.

[8] B. Bouchon, M. Rifqi, S. Bothorel, *Towards general measures of comparison of objects*, Fuzzy Sets and Systems **84** (2) (1996), 143–153.

[9] P. Ceravolo, E. Damiani, *Mining class hierarchies from XML data: representation techniques*, in: FUZZY DAYS, 2004.

[10] P. Ceravolo, E. Damiani, B. Oliboni, *Fuzzy techniques for metadata construction*, in: Proceedings of the Information Processing and Management of Uncertainty in Knowledge-Based Systems (IPMU 2004). Special Session "Fuzzy Logic in the Semantic Web, a New Challenge", 2004, pp. 1019–1026.

[11] P. Ceravolo, M.C. Nocerino, M. Viviani, *Knowledge extraction from semi-structured data based on fuzzy techniques*, in: Proceedings of the 8th International Conference on Knowledge-Based Intelligent Information & Engineering Systems (KES 04), 2004, pp. 328–334.

[12] C.Y. Chung, M. Gertz, N. Sundaresan, *Reverse engineering for Web data: from visual to semantic structure*, in: Proceedings of the 18th International Conference on Data Engineering (ICDE 02), 2002, pp. 53–63.

[13] D. Florescu, A. Levy, A. Mendelzon, *Database techniques for the world wide web: a survey*, ACM SIGMOD Record **27** (3) (1998), 59–74.

[14] A. Halevy, Z. Ives, P. Mork, I. Tatarinov, *Piazza: data management infrastructure for semantic web applications*, in: Proceedings of the 12th International World Wide Web Conference (WWW 03), 2003, pp. 556–567.

[15] A. Halevy, Z. Ives, D. Suciu, I. Tatarinov, *Schema mediation in peer data management systems*, in: proceedings of the 19th International Conference on Data Engineering (ICDE 03), 2003, pp. 505–515.

[16] P. Kilpelinen, *Tree matching problems with applications to structured text databases*, PhD thesis, University of Helsinski, November 1992.

[17] G.J. Klir, T.A. Folger, *Fuzzy Sets, Uncertainty, Information*, Prentice-Hall, 1988.

[18] L.J. Kohout, *Defining homomorphisms and other generalized morphisms of fuzzy relations in monoidal fuzzy logics by means of bk-products*, in: Proceedings of the 7th Joint Conference on Information Sciences (JCIS 03) (Subsection: 9th International Conference on Fuzzy Theory and Technology), 2003, p. 13.

[19] H.P. Leung, F.L. Chung, S.C. Chan, *On the use of hierarchical information in sequential mining-based XML document similarity computation*, Knowledge and Information Systems **7** (4) (2005).

[20] S.K. Pal, V. Talwar, P. Mitra, *Web mining in soft computing framework: relevance, state of the art and future directions*, IEEE Trans. on Neural Networks **13** (5) (2002), 1163–1177.

[21] D. Parry, *A fuzzy ontology for medical document retrieval*, in: Proceedings of the 2nd Australasian Workshop on Data Mining and Web Intelligence (DMWI 04), Australian Computer Society, 2004, pp. 121–126.

[22] A. Perchant, I. Bloch, *Fuzzy morphisms between graphs*, Fuzzy Sets and Systems **128** (2) (2002), 149–168.

[23] I. Tatarinov, Z. Ives, J. Madhavan, A. Halevy, D. Suciu, N. Dalvi, X. Dong, Y. Kadiyska, G. Miklau, P. Mork, *The Piazza peer data management project*, SIGMOD Record **32** (3) (September 2003), 47–52.

[24] A. Termier, *Extraction of frequent trees in an heterogeneous corpus of semi structured data: application to XML documents mining*, PhD thesis, Paris South University, April 2004.

[25] A. Termier, M.C. Rousset, M. Sebag, *TreeFinder: a first step towards XML data mining*, in: Proceedings of the IEEE International Conference on Data Mining (ICDM 02), Maebashi City, Japan, 2002, pp. 450–457.

[26] A. Termier, M.C. Rousset, M. Sebag, *Dryade: a new approach for discovering closed frequent trees in heterogeneous tree databases*, in: Proceedings of the IEEE International Conference on Data Mining (ICDM 04), Bringhton, UK, 2004, pp. 543–546.

[27] K. Wang, H. Liu, *Discovering typical structures of documents: a road map approach*, in: Proceedings of the 21 ACM SIGIR International Conference on Information Retrieval, Melbourne, Australia, August 1998, pp. 146–154.

[28] K. Wang, H. Liu, *Discovering structured association of semistructured data*, IEEE Trans. on Knowledge and Data Engineering **12** (3) (2000), 353–371.

[29] Q.Y. Wang, J.X. Yu, K.F. Wong, *Approximate graph schema extraction for semi-structured data*, in: Proceedings of the 7th International Conference on Extending Database Technology (EDBT 00), 2000, pp. 302–316.

[30] L. Xyleme, *A dynamic warehouse for XML data of the web*, IEEE Data Engineering Bulletin **24** (2) (June 2001), 40–47.

[31] R. Yager, *Families of owa operators*, Fuzzy Sets and Systems **57** (3) (1993), 125–148.

[32] X. Yan, J. Han, *gSpan: graph-based substructure pattern mining*, in: Proceedings of the IEEE International Conference on Data Mining (ICDM 02), Maebashi City, Japan, December 2002, pp. 721–724.

[33] X. Yan, J. Han, *CloseGraph: mining closed frequent graph patterns*, in: Proceedings of the ACM SIGKDD International Conference on Knowledge Discovery and Data Mining (KDD 03), Washington, DC, USA, August 2003, pp. 286–295.

[34] M. Ying, B. Bouchon, *Quantifiers, modifiers and qualifiers in fuzzy logic*, J. Appl. Non-classical Logics **7** (3) (1997), 335–342.

[35] L.A. Zadeh, *A computational approach to fuzzy quantifiers in natural languages*, Comput. Math. Appl. **9** (1) (1983), 149–184.

[36] M. Zaki, *Efficiently mining frequent trees in a forest*, in: Proceedings of the 8th ACM SIGKDD International Conference on Knowledge Discovery and Data Mining (KDD 02), Edmonton, Canada, July 2002, pp. 386–395.

[37] M. Zaki, *Efficiently mining frequent embedded unordered trees*, Fund. Inform. **65** (2005), 1–20.

[38] M.J. Zaki, C. Aggarwal, *Xrules: an effective structural classifier for XML data*, in: Proceedings of the 9th International Conference on Knowledge Discovery and Data Mining (ACM SIGKDD 03), 2003, pp. 316–325.

CHAPTER 13

Bottom-up Extraction and Maintenance of Ontology-based Metadata

Paolo Ceravolo[a], Angelo Corallo[b], Ernesto Damiani[a], Gianluca Elia[b], Marco Viviani[a] and Antonio Zilli[b]

[a] *Università degli Studi di Milano, Dipartimento di Tecnologie dell'Informazione, Via Bramante 65, 26013 Crema (CR), Italy*
E-mail: {ceravolo,damiani,viviani}@dti.unimi.it

[b] *Università degli Studi di Lecce–ISUFI, at eBusiness Management School, Via per Monteroni, 73100 Lecce (LE), Italy*
E-mail: {angelo.corallo,gianluca.elia,antonio.zilli}@ebms.unile.it

Abstract

In this chapter, several flexible techniques aimed at extracting, maintaining and enriching semantic-web style metadata are discussed. Such techniques were designed for being applied in the framework of dynamic *Communities of Practice* (CoP) interactions. Namely, we present a way of building ontologies that proceeds in a *bottom-up* fashion, defining concepts as clusters of concrete objects. Unlike huge, "supply-side" normative ontologies, our bottom-up ontologies are based on use of implicit and, therefore, parsimonious `part-whole` and `is-a` relations. This makes them suitable for the ad-hoc style of conceptualization used within communities of practice and peer-to-peer (P2P) communities. Also we discuss how metadata based on bottom-up ontologies can be associated with a flexible *degree of trust* by collecting user feedback. Our bottom-up extraction method complements current practice, where, as a rule, ontologies are built top-down. It is not claimed that bottom-up construction is a generally valid recipe; rather, the approach is intended to enrich the ontology developer's palette when designing and implementing Semantic Web applications.

Keywords

community of practice, semantic-web, bottom-up ontology, ad-hoc conceptualization, metadata extraction and maintaining, fuzzy techniques, clustering techniques, trusted metadata

FUZZY LOGIC AND THE SEMANTIC WEB
Edited by Elie Sanchez
© 2006 Elsevier B.V. All rights reserved

1. Introduction

In the emerging scenarios of the knowledge-based economy, the concept of *community* is emerging as a new organizational form, that complements existing structures, supporting knowledge sharing, diffusion and application processes. Communities do not operate in a vacuum; rather they have to confront with a huge amount of digital information such as text or semi-structured documents in the form of web pages, reports, papers and e-mails. In order to improve the effectiveness of community-wide co-operation, the capability of extracting sharable knowledge from heterogeneous documents produced by multiple sources and in a timely fashion is therefore crucial. In turn, knowledge extraction relies on a technological platform supporting rich conversations and fostering flexible knowledge access to the whole community. Availability of shared knowledge can help companies to use their intellectual capital more effectively and contribute to business results in several ways [22]:

- *increasing reuse of knowledge assets* through the creation of mutual trust and sense of obligation among community's members;
- *decreasing delay in answering to customer and stakeholder needs*, through real time connections among community's members, which are independent from time and space;
- *decreasing the learning curve for new employees* through a sort of mentorship led by community experts;
- *increasing organizational learning* through the creation of an informal learning environment in which both novices and experts may interact, share experiences and learn from each other;
- *generating new ideas for new products and services*, providing a safe environment for people to exchange ideas with others and ask difficult questions without the normal pressures of day-to-day performance.

In this chapter, several flexible techniques aimed at dynamically extracting, maintaining and enriching semantic-web style metadata in the framework of cooperating communities are discussed. Specifically, we present a way of building specification of a shared conceptualization [16] proceeding in a *bottom-up* fashion, i.e. defining concepts as clusters of concrete objects. Unlike huge, "supply-side" normative ontologies, our bottom-up ontologies are based on discovery of implicit `part-whole` and `is-a` relations. This makes them suitable for the ad-hoc, human-centered style of conceptualization [10] used within dynamic communities. Also, we discuss how metadata based on bottom-up ontologies can be associated with a flexible *degree of trust* by collecting user feedback. As we shall see, our bottom-up method nicely complements current practice, where, as a rule, ontologies are designed top-down. The chapter is organized as follows: Section 2 discusses the need for dynamic ontology building and maintenance in the framework of distributed communities, while Section 3 outlines the application of fuzzy techniques to efficient structure-based clustering of XML data. Section 4 describes how content-based clustering can be coupled with structure-based techniques to produce a finer-grained classification, suitable to be converted into an ontology. Section 5 contains a worked-out example of XML clustering, while Section 6 and 7 describe how clusters can be used for ontology extraction and maintenance.

2. Communities of practice (CoP) and the semantic web

In recent years, the term *community* has been widely used to describe several and different organizational entities related to knowledge management and competitiveness dynamics. Main examples are *community of interest* [1], *team* [25], *network* [6], *knowledge community* [2,26] and *virtual community* [1,27]. Generally speaking, we can say that a community can be conceived as a set of relations among nodes – where a node can be represented by a person, but also by an organization or by another community – activities and social world, ever changing over time and space, in which interests, work practices and objectives are common and shared, and participation can be more or less spontaneous. A particularly interesting form of community is a *Community of Practice* (CoP). The concept of CoP was introduced in 1991 by Lave and Wenger [21] and other researchers have continued to develop it during recent years. CoPs exist within businesses and across business units and company boundaries; even though they are informally constituted, these self-organizing systems share the capability of creating and using organizational knowledge through informal learning and mutual engagement [31], putting users at the center of the space of interactions.

According to the value of a set of specific parameters – such as purpose, membership, sponsorship, mandate, evolution, main outcomes, accountability, organizational support, infrastructure, visibility – CoPs can be classified into three different types [28]:

- *Informal CoP*: configured as grass-roots structures, loosely organized, and composed of people highly motivated to discuss their work;
- *Supported CoP*: characterized by an internal sponsorship and focused on developing new knowledge;
- *Structured CoP*: aligned with strategic goals and strongly oriented towards organizational performance. Its members are highly motivated and they utilize new capabilities (generated through collaboration and learning) to create new solutions and business value.

Besides the common purpose, the other fundamental elements [24,23] that characterize a community of practice are: *identity* – that allows the members of the community to feel themselves as an important and potentially active node of the community structure; *trust* – that represents the basic element to activate knowledge sharing and knowledge exchange dynamics; *participation* – that concerns active and proactive involvement in the community's dynamics, overcoming the simple "membership" focused just on the formal adhesion to the community; *leadership* – that gives information about the attitude and the capacity of each member, at each level of the community, to activate and sustain innovations processes and knowledge creation practices (diffuse leadership); *culture* – that constitutes an important aspect of members' "soft" profile; *interests* – that express the basic motivation of the participation; *experiences* – that contribute to shape the knowledge background of each member, with their personal attitudes and skills, creating a unique personal competence profile.

Nowadays, communication technologies have created a space of flows that substitutes for the space of places: geographical proximity is giving way to relational proximity. These phenomena greatly impact on the architecture of a community of practice, giving to it a distributed extension of the internal processes and dynamics. Moreover, in current business contexts that require multi-disciplinary approaches and competence, the usage of a common language (shared with all the involved members) represents the basic prerequisite to activate collaboration, capabilities creation processes, knowledge sharing dynamics, innovative solutions development. In our approach, the common machine-readable and machine-understandable language shared within a CoP is expressed by a dynamically created and maintained bottom-up ontology to be used as the community's *lingua franca*. The use of ontology for describing domains of competence and the knowledge base of collaborative groups like CoPs promises higher speed and effectiveness in the process of building knowledge; furthermore, it makes it easier for new member to understand the expertise of the community as a whole.

Another distributed community of paramount interest for our approach is the virtual marketplace hosting e-commerce transactions [10]. Current e-commerce platforms for selling products or services reinvented "customer care" traducing in the virtual context the human touch that customers find in their preferred retail store. In this context, language is the paramount element for interaction, as in a e-commerce transaction we only get what we say we need. Indeed, language represents the basic common infrastructure for cumulative, knowledge-based communication: "we do not see reality as it is, but as our languages are" [4]. But if in a real-world retail store we can speak until we understand each other, in a virtual web-based store customers have no patience and leave quickly if they cannot find the product they look for. The success of web-based e-commerce, for instance, lies in the effectiveness of mutual understanding: if the language used by the customer (or Internet user) is understood by the store owner then *probably* the customer will buy something, otherwise *surely* she will not buy. To improve the language based interaction, controlled vocabularies or ontologies can be used. An ontology-based description enables powerful search functionalities giving to the customer the possibility of focusing on the properties of products he looks for, discarding parameters she is not interested in. For instance, the Semantic Web approach uses descriptions of web services written in special-purpose language like OWL-S [33] while the Google search engine proposed something like these with its new Froogle facility, whose user can look for everything that is sold on American Internet stores. Visualization of Froogle search results can be adapted to the interest and necessity of the user, putting her in a central position with respect to the action of buying; also, a world-wide ontology of products and services can be utilized so a virtual promenade visiting many stores is possible.

2.1. *Building ontologies for CoPs*

Traditionally, the natural approach for community-oriented ontology building has been a collaborative one. The aim of this approach is to valorize the contribution of each member in the knowledge creation activity with his daily work. An example is Ontoligua Server [15], a web space where members of the community (a community of ontology developers) can access ontologies, browse them, edit them and propose modification. This process enables the introduction in the ontology of an interdisciplinary vision as each member of the community contributes with her own background. More importantly, cooperatively built ontologies can quickly incorporate new "operative knowledge" deriving from job experience. However, some drawbacks do exist. Cooperatively building an ontology is hard work; discussions and checkpoints are continuous; total time construction is usually long, even in presence of a careful process management and updates are very frequent. Each member of the community must commit a considerable amount of time to the ontology building activity, which is simply not realistic as far as distributed communities are concerned.

In order to adapt the ontology building process to the dynamic cooperation style of distributed communities, one wonders whether an automatic approach can be taken. Historically, automatic knowledge extraction applications working on free text documents started by indexing documents via vectors of (normalized) keyword occurrences [19]. This encoding gives rise to a *vector space model* where every document d, is considered to be a vector $\boldsymbol{d}$, in the term space (set of document words). In the simplest form, each document is represented by the (*TF*) vector $d_{f_t} = (f_{t_1}, f_{t_2}, \ldots, f_{t_n})$ where f_{t_i} is the frequency of the ith term in the document. A variant of this model includes an additional weighting of each term based on its inverse document frequency. Finally, in order to account for documents of different lengths, each document vector is normalized so that it is of unit length.

Documents were grouped or clustered into (approximations of) concepts by means of a suitable *distance* function computed on vectors, e.g., Euclidean or scalar product based ones. The most common similarity measure used for computing the similarity φ between documents, is defined as $\varphi(d_1, d_2) = \cos(\boldsymbol{d}_1, \boldsymbol{d}_2) = (\boldsymbol{d}_1 \bullet \boldsymbol{d}_2)/\|\boldsymbol{d}_1\| \cdot \|\boldsymbol{d}_2\|$, where $\bullet$ denotes the scalar product and $\|\boldsymbol{d}\|$ the Euclidean length.

These classical document classification algorithms present three main problems. First, they need a high number of documents (typically, many thousands) to work properly. Secondly, they hardly take into account document structure and are therefore unsuitable for semi-structured document formats used on the Web. Thirdly, the final step of identifying document clusters with concepts often gives raw results that contrast with the intuition of the human users. The conceptualization step can be significantly improved only through the effort of a domain expert, which introduces a delay not compatible with community-style Web interaction. In Section 3, we shall describe an automatic approach designed to deal with for Web-based data formats.

3. A bottom-up fuzzy ontology construction and maintenance engine

In our approach, ontology represents both the type and form of knowledge to be extracted. As far as data are concerned, however, it is fundamental having a simple and widely accepted semi-structured format capable of expressing them. The *eXtensible Markup Language* (XML) is today the most widespread format for exchanging and representing information on the Internet; also, many datatypes can be mapped on XML data model and represented as instances by using wrappers. Therefore, without loss of generality, we shall assume data to be represented by XML documents. Similarity between XML documents has been defined in different ways, taking into consideration only documents' structure, their content, or both [8]. Here we adopt a structure-based approach based on a fuzzy representation of XML documents' structure. Our fuzzy encoding allows to compute similarity values between documents and use them to create similarity classes representing the domain of competence, giving to the Community of Practice an automatic means to extract and to exchange knowledge.

The last task of the approach consists in the use of a hybrid approach based on k-Means and k-NN algorithms, applied in every structure-based similarity class to take into consideration the content to extract restriction and constraints previously unconsidered (Figure 1).

3.1. *Encoding XML data*

XML (*eXtensible Markup Language*) has its own standard data model, called *Information Set* (*Infoset*) [34]. In this chapter we rely on a simplified, abstract representation of the Infoset data model [13] as a *multi-labeled, multi-sorted directed graph* where, for the sake of simplicity, only two types of nodes (element and attributes) are considered and labels correspond to element and attribute names. Our simplified Infoset can be easily extended to coincide with the standard XML data model simply by increasing the number of sorts and adding some additional constraints [13]. Also, in our representation of the XML Infoset the sub-graph representing (element- and attribute-) containment is always a tree, where leaf nodes represent content and values, while non-leaf nodes correspond to tags. Under this assumption, here we shall consider every XML document as an instance of the simplified XML Infoset model described in [13] (Figure 2).

The first step of our fuzzy extraction process consists of encoding XML documents (i.e. instances of our simplified XML data model) as *fuzzy bags* (i.e. fuzzy multi-sets).

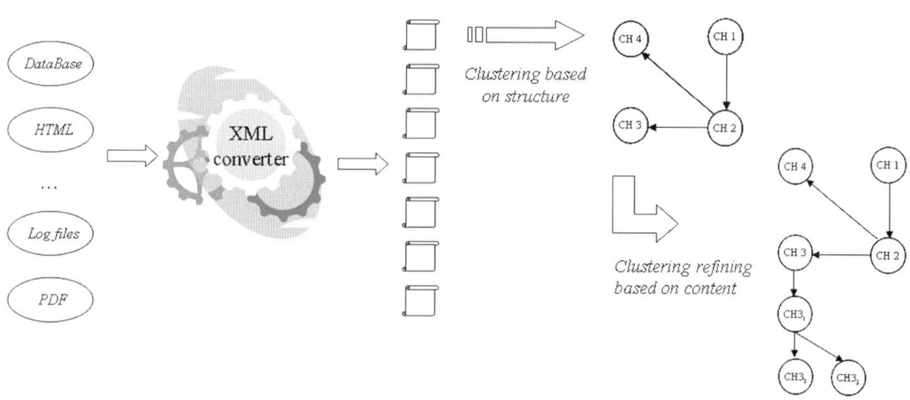

Fig. 1. The complete process of clustering an heterogeneous data flow based on structure and content information.

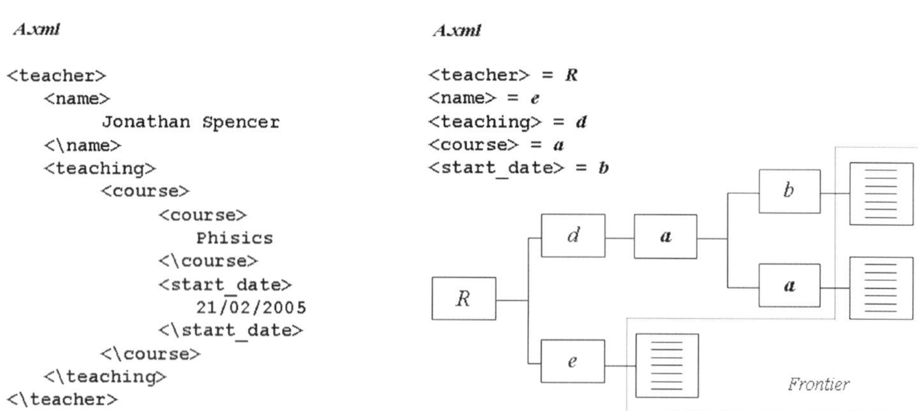

Fig. 2. The tree representation of an XML fragment.

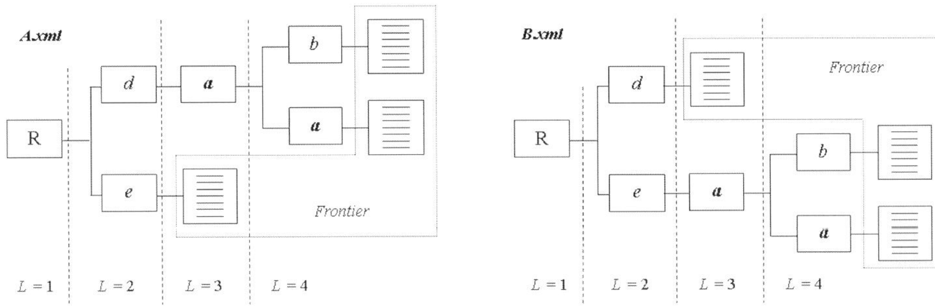

Fig. 3. The encoding process.

A *bag* is a collection of elements which may contain duplicates. As we shall see, bags provide a good *flat encoding* for XML documents, especially when duplicates are significant, since it can express the fact that the same tag can be repeated at different levels and/or with different cardinalities at the same level. Intuition tells us that adopting a model based on bags rather than sets will make our technique sensitive to differences in cardinality between tags of the same name. Again intuitively, similarity between bags must decrease when differences in elements' cardinalities increase; but the amount of this decrease may selectively depend on the specific tags involved, allowing for fine-tuning similarity values. Our encoding algorithm works as follows: for each document, tag content is discarded and a fuzzy membership degree is associated to tag names, obtaining a *fuzzy bag*. More specifically, the encoding of a XML tree as a fuzzy bag is performed as follows: supposing that every tag x_i has associated a membership degree $\mu(x_i)$, initially equal to 1 ($\mu(x_i) = 1$), we divide every membership degree by its nesting level L (see Figure 3), obtaining a "lossy" representation of the XML tree that takes into consideration the original nesting level of tags, but not the brotherhood relations.

The content of the leaf tags (according to our data model assumptions, XML documents are well-behaved, i.e. only the frontier of the XML tree contains data) is contextually saved independently from the fuzzy bag representation. Applying the method described before to the tree representation of generic XML documents A.xml and B.xml (Figure 3), we obtain the same fuzzy bag:

$$A = B = \{R/1, e/0.5, d/0.5, a/0.33, a/0.25, b/0.25\}.$$

We use the following representation to indicate the presence of content information connected to leaf nodes:

$$A = \bigl\{R/1, e[data]/0.5, d/0.5, a/0.33, a[data]/0.25, b[data]/0.25\bigr\}$$

and

$$B = \bigl\{R/1, e/0.5, d[data]/0.5, a/0.33, a[data]/0.25, b[data]/0.25\bigr\}.$$

Note that, thanks to fuzzification, our approach is also sensitive to differences in structure between XML documents containing the same group of tags in different position. The more the topological structure of two XML documents differs, the more the similarity between the corresponding bags decreases.

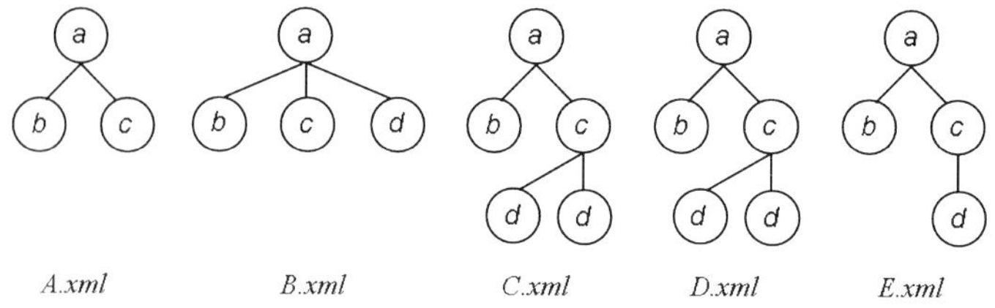

Fig. 4. Tree representation of XML documents in a cluster.

3.2. *Building document classes using structural information*

The main feature of our encoding technique is that, unlike XML trees, our fuzzy bags lend themselves to fast and efficient clustering. When comparing fuzzy bags representing XML documents several measures of object similarity can be used, e.g., those studied in [3]; choosing the "right" similarity is usually based on domain- and even dataset-dependent criteria. (For data-dependent criteria for choosing between non-symmetrical or symmetrical measures, see for instance [12].) Applying the extension principle, it is rather straightforward to extend most of the object comparison measures originally defined using fuzzy sets to fuzzy bags [5]. Once similarity values between fuzzy bags encoding XML documents have been computed, it is not difficult to discover frequent patterns and use them to build a classification. The rationale behind our approach is that we prefer to use object comparison measures rather than distance functions, the latter being too restrictive to represent complex objects such as XML documents and relationships between them [12]. More specifically, we compute a collection of blocks composed of documents close to a candidate pattern, again based on similarity (although our blocks are not the result of proper clustering, since object comparison measures lack the mathematical property of a distance, we shall loosely call them *clusters* in the following). Intuitively, bags within a cluster are more similar to each other than they are to the ones belonging to a different cluster. The next step is to assign meaning to discovered blocks, first of all choosing a good representative for each block. While it is possible to adopt "quick-and-dirty" solutions, e.g., selecting the smallest block member, we build a representative that does not correspond to any particular bag belonging to the cluster. Namely, we construct a fuzzy bag containing the entire vocabulary of tags in the cluster. Each tag is taken with the same cardinality and the membership degrees it has in the bag where it appears with the highest cardinality. Looking at Figure 4, where a set of XML documents is grouped in a cluster, encoded as:

$$A = \{a/1, b/0.5, c/0.5\},$$
$$B = \{a/1, b/0.5, c/0.5, d/0.5\},$$
$$C = \{a/1, b/0.5, c/0.5, d/0.3, d/0.3\},$$
$$D = \{a/1, b/0.5, c/0.5, d/0.3, d/0.3\},$$
$$E = \{a/1, b/0.5, c/0.5, d/0.3\}.$$

The cluster-head would be:

$$CH = \{a/1, b/0.5, c/0.5, d/0.5, d/0.3, d/0.3\}.$$

4. Content-based document classification

Once we have clustered XML documents based on their structure, it is time to take their content into account. XML content can belong to basic XML Schema datatypes (URLs, numbers, dates, etc.) or be a text blob. In the first case, it is useful to apply *type-driven matching*, while in the second hypothesis, we have to use traditional, keyword-based text categorization techniques. Following the ideas expressed in [11], XML data similarity can be computed by taking into account their estimated type T. Text comparison hardly captures the real similarity between typed values such as dates, numerical types and even strings, etc. For string edit distance and URI/dates comparison, details are given in [7] and [11].

Building a bottom-up ontology (and in particular in our case, where a content analysis of data follows a structural analysis), comparison between numerical data and equality between strings meaning assume a fundamental importance in expressing constraints on classes inferred after the phase of similarity comparison based on the structure of the documents. An example is given in Section 5. Approaches to content-based clustering (see [29] for details) can be classified as follows:

- *Hierarchical Algorithms*. They create a tree of node subsets by successively subdividing or merging the graph's nodes. Typical examples are k-Nearest-Neighbor (k-NN), linkage or Ward methods.
- *Iterative Algorithms*. The simplest and most well-known amongst these algorithms is k-Means, which resembles a self-organizing Kohonen network whose neighborhood function is set to size 1.
- *Meta-Search Algorithms*. They treat a clustering task as an optimization problem where a given goal criterion is to be minimized or maximized (genetic algorithms, simulated annealing, two-phase greedy strategy).
- *Exclusive* (they assign every node to exactly one cluster) *vs. Non-exclusive Algorithms* (they assign a membership value to a node with respect to each cluster, ex. fuzzy clustering algorithms).

In particular, k-Nearest-Neighbor [30,18] is one of the most used techniques for text categorization, due to its good performance. It is a supervised classification task: given a set of labeled prototypes (i.e. categories) and a test document, the k-NN method finds its k nearest neighbors among the training documents. The categories of the k neighbors are used to select the nearest category for the test document: each category gets the sum of votes of all the neighbors belonging to it and that one with the highest score is chosen. Other strategies calculate these scores taking into account the distances between the k neighbors and the test document or, alternatively, using a similarity measure like the scalar product.

On the other hand, the k-Means [30] algorithm is an unsupervised technique (no predefined classes prior process) often used in document clustering applications for grouping objects based on similarity. The algorithm is based on the mean value of the object in the cluster, which can be viewed as the clusters *center of gravity*. As input it takes the number of clusters k and a database containing n objects. It returns a set of k clusters that minimizes the squared-error criterion, which is defined for a clustering $\mathcal{L}$ of a data set $\mathcal{H}$, as $e^2(\mathcal{H}, \mathcal{L}) = \sum_{j=1}^{k} \sum_{i=1}^{n_j} \|\boldsymbol{d}_i^{(j)} - \boldsymbol{c}_j\|^2$ where $\boldsymbol{d}_i^{(j)}$ is the ith object belonging to the jth cluster and $\boldsymbol{c}_j$ is the centroid of the jth cluster, defined as $\boldsymbol{c} = \frac{1}{|\mathcal{H}|} \sum_{d \in \mathcal{H}} \boldsymbol{d}$.

Hierarchical Clustering algorithms (including k-Nearest-Neighbor) are generally considered better, although slower, than k-Means [14,17]. For this reason *hybrid approaches*

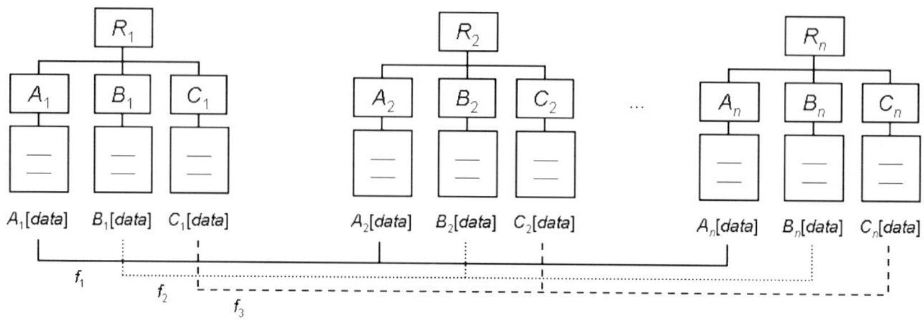

Fig. 5. The aggregation between data belonging to the same tag in different documents.

involving both k-Means and hierarchical methods have been proposed. Our approach uses an hybrid approach based on the execution of an improved version of the k-Means algorithm on a subset of the initial document set; this step provides the initial categories to the k-NN algorithm.

4.1. *The aggregation problem*

Our clustering techniques work on a collection of semi-structured documents that contain data in particular positions counter signed by tags. Therefore, content-based similarity *at the tag level* is computed comparing the contents of the same tag in two different documents, while content-based similarity *at the document level* is computed by aggregating a document pair's tag-level similarity values.

Referring to Figure 5 it is necessary to choose two different functions: a function f to compare data belonging to tags with the same name in different documents:

$$f_1\big(A_1[data], A_2[data], \ldots, A_n[data]\big),$$
$$f_2\big(B_1[data], B_2[data], \ldots, B_n[data]\big),$$
$$f_3\big(C_1[data], C_2[data], \ldots, C_n[data]\big),$$

and a function F to aggregate the individual fs:

$$F(f_1, f_2, f_3).$$

Occurrences of terms have distinct informative role depending on the tags they belong to. So, it is possible either to define a different function f for each group of data of the same tag in different documents, or choosing a function considering the membership value $\mu(x_i)$ associated to the ith tag.

5. An example of class subdivision and refinement

As we have seen, our clustering method consists of two phases: a structural analysis and a content-based one. The first phase works particularly well in those cases where documents present an high variety of tags cardinality. This happens especially in those situations where tags are optional. Comparing a large set of data, it is reasonable that some documents may

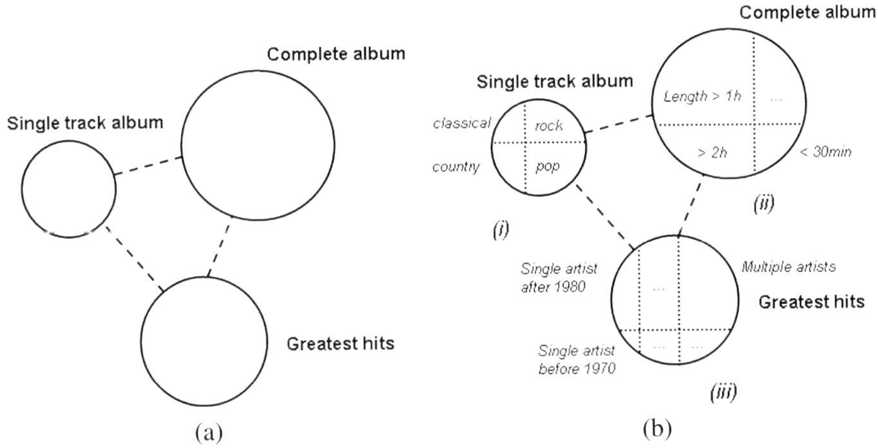

Fig. 6. Domain class subdivision based on structure (a) and refinement based on content (b). The outline lines between classes indicates future relations to be extracted building the ontology.

have common features. So we start by processing a subset of the available document set. This subset can be computed automatically, or selected by a domain expert. In the first case we start from a *tag cardinality histogram* of the document set (i.e. a simple histogram showing the number of documents for each given number of distinct tags) and we pick a sample of documents so that the relative proportions of tag cardinalities of the whole document set are exactly reproduced in the sample. Similarity values obtained from the comparison of this subset of documents are then used to cluster them obtaining tentative classes expressing a particular domain. In this section, we describe an experimental evaluation carried out on documents belonging to a musical domain. We rely on the fact that multimedia files have associated a certain number of fields containing optional information. For example, `mp3` files have a fixed (now growing) number of predefined tags (`author`, `title`, `genre`, ...). This information can be organized in XML fragments available to a distributed community interested in the musical domain. The structural analysis phase will cluster fragments describing single titles (i.e. CDs containing a single track) and distinguish them with respect to other fragments describing multitrack titles (i.e. CDs containing multiple tracks) (Figure 6(a)).

For every class forming the domain Figure 6(b) shows possible finer-grained divisions based on content: in Figure 6(b)(i), an operation of *strings comparison* is carried out on the tag <genre>); in Figure 6(b)(ii) the refinement is obtained considering the length of an album, operating on the *numerical content* of the tag <length>, in particular in this case we are in presence of "date/time" format; finally, in Figure 6(b)(iii) we combine string (the name of the artist <artist_name>) and numerical types (the year of publication <pub_year>) in order to obtain detailed classifications.

6. Ad-hoc ontology construction

In the previous section we have described how XML fragments can be clustered; for each cluster, a *cluster-head* is chosen that represents a description of typical instances in our

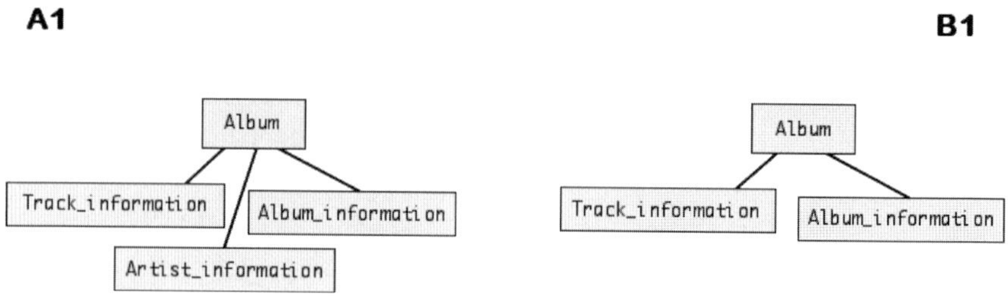

Fig. 7. A sample of an is-a relation.

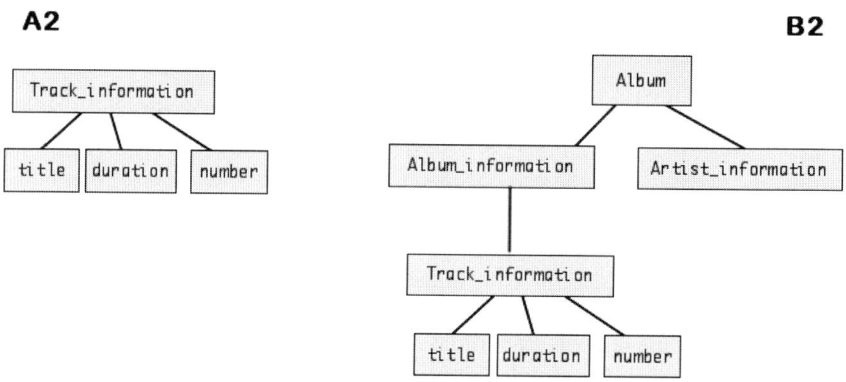

Fig. 8. A sample of a part-of relation.

domain. The next step is to use these descriptions for designing a hierarchy of classes. For this purpose our system can provide a list of *candidate relations* interconnecting cluster-heads.

Basically, we deal with two types of relations: `is-a` and `part-of`. A class A `is-a` a class B if it is a specialization of B. Therefore a description of A must contain the same attributes than B and at least one more. On the contrary A is a `part-of` B if it is (strictly) included in B, i.e. A has a subset of the attributes of B. Nesting is also important: as illustrated in Figure 7, in case of `is-a` relation, we expect nodes in A to have the same nesting level as the corresponding ones in B. In case of a `part-of` relation the nodes of A appear in B at a lower level, as shown in Figure 8.

In our application documents are encoded as fuzzy bags, and each node is associated to a weight (Figure 3). As illustrated in Figure 7, in case of `is-a` relation, we expect nodes of A to have the same weights as those in B. On the other hand, in case of a `part-of` relation, the nodes of A have lower weights as those in B, as shown in Figure 8.

We are now ready to formalize our conditions for detecting `is-a` and `part-of` relations as simple first-order logic formulae. Namely, we have:

- `is-a`:

$$x \in (A \cap B) \to \mu_A(x) = \mu_B(x); \qquad (1)$$

- `part-of`:

$$|A| > |B|;$$

$$x \in (A \cap B) \rightarrow \mu_A(x) > \mu_B(x). \tag{2}$$

Here we are not interested in formally evaluating Equation (1) and Equation (2) in order to provide the community with a reasonable approximate estimate of the degree of `is-a` and `part-of` relationships among clusters. Rather, we rely on the fact that our knowledge extraction process grouped documents according to a fuzzy criterion to identify suitable heuristics. Firstly, we need a value expressing how much two fuzzy bags (encoding cluster-heads) are included into each other. Following Equation (1) we can use $\frac{|A \cap B|}{|B|}$ for evaluating the `is-a` relation, and following Equation (2) $\frac{|A \cap B|}{|A|}$ for evaluating the `part-of` relation.

Then, we need to take into account the difference between the root nodes in A and in B. In case of `is-a` we expect them to coincide. In case of `part-of` the two roots must be distinct. We model this situation by a "degree of separation" $\Delta(B_r, A_r)$, where B_r is the root node of the object B, A_r is the root node of the object A and Δ is evaluated referring to B. As far as measuring `is-a` is concerned, we want our measure to decrease when the distance between the roots' increases. When measuring `part-of` we want the opposite, i.e. increasing the root nodes distance should bring our measure towards 1. So, we start with the following tailored versions of the Jackard similarity index [20]:

$$\frac{|A \cap B|}{|B|} \cdot \frac{1}{\Delta(B_r, A_r) + 1} \tag{3}$$

for `is-a` and

$$\left| \frac{|A \cap B|}{|A|} - \left(\frac{1}{\Delta(B_r, A_r) + 1} \right)^2 \right| \tag{4}$$

for `part-of`.

In order to make our similarity measures more precise, we need to take a closer look at the intersection $A \cap B$. More specifically, we do not want to know just how many nodes A and B share, but also if the nodes they have in common are in the same position. Therefore, we need to evaluate if (and how much) elements of $A \cap B$ in the subtree rising from A_r have the same distance from A_r in A as they have from B_r in B. Intuitively, if all the elements of the subtree are at the same distance from A_r in A as they have from B_r in B, their contribution to the total degree of `is-a` must coincide with the cardinality of $|A \cap B|$. For each node of $A \cap B$, we consider the ratio between the maximum and the minimum value of the difference between its distances from A_r and B_r. Note that the reason for being more than one value of this distance is that A and B are modeled as fuzzy bags rather than sets. Designating by A_m the generic node of $A \cap B$ in the subtree whose root is A_r, we get:

$$\frac{|A \cap B| - 1}{\sum_m \left(\frac{\max(\Delta(A_r, A_m),\ \Delta(B_r, A_m)) + 1}{\min(\Delta(A_r, A_m),\ \Delta(A_r, A_m)) + 1} \right)}. \tag{5}$$

In summary, we obtain two estimates respectively for the degree of `part-of` and `is-a` relations as follows (remember $A_m \in A \cap B$):

- `is-a`:

$$\left(\frac{|A \cap B|}{|B|} \cdot \frac{1}{\Delta(B_r, A_r) + 1}\right) \cdot \left(\frac{|A \cap B| - 1}{\sum_m \left(\frac{\max(\Delta(A_r, A_m), \Delta(B_r, A_m)) + 1}{\min(\Delta(A_r, A_m), \Delta(B_r, A_m)) + 1}\right)}\right). \tag{6}$$

- `part-of`:

$$\left(\left|\frac{|A \cap B|}{|A|} - \left(\frac{1}{\Delta(A_r, B_r) + 1}\right)^2\right|\right) \cdot \left(\frac{|A \cap B| - 1}{\sum_m \left(\frac{\max(\Delta(A_r, A_m), \Delta(B_r, A_m)) + 1}{\min(\Delta(A_r, A_m), \Delta(B_r, A_m)) + 1}\right)}\right). \tag{7}$$

For example, we can compute the degree of `is-a` of the two objects A1 and B1 in Figure 7 as follows:

$$\frac{3}{3} \cdot \frac{1}{0+1} \cdot \frac{3-1}{\frac{1+1}{1+1} + \frac{1+1}{1+1}} = 1.$$

The `part-of` degree is equal to:

$$\left|\frac{3}{4} - \left(\frac{1}{1}\right)^2\right| \cdot \frac{3-1}{\frac{2}{2} + \frac{2}{2}} = 0.25.$$

In a similar way, the degree of `is-a` linking A2 and B2 in Figure 8 can be computed as follows:

$$\frac{4}{7} \cdot \frac{1}{3} \cdot \frac{4-1}{\frac{2}{2} + \frac{2}{2} + \frac{2}{2}} = 0.19.$$

The degree of `part-of` is equal to:

$$\left|\frac{4}{4} - \left(\frac{1}{2+1}\right)^2\right| \cdot \frac{4-1}{\frac{2}{2} + \frac{2}{2} + \frac{2}{2}} = 0.91.$$

7. An example of ad-hoc ontology construction

We are now ready to outline our bottom-up procedure for ontology building; referring to cluster-heads A1, B1, A2 and B2 in Figures 7 and 8, we can connect them each other in a cyclic graph, labeling edges with the degrees of `is-a` and `part-of` relations obtained applying Equation (6) and Equation (7) (results are shown in Figure 9); then, we represent classes as vertices of a directed graph. In other words, the original cyclic graph is converted into a Directed Acyclic Graph (DAG) expressing a partial order. In Figure 10 the result of this process is shown.

Is-a	A1	B1	A2	B2
A1	1	1	0	0.42
B1	0.75	1	0	0.32
A2	0	0	1	0.14
B2	0.75	0.75	0.33	1

Part-of	A1	B1	A2	B2
A1	0	0.25	0	0
B1	0	0	0	0
A2	0	0	0	0.88
B2	0.31	0.38	0.46	0

Fig. 9. `is-a` and `part-of` values between cluster-heads.

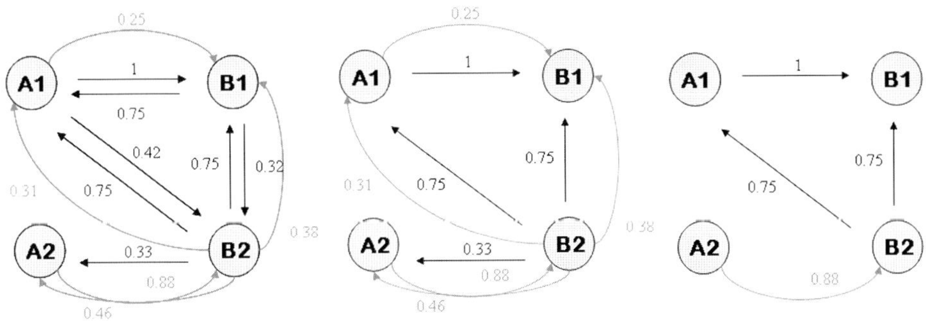

Fig. 10. The direct cyclic graph representing classes and relations between them is converted initially in a DAG and then in a tree.

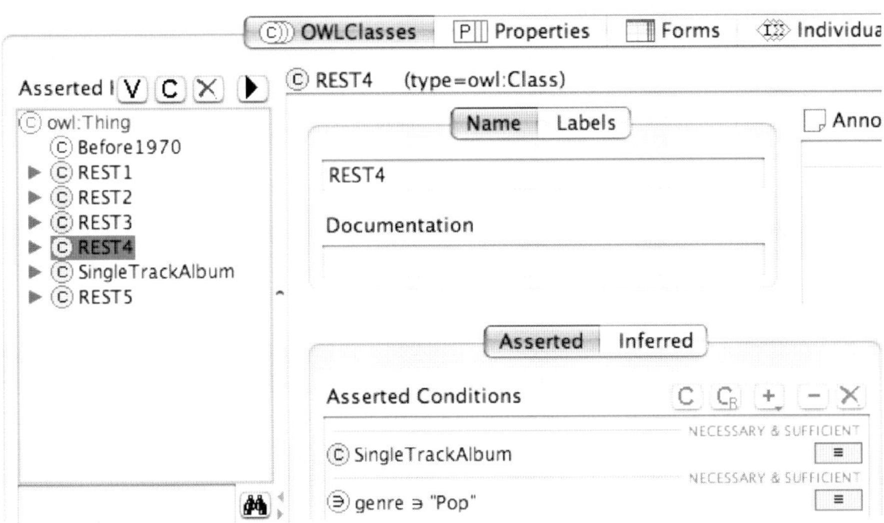

$$\mathtt{REST1} \equiv \mathtt{SingleTrackAlbum}$$

$$\mathtt{genre} \ni \mathtt{Rock}$$

$$\mathtt{REST2} \equiv \mathtt{SingleTrackAlbum}$$

$$\mathtt{genre} \ni \mathtt{Classical}$$

$$\mathtt{REST3} \equiv \mathtt{SingleTrackAlbum}$$

$$\mathtt{genre} \ni \mathtt{Country}$$

$$\mathtt{REST4} \equiv \mathtt{SingleTrackAlbum}$$

$$\mathtt{genre} \ni \mathtt{Pop}$$

Fig. 11. Four `SingleTrackAlbum` restrictions.

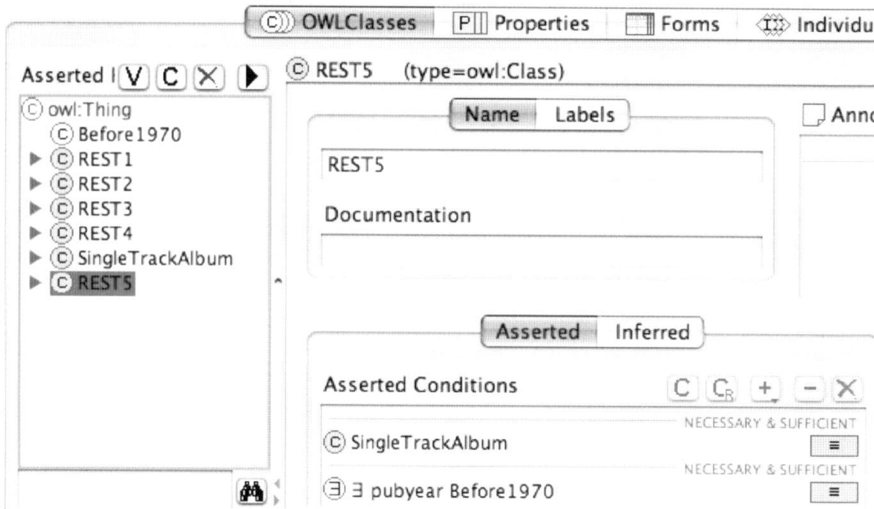

$\texttt{REST5} \equiv \texttt{GreatestHits}$

$\exists\ \texttt{pubyear Before1970}$

Fig. 12. A special type value restriction.

7.1. *Class constraints definition*

Section 5 described clustering on content values as a refining of the structural analysis phase. When building our bottom-up ontology, we express content-based partition of a structure cluster C in terms of constraints on the class corresponding to C. In fact, all instances of this class can be described using the set of attributes belonging to it; but documents belonging to the class can be grouped in subsets according to their content. This information is incorporated into the ontology by defining suitable constrained classes. In Section 5, the cluster *single track album* was split in four subset according to the values of the `<genre>` tag. Using OWL [32] we can define a general class `SingleTrackAlbum` and four sub-classes of `SingleTrackAlbum` carrying restrictions on the allowed values for the attribute `genre` (Figure 11). It is left to the community (or to the ontology engineer) to find meaningful names for constrained classes. Dealing with clusters like `GreatestHits`, we have to implement a more complex transformation. In these cases values restrictions are expressed by means of special data types. Our example includes a time-related data type that cannot be expressed in OWL. For this reason, we are forced to define a class whose sub-classes have the allowed length values and refer to these subclasses for expressing a value range. For instance, referring to Figure 6 we can define a restriction of the class `GreatestHits` as a class whose attribute `pubyear` is constrained to one of the allowed values in the class `Before1970` (Figure 12).

8. Conclusion

This chapter has discussed the role of fuzzy techniques in bottom-up generation of ontologies suitable for distributed peer-to-peer communities of practice. Admittedly, our bottom-up approach to ontology construction produces approximate, "quick and dirty" ontologies; therefore, it requires gathering the community's feedback for assessing the ontology correctness, following the evolution of the community's view of the domain. In our approach, ontology assertions are associated with a *trust level*. Our trust level evolves according to the community views throughout the ontology life-cycle by means of suitable voting algorithms [9].

A future development of this work includes a complete framework for community-based representation and update of trust-related metadata expressing a community's view over an automatically constructed ontology.

Acknowledgements

This work was partly funded by the Italian Ministry of Research Fund for Basic Research (`FIRB`) under project `RBAU01CLNB_001` "Knowledge Management for the Web Infrastructure" (`KIWI`).

References

[1] A. Armstrong, J. Hagel, *The Real Value of On-Line Communities*, Harvard Business Review **74** (1996), 134–141.

[2] R. Borowsky, J. Hofer-Alfeis, S. Klabunde, A. Schneider, R. Schmidt, S. Schoen, *Lösungsbausteine für Wissensmanagement-Prozesse*, Ergebnisbericht aus den Projektgemeinschaften, Siemens, 1998.

[3] B. Bouchon-Meunier, M. Rifqi, S. Bothorel, *Towards general measures of comparison of objects*, Fuzzy Sets and Systems **84** (2) (1996), 143–153.

[4] M. Castells, *The Rise of the Network Society*, Blackwell Publishers, 2004.

[5] P. Ceravolo, M.C. Nocerino, M. Viviani, *Knowledge extraction from semi-structured data based on fuzzy techniques*, in: Knowledge-Based Intelligent Information and Engineering Systems, Proceedings of the 8th International Conference, KES 2004, Part III, 2004, pp. 328–334.

[6] R. Charan, *How networks reshape organizations for results*, Harvard Business Review **69** (5) (1991), 104–115.

[7] G. Cormode, S. Muthukrishnan, *The string edit distance matching problem with moves*, in: SODA '02: Proceedings of the Thirteenth Annual ACM-SIAM Symposium on Discrete Algorithms, San Francisco, California, 2002, pp. 667–676.

[8] D.R. Cutting, D.R. Karger, J.O. Pedersen, J.W. Tukey, *Scatter/Gather: a cluster-based approach to browsing large document collections*, in: SIGIR '92, 1992, pp. 318–329.

[9] E. Damiani, S. De Capitani di Vimercati, S. Paraboschi, P. Samarati, *Managing and sharing servents' reputations in P2P systems*, IEEE Trans. on Knowledge Data Engineering **4** (15) (2003), 840–854.

[10] E. Damiani, R. Khosla, W. Grosky, *Human Centered e-Business*, Kluwer Academic Publishers, 2003.

[11] E. Damiani, N. Lavarini, B. Oliboni, L. Tanca, *An approximate querying environment for XML data*, in: V. Loia, M. Nikravesh, L.A. Zadeh (Eds.), Fuzzy Logic and the Internet, Series on Studies in Fuzziness and Soft Computing, Vol. 137, 2004.

[12] E. Damiani, M.C. Nocerino, M. Viviani, *Knowledge extraction from an XML data flow: building a taxonomy based on clustering technique*, in: Current Issues in Data and Knowledge Engineering, Proceedings of EUROFUSE 2004: 8th Meeting of the EURO Working Group on Fuzzy Sets, 2004, pp. 133–142.

[13] E. Damiani, L. Tanca, *Blind queries to XML data*, in: Proceedings of Database and Expert, in: Systems Applications: 11th International Conference, DEXA 2000, 2000, pp. 345–356.

[14] R.C. Dubes, A.K. Jain, *Algorithms for Clustering Data*, Prentice-Hall, 1988.

[15] A. Farquhar, R. Fikes, J. Rice, *The Ontolingua server: a tool for collaborative ontology construction*, Technical Report 96-26, Knowledge Systems Laboratory, Stanford University, 1996.

[16] T.R. Gruber, *Toward principles for the design of ontologies used for knowledge sharing*, Technical Report KSL 93-04, Knowledge Systems Laboratory, Stanford University, 1993.

[17] G. Hamerly, C. Elkan, *Alternatives to the k-means algorithm that find better clusterings*, in: CIKM '02: Proceedings of the Eleventh International Conference on Information and Knowledge Management, McLean, VA, USA, 2002, pp. 600–607.

[18] E. Han, G. Karypis, V. Kumar, *Text categorization using weight adjusted k-Nearest Neighbor classification*, in: PAKDD '01: Proceedings of the 5th Pacific-Asia Conference on Knowledge Discovery and Data Mining, Springer-Verlag, 2001, pp. 53–65.

[19] A.K. Jain, M.N. Murty, P.J. Flynn, *Data clustering: a review*, ACM Computing Surveys **31** (3) (1999), 264–323.

[20] G. Klir, T. Folger, *Data Clustering: Fuzzy Sets, Uncertainty and Information*, Prentice-Hall, 1988.

[21] J. Lave, E. Wenger, *Situated Learning: Legitimate Peripheral Participation*, Cambridge University Press, 1991.

[22] E. Lesser, K. Everest, *Using communities of practice to manage intellectual capital*, Ivey Business Journal **65** (4) (2000), 37–41.

[23] E.L. Lesser, M.A. Fontaine, J.A. Slusher, *Knowledge and Communities*, Butterworth Heinemann, 2000.

[24] R. Maier, *Knowledge Management Systems – Information and Communication Technologies for Knowledge*, Springer-Verlag, 2002.

[25] R. McDermott, *Knowing in community: ten critical success factors in building communities of practices*, IHRIM J. **4** (2000).

[26] C. North, B. Shneiderman, C. Plaisant, *User controlled overviews of an image library: a case study of the visible human*, in: Proceedings of the First ACM International Conference on Digital Libraries, 1996, pp. 74–82.

[27] H. Rheingold, *The Virtual Community*, 2nd ed., The Millennium Whole Earth Catalog, Harper, San Francisco, 1994.

[28] H. Saint-Onge, D. Wallace, *Leveraging Communities of Practices for Strategic Advantage*, Butterworth Heinemann, 2003.

[29] B. Stein, S. Meyer zu Eissen, *Document categorization with MajorClust*, in: Proceedings of the 12th Workshop on Information Technology and Systems (WITS '02), Barcelona, Spain, Technical University of Barcelona, 2002.

[30] M. Steinbach, G. Karypis, V. Kumar, *A comparison of document clustering techniques*, KDD Workshop on Text Mining, 2000.

[31] E. Wenger, *Communities of practice: the key to a knowledge strategy*, Knowledge Directions **1** (2) (1999), 48–63.

[32] http://www.w3.org/OWL/

[33] http://www.daml.org/services/owl-s/

[34] http://www.w3.org/XML/

CHAPTER 14

Approximate Knowledge Graph Retrieval: Measures and Realization

T.H. Cao and Dat T. Huynh

Faculty of Information Technology, Ho Chi Minh City University of Technology, Viet Nam
E-mail: {tru,htdat}@dit.hcmut.edu.vn

Abstract

The World Wide Web is moving to its next generation called Semantic Web. Managing and searching for web information goes beyond the conventional relational database model, as the data are semi-structured and inexact answers are often the case. This research work studies similarity and subsumption measures for matching knowledge graphs based on the non-directed and directed semantic distances between their entity types and attribute values. A knowledge base represented in RDF is constructed on an ontology that defines its concept types and relation types with their signatures. A query engine is developed and implemented to query the knowledge base, returning approximate answers with similarity and subsumption degrees. On the one hand, for a user-friendly interface and easily readable query expressions, conceptual graphs are employed at the front-end. On the other hand, in order to take the advantage of the existing platform of Sesame, the query modification tactic is first used to retrieve the knowledge graphs that are close to a query graph, before the matching degrees of those answer graphs are calculated.

Keywords

semantic distance, similarity degree, subsumption degree, query modification, graph matching

1. Introduction

The explosion of information on the World Wide Web is challenging the conventional relational database model at least in two aspects. Firstly, such information is no longer

FUZZY LOGIC AND THE SEMANTIC WEB
Edited by Elie Sanchez
© 2006 Elsevier B.V. All rights reserved

as simple as to be easily and naturally represented in relational tables. Secondly, exact answers are hardly obtained when searching for information on the Web.

For the first problem, the Resource Description Framework (RDF) has provided a common language for representing Web information, whose statements can be viewed as graphs. New technologies for management of knowledge and information, in particular on the Web, have also emerged. For example, Sesame [16] with its query language SeRQL [24] is for storing and querying RDF graphs. Those are open sources and have been used in many large-scale systems.

Meanwhile, the second problem is mainly concerned with the semantic distance between concepts and relations in knowledge graphs. It has attracted significant research effort whose methods can be classified into two main approaches, namely, corpus-based and ontology-based. In the former approach, the semantic distance between words representing concepts or relations is determined on the basis of their contexts of occurrence in a linguistic corpus. In the latter approach, it is determined on the basis of the semantic nets involving the words. However, those developed methods do not work for un-interpreted type labels, because they rely on the actual occurrence and meanings of involved words.

In this work, since our knowledge bases of concern are built on ontology, for concept and relation type labels, we adapt the ontology-based semantic distance formula for words given by Jiang and Conrath [15]. Instead of calculating the occurrence probabilities of words as required in that formula based on a linguistic corpus, we derive the occurrence probabilities of concept and relation types from the populations of the types in a knowledge base of discourse. For fuzzy set-based attribute values, we propose a measure of their similarity that uses fuzzy conditional probability.

While symmetric measures like similarity have been discussed much in literature, research on asymmetric measures appears to be sporadic. In particular, in practice one may want to measure how much a query graph subsumes an answer graph. We define such a asymmetric subsumption measure for both entity types and attribute values, also in the probabilistic framework of Jiang–Conrath's semantic distance. Therefore, our defined matching degrees between entity types and attribute values of two knowledge graphs can be combined to give an overall matching degree between the graphs.

Regarding the implementation aspect, realizing an approximate retrieval system of knowledge graphs from scratch is a formidable task. Therefore, we choose the query modification approach as an alternative. In particular, we exploit the infrastructure of Sesame and SeRQL for management and exact matching of knowledge graphs. That is, knowledge is stored as RDF graphs in Sesame. A query graph is first generalized to retrieve RDF graphs via SeRQL. Those exact answer graphs, which are subsumed by the generalized query graph, are then matched to the original query graph to return their similarity and subsumption degrees.

However, SeRQL is for machine rather than human consumption. Meanwhile, Conceptual Graphs (CG) have been shown to have a smooth mapping to and from natural language [6,25]. Significantly, while Zadeh has regarded a methodology for computing with words as the main contribution of fuzzy logic [30], Sowa has regarded a smooth mapping between logic and natural language as the main motivation of conceptual graphs [26]. It shows that fuzzy logic and conceptual graphs are two logical formalisms that emphasize the same target of natural language. Their combination can offer a formalism to represent and reason with linguistic information. So, we employ conceptual graphs for query expressions at the interface layer, which are translated to SeRQL queries to be executed by the Sesame query engine.

The rest of the paper is organized as follows. Section 2 reviews previous definitions of similarity measures for words, and proposes our similarity and subsumption measures for concept and relation types. Section 3 defines our matching degrees for fuzzy attribute values. Section 4 summarizes the basic notions of Sesame, SeRQL, and conceptual graphs for storing and querying knowledge graphs. Sections 5 and 6 present our query modification tactic and CG-SeRQL mapping algorithm, and their use in the semantic web knowledge and information management system VN-KIM that we are developing. Finally, Section 7 concludes the paper and suggests future work.

2. Matching measures for entity types

2.1. *Similarity between words*

The similarity between expressions, such as words, concepts, or relations, measures how close their semantics or meanings are. Previous methods for words could be classified into corpus-based or ontology-based approaches, which were also called knowledge-poor or knowledge-rich, respectively.

In the corpus-based approach, relations between words from a linguistic corpus were investigated. Its main assumption was that similar words were used in similar contexts, and similar contexts used similar words. So, a distribution of co-occurrences of words within different contexts could be used to calculate the similarity between the words [8,11]. The advantage of this approach is that it does not require domain knowledge. As such, however, it misses the inheritance of the semantics of words in natural language.

Meanwhile, in the ontology-based approach, predefined semantics, in particular taxonomy, of concepts was taken into account, where lexical resources were built into a network or directed graph. The similarity between concepts was then calculated based on the properties of the paths connecting them. So, the approach is intuitive and easy to understand. It could also be classified further into the dictionary-based and taxonomy-based approaches.

In dictionary-based methods, a monolingual dictionary was turned into a network by creating a node for every headword, and edges for relations between a headword and the words in its definition. The similarity between words was then calculated by spreading activation on this network [14,18].

In contrast, the simplest taxonomy-based method computed the similarity between two concepts on the basis of the number of edges on the shortest path between them in a hierarchy, in particular WordNet [19,28]. In [13], it was suggested that a concept was close in semantics to another if the path connecting them was not too long and did not change its direction too often.

Unlike those edge-based methods, the node-based method in [23] combined ontology-based and corpus-based approaches to measure the similarity between a pair of concepts c_1 and c_2 as the extent to which they shared common information. That shared information was determined by the information content of the least common super-concept, denoted by $c_1 \vee c_2$, which subsumed both c_1 and c_2:

$$Sim_R(c_1, c_2) = -\log Pr(c_1 \vee c_2) \quad (1)$$

where $Pr(c)$ was the occurrence probability of the word representing a concept c in a corpus of discourse. The drawback of this method is that many pairs of concepts may have the same similarity degree if they have the same least common super-concept.

In order to overcome the shortcoming of [23], in [15] Jiang and Conrath used both edge-based and node-based measures, assuming that the semantic distance of the link connecting a child concept c to its parent concept $Par(c)$ was the information content of c that is not in $Par(c)$. Formally, one had:

$$Dist_{JC}\big(c, Par(c)\big) = -\log Pr\big(c \mid Par(c)\big) = -\log \frac{Pr(c \wedge Par(c))}{Pr(Par(c))}. \tag{2}$$

Since $c \wedge Par(c) = c$, one obtained:

$$Dist_{JC}\big(c, Par(c)\big) = -\log \frac{Pr(c)}{Pr(Par(c))} = -\log Pr(c) + \log Pr\big(Par(c)\big) \tag{3}$$

where $-\log Pr(c)$ represents the information content of a concept c.

Given the semantic distance from a node to its parent, the semantic distance between any two nodes in a hierarchy was the sum of the distances along the shortest path that connected the nodes:

$$Dist_{JC}(c_1, c_2) = \sum_{c \in Path(c_1, c_2) \setminus \{c_1 \vee c_2\}} Dist_{JC}\big(c, Par(c)\big) \tag{4}$$

where $c \in Path(c_1, c_2)\setminus\{c_1 \vee c_2\}$ denoted the set of all the concepts along the path connecting c_1 and c_2 except for $c_1 \vee c_2$. After elimination of opposite terms, one obtained:

$$\begin{aligned} Dist_{JC}(c_1, c_2) &= Dist_{JC}(c_1, c_1 \vee c_2) + Dist_{JC}(c_2, c_1 \vee c_2) \\ &= -\big(\log Pr(c_1) + \log Pr(c_2)\big) + 2\log Pr(c_1 \vee c_2). \end{aligned} \tag{5}$$

The measure proposed in [20] used the same elements as in $Dist_{JC}()$, but in a different formula. However, as evaluated in [4], the Jiang–Conrath method is overall the best among the above-mentioned methods.

2.2. *Similarity and subsumption between entity types*

The methods reviewed above measure similarity between actual words. In the Jiang–Conrath method, for instance, the information content of a concept is calculated via the occurrence probability of the word representing that concept in a corpus. However, a word representing a type in a hierarchy may be just a dummy identifier for the type, not necessarily attached to any linguistic corpus nor having any meaning. So, instead of relying on the occurrence probabilities of type labels, in this work we propose to use those of type entities. That is, we define:

$$Pr(t) = Pop(t)/N \tag{6}$$

where $Pop(t)$ is the number of entities of a type t and N is the total number of entities in a knowledge base of discourse. Adapting Equation (5), one obtains:

$$\begin{aligned} Dist(t_1, t_2) &= -\big(\log\big(Pop(t_1)/N\big) + \log\big(Pop(t_2)/N\big)\big) + 2\log\big(Pop(t_1 \vee t_2)/N\big) \\ &= -\big(\log Pop(t_1) + \log Pop(t_2)\big) + 2\log\big(Pop(t_1 \vee t_2)\big). \end{aligned} \tag{7}$$

The similarity degree between two concept types can be defined as the inverse of their semantic distance, and scaled into [0, 1], as follows:

$$Sim(t_1, t_2) = \frac{1}{1 + Dist(t_1, t_2)} \tag{8}$$

This *population-based* measure can also be applied to the semantic distance between two relation types in a hierarchy, where the population of a relation type is the total number of entity tuples that are involved in relations of that type.

One can notice that the above-mentioned measures are symmetric, for both words and entity types, as they should be with respect to the notion of similarity. For example, the similarity degree between a vehicle in the query "Find a vehicle" and a car in the answer "A certain car" is the same as the similarity between a car in the query "Find a car" and a vehicle in the answer "A certain vehicle", which is the similarity degree between the entity types "vehicle" and "car".

However, in practice, looking at the subsumption relationship, one may consider the answer in the former case as an exact answer, because every car is a vehicle, while the answer in the latter case is just an approximate one, with respect to the queries. In general, one may want to evaluate how much a query subsumes an answer.

As such, unlike similarity functions, which are symmetric, subsumption functions are asymmetric. In other words, similarity is based on non-directed semantic distances, while subsumption is based on directed semantic distances. Equation (2) above gives an idea to define a subsumption distance from a type t_1 to a type t_2 as follows:

$$Dist_S(t_1 \rightarrow t_2) = -\log Pr(t_1 \mid t_2) = -\log \frac{Pr(t_1 \wedge t_2)}{Pr(t_2)} \tag{9}$$

where $t_1 \wedge t_2$ denotes the greatest common subtype of t_1 and t_2, and $Pr(t)$ is defined as in Equation (6) for a type t.

Then the subsumption degree of a type to another can be defined as the inverse of the directed semantic distance from t_1 to t_2, and scaled into [0, 1], as follows:

$$Sub(t_1 \rightarrow t_2) = \frac{1}{1 + Dist_S(t_1 \rightarrow t_2)}. \tag{10}$$

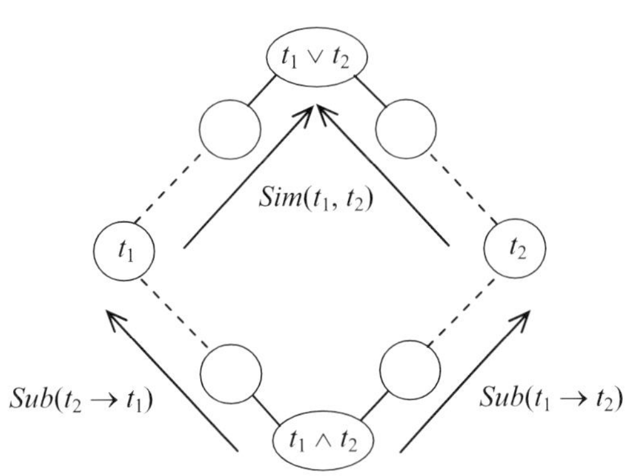

Fig. 1. Subsumption versus similarity.

Table 1
Voting pattern for *high* dice values.

Scores \ Voters	P_1	P_2	P_3	P_4	P_5	P_6	P_7	P_8	P_9	P_{10}
1										
2										
3	x	x								
4	x	x	x	x	x					
5	x	x	x	x	x	x	x	x	x	
6	x	x	x	x	x	x	x	x	x	x

This definition satisfies intuition in the following special cases:

- If t_1 is a super-type of t_2, i.e., $t_1 \wedge t_2 = t_2$, then $Sub(t_1 \rightarrow t_2) = 1$ while $Sub(t_2 \rightarrow t_1) < 1$.
- If t_1 and t_2 are disjoint, i.e., $t_1 \wedge t_2 = \varnothing$, then $Sub(t_1 \rightarrow t_2) = Sub(t_2 \rightarrow t_1) = 0$.

Figure 1 illustrates the difference between similarity and subsumption measures.

3. Matching measures for attribute values

3.1. *Voting model of fuzzy sets*

In this work, for defining matching measures on fuzzy set-based attribute values, we apply the voting model interpretation of fuzzy sets [2,10]. That is, given a fuzzy set A on a domain U, each voter has a subset of U as his/her own crisp definition of the concept that A represents. For example, a voter may have the interval [0, 35] representing human ages from 0 to 35 years as his/her definition of the concept *young*, while another voter may have [0, 25] instead.

The membership function value $\mu_A(u)$ is then the proportion of voters whose crisp definitions include u. This model defines a *mass assignment* (i.e., probability distribution) on the power set of U, where the mass (i.e., probability value) assigned to a subset of U is the proportion of voters who have that subset as a crisp definition for the fuzzy concept A. As such, this mass assignment corresponds to a family of probability distributions on U.

Let us take the Dice example in [2]. Given the dice values from the set $\{1, 2, 3, 4, 5, 6\}$, suppose that a score *high* is defined by the discrete fuzzy set {3:0.2, 4:0.5, 5:0.9, 6:1}, i.e., the membership of value 3 is 0.2, and so on. The voting pattern of a group of 10 persons for this score could be as in Table 1.

That is, all voters, P_1 to P_{10}, vote for value 6 as a *high* score, while only two of them, P_1 and P_2, vote for 3 as a *high* score, and so on. In other words, the crisp definition of P_{10} for the *high* score is $\{6\}$, while that of P_1 and P_2 is $\{3, 4, 5, 6\}$, for instance. An assumption made in this voting model is that any person who accepts a value as a *high* score also accepts all values that have higher membership grades in the fuzzy set *high*. This model defines the following mass assignment on the power set of $\{1, 2, 3, 4, 5, 6\}$:

$$\{6\}:0.1 \ \{5, 6\}:0.4 \ \{4, 5, 6\}:0.3 \ \{3, 4, 5, 6\}:0.2.$$

3.2. *Conditional probability on fuzzy sets*

Using the voting model, given two fuzzy sets A and B on a domain U, the conditional probability $Pr(A \mid B)$ is defined by [1]:

$$Pr(A \mid B) = \sum_{S,T \subset U} Pr(u \in S \mid u \in T) \cdot m_A(S) \cdot m_B(T) \tag{11}$$

where $m_A(S)$ and $m_B(T)$ are the masses of subsets S and T of U, respectively. The intuitive meaning of $Pr(A \mid B)$ is that it is the probability for a fuzzy proposition $x \in A$ being true given $x \in B$ being true. In other words, it is the fuzzy conditional probability of $x \in A$ given $x \in B$. The above definition can also be adapted for fuzzy sets on continuous domains, using integration instead of addition.

For example, suppose that the dice value *about_5* is defined by the fuzzy set {6:0.3, 5:1, 4.0.3}, whose mass assignment is {5}:0.7 {4, 5, 6}:0.3. Then, one has:

$$\begin{aligned}
&Pr(high \mid about_5)\\
&\quad = Pr\big(u \in \{5\} \mid u \in \{6\}\big) \cdot m_{high}\big(\{6\}\big) \cdot m_{about_5}\big(\{5\}\big)\\
&\qquad + Pr\big(u \in \{5\} \mid u \in \{5,6\}\big) \cdot m_{high}\big(\{5,6\}\big) \cdot m_{about_5}\big(\{5\}\big)\\
&\qquad + Pr\big(u \in \{5\} \mid u \in \{4,5,6\}\big) \cdot m_{high}\big(\{4,5,6\}\big) \cdot m_{about_5}\big(\{5\}\big)\\
&\qquad + Pr\big(u \in \{5\} \mid u \in \{3,4,5,6\}\big) \cdot m_{high}\big(\{3,4,5,6\}\big) \cdot m_{about_5}\big(\{5\}\big)\\
&\qquad + Pr\big(u \in \{4,5,6\} \mid u \in \{6\}\big) \cdot m_{high}\big(\{6\}\big) \cdot m_{about_5}\big(\{4,5,6\}\big)\\
&\qquad + Pr\big(u \in \{4,5,6\} \mid u \in \{5,6\}\big) \cdot m_{high}\big(\{5,6\}\big) \cdot m_{about_5}\big(\{4,5,6\}\big)\\
&\qquad + Pr\big(u \in \{4,5,6\} \mid u \in \{4,5,6\}\big) \cdot m_{high}\big(\{4,5,6\}\big) \cdot m_{about_5}\big(\{4,5,6\}\big)\\
&\qquad + Pr\big(u \in \{4,5,6\} \mid u \in \{3,4,5,6\}\big)\\
&\qquad \times m_{high}\big(\{3,4,5,6\}\big) \cdot m_{about_5}\big(\{4,5,6\}\big)\\
&\quad = 0.53.
\end{aligned}$$

3.3. *Similarity and subsumption between fuzzy sets*

Since fuzzy attribute values were introduced into knowledge and data bases, a measure of similarity between fuzzy sets has become a natural demand. Several measures have been proposed for different purposes [17]. For instance, in [21] for object matching, the similarity degree between two fuzzy sets A and B on a domain U was defined by:

$$e(A, B) = \frac{Car(A \cap B)}{Car(A \cup B)} \tag{12}$$

where $Car(C)$ was the cardinality of a fuzzy set C. Meanwhile, in [27] for image retrieval, the fuzzy distance between A and B was defined as follows:

$$d(A, B) = \sum_{u \in U} \mu_{(A \setminus B) \cup (B \setminus A)}(u) = \sum_{u \in U} \big|\mu_A(u) - \mu_B(u)\big| \tag{13}$$

assuming that U was a discrete domain. For the continuous case of U, integration was used instead of addition.

In this work, our matching measures for fuzzy sets are defined in the probabilistic framework for words and entity types as presented above, so that they can be combined in integrated matching measures between knowledge graphs. We note that the set of all fuzzy sets on a domain U also forms a lattice with the fuzzy subset relation as the partial order, as compared to entity type lattices [5]. In particular, for two fuzzy sets A and B on U, their least common fuzzy super-set is $A \cup B$, and their greatest common fuzzy subset is $A \cap B$. Therefore, Equation (5) can be adapted to define the similarity distance between fuzzy sets as follows:

$$\begin{aligned} Dist(A, B) &= -\big(\log Pr(A) + \log Pr(B)\big) + 2 \log Pr(A \cup B) \\ &= -\log Pr(A \mid A \cup B) - \log Pr(B \mid A \cup B). \end{aligned} \tag{14}$$

Meanwhile, the subsumption distance from A to B is adapted from Equation (9):

$$Dist_S(A \to B) = -\log Pr(A \mid B). \tag{15}$$

The proposed measures employ the definition of fuzzy conditional probability introduced above.

4. Storing and querying knowledge graphs

4.1. *Sesame*

In order to effectively manage and exploit knowledge and information represented in RDF and RDFS, a system for efficient storage and expressive queries of large quantities of RDF and RDFS statements is a demand. Sesame is such an open framework meeting that need. The most essential design principle of Sesame is to make it extendable and adaptable to different operating environments. That idea has been realized in a generic modular architecture that enables Sesame to be installed on different database management systems (DBMS) and accessed via various communication protocols, as shown in Figure 2.

For achieving it, Sesame's developers have introduced storage and inference layers (SAIL) between upper modules and lower data repositories. They provide a set of basic methods for storing, querying, and manipulating RDF data provided, while optimized implementation of those methods is handled by a particular DBMS in use. In addition to memory and file storage mechanisms, Sesame currently supports the three DBMSs PostgreSQL, MySQL and Oracle 9i. Moreover, SAILs can also be stacked on top of each other, whence the developers can add in new SAILs for supporting further operations such as caching, concurrency handling, or inference.

Since clients can be on very different operational environments such as a Web server or a standalone application, Sesame offers a flexible communication mechanism. By placing all common communication methods at the module Request Router and putting all protocol-specific methods at the protocol handlers, Sesame can support different communication protocols such as HTTP, RMI, and SOAP.

Besides the above-mentioned essential features for storage and communication, Sesame also provides modules for querying, administration, security and versioning. In brief, it is a generic architecture that can be installed on a variety of different DBMSs, and a good candidate for storing, managing, and retrieving large-scale knowledge bases on the Web.

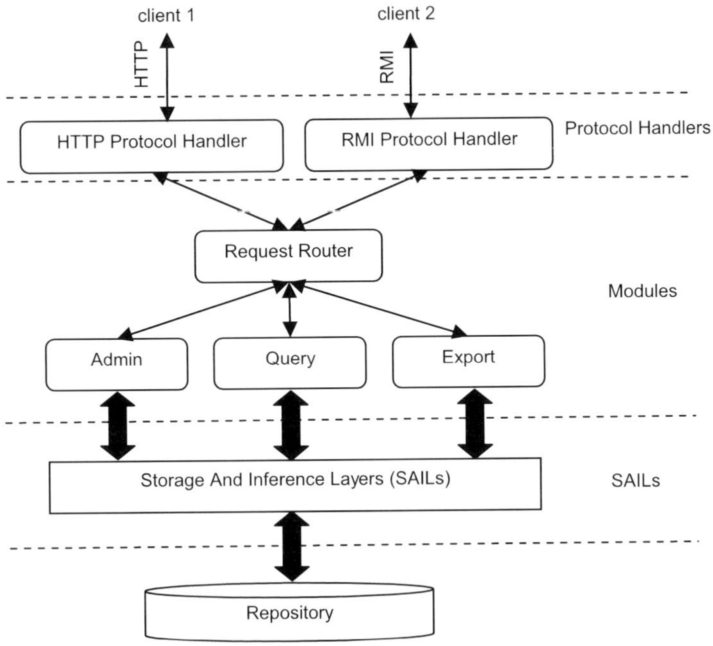

Fig. 2. Sesame architecture.

4.2. *SeRQL*

Along with the emergence of RDF/RDFS standard, several RDF/RDFS query languages have been proposed, such as RQL and RDQL. However, they still lack a strong path expression ability, which helps to simplify query formulation. Therefore, Sesame offers SeRQL (Sesame RDF Query Language) as a new RDF/RDFS query language that inherits the best features of RQL and RDQL, and equipped with some new features. SeRQL also supports subsumption, where a concept or relation type in a query can match with its subtypes in a knowledge base.

The basic building blocks of SeRQL are universal resource identifiers (URI), literals, and variables. An URI in Sesame can be written in two formats, either as a full URI, e.g., <http://www.dit.hcmut.edu.vn/vnkim/vnkimkb.rdf#Country_1>, or an abbreviated URI, e.g., <vnkimkb_rdf: Country_1>. In this case, the prefix vnkimkb_rdf will be mapped to the full string http://www.dit.hcmut.edu.vn/vnkim/vnkimkb.rdf by the Sesame query engine. RDF literals represent properties of objects and can optionally be attached with language and data type tags. Variables are URIs or properties that are wanted, and identified in queries by names.

All RDF statements in a repository can be regarded as a graph, and searching can be performed by matching a query graph, consisting of path expressions, to that RDF graph. One outperforming feature of SeRQL as compared to other RDF/RDFS query languages is that it allows path expressions of arbitrary length. For example, path expressions of length 1 in SeRQL for the query "Find all cities that are located in Vietnam" could be as follows:

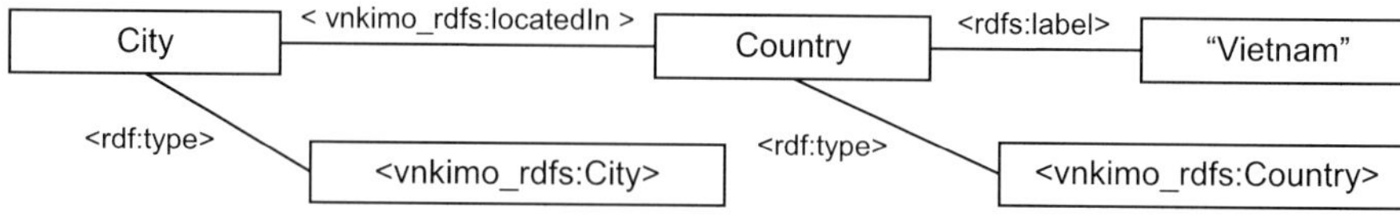

Fig. 3. Graph representation of a SeRQL query.

```
{City} rdf:type {<vnkimo_rdfs:City>},
{City} vnkimo_rdfs:locatedIn {Country},
{Country} rdf:type {<vnkimo_rdfs:Country>},
{Country} rdfs:label {"Vietnam"}
```

The second and the fourth path expressions, for instance, can be combined and rewritten as the following path expression of length 2:

```
{City} vnkimo_rdfs:locatedIn {Country} rdfs:label {"Vietnam"}
```

In SeRQL, nodes and edges in path expressions can be variables or URIs, while nodes can also be literals. In this example, City and Country are variables, "Vietnam" is a literal, and <vnkimo_rdfs:City> and <vnkimo_rdfs:Country> are URIs. This SeRQL query can be illustrated as a graph as shown in Figure 3.

A SeRQL selection query is built up from the four main clauses SELECT, FROM, WHERE, and USING NAMESPACE. The SELECT clause is used to determine which variables must be returned and in which order. The FROM clause contains path expressions. The WHERE clause expresses constraints on the values of variables. Finally, the USING NAMESPACE clause declares namespaces that are used for the mapping of abbreviated URIs.

For example, the full SeRQL query for "Find all cities in Vietnam whose population is over one million" is:

```
SELECT City
FROM
    {City} rdf:type {<vnkimo_rdfs:City>},
    {City} vnkimo_rdfs:locatedIn {Country},
    {City} vnkimo_rdfs:hasPopulation {Population},
    {Country} rdf:type {<vnkimo_rdfs:Country>},
    {Country} rdfs:label {"Vietnam"}
WHERE
    Population > "1000000"^^ <xsd:positiveInteger>
USING NAMESPACE
    rdf = <http://www.w3.org/1999/02/22-rdf-syntax-ns#>
    rdfs = <http://www.w3.org/2000/01/rdf-schema#>
    vnkimo_rdfs = <http://www.dit.hcmut.edu.vn/vnkim/vnkimo-rdf-schema#>
```

SeRQL also provides some short cuts to simplify query formulation. One of those short cuts is the notation for branches in path expressions. That is useful for the case when one subject has several relations with other objects. Instead of repeating the subject for each path expression, those path expressions can share the common subject by using a semicolon. For example, the above path expressions could be shortened as follows:

Fig. 4. An example CG.

{City} rdf:type {<vnkimo_rdfs:City>}; vnkimo_rdfs:locatedIn {Country},
{Country} rdf:type {<vnkimo_rdfs:Country>}; rdfs:label {"Vietnam"}

Another useful short cut is for reified statements, where a node is itself a statement. For example, one can have a statement in the form below:

{ {reifiedSubj} reifiedPred {reifiedObj} } pred {obj}

This would be equivalent to the following set of path expressions:

{reifiedStatement} rdf:type {rdf:Statement},
{reifiedStatement} rdf:subject {reifiedSubj},
{reifiedStatement} rdf:predicate {reifiedPredicate},
{reifiedStatement} rdf:object {reifiedObj},
{reifiedStatement} pred {obj}

Finally, it is worth mentioning that property constraints can be combined using the logical *not*, *and*, *or* operators in the WHERE clause. That makes SeRQL be an expressive query language for RDF knowledge bases, as SQL for relational databases.

4.3. *Conceptual graphs*

A conceptual graph is a bipartite graph of *concept* vertices alternate with (conceptual) *relation* vertices, where edges connect relation vertices to concept vertices. Each concept vertex, drawn as a box and labeled by a pair of a *concept type* and a *concept referent*, represents an entity whose type and referent are respectively defined by the concept type and the concept referent in the pair. Each relation vertex, drawn as a circle and labeled by a *relation type*, represents a relation of the entities represented by the concept vertices connected to it. Concepts connected to a relation are called *neighbor concepts* of the relation. For example, the CG in Figure 4 says "John studies a subject in Computer Science".

In this example, [STUDENT: John], [SUBJECT: *], [FIELD: Computer Science] are concepts with STUDENT, SUBJECT and FIELD being concept types, whereas (STUDY) and (IN) are relations with STUDY and IN being relation types. The referents John and Computer Science of the concepts [STUDENT: John] and [FIELD: Computer Science] are *individual markers*. The referent * of the concept [SUBJECT: *] is the *generic marker* referring to an unspecified entity. It is assumed that two concepts with two different individual markers are assumed to refer to two different entities, while concepts with the same individual marker are assumed to refer to the same entity.

To denote that two concepts with the generic marker refer to the same unspecified entity, a *coreference link* drawn as a dotted line connecting them is used. For example, the CG in Figure 5 adds to the CG in Figure 4 a fact that "There is a person who studies the same subject as John does". In this example, the CG can be transformed into a semantically equivalent one without a coreference link by joining the two concepts [SUBJECT: *]

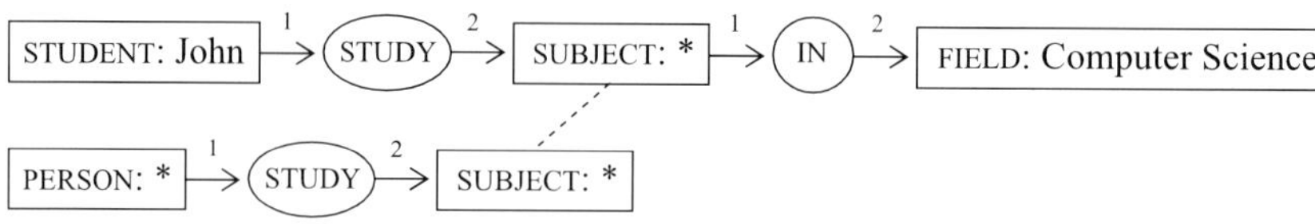

Fig. 5. A CG with a coreference link.

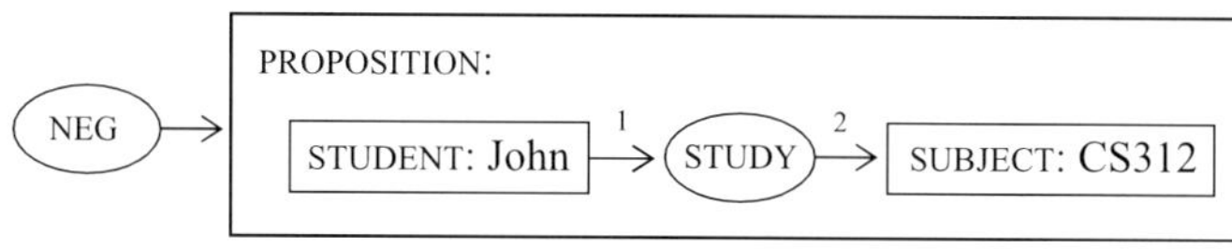

Fig. 6. A negated CG.

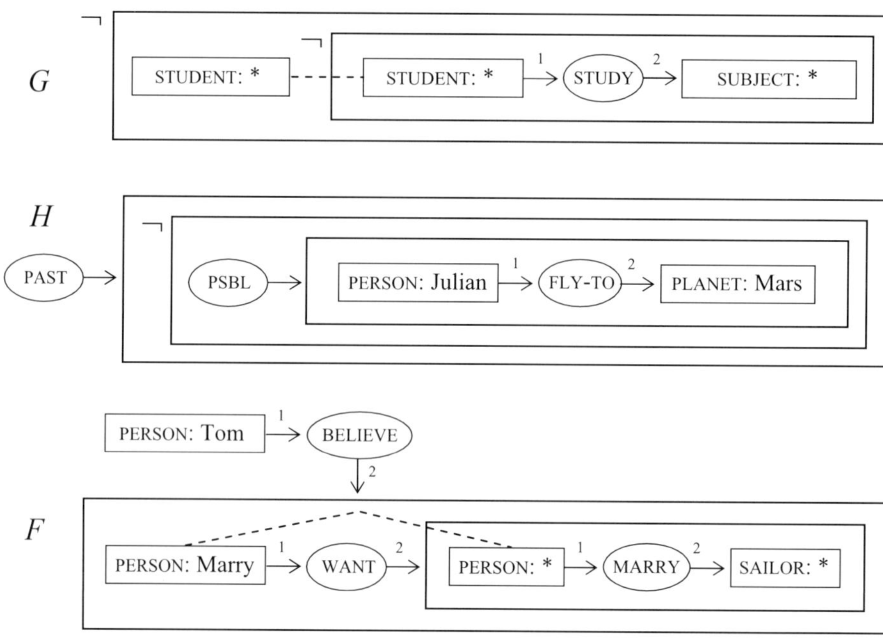

Fig. 7. Nested CGs.

into a single one. However, coreference links are more appealing when concepts occur in different contexts, such as *negative contexts*, where the concepts cannot be joined.

Corresponding to the notion of sorts in order-sorted predicate logic, concept types are partially ordered by the concept subtype order. For example, one may have STUDENT as a subtype of PERSON. Relation types can also be partially ordered. For example, with the CG in Figure 4, one may have STUDY as a subtype of a relation type ACT.

Meanwhile, each relation type has a signature defining its arity and the least specific concept type for each of its arguments. A relation type can be viewed as a type each entity of which is a tuple of entities of the argument types (being concept types) of that relation type. Thus, an argument type of a relation type is a subtype of the respective argument type of any super-type of that relation type.

For example, the signature of the relation type STUDY in the CG in Figure 4 can be (STUDENT, SUBJECT), with arity 2 and argument types STUDENT and SUBJECT for the edge labeled 1 and the edge labeled 2, respectively. Then the CG [PHD_STUDENT: John] → (STUDY) → [SUBJECT: *] is *well-typed*, but not the CG [PERSON: John] → (STUDY) → [SUBJECT: *], provided that PHD_STUDENT is a subtype of STUDENT, which is in turn a subtype of PERSON.

A proposition can be represented as a concept of type PROPOSITION whose referent is a CG describing that proposition. The negation of a proposition is then represented by a relation of type NEG connected to the concept representing that proposition, as illustrated in Figure 6.

Nested CGs can be used to express arbitrary first-order logic sentences and, furthermore, modal and higher-order logic ones. For examples, in Figure 7, G says "Every student studies some subject", H says "Julian couldn't fly to Mars", and F says "Tom believes that Mary wants to marry a sailor", where $\neg$ denotes a NEG relation [25].

5. Knowledge graph retrieval

5.1. *Query modification*

As shown above, SeRQL is a powerful query language for RDF graphs, but it is not user-friendly. With the graphical knowledge representation, CGs are more readable. It has also been shown that there is a close mapping between CGs and RDF graphs [29]. In order to exploit the Sesame infrastructure, we use non-nested CGs extended with the queried referent at the interface layer and map them to SeRQL for knowledge querying.

We define a *query CG* as a non-nested CG whose concept referents can be either an individual referent, the generic referent, denoted by '*', or the queried referent, denoted by '?'. The generic referent means that it does not care about a matched individual referent. The queried referent represents the referent of an entity in question. Each query CG is associated with value constraints on the properties of its concepts.

As mentioned before, a query CG is first generalized by replacing its concept and relation types by their immediate super-types. This generalized query is mapped to an equivalent SeRQL query to retrieve the knowledge graphs that it subsumes, using the Sesame exact matching engine. However, such a generalization must conform to the signatures of involved relation types. We recall that, if a CG [C_1] → (R) → [C_2] is well-typed, then so is [C_1] → (R$'$) → [C_2] for any super-type R$'$ of R.

From this observation, given a query CG, all of its relation types can be replaced by their immediate super-types first. Then for each new relation, its neighbor concept types can be replaced by their immediate super-types if and only if the resulting CG is still well-typed. We note that, if the referent of a concept is an individual referent, then its concept type is not changed. For an example, suppose that one wants to find all public companies that are registered in a country of North America and have their stock exchange names beginning with "AB". This query can be posed by the query CG Figure 8, which also shows its generalized CG, where COMPANY, LOCATED-IN, and POLITICAL-REGION are immediate super-types of PUBLIC-COMPANY, REGISTERED-IN, and COUNTRY, respectively.

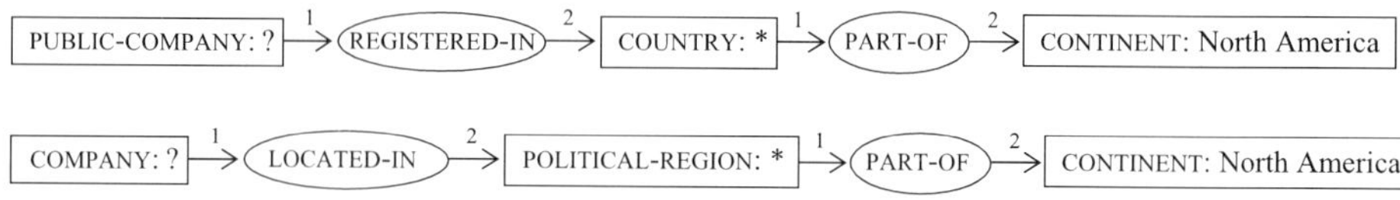

Fig. 8. A query CG and its generalization.

For attribute values in a query CG, they are first replaced by '?' to retrieve corresponding attribute values in answer graphs. Then both entity types and attribute values in those answer graphs are matched with the corresponding ones in the original query, giving overall matching degrees between the graphs.

5.2. *CG-SeRQL mapping*

The generalized query CG in Figure 8 can be mapped to the following SeRQL query:

```
SELECT x1, z1
FROM
    {x1} < rdfs:label> {z1},
    {x1} <rdf:type> {<www.dit.hcmut.edu.vn/vnkim/vnkimo_rdfs#Company>},
    {x1} <www.dit.hcmut.edu.vn/vnkim/vnkimo_rdfs#stockExchangeIndex> {p1},
    {y1} <rdf:type> {<www.dit.hcmut.edu.vn/vnkim/vnkimo_rdfs#PoliticalRegion>},
    {x1}< www.dit.hcmut.edu.vn/vnkim/vnkimo_rdfs#locatedIn> {y1},
    {y1} <www.dit.hcmut.edu.vn/vnkim/vnkimo_rdfs#partOf>
            {<www.dit.hcmut.edu.vn/vnkim/vnkimkb_rdf#Continent_34>}
WHERE p1 like "AB*"
```

In this example, the queried concept [Company: ?] is translated into the pair of variables (x1, z1) in the SELECT clause of the SeQRL query, where x1 represents the identifier and z1 represents the label of a result entity. The generic concept [PoliticalRegion: *] is translated into the variable y1. The individual concept referent North America is translated into its identifier. The relations, including the implicit properties like label, type, and stockExchangeIndex, in the query CG, are translated into RDF statements in the FROM clause. The constraint on the stock exchange names is specified in the WHERE clause.

Following is pseudocode of our algorithm to generalize and translate a query CG into its equivalent SeRQL clauses:

```
Generalize the query CG and replace its attribute values by '?'
For each relation in the generalized query CG do
{
   Get all neighbor concepts of the current relation
   // process each concept sequentially
   For each concept do
   {
      If the concept is new // i.e. not processed before
      {
         If the concept referent is '?' or '*'
         {
            If the concept referent is '?'
```

```
            {
               Assign to it a variable starting with 'x' // e.g., x1, x2, x3, ...
               Assign to it a label variable starting with 'z' // e.g., z1, z2, z3, ...
               Append these variables to the SELECT clause
               Append the "rdfs:label" statement for this concept to the FROM clause
            }
            If the concept referent is '*'
            {
               Assign to it a variable starting with 'y' // e.g., y1, y2, y3, ...
            }
            Append the "rdf:type" statement for this concept to the FROM clause
            Append the constrained property statements for this concept to the FROM clause
            Assign to the constrained properties variables starting with 'p' // e.g., p1, p2, p3, ...
            Append the property constraints to the WHERE clause
         }
         If the concept referent is specific, assign to it its identifier as a dummy variable
         Remember this concept having been processed
      }
   } // end of process concepts
   Assign to the current relation its URI
   Append the statement for the current relation with its neighbor concepts to the FROM clause
}
// process the disconnected concept nodes
For each concept that has not been visited yet, process it as above.
```

5.3. *Approximate knowledge graph matching*

The answer knowledge graphs for a generalized query CG are those that are subsumed by it. They are then matched to the original query CG to compute their matching degrees, using the similarity and subsumption measures defined in Section 3. The similarity/subsumption degree between two knowledge graphs is the average of the similarity/subsumption degrees between their matched concepts, relations, and attribute values.

For example, suppose that an answer CG for the generalized query CG in Figure 8 is as depicted in Figure 9, and $Pop(\text{PUBLIC-COMPANY}) = 5069$, $Pop(\text{COMPANY}) = 7861$, $Pop(\text{RIGISTER-IN}) = 10816$, $Pop(\text{LOCATED-IN}) = 14545$. Then the similarity degree between the original query CG and the answer CG is computed as follows:

$$\begin{aligned}
&Dist(\text{PUBLIC-COMPANY, COMPANY}) \\
&\quad = -\log\bigl(Pop(\text{PUBLIC-COMPANY})/Pop(\text{COMPANY})\bigr), \\
&Dist(\text{REGISTERED-IN, LOCATED-IN}) \\
&\quad = -\log\bigl(Pop(\text{RIGISTERED-IN})/Pop(\text{LOCATED-IN})\bigr),
\end{aligned}$$

$$\begin{aligned}
SimDegree = {} & \bigl(Sim(\text{PUBLIC-COMPANY, COMPANY}) \\
& + Sim(\text{REGISTERED-IN, LOCATED-IN}) + 1 + 1 + 1\bigr)/5 \\
= {} & (0.612 + 0.701 + 1 + 1 + 1)/5 \\
= {} & 0.863.
\end{aligned}$$

Fig. 9. An approximate answer CG.

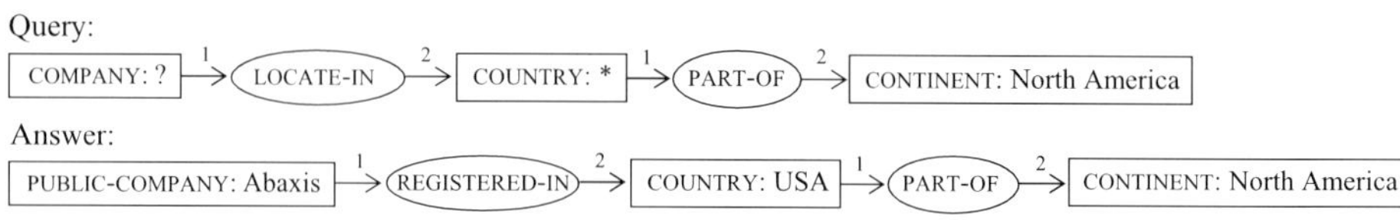

Fig. 10. Another example of query and answer CGs.

Meanwhile, for the subsumption degree of the original query CG to the answer CG, one has:

$$\begin{aligned}
&Dist_S(\text{PUBLIC-COMPANY} \rightarrow \text{COMPANY})\\
&\quad = -\log\big(Pop(\text{PUBLIC-COMPANY})/Pop(\text{COMPANY})\big),\\
&Dist_S(\text{REGISTERED-IN} \rightarrow \text{LOCATED-IN})\\
&\quad = -\log\big(Pop(\text{RIGISTERED-IN})/Pop(\text{LOCATED-IN})\big),
\end{aligned}$$

$$\begin{aligned}
SubDegree &= \big(Sub(\text{PUBLIC-COMPANY} \rightarrow \text{COMPANY})\\
&\qquad + Sub(\text{REGISTERED-IN} \rightarrow \text{LOCATED-IN}) + 1 + 1 + 1\big)/5\\
&= (0.612 + 0.701 + 1 + 1 + 1)/5\\
&= 0.863.
\end{aligned}$$

One can observe that, when t_1 is a subtype of t_2, then $Dist(t_1, t_2) = Dist_S(t_1 \rightarrow t_2)$, and thus $Sim(t_1, t_2) = Sub(t_1 \rightarrow t_2)$. Now, suppose a query CG similar to the answer CG in Figure 9 and an answer CG similar to the query CG in Figure 8, as shown in Figure 10. Then the similarity between the two graphs is still the same as in the above example, but the subsumption degree from the query graph to the answer graph is 1.

6. VN-KIM query module

6.1. *Ontology and Knowledge Base*

World Wide Web is moving to its next generation, called Semantic Web [3], in which not only humans can read and understand web pages, but computers can also understand and process them automatically. A clear disadvantage of the current web can be seen in search engines such as Google, which often return many useless documents that have nothing to do with what a user wants to find. The main reason is that no machine-understandable semantics is given to words in documents and queries [7].

KIM [22] is a "killer application" for Semantic Web. It is firstly a knowledge-based system of popular named entities in the world. It can automatically extract the class of a named entity in a web page and annotate that information in the web page. Those semantics-enhanced web pages are then stored for efficient searching using entity names and their

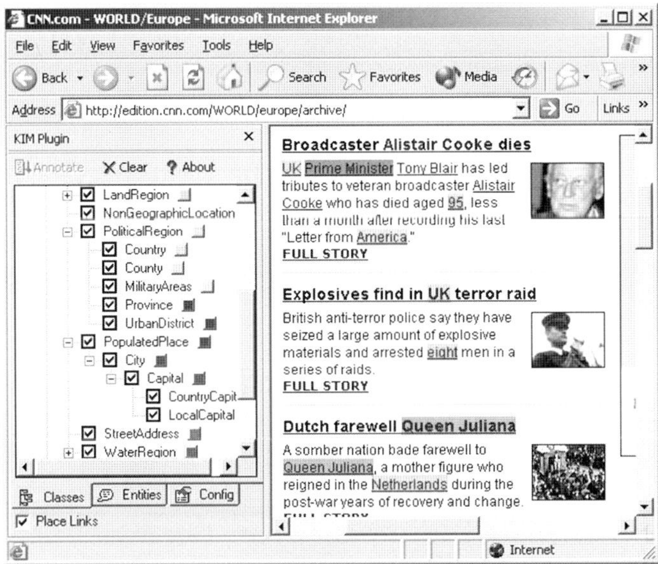

Fig. 11. Information extraction by KIM.

classes. For example, one can search for documents about "Washington" as a person, while current search engines may return any document that contains "Washington" though it is the name of a press or a university. Figure 11 shows a web page with highlighted named entities recognized by KIM.

Currently KIM ontology consists of about 250 classes, grouped into three upper-level concepts: Entity, EntityResource, and LexicalResource. Among them, Entity is the most important concept and specialized into Object (for people, locations, ...), Happening (for events, situations, ...), and Abstract (for addresses, professions, ...). It also includes about 100 relations and attributes, such as subRegionOf for the class Location or hasPosition for the class Person. KIM knowledge base contains about 80,000 entities of the most important and popular people, cities, organizations, ..., in the world. For storing and managing the ontology and knowledge base, KIM employs Sesame. Annotated documents are indexed and retrieved using Lucene [12], by entity names and classes rather than keywords. KIM information extraction engine is based on GATE [9], an architecture for developing natural language processing applications.

We are currently developing a similar system as KIM, named VN-KIM, but for Vietnamese language and domain, and being to overcome the disadvantages of KIM. One of the shortcomings of KIM is its usage of fixed patterns for knowledge querying. For example, Figure 12 shows such a pattern to express a query involving three entities, namely X, Y, and Z. In this pattern, one can express a relation between X and Y, and a relation between Y and Z, and the name and one property value restriction for each of the three entities, but no more.

While KIM query patterns are simple for usage and helpful in some cases, they are not sufficient and flexible enough for more expressive queries, which allow an arbitrary number of entities involved, multiple relations between them, and several property value constraints. On the other hand, SeRQL is much more expressive but not easy to use for end-users, who are not familiar to the language syntax. Below, we present our VN-KIM

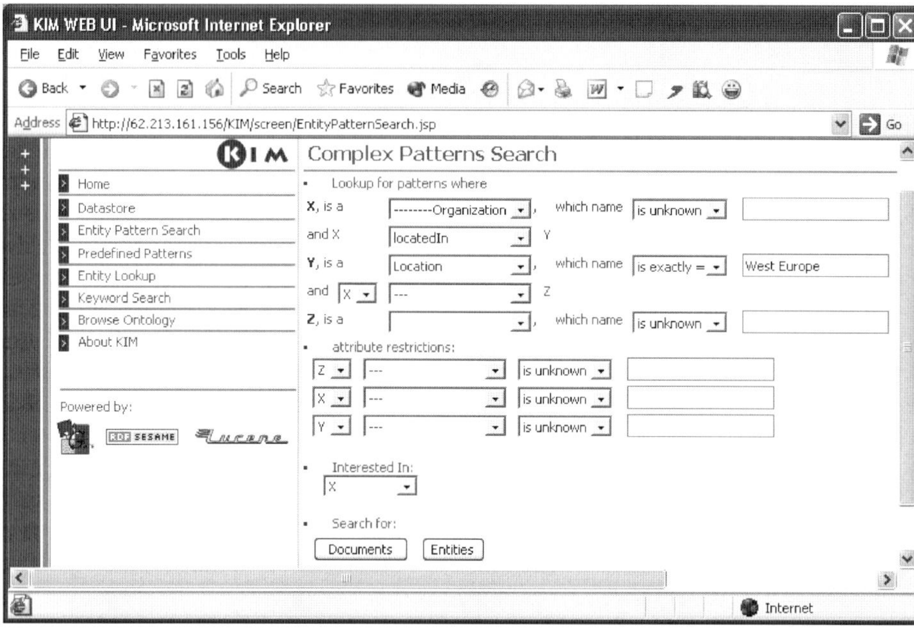

Fig. 12. A fixed query pattern in KIM.

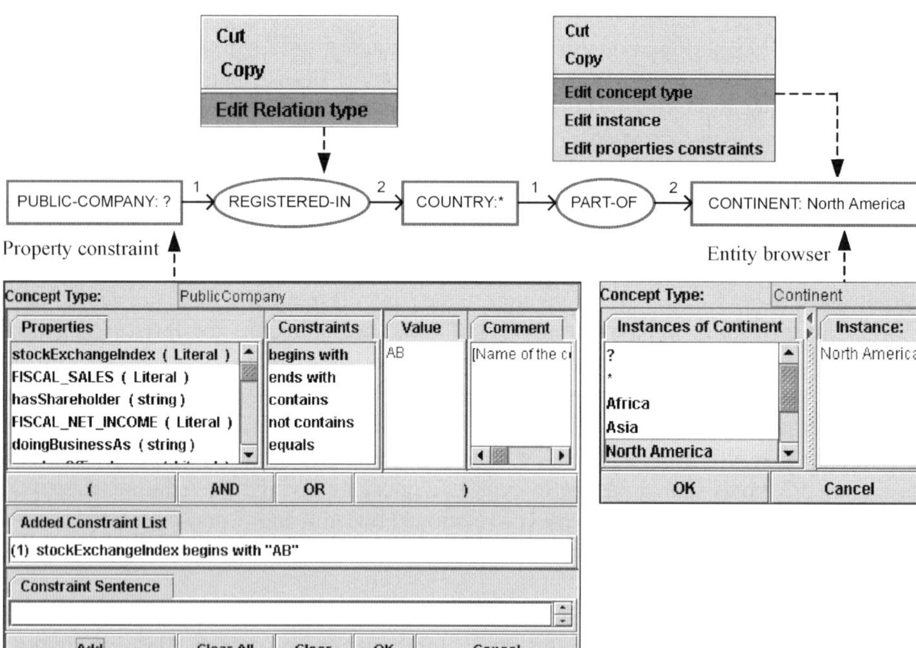

Fig. 13. Editing a query CG.

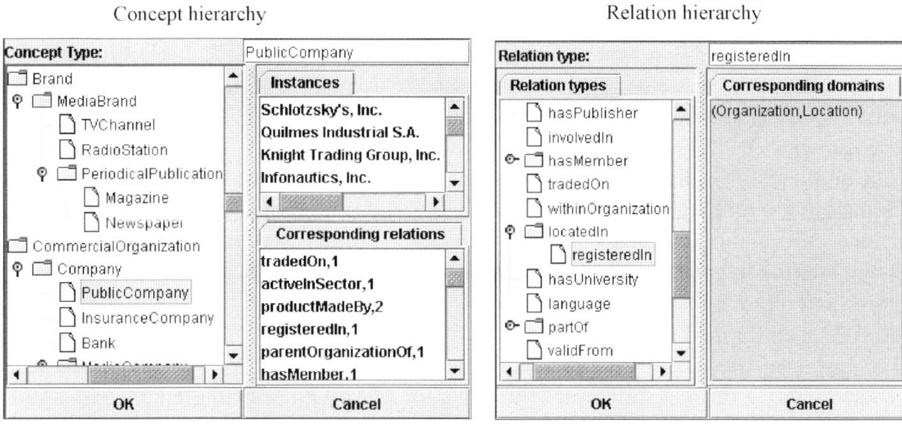

Fig. 14. Partial concept and relation ontology.

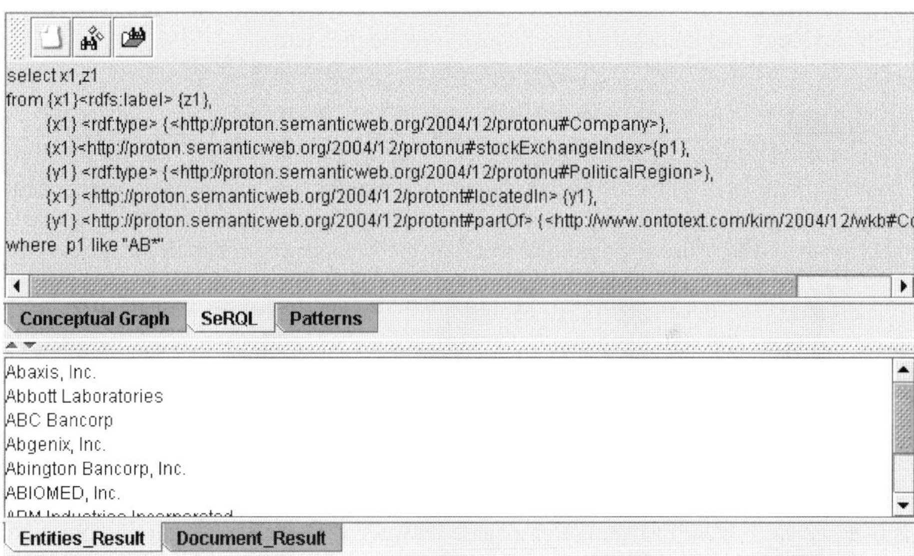

Fig. 15. Query results.

query editor and answering mechanism, which employs simple CGs for expressive and user-friendly knowledge querying.

6.2. *Query editor*

VN-KIM query editor allows a user to pose queries in fixed patterns, CGs, or SeRQL clauses. It provides drawing tools for editing query CGs with respect to its ontology and knowledge base. Concept editing is assisted by browsing the concept type hierarchy and existing entities of a particular concept type. The relation type hierarchy can also be browsed for relation editing, and consistency of a relation and its neighboring concepts in a CG is checked after the CG is drawn. Conditions on properties values of an entity can be expressed using the constraint relations of the value domains as specified in the ontology. Query CGs are then mapped into SeRQL clauses as presented in Section 5 for knowledge and document retrieval.

The query editor is also equipped with functions for saving a query CG to an XML file, and loading a query CG from an XML file. The XML file contains both the logical and graphical information of each element in the query graph. The logical properties of one element are its category (i.e., concept, relation, or edge) and other specific properties (e.g. the type and referent of a concept element). The graphical properties such as positions and colors are used for displaying the graph.

For instance, the example query in Figure 8 can be posed by the CG edited using VN-KIM query editor as shown in Figure 13, together with assisting commands and information. The sub-hierarchies showing the concept and relation types involved are given in Figure 14, and the equivalent SeRQL clauses and results of the generalized query are presented in Figure 15. Their matching degrees with the original query are evaluated using the defined similarity and subsumption measures. For readability, we have run the example on KIM ontology and knowledge base, which uses English terms.

7. Conclusion

Previous definitions of semantic distance were for words on the basis of their actual occurrence or semantics, which did not work for un-interpreted type labels. We have proposed a population-based measure as a solution, whereby the occurrence probability of a type label is computed as that of its entities in a knowledge base of discourse. For fuzzy set-based attribute values, we have used fuzzy conditional probability instead. The Jiang–Conrath method has been adapted accordingly, for measuring the similarity between two knowledge graphs.

In contrast to those similarity measures, which are symmetric, we additionally define a asymmetric subsumption measure for both entity types and attribute values, also in the probabilistic framework of Jiang–Conrath's semantic distance. This way provides a basis to combine subsumption degrees between entity types and attribute values of two knowledge graphs to give an overall subsumption degree between the graphs, measuring how much the query graph subsumes the answer graph.

In realization of an approximate knowledge graph retrieval system, we have employed the exact query engine of Sesame with its query language SeRQL. A query graph is first generalized by replacing its concept and relation types with their immediate super-types, conforming to the relations' signatures. The answer graphs that are subsumed by the generalized query graph are then matched to the original query graph to measure their similarity and subsumption degrees. For user-friendly query expressions at the front-end of VN-KIM, we have presented an algorithm to map conceptual graphs to SeRQL clauses.

Although measures of similarity and subsumption between fuzzy attribute values have been defined as presented in this paper, we have not implemented attribute value matching in VN-KIM yet. That would require a major change of Sesame and SeRQL to accommodate fuzzy sets. Also, query CGs in the current form are just simple ones, while nested CGs would be useful for more expressive queries. These are among the topics that we are currently investigating.

References

[1] J.M. Baldwin, J. Lawry, T.P. Martin, *A note on probability/possibility consistency for fuzzy events*, in: Proc. 6th International Conference on Information Processing and Management of Uncertainty in Knowledge-Based Systems, Granada, Spain, 1996, pp. 521–525.

[2] J.F. Baldwin, T.P. Martin, B.W. Pilsworth, *Fril – Fuzzy and Evidential Reasoning in Artificial Intelligence*, Research Studies Press, New York, 1995.

[3] T. Berners-Lee, *Semantic web roadmap*, first draft, available at http://www.w3.org/DesignIssues/CG.html, last modified: February 2001.

[4] A. Budanitsky, G. Hirst, *Semantic distance in WordNet: an experimental, application-oriented evaluation of five measures*, in: Proc. Workshop on WordNet and Other Lexical Resources, 2nd Meeting of the North American Chapter of the Association for Computational Linguistics, Pittsburgh, Pennsylvania, USA, 2001.

[5] T.H. Cao, *A formalism for representing and reasoning with linguistic information*, Int. J. Uncertainty, Fuzziness and Knowledge-Based Systems **10** (2002), 281–307.

[6] T.H. Cao, *Conceptual graphs for modelling and computing with generalized quantified statements*, in: J. Lawry, J. Shanahan, A. Ralescu (Eds.), Modelling with Words: Learning, Fusion, and Reasoning within a Formal Linguistic Representation Framework, Lecture Notes in Artificial Intelligence, vol. 2873, Springer, Berlin, 2003, pp. 80–101.

[7] T.H. Cao, T.H.D. Nguyen, T.C.T. Qui, *Searching the web: a semantics-based approach*, in: H.G. Bock, et al. (Eds.), Modelling, Simulation and Optimization of Complex Processes, Springer, Berlin, 2005, pp. 57–68.

[8] K.W. Church, P. Hanks, *Word association norms, mutual information, and lexicography*, in: Proc. 27th Annual Meeting of the Association for Computational Linguistics, 1989, pp. 76–83.

[9] H. Cunningham, et al., *Developing Language Processing Components with GATE*, GATE User's Guide version 2.1, 2003, available at http://gate.ac.uk/.

[10] B.R. Gaines, *Fuzzy and probability uncertainty logics*, J. Inform. and Control **38** (1978), 154–169.

[11] S. Gauch, J. Wang, S.M. Rachakonda, *A corpus analysis approach for automatic query expansion and its extension to multiple databases*, in: Proc. International Conference on Information and Knowledge Management, 1997.

[12] O. Gospodnetic, *Parsing, indexing, and searching XML with Digester and Lucene*, IBM DeveloperWorks, 2003, available at http://jakarta.apache.org/lucene/.

[13] G. Hirst, D. St-Onge, *Lexical chains as representations of context for the detection and correction of malapropisms*, in: C. Fellbaum (Ed.), WordNet: An Electronic Lexical Database, MIT Press, Cambridge, 1998, pp. 305–332.

[14] N.D. Ho, F. Cédirick, *Lexical similarity based on quantity of information exchanged – synonym extraction*, in: Proc. Research Informatics Vietnam-Francophone Conference, Hanoi, Vietnam, 2004, pp. 193–198.

[15] J. Jiang, D.W. Conrath, *Semantic similarity based on corpus statistics and lexical taxonomy*, in: Proc. International Conference on Research in Computational Linguistics, Taiwan, 1997.

[16] A. Kampman, F. Harmelen, J. Broekstra, *Sesame: a generic architecture for storing and querying RDF and RDF Schema*, in: Proc. 1st International Semantic Web Conference, 2002, available at http://www.openRDF.org/, last visited: January 2005.

[17] F. Klawonn, J.L. Castro, *Similarity in fuzzy reasoning*, Mathware Soft Computing **3** (1995), 197–228.

[18] H. Kozima, A. Ito, *Context-sensitive word distance by adaptive scaling of a semantic space*, in: R. Mitkov, N. Nicolov (Eds.), Recent Advances in Natural Language Processing, John Benjamins Publishing Company, Amsterdam, 1997, pp. 111–124.

[19] C. Leacock, M. Chodorow, *Combining local context and WordNet similarity for word sense identification*, in: C. Fellbaum (Ed.), WordNet: An Electronic Lexical Database, MIT Press, Cambridge, 1998, pp. 265–284.

[20] D. Lin, *An information-theoretic definition of similarity*, in: Proc. 15th International Conference on Machine Learning, Madison, WI, USA, 1998.

[21] S. Mitaim, B. Kosko, *Neural fuzzy agents for profile learning and adaptive object matching*, Presence **7** (1998), 617–637.

[22] B. Popov et al., *KIM – semantic annotation platform*, in: Proc. 2nd International Semantic Web Conference, 2003.

[23] P. Resnik, *Using information content to evaluate semantic similarity in taxonomy*, in: Proc. 14th International Joint Conference on Artificial Intelligence, Montreal, Canada, 1995, pp. 448–453.

[24] *SeRQL Manual*, available at http://www.openRDF.org/.

[25] J.F. Sowa, *Conceptual Structures – Information Processing in Mind and Machine*, Addison-Wesley Publishing Company, Massachusetts, 1984.
[26] J.F. Sowa, *Matching logical structure to linguistic structure*, in: N. Houser, D.D. Roberts, J. Van Evra (Eds.), Studies in the Logic of Charles Sanders Peirce, Indiana University Press, 1997, pp. 418–444.
[27] C. Vertan, N. Boujemaa, *Using fuzzy histograms and distances for color image retrieval*, in: Proc. 3rd UK Conference in Image Retrieval, Brighton, UK, 2000.
[28] Z. Wu, M. Palmer, *Verb semantics and lexical selection*, in: Proc. 32nd Annual Meeting of the Association for Computational Linguistics, New Mexico, USA, 1994, pp. 133–138.
[29] H. Yao, L. Etzkorn, *Conversion from the Conceptual Graph (CG) model to the Resource Description Framework (RDF) model*, in: Proc. 12th International Conference on Conceptual Structures, 2004, pp. 98–114.
[30] L.A. Zadeh, *Fuzzy logic = computing with words*, IEEE Trans. on Fuzzy Systems **4** (1996), 103–111.

Integration of Processing of Fuzzy Information

CHAPTER 15

Soft Integration of Information with Semantic Gaps

Trevor Martin[*]

Artificial Intelligence Group, Department of Engineering Mathematics, University of Bristol, Bristol BS8 1TR, UK
E-mail: trevor.martin@bris.ac.uk

Ben Azvine

Intelligent Systems Lab, BT Research and Venturing, Adastral Park, Ipswich IP5, UK
E-mail: Ben.Azvine@bt.com

Abstract

The combination of information from multiple semi-structured (or structured) sources is a widespread problem. Given two sources $s1$ and $s2$ which refer to approximately the same sets of real world entities, the information fusion task is to determine whether an object from source $s1$ refers to the same real-world entity as an object from source $s2$, and how the properties correspond (e.g., *author* and *composer* should correspond almost exactly to *creator*, *business-name* should correspond to *company-name*, etc.).

It is rare for two sources to adhere to precisely the same conventions. Even where agreed conventions exist, interpretations may differ – for example, consider an electronic order for goods with a piece of information marked as *shippingDate*. It is not clear whether *shippingDate* refers to the time when goods are delivered to the customer or the time when they leave the supplier. This is an example of a semantic gap – a term which can be interpreted in multiple and possibly inconsistent ways.

The semantic web vision is for "islands of standardisation" in which there are small, agreed ontologies for particular sets of users or communities. Wherever two communities meet or overlap, integration of information is a potential problem.

In this paper we briefly outline the background to this problem, and argue that it is essential to account for the uncertainty inherent in the process. We go on to describe a framework for integrating information sources, given a number of approximate mappings between their attributes.

[*] Currently BT Senior Research Fellow, Intelligent Systems Lab, Adastral Park, Ipswich IP5, UK

FUZZY LOGIC AND THE SEMANTIC WEB
Edited by Elie Sanchez
© 2006 Elsevier B.V. All rights reserved

Keywords

fuzzy, information integration, instance matching, schema mapping, semantic interoperability, ontology alignment

1. Introduction

A key feature of a human language is the association of particular symbols (or sounds) with particular objects in the world. Because different individuals use the same symbol to denote the same objects, communication is possible. Thus, specific symbols ("Queen Elizabeth II", "Buckingham Palace") denote specific objects. However, the association between symbols and objects is not necessarily unique or one-to-one. For example, "Elton John", "Reg Dwight" and "the writer of the music of 'Candle in the Wind' " all denote the same person. The word "Europe" denotes a collection of countries, but its precise definition is elusive – is it a set of countries marked as Europe on a particular map, the members of the European union (now? in 1970? in 1975? in 2010?), the countries eligible to enter European Championship football, countries eligible to enter the Eurovision song contest . . .?

Such questions may be dismissed as belonging to philosophy and linguistics, but have direct relevance to computerised information systems and the flexible management of information. Most information systems are built (explicitly or implicitly) using crisply defined categories and rely on the underlying theory of databases, which requires a unique identifier for every individual entity. For example

- a person may be identified by national insurance number, employee identifier, passport number, or a variety of other methods;
- publications are identified by ISBN, by document-identifier, by (author + date + publisher), etc.;
- supermarket products are identified by barcode, by (supplier + itemname + size), etc.

Within a single database or information source, this can work adequately (although there are still problems if an object is assigned more than one "unique" identifier). The problem becomes more serious when attempting to combine information from multiple sources – we face the problem of not knowing whether the different sources are referring to the same object or not. Additionally data may be in slightly different formats, increasing the problem of matching.

To illustrate, take a simple request to *"find the phone number of, and directions to, an indian restaurant capable of hosting a meal for* 25 *vegetarians near the centre of Ipswich"*. Table 1 shows information available from web sources (Feb 2005).

No single source has the complete answer, but collectively they contain the relevant information. At present, it is difficult for a computer to recognise that the data refer to the same restaurant, let alone fuse the data to provide an integrated view. There is sufficient information to answer the query, but the process requires a human to interpret the data. The four entries in the first row of data contain four different names, two telephone numbers and (syntactically) three different addresses but clearly refer to the same restaurant. Similarly, the second row contains different postcodes and telephone numbers but obviously denotes a second restaurant, distinct from the first. A human can recognise that different parts of each entry correspond to restaurant name, address, etc., either because of the form of the text or because the text is labelled.

Table 1
Overlapping information from multiple sources (January 2005).

Scoot	yell.com	bt.com directory enquiries	g2i.co.uk
Maharani 46 Norwich Road, Ipswich, IP1 2NJ **Tel:** 01473 213388	**Maharani Indian Cuisine** 46, Norwich Rd Ipswich Suffolk IP1 2NJ **Tel:** 01473 232266 **Classification:** Restaurants – Indian **Options:** <Map/Directions> <Send to mobile> <Save to address book>	**Maharani Indian Cuisine & Takeaway** 46 Norwich Road, Ipswich, IP1 2NJ **Tel:** 01473 213388	**Maharani Indian Restaurant** **Tel:** 01473 232266 **Address:** 46 Norwich Rd Ipswich IP1 2NJ Suffolk **Location** Town <map> **Cuisine** Indian Vegetarian **Seating Capacity** Over Sixty Licenced Take Away Available
TAJ Mahal 40-42 Norwich Rd, Ipswich, IP11 23 **Tel**: 01473 210722	**Taj Mahal** 40/42, Norwich Rd Ipswich Suffolk IP1 2NJ **Tel**: 01473 257712 **Classification:** Restaurants – Indian **Options:** <Map/Directions> <Send to mobile> <Save to address book>	**TAJ Mahal Indian Restaurant** 40-42 Norwich Rd, Ipswich, IP11 2NJ **Tel**: 01473 210722	**etc.**

The problem of integrating data from multiple sources is becoming more acute, with the increasing spread of electronic data storage. Data may be stored in databases or non-database sources (semantic web, XML, web, semi-structured repositories, etc.) or as free text in a document base. Electronically stored data is usually classified to some degree at least, whether by the database schema, the nature of the storage (e.g., a hierarchical structure) or in more recent systems by some sort of taxonomy or ontology. We can regard this classification as meta-data, together with headings, semi-structured components and any explicit meta-data that may be present.

The classification structure and attributes (properties) of the objects (i.e. the values associated with meta-data tags) can be used to guide the integration of information. Even if different hierarchies use different categories, there is likely to be a degree of correspondence, and objects placed within similar categories are likely to have similar properties. For example, the automotive industry and chemical industry have separately agreed standards for electronic data exchange (www.starstandard.org, www.cidx.org) which are almost identical in some categories, overlap in some and are disjoint in others. However it is perfectly possible that a third party might wish to merge information formatted according to the two different standards. Similarly, a digital library and an online bookseller refer to the same (structured) objects but may differ in categorisation and details stored about each book.

For example, documents in a digital library or online bookshop may have some or all of the properties such as author, publication date, title, publication format, ISBN, publisher, etc. Restaurants in a classified directory might have a name, address, telephone number, as well as an optional brief description and other properties such as location, menu details, etc. Entries in product catalogues might have names, part numbers, prices, categories and short descriptions. These are all examples of structured objects that refer to real-world entities. The properties can be used to group the objects into classes, which may in turn form some sort of hierarchical structure. This leads to the "ontology alignment" or "schema matching" problem; the method outlined here is focused on "instance-matching" but could be extended to use and/or predict the hierarchical classification.

In the absence of a centrally agreed standard, it is rare for two sources to adhere to precisely the same conventions. Even where agreed conventions exist, interpretations may differ – for example, consider an electronic order for goods with a tag *<shippingDate>*. It is not clear whether *<shippingDate>* refers to the time when goods are delivered to the customer or the time when they leave the supplier. Additional problems arise from subtle differences in the information stored – for example, a classified directory may list phone numbers as *<area-code><local-number>* whereas another might just list the *<local-number>*. An author may be listed as *<surname>*, *<initials>* or as *<forename><surname>*. Thus even when the information from different sources is "identical", a straightforward syntactic match might not reveal this.

We consider two information sources referring to discrete sets of objects with associated properties. The aim of the work described here – a Structured Object Fusion Toolkit (SOFT) – is to identify when sources are referring to the same entity, and to enable merging of the common attributes and integration of the unique attributes in order to provide a reliable fused information source.

The method is based on the identity of indiscernibles, proposed by Leibniz – if we cannot tell ("discern") two objects from each other, then they are the same object. Here, we can only tell two objects apart on the basis of their attribute values. Let us assume that sources $s1$ and $s2$ have the same sets of attributes. Then, if an object from $s1$ has identical attribute values to an object from $s2$, we must conclude that they refer to the same real-world entity. If the attributes differ but the sets of possible attribute values can be related in some way, then we can find an approximation to the true mapping from $s1$ to $s2$.

2. Background

The problem has a long history and has been considered from two perspectives – instance (or record) -based and schema (or ontology) -based approaches. Additionally, we can classify previous approaches to the problem according to the background of the research – semantic web researchers tend to concentrate on ontologies whereas database developers focus on ways of combining database schemas.

The problem of *record linkage* was identified in the USA public health area, when combining different records that (possibly) referred to the same patient. Newcombe [25] proposed a frequency-based approach which was later formalised by Fellegi and Sunter [17]. These approaches assume that the two data sources have common attributes, and are commonly applied to the so-called "merge/purge" problem in business databases to filter out duplicate entries. The methods focus on calculating a weight for each attribute in the database, according to the likelihood of finding matching values within that attribute's domain (i.e. the set of all values appearing in the column).

The initial formulation treated binary matches (true/false) but was extended to categorical matches (one of a small set of values) and continuous matches (e.g., a number in the interval [0, 1]). By assuming conditional independence between records matching on different attributes it is possible to estimate the conditional probabilities for each attribute matching, given that the records are (or are not) identical, and hence to find thresholds for classifying two records as matching or not according to the weighted sum of matches. The estimation can be on the basis of minimum error probabilities, expectation maximisation, utility (cost of incorrect decision), etc. – see [13] for an overview.

These methods implicitly take into account knowledge of the database schema, as they assume each record consists of the same set of attributes.

The record linkage problem was extended to *analytic linkage* (also referred to as *entity matching*) by considering the combination of data taken from two or more sources, e.g., the integration of heterogeneous databases. Dey et al. [9] give a summary of probabilistic approaches, based on the same framework as the record linkage work outlined in the previous paragraph. Again, knowledge of the schema is assumed in that matching pairs of attributes are known.

These methods use several techniques to try to match attributes, such as standardising the form of names and addresses, and applying heuristics (for example first-n-characters match, common substrings, edit distance is below a specified threshold). Bilenko, Mooney et al. [6] describe *SoftTF-IDF*, an adaptive matching function, which takes account of the frequencies of similar and identical words within a domain.

The problem can also be approached at the schema level, by looking at labels (i.e. attribute names) and constraints associated with allowed values.

Rahm and Bernstein [27] provide a useful survey of approaches to the automation of schema matching, including both database and semantic web problems. They identify three main groups, with methods arising from the fields of:

- information retrieval – using distance-based matching techniques such as the edit distance to overcome the inadequacy of exact, "keyword-based" matching. These assume the use of fairly simple mappings between attribute domains.
- machine learning – using algorithms to create a mapping between attributes based on the similarity among their associated values. Bayesian classifiers are the most common approaches (e.g., GLUE [12] and Autoplex [5]).
- graph theory – by representing schemata in tree or graph form, e.g., the TreeMatch algorithm [21] which estimates the similarity of leaf nodes in an XML DTD by estimating the similarity of their ancestors.

They also present a taxonomy covering many existing approaches based on the split between metadata matching and content (instance) matching as mentioned above, but also explicitly considering other aspects of matching such as

- structure matching (e.g., an attribute such as address in one source could map to several attributes such as street, town and postcode in another).
- hierarchical and similarity-based matching of terms (e.g., book is a type of publication, make and brand are roughly synonymous).
- matching based on data constraints (e.g., DoB is of type Date and so could match an attribute BirthDate in another source, which is also of type Date).

Examples listed include:

- SemInt (Northwestern University) [19] which extracts features from attributes (such as data type, length, key information) and uses a neural network to cluster possible matching candidates.
- Learning Source Descriptions (LSD, University of Washington) [11] which uses machine learning techniques to map a new schema into a predefined global schema.
- SKAT (Semantic Knowledge Articulation Tool, Stanford) [23] is a prototype which focuses more on ontologies, allowing the user to specify first-order logic rules to translate between ontologies.
- Cupid (Microsoft Research) [21] matches at both element and structure levels, using thesauri to determine similarities between data and a tree-matching algorithm based on node similarities.

Almost all of these systems include some measure of uncertainty in the matching process, such as Cupid's "plausibility factor". The need for uncertainty has also been noted by others. For example, Chang and García-Molina [8] developed an approach for the precise translation of Boolean queries across different information sources. In a subsequent paper [7] they presented a real-world case study (combining book searches from four different web sites), and found that it was only possible to make exact mappings in 30% of the rules – the remaining 70% required approximation.

Other tools include

- Glue [12]
- OntoBuilder [24]
- Prompt [26]

Gal et al. [18] recognised the need to include uncertainty in the matching process, and outlined a fuzzy framework for schema integration based on the notion of similarity relations. Gal has also looked at the problem of evaluating the matching between schemata, compared to a notional "ideal" matching that would be produced by a human.

Fagin [14–16] recognised the need to handle uncertainty in real-world databases and investigated the combination of information drawn from standard relational databases with that taken from fuzzy (graded) sources, not necessarily relational. He analysed algorithms for evaluating conjunctive queries efficiently, concluding that in most cases this could be done with complexity better than linear in the database sizes.

Ding and Foo [10] reviewed the problem from a semantic web perspective, noting that

> current researches on semi-automatic or automatic ontology research in all three aspects of generation, mapping and evolving have so far achieved limited success.

Madhavan, Bernstein et al. [20] outlined a way of defining languages to represent mappings between schemas and ontologies (domain models in their paper), and concluded by saying that

> the next step in this work is to develop an appropriate probabilistic representation of mappings that enables capturing both inaccuracy and uncertainty.

The integration of semi-structured information from heterogeneous sources is an unsolved and important problem which is ideally suited to the techniques of soft computing – see also [22].

3. SOFT – a Structured Object Fusion Toolkit

3.1. *Overview*

We focus here on the problem of matching at the instance level, using relatively straightforward mappings from one source attribute to another; this excludes mappings which involve composition of two or more attributes – for example where one source uses <customer first name> and <customer last name> which could correspond to <customer name> in another source.

Assume we have two sets of objects $A = \{a_1 \dots a_n\}$ and $B = \{b_1 \dots b_m\}$, and that we wish to establish an approximate mapping

$$h : A \to B.$$

We can envisage an underlying universe U (the "real world") which is modelled by the objects in sets A and B, i.e. there are functions

$$f : U \to A$$

and

$$g : U \to B$$

which are unknown and in practise may be incomplete and/or "noisy" in some way, meaning that some or all of the mappings f and g may be incorrect. We seek the best approximation to h such that

$$h(f(x)) = g(x),$$

i.e. we seek to establish a correspondence between the objects in A and B, based on the properties (attributes) of the objects.

Let the objects in A and B have attribute values taken from $C_1, C_2, \dots, D_1, D_2, \dots$ with relations defined as

$$R_i : A \to C_i, \quad i = 1 \dots n_A,$$

$$S_j : B \to D_j, \quad j = 1 \dots n_B.$$

Note that these are relations, i.e. they can be single- or multi-valued. Examples would be height, weight, nationality, name, passport number, phone number(s), etc.

We do not assume that the information about A and B in relations R_i, S_j is identical or completely consistent, but we do assume that some of these relations reflect similar or identical properties of the objects in A and B. Thus for some choices of pairs of codomains (C_i, D_j) we assume an exact or approximate matching function h_{ij} which for each element of C_i returns a (possibly fuzzy) subset of D_j. As shown below, this can be converted to a mass assignment giving an estimate of the probability that the element corresponding to some $c \in C_i$ lies in a subset $\{d_1 \dots d_k\} \subseteq D_j$. (We will refer to h_{ij} as a function even though its output is not necessarily a single value.)

The h_{ij} can be obvious mappings from codomain to codomain, involving exact matches or small permutations, truncations, etc.; alternatively they can be more sophisticated functions, possibly the product of a machine learning process. The proportion of a domain that matches under such a mapping gives an indication of the "overlap" between domains and hence the possibility that two attributes correspond.

3.2. *How do we derive h?*

This is obvious if the functions h_{ij} are exactly known – for each object a_k in A,

$$h(a_k) = S_j^{-1}\big(h_{ij}\big(R_i(a_k)\big)\big),$$

i.e. we obtain a subset of objects in B which possibly correspond to the object a_k. To illustrate, consider a set of authors identified by *author-id* (A) and the relations

R_1 (author-id → name)

author-id	name
author-1	Charles Dodgson
author-1	Lewis Carroll
author-2	Herbert Wells
author-3	I.M. Copi
author-4	Michael Coren

R_2 (author-id → book title)

author-id	book title
author-1	Symbolic Logic
author-1	Alice in Wonderland
author-2	The Invisible Man
author-3	Symbolic Logic
author-4	The Invisible Man

R_3 (author-id → date-of-birth)

author-id	date-of-birth
author-1	27 January 1832
author-2	21 September 1866
author-5	30 December 1865

and (from another source) a set of writers identified by real name (B) with one-to-many relations

S_1 (writer-real-name → ISBN)

writer (real name)	ISBN
C.L. Dodgson	0486204928
C.L. Dodgson	0091737648
C.L. Dodgson	0573150060
H.G. Wells	0809596369
H.G. Wells	0785731547
I.M. Copi	0023249803
M.Coren	0747511586

S_2 (writer-real-name → pseudonym)

writer (real name)	pseudonym
C.L. Dodgson	Lewis Carroll
H.G. Wells	Walker Glockenhammer
H.G. Wells	Septimus Browne

S_3 (writer-real-name → year-of-birth)

writer (real name)	year-of-birth
C.L. Dodgson	1832
H.G. Wells	1866
A.C. Doyle	1859
J.R. Kipling	1865
M. Coren	1959

We assume there is a function h_{21} which will return the set of ISBNs corresponding to a particular title

h_{21}

BOOK TITLE	ISBN
Alice in Wonderland	0091737648 0573150060
The Invisible Man	0809596369 0785731547 0747511586
Symbolic Logic	0486204928 0023249803

and the obvious function h_{33} which matches *date-of-birth* with *year-of-birth*.

In order to link *author-2* from set A to an *author name* from set B, we could look for the set of book titles written by that author using R_1, find the corresponding ISBNs using h_{21} and then work backwards to *writer (real name)* via S_1.

Thus

$$R_1(author_2) = \{The\ Invisible\ Man\},$$

$$h_{21}\big(R_1(author_2)\big) = \{0809596369, 0785731547, 0747511586\},$$

$$S_2^{-1}\big(h_{21}\big(R_1(author_2)\big)\big) = \{H.G.\ Wells, M.\ Coren\}.$$

Given complete and certain information, we can choose additional pairs of attributes to refine this mapping. Taking the obvious mapping h_{33} from date-of-birth to year-of-birth would give a second piece of evidence

$$S_3^{-1}\big(h_{33}\big(R_3(author_2)\big)\big) = \{H.G.\ Wells\}$$

so that we could conclude that *author-2* in set A corresponds to *H.G. Wells* in set B as this is the intersection of the two results.

This is only possible in the case of complete and certain information – if there is a possibility that the entry in Table S_3 is incorrect, then we cannot exclude the possibility of *author-2* being M. Coren, i.e. taking the intersection is not correct.

In general, we have to allow for the fact that h_{ij} are approximate, and may be inconsistent. For example, year of birth might be missing, inaccurate or not known for sure ("*born in the mid* 1860'*s*"). Taking the crisp information in the example above and allowing flexibility in the match function h_{33} could give

$$S_3^{-1}\big(h_{33}\big(R_3(author_5)\big)\big) = \{H.G.\ Wells/\mu_1, J.R.\ Kipling/\mu_2, A.C.\ Doyle/\mu_3\}$$

where the memberships μ_i reflect the degree of match between 30 *December* 1865 and 1866, 1865, 1859 respectively.

3.3. *Representation of approximate correspondences*

Clearly it is not sensible to attach too much importance to the precise membership values in a situation where we are not sure that two attribute sets correspond, or that two particular objects or attribute values are "the same". We interpret memberships as indicating an ordering over the elements of a set, so that higher membership indicates a better match. We treat fuzzy sets using the mass assignment theory of Baldwin [2,4], and use the least prejudiced distribution in cases where a distribution is required – see also the pignistic probability function of Smets [1,28].

A mass assignment M on a set B is a distribution over the power set of B

$$M = \{B_i, m(B_i)\}$$

where $m(B_i) \geqslant 0$ for all $B_i \subseteq B$ and $\sum_{B_i \subseteq B} m(B_i) = 1$.

This is related to a fuzzy subset of B

$$F = \{b_i/\mu(b_i)\}$$

by

$$\mu(b_i) = \sum_{b_i \in B_k} m(B_k).$$

The corresponding least prejudiced distribution is calculated by

$$LPD(b_i) = \sum_{b_i \in B_k} \frac{m(B_k)}{|B_k|}.$$

The mass assignment – fuzzy set transformation relies on a voting (or possible worlds) model. For example, suppose that 10 people are asked which dice values they would accept as being *large*. If two will only accept {6} as large, five will accept {5, 6} and the remaining three will accept {4, 5, 6} then we obtain a mass assignment

$$\{6\} : 0.2, \{5, 6\} : 0.5, \{4, 5, 6\} : 0.3$$

from which we can extract a fuzzy set by treating the memberships as the proportion of voters who accept a particular value as satisfying "large"

– all accept 6 so its membership is 1,
– 80% accept 5, so its membership is 0.8,
– 30% accept 4, so its membership is 0.3.

Thus the fuzzy set *large* is $\{6/1, 5/0.8, 4/0.3\}$.

The reverse process is possible, i.e. we can go from a fuzzy set to mass assignment by putting elements in order of decreasing membership; probability mass on the sets is calculated by subtracting one membership from the next. Taking the fuzzy set

$$\{6/1, 5/0.8, 4/0.3\}$$

we calculate

on $\{6\}$ a mass of $1 - 0.8 = 0.2$,
on $\{6, 5\}$ a mass of $0.8 - 0.3 = 0.5$,
and on $\{6, 5, 4\}$ a mass of $0.3 - 0 = 0.3$.

The probability mass of 0.3 on the set $\{4, 5, 6\}$ could be split between the elements 4, 5, 6 in any way – this means a mass assignment corresponds to a *family* of probability distributions. For example, at one extreme we could assign all the mass to 6, i.e. $Pr(6) = 1$, or we could maximise the mass on 5, i.e. $Pr(6) = 0.2$, $Pr(5) = 0.8$ since the mass on $\{6, 5\}$ and on $\{6, 5, 4\}$ could all be assigned to the element 5.

A more reasonable choice, in the absence of any other information, is to split the probability mass evenly between elements in a set. (For example, if you are told a dice value is even and then asked to bet on the specific value, the safest assumption is that 2, 4 and 6 are equally likely.)

In the example above, the mass of 0.5 on $\{6, 5\}$ would be split 0.25 on 6 and the same on 5; mass of 0.3 on $\{6, 5, 4\}$ would be split 0.1 on each.

This would give a probability distribution of

$$Pr(6) = 0.2 + 0.25 + 0.1 = 0.55,$$

$$Pr(5) = 0.25 + 0.1 = 0.35,$$

$$Pr(4) = 0.1.$$

Because this distribution involves no bias towards any of the elements, it is called the *least prejudiced distribution* corresponding to the mass assignment. This also gives a way of transforming a distribution into a fuzzy set – by assuming the distribution is the least prejudiced distribution, the related mass assignment can be determined and hence a fuzzy set.

To convert the (least prejudiced) distribution to a fuzzy set, put the probabilities into decreasing order and number them from $1 - m$. Denote these probabilities by

$$p_1, p_2, \ldots, p_m.$$

Then, for the nth element in the sequence:

$$\mu_n = \mu_{n+1} + n \times (p_n - p_{n+1}).$$

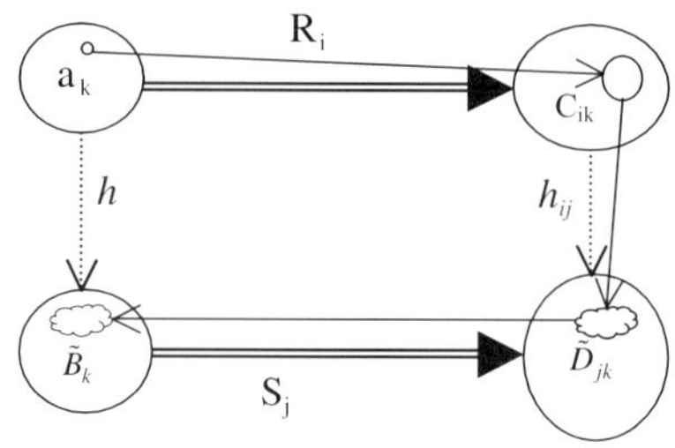

Fig. 1. Diagrammatic representation of the approximate mapping h.

3.4. *Updating h with approximate correspondences*

We treat each selected h_{ij} as an approximate piece of evidence or observation of the elements in B that possibly correspond to elements in A. Because the evidence is not certain, we cannot completely rule out (or rule in) any elements – all we can do is update the probability that an element in A corresponds to a given subset of elements in B.

To start with, the most general statement we can make for any element a_k is that it corresponds to some element in B, i.e. (for all k)

$$h(a_k) = \{b_1/1, b_2/1, \ldots, b_m/1\},$$

i.e.

$$Pr\big(h(a_k) \in B\big) = 1$$

so that $h(a_k) \in B$ has a probability of 1, or if we represent it as a mass assignment

$$h(a_k) = \{B : 1\}$$

(note that incompleteness, i.e. assignment of mass to the empty set means that a_k corresponds to an element outside B). This is our initial (and uninformative) approximation to h.

If we can establish a reasonably specific matching h_{ij} between two codomains C_i and D_j, we can use this to refine the current approximation to h, since if

$$R_i(a_k) = C_{ik}$$

and

$$h_{ij}(C_{ik}) = \widetilde{D}_{jk} \quad \text{where } \widetilde{D}_{jk} \text{ denotes a fuzzy subset of } D_j$$

and

$$S_j(\widetilde{B}_k) = \widetilde{D}_{jk}$$

then

$$h(a_k) = \widetilde{B}_k,$$

i.e. a_k corresponds to a fuzzy subset $\widetilde{B}_k$ (see Figure 1).

Given a possible match function h_{ij} we can transform each $\widetilde{B}_k$ to a mass assignment and then update the probability that

$$h(a_k) \in B_l$$

for each B_l based on the "observed" values derived from $h(a_k)$.

How should we combine the masses m_{ij} from different i, j pairs? Consider a number of observations which are represented as mass assignments over the universe of possible values B.

Let M_n be the mass assignment on B that makes the observed values most likely after n observations, i.e. choose the masses to maximise

$$Pr(M_n|o_1, o_2, \ldots, o_n).$$

This gives a way of updating M after each observation.

Using a naive Bayes assumption

$$Pr(M_n|o_1, o_2, \ldots, o_n) = \frac{Pr(o_1, o_2, \ldots, o_n|M_n) \times Pr(M_n)}{Pr(o_1, o_2, \ldots, o_n)},$$

$$Pr(o_1, o_2, \ldots, o_n|M_n) = Pr(o_1|M_n) \times Pr(o_2|M_n) \times \cdots \times Pr(o_n|M_n).$$

Assuming each possible mass assignment M_n is equally likely,

$$M_n(B_k) = \frac{N_n(B_k)}{\sum_{X \subseteq B} N_n(X)}$$

where $N_n(X)$ is number of times the subset X has been observed.

3.5. *Choice of pairs R_i, S_j*

Clearly it is not useful to update with a pair of relations (R_i, S_j) whose codomains hardly match each other. The possible pairs of relations are ordered according to the maximum probability of matching (from the least prejudiced distribution), averaged over all elements of the codomain C_i:

$$AvMatch(h_{ij}) = \frac{\sum_{x \in C_i} \max_{y \in D_j}(Pr(y \in LPD(h_{ij}(x))))}{|C_i|}.$$

We use this in preference to the average maximum probability of matching, defined as

$$AvMaxMatch(h_{ij}) = \frac{\sum_{x \in C_i} \max_{y \in D_j}(Pr(y \in h_{ij}(x)))}{|C_i|}$$

because the latter measure is not necessarily helpful if there is a large amount of uncertainty in the approximate mapping. For example, if

$$C_i = \{c_1, c_2\},$$

$$D_j = \{d_1, d_2, d_3\}$$

then the universal matching function,

$$h1_{ij}(c_1) = \{d_1/1, d_2/1, d_3/1\},$$

$$h1_{ij}(c_2) = \{d_1/1, d_2/1, d_3/1\}$$

has an *AvMaxMatch* of 1 (since it definitely links every element of C_i to *something* in D_j) whereas

$$h2_{ij}(c_1) = \{d_1/0.9\},$$

$$h2_{ij}(c_2) = \{d_2/1, d_3/0.4\}$$

would only have an *AvMaxMatch* of 0.95 although it is much more specific than $h1_{ij}$. For the cases above

$$AvMatch(h1_{ij}) = 1/3,$$

$$AvMatch(h2_{ij}) = 0.85.$$

This discriminates against matching functions h_{ij} that are very unspecific.

It also makes little sense to choose A and B unless they are close to being key domains, i.e. uniquely identifying an object. This can be estimated by looking at the cardinality of the domain.

3.6. *Efficiency considerations*

The discussion above assumes we work with mass assignments over the universe B rather than a distribution on B since this avoids having to make assumptions about how to divide the mass on non-singleton sets. However, there is clearly an implementation problem since the power set of B has $2^{|B|}$ elements which is not feasible for large universes. We can constrain the size of our work space by requiring M_i to correspond to a fuzzy set, i.e. to be a nested mass assignment.

This simplifies both the representation and the processing involved. It turns out that we just need to consider the mean membership of each element; this is equivalent to taking an average mass assignment and then restricting it (by means of type-2 restrictions [3]) until it is nested and hence corresponds to a fuzzy set.

3.7. *When to stop*

We measure the effectiveness of each attribute pairing ($R_i \sim S_j$) by monitoring the total entropy of h. We have, for each element a_k, a corresponding fuzzy set $\widetilde{B}_k$ which is the current best estimate of the elements in B which are "identical to" a_k. The average over all elements a_k of the entropies of the corresponding fuzzy sets $\widetilde{B}_k$ is monitored.

We calculate the entropy of a nested mass assignment (fuzzy set) by creating the corresponding least prejudiced distribution ($b_i : p_i$) and finding $\sum_i p_i \log_2(1/p_i)$.

This can be done quickly by putting the N elements in increasing order $i = 0 \ldots N-1$

$$p_i = p_{i-1} + \frac{\mu_i - \mu_{i-1}}{N-i} \quad \text{for } i = 1 \ldots N-1, \text{ with } p_0 = \frac{\mu_0}{N}.$$

In the case of an incomplete mass assignment, it is assumed the mass is spread equally amongst all elements in the domain – this increases the entropy considerably.

This value is calculated for all fuzzy sets B_k, and the total represents the entropy of the current h. If this increases, then the last attribute pairing $(R_i \sim S_j)$ has made the approximate h less specific and so is rejected. It is assumed that any lower ranked attribute pairs will also make the approximate h less specific and so the process stops.

ALGORITHM (informal outline).
Inputs

- two chosen domains A, B
- sets of relations R_i, S_j and associated codomains C_i, D_j
- set of approximate matching functions h_{ij} which (for some or all i–j pairs) associate each element in C_i with a fuzzy set of elements in D_j

Output – a partial fuzzy mapping between A and B
Optional: verify that A and B are key domains
foreach a_k in A
    initialise $B_k = B$, i.e. for all b, $\mu_k(b) = 1$
end for
foreach pair R_i–S_j
    if background knowledge doesn't rule out this pair (e.g., from schema matching)
        find $AvMatch(R_i, S_j)$
    store in table if non-zero
end for
Sort table by decreasing *AvMatch* score (i.e. probability of correspondence)
$n := 0$
while next $(R_i$–$S_j)$ in table decreases the overall entropy of h
    foreach $a_k \in A$

$$\mu_k(b) := \frac{\mu_k(b) \times n + \mu_{ij}(b)}{n+1}$$

    end for
    $n := n + 1$
endwhile
output $h : a_k \to \mu_k(b)$ for all k and b

4. Implementation and testing

The algorithm has been implemented as a prototype and a number of tests have been performed on small datasets describing restaurants (3 datasets) and films (2 datasets). For illustration purposes we have used two xml datasets describing restaurants (denoted dbX and dbY below) to test the method. These are derived from a classified directory and an online source, and have the following structure:

dbX	Name, TelNo, Addr, FoodType, Meal, TextLine (optional)	204 entries
dbY	name, phone, heading, textline (optional, may be multiple lines)	426 entries

Manual comparison suggests there are about 150 common entries, e.g.,

dbX129

Name	the george
TelNo	01473 652408
Addr	george st hintlesham ipswich suffolk ip8 3nh
FoodType	pubs european english
Meal	dinner breakfast lunch
TextLine	english pub

dbY165

name	the george at hintlesham
phone	652408
textline	george the, george st hintlesham
textline	traditional ales & fine food
heading	public houses

The chosen mapping functions were straightforward

Phone/TelNo
<*area code*> <*number*> matches <*area code*> <*number*>
<*area code*> <*number*> matches <*number*> and vice versa
<*number1*> partially matches <*number2*> if they differ by permutation of two digits
degree of match = proportion of digits in the "correct" position

Text strings
String *STR1* is an approximate substring of *STR2* if *STR1* is shorter than *STR2* and most words in *STR1* are also in *STR2*
STR1 is an approximate permutation of *STR2* if they have a high proportion of common words
degree of match = proportion of common words, must be at least two
Both ignore "stop" words such as *the*, *and*, etc.

5. Initial results

Mapping from *dbX* to *dbY*, the average maximum matches between domains are

dbX attribute	dbY attribute	average match using max in LPD
TelNo	phone	65%
Name	name	60%
Name	textline	55%
Addr	textline	39%

On the basis of telephone number matching, 72 entries out of the 204 in *dbX* are unassigned, 131 appear to have a perfect match and one appears to have a perfect match with two entries in the *dbY* database (this is actually correct, as the *dbY* database contains a duplicate entry). When names are also matched, 44 are unassigned, 18 partially match two or more *dbY* entries and 142 match a single entry. Figure 2 plots the number of definite matches (taken as the element in B_k with highest membership correctly corresponding to a_k according to a manually determined ground truth), and Figure 3 shows the entropy of each B_k as more attribute pairs are used. The average entropy decreases and then starts to increase with the fourth pair, indicating that no more attribute pairs should be used.

Some examples of matches are given below – first, some examples of matches which are identified as the best possibility but without being certain:

dbX128

Name	moon & mushroom inn
TelNo	01473 785320
Addr	high rd swilland ipswich suffolk ip6 9lr
FoodType	pubs european english
Meal	dinner breakfast lunch
TextLine	english pub

dbY171

name	half moon inn
phone	785320
textline	half moon inn high rd swilland witnesham
heading	public houses

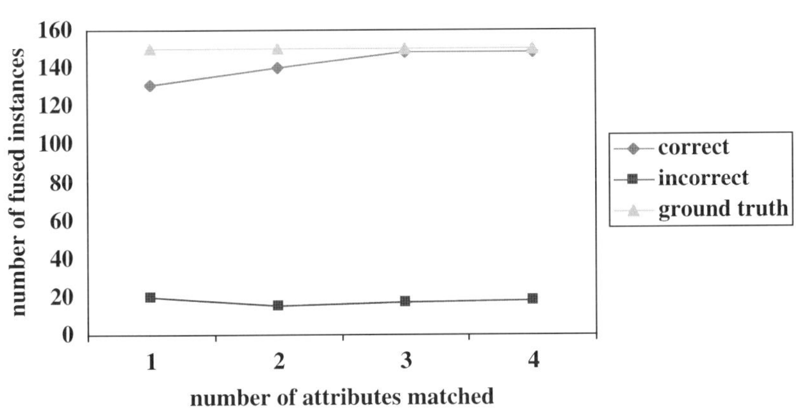

Fig. 2. Number of correctly and incorrectly fused entities in restaurant data, with increasing number of matched attributes.

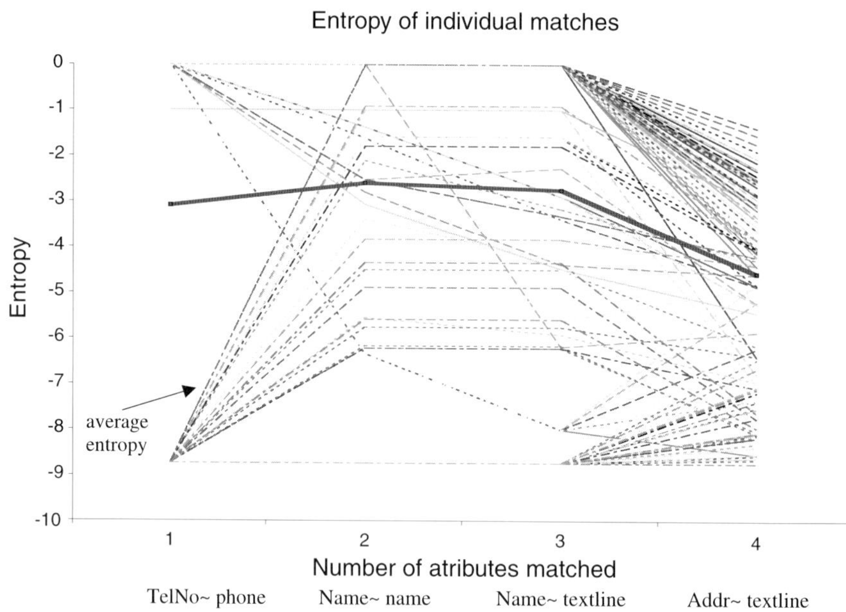

Fig. 3. Entropy of individual fuzzy sets of possible matches Bk in restaurant data, with increasing number of matched attributes. The thick line gives the average entropy.

dbX86

Name	pizza hut uk ltd
TelNo	01473 604770
Addr	upper brook st ipswich suffolk ip4 1du
FoodType	pizza european
Meal	lunch dinner
TextLine	pizza

dbY51

name	pizza hut
phone	216922
heading	food - delivered
textline	pizza hut uk ltd 45-49 upper brook st ipswich

In this case, the system detects a match even though the most reliable indicator (TelNo → phone) is wrong. The system appears to be reasonably robust against errors, e.g., the two entities

dbX8

Name	hilton fish bar
TelNo	01473 727510
Addr	378 bramford rd ipswich suffolk ip1 5ay
FoodType	european fish and chips
Meal	lunch dinner
TextLine	

dbY28

name	fish net the
phone	727510
heading	fish & chip shops & restaurants
textline	fish net the 385 spring rd ipswich

are (correctly) found not to match in spite of having the same phone number.

6. Further development

Although initial results are promising, further testing is necessary to fully establish the validity of this method, involving a variety of datasets. Future developments include

- multi-attribute matching, so that (for example) we could combine <*first name*> and <*last name*> from one source to match <*full name*> in the second source
- use of constraints and schema – level knowledge to find likely attribute matches
- induction and matching of hierarchical structure where one or both sources have a hierarchical classification

Acknowledgements

This work was partially supported by BT Intelligent Systems Lab.

References

[1] A. Ayoun, P. Smets, *Data association in multi-target detection using the transferable belief model*, Internat. J. Intelligent Systems **16** (10) (2001), 1167–1182.

[2] J.F. Baldwin, *The management of fuzzy and probabilistic uncertainties for knowledge based systems*, in: S.A. Shapiro (Ed.), Encyclopedia of AI, Wiley, 1992, pp. 528–537.

[3] J.F. Baldwin, *A calculus for mass assignments in evidential reasoning*, in: M. Fedrizzi, J. Kacprzyk, R.R. Yager (Eds.), Advances in the Dempster–Shafer Theory of Evidence, Wiley, New York, 1994.

[4] J.F. Baldwin, T.P. Martin, B.W. Pilsworth, *FRIL – Fuzzy and Evidential Reasoning in AI*, Research Studies Press (John Wiley), 1995, p. 391.

[5] J. Berlin, A. Motro, *Autoplex: automated discovery of content for virtual databases*, in: Lecture Notes in Comput. Sci., vol. 2172, 2001, pp. 108–122.

[6] M. Bilenko, R. Mooney, W. Cohen, P. Ravikumar, S. Fienberg, *Adaptive name matching in information integration*, IEEE Intelligent Systems **18** (2003), 16–23.

[7] K.C.C. Chang, H. Garcia-Molina, *Approximate query mapping: accounting for translation closeness*, VLDB Journal **10** (2–3) (2001), 155–181.

[8] K.C.C. Chang, H. Garcia-Molina, A. Paepcke, *Boolean query mapping across heterogeneous information sources*, IEEE Trans. on Knowledge and Data Engineering **8** (4) (1996), 515–521.

[9] D. Dey, S. Sarkar, P. De, *A distance-based approach to entity reconciliation in heterogeneous databases*, IEEE Trans. on Knowledge and Data Engineering **14** (3) (2002), 567–582.

[10] Y. Ding, S. Foo, *Ontology research and development. Part 2 – a review of ontology mapping and evolving*, J. Inform. Sci. **28** (5) (2002), 375–388.

[11] A. Doan, P. Domingos, A.Y. Halevy, *Reconciling schemas of disparate data sources: a machine-learning approach*, SIGMOD Record **30** (2) (2001), 509–520.

[12] A. Doan, P. Domingos, A. Halevy, *Learning to match the schemas of data sources: a multistrategy approach*, Machine Learning **50** (3) (2003), 279–301.

[13] M.G. Elfeky, V.S. Verykios, A.K. Elmagarmid, *TAILOR: a record linkage tool box*, in: Proc. International Conference on Data Engineering, San Jose, CA, IEEE Computer Society, 2002, pp. 17–28.

[14] R. Fagin, *Combining fuzzy information from multiple systems*, J. Comput. System Sci. **58** (1) (1999), 83–99.

[15] R. Fagin, *Combining fuzzy information: an overview*, SIGMOD Record **31** (2) (2002), 109–118.

[16] R. Fagin, *Fuzzy queries in multimedia database systems*, in: Proceedings of the ACM SIGACT SIGMOD SIGART Symposium on Principles of Database Systems, 1998, pp. 1–10.

[17] I.P. Fellegi, A.B. Sunter, *A theory for record linkage*, J. Amer. Statist. Assoc. **64** (1969), 1183–1210.

[18] A. Gal, A. Trombetta, A. Anaby-Tavor, D. Montesi, *A model for schema integration in heterogeneous databases*, in: Proc. Seventh International Database Engineering and Applications Symposium (IDEAS'03), Hong Kong, IEEE Press, 2003, pp. 2–11.

[19] W.S. Li, C. Clifton, *SEMINT: a tool for identifying attribute correspondences in heterogeneous databases using neural networks*, Data and Knowledge Engineering **33** (1) (2000), 49–84.

[20] J. Madhavan, P.A. Bernstein, P. Domingos, A.Y. Halevy, *Representing and reasoning about mappings between domain models*, in: Proceedings of the National Conference on Artificial Intelligence, 2002, pp. 80–86.

[21] J. Madhavan, P.A. Bernstein, E. Rahm, *Generic schema matching with cupid*, in: Proceedings of the International Conference on Very Large Data Bases, 2001, pp. 49–58.

[22] T.P. Martin, *Searching and smushing on the semantic web – challenges for soft computing*, Studies in Fuzziness and Soft Computing **139** (2004), 167–186.

[23] P. Mitra, G. Wiederhold, M. Kersten, *A graph-oriented model for articulation of ontology interdependencies*, in: Lecture Notes in Comput. Sci., vol. 1777, 2000, pp. 86–100.

[24] G. Modica, A. Gal, H.M. Jamil, *The use of machine-generated ontologies in dynamic information seeking*, in: Lecture Notes in Comput. Sci., vol. 2172, 2001, pp. 433–448.

[25] H.B. Newcombe, J.M. Kennedy, S.J. Axford, A.P. James, *Automatic linkage of vital records*, Science **130** (1959), 954–959.

[26] N.F. Noy, M.A. Musen, *The PROMPT suite: interactive tools for ontology merging and mapping*, Internat. J. Human Computer Studies **59** (6) (2003), 983–1024.

[27] E. Rahm, P.A. Bernstein, *A survey of approaches to automatic schema matching*, VLDB Journal **10** (4) (2001), 334–350.

[28] P. Smets, *Constructing the pignistic probability function in a context of uncertainty*, in: M. Henrion, et al. (Eds.), Uncertainty in Artificial Intelligence, vol. 5, Elsevier, 1990, pp. 29–39.

CHAPTER 16

Processing Fuzzy Information in Semantic Web Applications

Sebastian Kloeckner and Klaus Turowski

Chair of Business Informatics and Systems Engineering, Business Faculty, University of Augsburg, Universitätsstraße 16, 86135 Augsburg, Germany
E-mail: {sebastian.kloeckner,klaus.turowski}@wiwi.uni-augsburg.de

Uwe Weng

Gladbacher Straße 118, 40219 Düsseldorf, Germany
E-mail: uwe.weng@web.de

Abstract

The actual "Version" of the World Wide Web is surely one of the most important achievements of the last decades. But it suffers from the intrinsic deficit to include the semantic structure of the offered information or knowledge as HTML was developed to format the content instead of describing the logical structure. While the actual research on semantic web and RDF offer improvements for this situation, it is likely that the problem will not be solved completely. This is mainly caused by the fuzziness of human language and the fact that the boundaries of ontologies cannot be defined exactly.

This article shows a description and a general methodology based on fuzzy inference in order to close this gap. The usage of fuzzy logic allows it to represent and process ambiguous facts making it possible to reconstruct the fuzziness of the real world and human way of thinking. Based on this representation and ability of processing it, better search algorithms can be developed and implemented in order to gain better results.

Keywords

fuzzy information, semantic web, XML, ontology, RDF, Dublin core, data exchange, ontology mapping

FUZZY LOGIC AND THE SEMANTIC WEB
Edited by Elie Sanchez
© 2006 Elsevier B.V. All rights reserved

1. Motivation

Due to ease of use especially the World Wide Web and web browsing have achieved a great popularity in recent years. During the first years most of the offered HTML web pages consisted of static content. The main goal was to offer information, which was relatively invariant over time. Therefore it was adequate for search engines to (re-)visit and index these pages in comparatively long periods of time.

Today a not neglectable amount of content is generated dynamically by deploying new web technologies like J2EE or ASP.NET. This dynamic generation of content allows it to integrate back office systems and databases. Consequently, the offered content and even the sources (URI) can change constantly implying that the concepts of the earlier mentioned search engines do not match these circumstances any more. Additionally, the offered HTML content, except from the standard meta data of the head-tag, does not contain any meta data about the offered content. This did not matter as far as the offered content did not change very often and was directly targeted at a human end user. But the appearance of automated processing of content, so-called web services, demands more information, especially meta information about the offered content.

2. World Wide Web

The Internet has become a huge data pool and it is difficult to find the desired information. For this reason search engines and catalogues are the most commonly visited web sites. Catalogues try to mirror the structure of the domains. In order to assign a web page to a particular category a registration is required. This results in high maintenance costs.

In contrast search engines perambulate the Internet and index the located web pages for a keyword-based search. But this indexing by keywords has a lot of disadvantages. First, different spellings or synonyms are normally not included while processing the request. Besides, the context, the keyword is used in, disappears. Hence, current search methods wrongly assume that the user enters the right words for getting the wanted information. The W3C states this as "Google et al. are great, but there are too many false hits" [3].

Additionally, because of the high volume of data one search engine is not able to index all pages on the Internet, even though the analysis is restricted to some HTML tags (e.g., <title> or <h1>). Therefore, meta crawlers have established. They send the search request to many engines and summarize the results (e.g., filtering out double Internet addresses).

As mentioned the search strategy is primary keyword-based. But, more and more dynamically generated web pages hamper a suitable and semantically correct indexing. The actual relevant content is stored in structured databases.

To improve the quality of the search result some engines additionally consult the links to other web sites or count the number of keyword occurrences for measuring the importance of a web page and classifying the probable content [12].

Furthermore, at the beginning of the Internet era indexing was reduced to content formulated in HTML. Meanwhile there are search possibilities about multimedia content and other resource types. There are two different index methods. The primitive way is to assign the semantic of the resource to the in HTML formulated and indexed content. The advanced stage is the interpretation of the resource itself.

3. Semantic web

As shown, the actual "Version" of the World Wide Web is surely one of the most important achievements of the last decades. But it suffers from several intrinsic problems. First, by the nature of the WWW, information and knowledge is stored distributedly.

Second, most of the information and knowledge offered by the WWW is only understandable for humans as the resources does not contain any (or only few) machine readable structures or semantics. HTML is unsuitable as representation language for knowledge because this protocol aims to format the content. This combination of distribution and missing meaning makes it impossible to combine or interpret the offered information and knowledge automatically.

In order to solve the problem of partial information and difficulties of interpretation each resource should provide information about itself, also called "metadata". Meta data are data about data. For instance, they are used to identify, describe, or locate resources. This metadata should be in a machine processable format in order to allow software-agents to process and analyze this metadata. Additionally, this metadata should be based on a common data model (schema) for expressing metadata and a common vocabulary as only a common understanding allows automatic processing of information and knowledge.

Such semantic interoperability is the goal of present W3C standardization efforts. The Resource Description Framework (RDF) is one result of its work and specifies a standard for describing data about (web) resources.

RDF is a declarative language, which expresses semantic content in form of subject–predicate–object triples. Hence, RDF provides a basic object–property–value data model for meta data. The subject of such triples is the resource which is identified by a Uniform Resource Identifier (URI) reference. Generally, URIs can link any pieces of information across the web. It is a basic concept of the Internet architecture. The W3C Recommendation emphasizes that RDF URIs can refer to any identifiable thing, including things that may not be directly retrievable on the web, because function calls hide them, or they are real objects which belongs to the domain of discourse, for instance [7]. Thus, RDF descriptions are not restricted to hypertext documents. The basic idea is that subjects are described in terms of properties and values. They build statements which are made about a resource. Both properties and their values can also be references to other resources identified by a URI. Instead of using constant literals it is even advisable to use URIs for referring to same things. Such links are important input for inference because the machine is able to recognize that the subject, predicate or value are the same object.

The favorite representation form is a conceptual graph consisting of nodes and directed arcs (Figure 1). Nodes represent a subject or value. A predicate is modeled as an arc, directed from the subject node to the object node. For a simpler exchange an XML-based syntax is also specified.

In contrast to pure XML, RDF gives the description about a document or resource in a general machine-processable meaning. Thus, it specifies the semantics of a formal language and not the content. From this point of view the semantic of a resource consists always of triples whose composition and meaning of every part is specified by RDF.

Furthermore, RDF allows a form of reification. RDF itself provides a vocabulary to describe a statement as statements referring to the parts of the triple as object. Build-in properties rdf:subject, rdf:predicate, and rdf:object are used to specify the role of the ob-

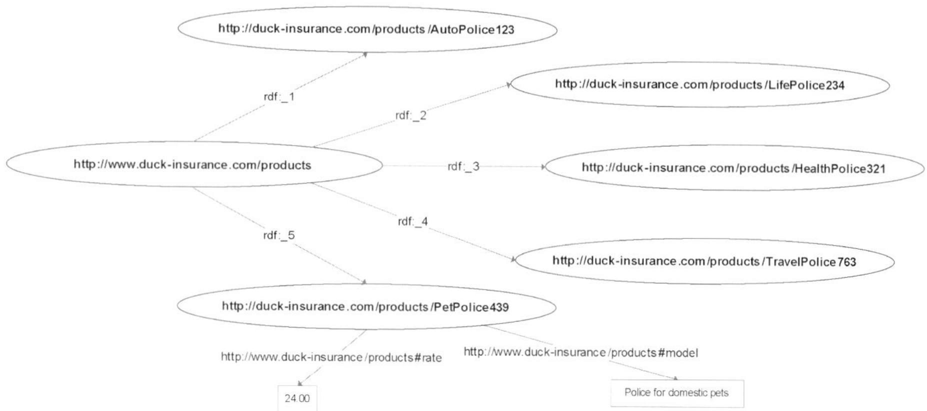

Fig. 1. Conceptual graph of a product structure.

```
<?xml version="1.0" ?>
<!DOCTYPE rdf:RDF (View Source for full doctype...)>
- <rdf:RDF xmlns:rdf="http://www.w3.org/1999/02/22-rdf-syntax-ns#" xmlns:p="http://www.duck-
   insurance.com/products/vocab#">
 - <rdf:Description rdf:about="http://www.duck-insurance.com/products">
  - <p:products>
   - <rdf:Bag>
       <rdf:li rdf:resource="http://www.duck-insurance.com/products/AutoPolice123" />
       <rdf:li rdf:resource="http://www.duck-insurance.com/products/LifePolice234" />
       <rdf:li rdf:resource="http://www.duck-insurance.com/products/HealthPolice321" />
       <rdf:li rdf:resource="http://www.duck-insurance.com/products/TravelPolice763" />
       <rdf:li rdf:resource="http://www.duck-insurance.com/products/PetPolice439" />
     </rdf:Bag>
   </p:products>
  </rdf:Description>
 - <rdf:Description rdf:ID="PetPolice439">
    <p:model rdf:datatype="http://www.w3.org/2001/XMLSchema#string">Police for domestic
      pets</p:model>
    <p:rate rdf:datatype="http://www.w3.org/2001/XMLSchema#decimal">24.00</p:rate>
  </rdf:Description>
</rdf:RDF>
```

Fig. 2. Semantic description of a product catalogue document.

ject within the original statement. It is also possible to make further statements about the original statement.

RDF supports inheritance of classes and properties. This kind of relationship facilitates inferences like "A legal person is a person" or "Every person has got an address which is a primary address for postal usage". In opposite to many object-oriented programming languages multiple derivation is allowed.

Additionally, statements can describe containers using predefined language elements. They do not construct a container like in programming languages. These elements express that the semantic of the subject is a container. For example, a company offers a set of products. A description about the product catalogue might look like this, cf. Figure 2.

An RDF Schema allows the definition of a particular vocabulary. It specifies the acceptable relations and valid constraints (e.g., attribute values) to the concepts of the domain model.

The meaning of an RDF Schema is comparable with an XML Schema. But, in contrast to RDF XML specifies only the syntactic conventions which can be checked by parsers while reading the document. The intended semantics of the elements remain outside the realm of XML. An XML document only obeys a grammar formulated in a Schema. For this reason, it is not sufficient to compare two documents by means of their XML Schemata. The contents of two documents are equal if their semantic is identical. In [2] the impacts are illustrated considering as example. Therefore, the RDF specification supports the semantic mapping between different models.

RDF is the basis for many vocabularies (e.g., OWL [9], DAML + OIL [1], etc.). Hence, an RDF Schema specification does not provide a particular vocabulary of application-specific classes and properties. Instead, it facilitates to describe such elements. Additional capabilities (e.g., introduction of cardinalities) are targets of ontology languages. They intend to provide further machine-processable semantics for resources.

In practice, the setup and maintenance of such ontologies are time-consuming and cost-intensive. Therefore, ontologies are primarily built for domain specific tasks. They are closed systems which return good results within the considered domain. But, they do not move the research very much towards the semantic web vision. Therefore, sharing knowledge over domain boundaries is a challenge which is addressed in [8].

4. Semantic web applications

The first popular applications of this standard may be the Dublin Core Standard and the RDF Site Summary (RSS). In the Primer document of the W3C recommendation further domain-specific examples are published [7]. The Dublin Core Standard is maintained by the Dublin Core Metadata Initiative. The objective of the organization is to develop an interoperable online meta data standard that supports a broad range of purposes and business models [4].

Among other things the result is an element set available for HTML and RDF. Some search engines use these descriptive elements for automated indexing. Figure 3 shows a simple example extended by Dublin Core elements.

Resource descriptions using the Dublin Core elements are encoded in HTML as meta tags. They give information about the owner and the content of the web document, for instance. Notice, the declaration of such tags is a manual step. The author is responsible for the correctness of the description. Sometimes homepage owners misapply the elements in order to get a higher ranking in search results. Hence, it is a question of trustworthiness, reliability, and seriousness (web of trust).

The Unified Resource Identifier for the appropriate Dublin Core vocabulary as RDF Schema is http://purl.org/dc/elements/1.1/.

The RDF Site Summary vocabulary is designed for describing short abstracts about any subjects. In addition to the headline a link to the full content is also possible. Figure 4 shows a feed example about a virtual insurance company.

A feed comprises a channel which has a title, link to the original place, description, and (optional) language, followed by a series of items. Each item element includes a title, link, and description. An optional module allows the combination with the Dublin Core Standard.

In opposite to HTML such feeds are reduced to the semantic content. The application operating a RSS file is responsible for the visualization. Hence, the application is able to understand the semantic meaning of the XML elements.

Modern web crawlers already support this kind of information fishing. Special RSS Feed readers are also available. Additionally, they can aggregate all feeds from the sites and show

```
<!DOCTYPE HTML PUBLIC "-//W3C//DTD HTML 4.01//EN"
     "http://www.w3.org/TR/html4/strict.dtd">
<html>
<head>
  <meta-http-equiv="expires" content="0">
  <!-- Dublin Core -->
  <meta name="DC.Title" content="Duck Insurance Company">
  <meta name="DC.Creator" content="Donald Duck">
  <meta name="DC.Subject" content="Insurance Consulting Rating">
  <meta name="DC.Description" content="We offer insurance products and services.">
  <meta name="DC.Date" content="2005-01-01">
  <meta name="DC.Type" content="Text">
  <meta name="DC.Format" content="text/html">
  <meta name="DC.Identifier" content="http://www.duck-insurance.com/">
  <meta name="DC.Source" content="http://www.duck-insurance.com/">
  <meta name="DC.Language" content="en">
  <meta name="DC.Coverage" content="USA">
  <meta name="DC.Rights" content="All rights reserved.">
  <title>Duck Insurance</title>
</head>
<body>
  <h1>Welcome to Duck Insurance</h1>
  We offer following products and services:<p/>
  <ul>
    <li>Auto insurance</li>
    <li>Life insurance</li>
    <li>Health insurance</li>
    <li>Travel insurance</li>
    <li>Pet insurance</li>
    </p>
    <li>Consulting</li>
    <li>Rating</li>
  </ul>
  <p/>
  <h3>Contact:</h3>
  <p/>
  Donald Duck Insurance<br/>
  400 Goofy Street. SW<br/>
  Mickeysota. MY 12344
  <p/>
  Phone: US 1-800-55-123
</body>
</html>
```

Fig. 3. Example document extended by Dublin Core.

them at a glance. In combination with an optional syndication module such applications can react to changes automatically. After registering, users do not visit all web sites personally in order to get the latest news. In the meantime, many companies (especially distributors of print media) offer such services. Pretty much anything that can be broken down into discrete items can be syndicated via RSS. The organization of web logs and discussion forums in RSS format are further application areas.

The Unified Resource Identifier for the RSS vocabulary as RDF Schema is http://web.resource.org/rss/1.0/schema.rdf.

But already these two schemas show that there is another problem: There are many ontologies on the web. Some of them are overlapping or their boundaries are fuzzy, as they are still under development. This implies that the equivalence of terms across ontologies can not be processes automatically.

5. Fuzzy control

It is indisputable that fuzzy approaches offer a high potential to improve decision making within business applications. There are many successful examples for software and hardware solutions. [10] gives a comprehensive overview of decision problems which are solved by fuzzy control. The ability to take the fuzziness of data over the deduction is particular interesting for many real world and linguistic terms related applications.

Nevertheless, fuzzy approaches mainly exist as self-contained solutions that are not designed to collaborate with other business application systems. There are still two obstacles that make it difficult to use fuzzy approaches for business applications:

```
<?xml version="1.0" encoding="iso-8859-1" ?>
- <rdf:RDF xmlns:rdf="http://www.w3.org/1999/02/22-rdf-syntax-ns#" xmlns="http://purl.org/rss/1.0/"
    xmlns:dc="http://www.duck-insurance.com/feeds/news.rss">
  - <channel rdf:about="http://www.duck-insurance.com/feeds/news.rss">
      <title>Duck Insurance</title>
      <link>http://www.duck-insurance.com</link>
      <description>News about company and products</description>
      <dc:publisher>Duck Insurance</dc:publisher>
      <image rdf:resource="http://www.duck-insurance.com/images/logo.jpg" />
    - <items>
      - <rdf:Seq>
          <rdf:li rdf:resource="http://www.duck-insurance.com/news/article.php?sid=20041223" />
          <rdf:li rdf:resource="http://www.duck-insurance.com/news/article.php?sid=20050115" />
        </rdf:Seq>
      </items>
    </channel>
  - <image rdf:about="http://www.duck-insurance.com/images/logo.jpg">
      <title>Duck Insurance News</title>
      <url>http://www.duck-insurance.com/images/logo.jpg</url>
      <link>http://www.duck-insurance.com</link>
      <dc:description>Logo of Duck Insurance</dc:description>
    </image>
  - <item rdf:about="http://www.duck-insurance.com/news/article.php?sid=20041223">
      <link>http://www.duck-insurance.com/news/article.php?sid=20041223</link>
      <title>Merry Christmas and a Happy New Year</title>
      <description>We wish all customers and interessents Merry Christmas and ...</description>
      <dc:identifier>http://www.duck-insurance.com/news/article.php?sid=20041223</dc:identifier>
    </item>
  - <item rdf:about="http://www.duck-insurance.com/news/article.php?sid=20050115">
      <link>http://www.duck-insurance.com/news/article.php?sid=20050115</link>
      <title>Duck Insurance wins Insurance award 2005</title>
      <description>The insurance association announces that the Duck Insurance Company ...</description>
      <dc:identifier>http://www.duck-insurance.com/news/article.php?sid=20050115</dc:identifier>
    </item>
  </rdf:RDF>
```

Fig. 4. News feed example of a virtual insurance company.

- The technical and logical integration of the fuzzy applications into the company-wide information system (caused by mismatched data types, for instance), and
- The development and programming of the fuzzy application itself.

The problem of technically integrating fuzzy approaches in existing business software was already addressed in [13]. A set of so-called fuzzy components was developed in the Java programming language in order to facilitate the technical integration of fuzzy approaches in existing business software. A fuzzy component is defined as a reusable piece of software which implements certain fuzzy functionality or fuzzy data, e.g., a fuzzy set, a fuzzy number or interval, a fact- or rule-base, or a fuzzy inference operator. The development of fuzzy applications with these components works fine as long as it takes place in the same development environment as the business application. When using different or even distributed environments there are two other integration problems: On the one hand, there is the problem of exchanging fuzzy information between applications. To solve this problem [14] defined a common interchange format for fuzzy data and rules by using the extensible markup language. XML makes it possible to represent and exchange information in an environment-neutral manner. An appropriate schema for the most important fuzzy data types was introduced to soften the lack of data integration. This schema contains the formal syntax for describing uncertainty information. The result is a standardized exchange of fuzzy information between applications and services that have reached agreement about the syntactic format before. On the other hand, there is the problem of distributed interoperability. [15] proposes to use web services in order to diminish this kind of integration problems. Hence, the implementation of fuzzy control is already available over the Internet and open to the public in a simple and standardized manner.

6. Fuzzy control in a semantic web

Existing simple, symbolic models are only approximations of complex real world concepts. They ignore the natural fuzziness of linguistic terms, for instance. Representation and control of uncertain information may result in a better picture of the domain of discourse.

The problem of searching and finding information itself has two faces: First, the information available on the web does not contain any semantic information or the used ontology is unknown or incompatible. Therefore, the information or knowledge of this resource cannot be categorized by crisp values. In order to generate the required semantic information the content has to be analyzed by the usage of fuzzy logic based on natural fuzziness of linguistic terms, if no semantic information is available at all. If at least some kind of semantic structure is available, this semantic structure has to be mapped to a known ontology in order to "understand" the offered information or knowledge. This could be realized based on fuzzy inference and equivalent or similar linguistic terms.

Second, the given information for the search process does not contain any context or semantic information. Starting point is always the search for certain information (web-content) or functionality (web-service). In one case the human user try to find information or knowledge concerning a certain topic, while in the other case machines try to get the required input or functionality. Based on this assumption several levels of interpretation are conceivable:

Given ontology and crisp search values: If the ontology and the crisp values for the search are given, the prime task of the search engine is to find information which maps with the given information best.

Given ontology with fuzzy search terms: If the domain of desired information is known, but the given search term is vague in relation to the given ontology, two search strategies are conceivable. One would be to use fuzzy inference and equivalent or similar linguistic terms in order to get the set of possible meaning of the search term. The other would be to find ontologies, which are comparable to the given ontology, and to map the given search term onto these ontologies. Both strategies would use fuzzy inference based on linguistic variables and terms in order to find equivalent or similar structures.

Crisp search values without given ontology: If the given search information does not contain any context or meaning at all, the process of finding information or knowledge must first estimate the possible context. This could be realized based on the fuzzy analysis of known ontologies resulting in an ordered set of applicable ontologies. Based on these ontologies the search itself could be executed.

Fuzzy values without given ontology: Without any crisp information, neither search nor ontology information, a successful search becomes very unlikely. As in the case of a given ontology and fuzzy search terms two search strategies are imaginable, but with a significantly lower chance of satisfying results. One strategy would be to find a set of possible meaning of the search term by usage of fuzzy inference and equivalent or similar linguistic terms. The other strategy would be to find known ontologies to which the search term could belong to. This would also be based on fuzzy inference concerning the known ontologies and the search term.

Some of the mentioned issues are candidates being solved by fuzzy logic:

The natural language is imprecise and vague. That is, the classification of attribute values are fuzzy (e.g., "when is a person an old person?"), or the membership to a type hierarchy is not distinct.

- The natural language supports different spellings.
- The syntactic constitution of semantic identical values differs between two sources. The challenge is to disagree semantic equality and similarity [8].
- The collected information is uncertain and differs between sources.
- The relationships between subjects themselves are fuzzy.
- The evaluation of trustworthiness and reliability is subjective. Therefore, information sources need a fuzzy weighting and classification.

Different integration levels for expressing semantic fuzziness are conceivable. The lowest level would be the use of existing elements like the meta-tags in Dublin Core Standard. For instance, the age of a person looks like the following Example 1. While the first line of the example offers a crisp value for the age of a person, the last line describes that the person is old with a right open interval beginning at the age of 50.

EXAMPLE 1. Abstract representation of fuzzy knowledge as meta information.

```
<meta name=''Person.age'' content=''65''> (crisp value)
<meta name=''Person.age'' content=''[65, 1.0]''> (singleton with peak at 65)
<meta name=''Person.age'' content=''[50, 0.0] [65, 1.0]''> (right open
interval for old persons)
```

The knowledge processor has to consider the syntax conventions in order to get a valid result.

The next integration level is the fuzzification of statements while reasoning. Origin crisp RDF statements are translated into linguistic terms or even fuzzy sets using rules. But, the fuzzification of particular values depends on the semantic of the concept. Not every subject is suitable to be fuzzified [8]. The rules could be embedded in the data, using the RDF-encoding of a rules language like [5]. They express additional relationships between concepts. The structure of the semantic net itself is the source for inference.

Full integration implicates a syntactic and semantic extension of the RDF language. One approach arranges the assignment of a membership grade (e.g., for expressing the trustworthiness of a statement) by using reification. In this case, the predicate would get a property "trust" in order to represent the membership degree of the couple (subject, object) to the extension of the predicate. Another form of integration on language level could be the definition of fuzzy data types. The currently supported data types in RDF are the XML Schema data types. For expressing linguistic terms which are based on fuzzy sets, an appropriate proposal is described in [15].

Fuzzy RDF does not aim to become a standard extension to plain RDF. It is conceived to be used only inside the application. Fuzzy data can be transferred as standard RDF using reification. The exact semantic can be brought back by defuzzification.

The extensible nature of RDF means that different meta data schemas can established by different groups. As already mentioned the standardization of ontologies for all real world concepts is an unreachable goal. Top-down standardization would not be successful, because of the diversity of domains, languages, and cultures around the world. It would also contradict the philosophy of the Internet as an independent communication and data

```
<?xml version="1.0" encoding="ISO-8859-1" ?>
<!DOCTYPE swrlx:Ontology (View Source for full doctype...)>
- <swrlx:Onotolgy xmlns:owlx="http://www.w3.org/2003/05/owl-xml"
    xmlns:swrlx="http://www.w3.org/2003/11/swrlx" xmlns:ruleml="http://www.w3.org/2003/11/ruleml"
    xmlns:xsl="http://www.w3.org/2001/XMLSchema-instance"
    xsl:schemaLocation="http://www.w3.org/2003/11/swrlx
    http://www.ruleml.org/swrl/xsd/modular/swrlx.xsd">
  - <owlx:Annotation>
      <owlx:Documentation>Classification of customers by age</owlx:Documentation>
    </owlx:Annotation>
    <ruleml:var>customer</ruleml:var>
    <ruleml:var>age</ruleml:var>
  - <!--
          If the age of a customer is greater then or equal to 65 years
          Then the customer is a pensioner.

    -->
  - <ruleml:imp>
      <ruleml:_rlab ruleml:href="#customerGroup" />
    - <ruleml:_body>
      - <swrlx:builtinAtom swrlx:builtin="http://www.w3.org/2003/11/swrlb#greaterThanOrEqual">
          <ruleml:var>age</ruleml:var>
          <owlx:DataValue
            owlx:datatype="http://www.w3.org/2001/XMLSchema#int">65</owlx:DataValue>
        </swrlx:builtinAtom>
      </ruleml:_body>
    - <ruleml:_head>
      - <swrlx:individualPropertyAtom swrlx:property="http://www.duck-
          insurance.com/ontology#hasStatus">
          <ruleml:var>customer</ruleml:var>
          <owlx:Individual owlx:name="http://www.duck-insurance.com/ontology#pensioner" />
        </swrlx:individualPropertyAtom>
      </ruleml:_head>
    </ruleml:imp>
  </swrlx:Onotolgy>
```


Fig. 5. Example for representing fuzzy knowledge in XML.

exchange platform being available for any purposes. For this reason it is important to examine the implications when combining different ontologies. At first, the representations might be different. For instance, properties of a subject have different names though the values have the same meaning.

At this point the representation of fuzzy knowledge is discussed. But, for an integrated conceptual solution the processing of uncertain information has also to be considered. Conventional RDF inference engines deduce new facts by value matching. A fuzzy inference engine transfers the fuzziness to the reasoning process by using fuzzy operators which are part of production rules. For instance, a MIN-operator would express a logical AND. [15] defines an XML syntax for formulating fuzzy rules. Hence, if a fuzzification makes sense semantically it is also required to transform values to a unique fuzzified spelling in order to discover similar words when approximate matching.

Though it can be shown that a fuzzy (RDF) inference engine is equivalent, both in its results and performance, to a conventional inference engine, the deduction of new facts should be limited to degree of memberships over a predefined value. Such threshold controls the relevance of a resource and affects the performance because it delimitates the search area at early time.

7. Technical, architectural solution/approach

In the following, technical solutions for certain problems which were addressed in the sections above are presented. It is a first architectural approach to enhance the Internet to a semantic web. Before, the determing factors are summarized:

Global and open base for communication and information exchange
Platform independence
Wide Area Network
Support of processing static and dynamic contents

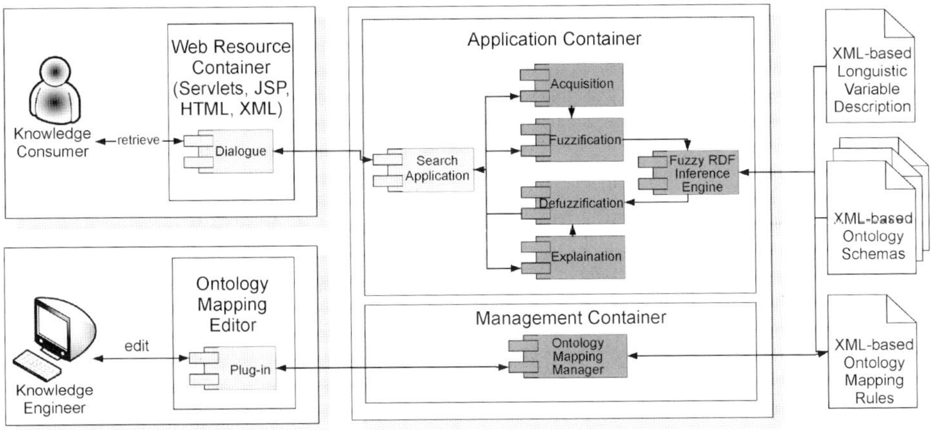

Fig. 6. Schematic architecture for processing fuzzy information in semantic webs.

The design of a semantic web may be similar to a knowledge-based system. Knowledge base and inference method are separated. The knowledge base contains domain specific information about resources of the real world, whereas the inference method is application neutral and can be reused in different web services. As representation form the Resource Description Framework (RDF) with its object–property–value data model is used. Building ontologies makes possible to describe knowledge as accurately as possible.

Additionally, the experience in knowledge engineering has shown that the more the knowledge base is domain-specific, the better the results in knowledge processing. It is absolutely comprehensible that current efforts in building ontologies are restricted to certain domains and applications and due to the globality of the web a standardization of ontologies is terminally impossible. The understanding of an ontology is always application specific. Therefore, the intersection of knowledge is a matter of particular interest. Hence, the aim of meta crawlers is to combine the ontologies of different concept-based (search) engines if required.

Because of the high volume of resources on the web ontologies have to be generated and integrated by machines. Hence, the acquisition like in knowledge based systems should be automated. For this aim clustering is a suitable data mining technique. Appropriate efforts are described in [6]. A fuzzy ontology generation framework is already available [11]. It is very important for the acceptance of a semantic web that the complexity of building web ontologies is hidden towards web site creators and other knowledge providers. Most documents are provided by less versed people in formal and logical languages. The know-how about knowledge representations are focused on people which are like to offer such services.

The result of knowledge acquisition is a semantic net with fuzzy data and relationships. For further asynchronous processing they have to be stored in an adequate database. Though conventional relational database management systems are able to store fuzzy sets in normalized form so-called fuzzy databases should be considered. They are optimized for handling fuzzy data and relations and support a fuzzy query language for information retrieval [16].

The problem solving component of a knowledge-based system provides search and inference methods. They operate on the knowledge base and produce new facts by deduction. For processing uncertain information a fuzzy inference engine is required. It matches the fuzzified request with the fuzzy data in the knowledge base by applying rules and predefined processing strategy (e.g., min–max-operator). The result is a list or subnet of resources (concepts) with different degrees of relevance or fuzzy information, in general. Both need possibly a defuzzification before further conventional processing. A first approach for such an inference component is shown in [13].

Another component is needed for dialogue between human and machine. Relating to the Internet a web browser is used for communication. The result of knowledge processing is displayed as HTML page. In connection with J2EE application server systems the dialogue component consists of the web browser front-end installed on the machine of the user, and the servlet engine which is responsible for generating new valid HTML pages.

A fourth component may explain the inference result. For instance, a search engine shows the search path through the semantic net in human readable form. Additionally, with the aid of the dialogue component the engine can ask the user for further details in order to constrain the search area. This example shows how important the generation of new web documents for knowledge processing itself is because inference is a dynamic process.

A fifth component would be an ontology mapping manager. This component allows a knowledge engineer to insert, alter and delete the ontology mapping rules. In order to get best inference results some of the rules have to be changed over time as ontologies can be modified or human understanding may change. Additionally, over time new ontologies will be developed. These new ontologies and their relationship to other ontologies have to be modeled as ontology mapping rules for usage during the inference process. And finally, some ontologies may become obsolete in the future. Consequently, they have to be removed from the ontology mapping rules as they could disturb the inference process.

Figure 6 shows a schematic diagram of the overall architecture for such a system.

8. Conclusion and future work

The shown generic idea offers a first approach for the use of fuzzy logic in order to improve the actual concepts of the semantic web. While solutions for the representation of fuzzy data types in XML are already available, a way of integrating such information into the offered content as well as the existing ontologies is still missing.

In order to get an efficient semantic web similarities and equalities of existing ontologies have to be represented. But these similarities and equalities cannot be precise as they always include some level of interpretation. Therefore fuzzy rules have to be developed in order to enable a successful inference. Additionally, further definitions of linguistic variables and terms have to be discussed due to the fact that they will be a part of the knowledge base used for inference. Furthermore, adequate inference rules (rule base) for evaluating and matching the different kinds of facts have to be defined. And finally, a content parsing engine, which is able to analyze content which does not contain any semantic or context information, has to be developed and implemented.

References

[1] D. Connolly, F. van Harmelen, I. Horrocks, D.L. McGuinness, P.F. Patel-Schneider, L.A. Stein, *DAML + OIL (March 2001) Reference Description*, W3C Note 18 December 2001, http://www.w3.org/TR/daml+oil-reference.

[2] S. Decker, F. van Harmelen, J. Broekstra, M. Erdmann, D. Fensel, I. Horrocks, M. Klein, S. Melnik, *The semantic web – on the respective roles of XML and RDF*, IEEE Internet Computing **4** (5) (2000), 63–74.

[3] I. Hermann, *Tutorial on semantic web technologies*, http://www.w3.org/Consortium/Offices/Presentations/RDFTutorial, 2005.

[4] D. Hillmann, *Using Dublin Core*, http://dublincore.org/documents/usaguide, 2003.

[5] I. Horrocks, P.F. Patel-Schneider, H. Boley, S. Tabet, B. Grosof, M. Dean, *SWRL: a semantic web rule language combining OWL and RuleML*, http://www.daml.org/2003/11/swrl/, 2003.

[6] A. Maedche, S. Staab, *Ontology learning for the semantic web*, IEEE Intelligent Systems **6** (2) (2001), 72–79.

[7] F. Manola, E. Miller, *RDF Primer*, W3C Recommendation 10 February 2004, http://www.w3.org/TR/rdf-primer.

[8] T.P. Martin, *Searching and smushing on the semantic web – challenges for soft computing*, in: Proc. FLINT 2001 – New Directions in Enhancing the Power of the Internet, Berkeley, CA, 2001, pp. 3–8.

[9] D.L. McGuinness, F. van Harmelen, *OWL web ontology language – overview*, W3C Recommendation 10 February 2004, http://www.w3.org/TR/owl-features/.

[10] H. Popp, *Einsatz von Fuzzy-Technik in Industrie und Dienstleistungsbereich – ein Überblick*, in: J. Biethahn, A. Hönerloh, J. Kuhl, V. Nissen (Eds.), Fuzzy Set-Theorie in betriebswirtschaftlichen Anwendungen, Vahlen Verlag, Munich, 1997, pp. 23–40.

[11] T.T. Quan, S.C. Hui, T.H. Cao, *FOGA: a fuzzy ontology generation framework for scholarly semantic web*, in: Proc. 15th European Conference on Machine Learning, Pisa, Italy, 2004.

[12] B.O. Szuprowicz, *Search engine technologies for the World Wide Web and Intranets*, Computer Technology Research Corporation, 1997.

[13] K. Turowski, U. Weng, *Building business applications with fuzzy components*, in: Proc. 3rd International ICSC Symposium on Intelligent Industrial Automation and Soft Computing, Genova, Italy, 1999, pp. 255–260.

[14] K. Turowski, U. Weng, *Representing fuzzy information using XML*, in: Second International ICSC Symposium on Engineering of Intelligent Systems, Paisley, Scotland, 2000, 6 pages.

[15] K. Turowski, U. Weng, *Processing fuzzy information with web services*, in: Proc. 6th WSEAS International Multiconference on Circuits, Systems, Communications and Computers, Crete, Greece, 2002, pp. 31–37.

[16] M. Umano, *Retrieval from fuzzy database by fuzzy relational algebra*, in: Proc. IFAC Symposium on Fuzzy Information, Knowledge Representation and Decision Analysis, Marseille, France, 1984, pp. 1–6.

CHAPTER 17

Fuzzy Logic Aggregation for Semantic Web Search for the Best (Top-k) Answer

Peter Vojtáš*

Department of Software Engineering, Faculty of Mathematics and Physics, Charles University, Prague and Institute of Computer Science, Czech Academy of Science, Prague

E-mail: peter.vojtas@mff.cuni.cz

Abstract

The paper is dealing with the problem of finding best (top-k) answers depending on scoring multicriterial user requirements which appears quite often on the web. Our solution is based on formal models of many valued logic (fuzzy, annotated) where user preference (score) is modeled by ordering of truth values – meaning best answer is the one with biggest truth value. The user global score is modeled by fuzzy aggregation (annotation) operators of all user attribute score. In this paper, we present several many valued definite logic programming based techniques with fuzzy aggregation. We present models for both deductive and inductive tasks and also an extension with similarities and finite domains. We advocate for an extension of web rule languages by fuzzy aggregation.

Keywords

fuzzy and annotated logic, semantic web, multicriterial querying, best answer, scoring, rating

1. Introduction

Today it is becoming important for Semantic web, browsers, relational and object database systems to be able to access (search, retrieve) multifeature data, such as video, images, audio, text and another objects with many different attributes (typically web resources like products, services, documents).

* Supported by Czech IT projects 1ET100300517 and 1ET100300419.

FUZZY LOGIC AND THE SEMANTIC WEB
Edited by Elie Sanchez
© 2006 Elsevier B.V. All rights reserved

Global score combining particular attribute score. R. Fagin, A. Lotem and M. Naor in [8] conclude:

> Early database systems were required to store only small character strings, such as the entries in a tuple in a traditional relational database. Thus, the data was quite homogeneous. Today, we wish for our database systems to be able to deal not only with character strings (both small and large), but also with a heterogeneous variety of multimedia data (such as images, video, and audio). Furthermore, the data that we wish to access and combine may reside in a variety of data repositories, and we may want our database system to serve as middleware that can access such data. One fundamental difference between small character strings and multimedia data is that multimedia data may have attributes that are inherently fuzzy (see L.A. Zadeh [28]). For example, we do not say that a given image is simply either "red" or "not red". Instead, there is a degree of redness, which ranges between 0 (not at all red) and 1 (totally red).

The most frequent source of fuzziness is our interpretation and/or representation of information and the user query. User query is typically vague and such attributes are interpreted fuzzy and we speak about user's ranking, rating, scoring, evaluation, preference, ... User is typically interested in the best (top-k) answers according to a global score combining particular attribute score. Human decision uses neither conjunctive nor disjunctive queries, it is rather like "the more required properties the object has, the better" [30].

There are many papers dealing with this scenario from different point of view.

Cost of web services/disc access. R. Fagin in [7] uses following example: the user wants to get information about restaurants. The user has a function that gives a score $g \in [0, 1]$ to each restaurant based on how good it is (the bigger g is the better the restaurant), $i \in [0, 1]$ how inexpensive it is, and c how close it is. In this example, the Zagat-Review web site gives ratings of restaurants, the NYT-Review web site gives prices, and the MapQuest web site gives distances. And there is a user function giving overall score to restaurants combining particular attribute scoring, a function monotone in all variables, e.g., a weighted average $(3g + 2i + c)/6$. As noted in [3], the Euclidean distance of (g, i, c) from the ideal point $(1, 1, 1)$ need not be a good overall score (same can hold for other measures used in Information Retrieval).

Fagin and others following this line of research (e.g., [10]) consider attribute scores are in different (distributed over the web) ordered lists. They consider two modes of access to data. The first mode of access is sorted (or sequential) access. Here the middleware system obtains the grade of an object in one of the sorted lists by proceeding through the list sequentially from the top. The second mode of access is direct (random) access. The task is to find algorithms (heuristics) for retrieving top-k object with highest overall score minimizing the price of sequential and direct access performed.

Similarly I.F. Ilyas et al. [13] consider a problem of finding k most similar video shots to a given image based on visual features as a ranked join optimization problem with respect to disc access. An interesting variant of multifeature join query optimization is discussed in [3], where computation of a candidate set (skyline) for best answer for arbitrary monotone combining function was studied, depending on the user preference ordering of attribute domain. A threshold version of ranked join optimization was also discussed in [20].

Declarative semantics of best answer. The above mentioned lines of research are concerned more with the *procedural* part of the problem and its efficiency. In this paper we would like to concentrate on the *declarative* and/or model theoretic part of the integration of particular score to a global one for extending querying capabilities of web solutions.

Rules become important. Quoting RuleML home page [25]:

> Rules in (and for) the Web have become a mainstream topic since inference rules were marked up for E-Commerce and were identified as a Design Issue of the Semantic Web, and since transformation rules were put to practice for document generation from a central XML repository. Moreover, rules have continued to play an important role in AI shells for knowledge-based systems and in Intelligent Agents, today both needing a Web interchange format, and such XML/RDF-standardized rules are now also usable for the declarative specification of Web Services.

Rules enhancing querying capabilities. The capabilities of ontology languages with respect to instances can be rather low [9,1], and even conjunctive queries – the least expressive query language usually considered in database research – are often not supported [2]. This area is a stronghold of rules, which offer extensive facilities for instance reasoning. In [6,9] consider combining DLs with the rule paradigm in order to state expressive instance queries with respect to terminological knowledge bases.

Concerning modeling scoring of user features and integrating them a many valued logic approach is suitable, especially that of generalized annotated programs of [14] and fuzzy definite logic programming [26]. Fuzzy description logic was extensively studied (see, e.g., [23]) nevertheless without fuzzy aggregation of concepts.

This paper. In this paper we are dealing only with the rule side, because we are convinced that rules and user query are main source of fuzziness (here we are not going to change the DL part of web modeling languages, although embedding of DL predicates similarly as in [6] is possible).

Main focus is in developing a rule system with fuzzy aggregation (annotation operator) and best answer (top-k) semantics.

In this paper we will develop a model of definite logic programming with fuzzy aggregation and/or annotation operators.

2. Motivating example, solution design, language

In what follows we give a motivating example, explain our approach which can serve as a design of a software solution. Single parts (often a public domain) were experimentally checked.

2.1. *Querying the web – web data standards and their translation*

Our motivating example is inspired by a hotel reservation system used in many papers, e.g., [18,3] and [20]. Context concerns winter resorts.

Query. Assume a user, say U1, is interested in a hotel that is cheap, new and is close to a downhill skiing slope.

Information on the web is usually in the RDF format. RDF – Resource Description Framework is a language that represents information not only about web resources but also things that can be identified on the web (e.g., items in online shopping and in our case about hotels), see the W3C web page [22].

According to [22] RDF statements are similar to a number of other formats for recording information, such as: entries in a simple record or catalog listing describing the resource in a data processing system, rows in a simple relational database, simple assertions in formal logic and information in these formats can be treated as RDF statements, allowing RDF to be used to integrate data from many sources.

Primary semantics of RDF is a labeled oriented graph (out-node a is called "subject", in-node b is "object" and the arrow is labeled by a property R).

$$a \xrightarrow{R} b$$

This graph structure is usually coded in a xml syntax, which simplified can look like

subject `<http://www.pv-FLSW-example.org/subject/a>`
predicate `<http://www.pv-FLSW-example.org/property/R>`
object `<http://www.pv-FLSW-example.org/object/b>`

For our many valued logic based methods it will be useful to think of RDF as a version of relational logic encoding the above information in a relational atom

$$R(a, b)$$

defining the extension of R as the set $\{\langle x, y\rangle : \langle x, y\rangle \in R\}$ and noting that $\langle a, b\rangle \in R$. This is precisely the denotation of R in the conventional Tarskian model theory and fits with relational database model (see [22]).

Query continued. Using an ontology (see, e.g., [24], we do not discuss details here) relating "cheap" to "price", "near" to "distance" and "new" to "year of construction (yoc)", results in RDF data like.

In our "good hotel for vacation" example it can look like (we use this as a running example)

subject `<http://www.pv-FLSW-example.org/hotelname/H1>`
predicate `<http://www.pv-FLSW-example.org/terms/price>`
object `<http://www.pv-FLSW-example.org/value/"2500">`

with following meaning (prefix abbreviated)

`exhotelname:H1` has a `exterms:price` which value is `exvalue: "2500"`

we can imagine a bigger amount of such data in triple notation

```
exhotelname:H1  exterms:price     exvalue: "2500".
exhotelname:H1  exterms:distance  exvalue: "200".
exhotelname:H1  exterms:yoc       exvalue: "1990".
exhotelname:H2  exterms:price     exvalue: "1000".
...             ...               ...
```

Corresponding relational tables can look like

Distance		Price		YOC	
Hotel	distance_value	Hotel	price_value	Hotel	yoc_value
H1	200	H1	2500	H1	1990
H2	300	H2	1000	H2	1999
H3	400	H3	500	H3	1989
H4	500	H4	1600	H4	2003
H5	1000	H5	750	H5	2000

Step 1. *Users scoring of attributes.* To model user's notion of close, cheap and new we use fuzzy sets, e.g.,

$$\text{fclose\_U1}(x) = \max\left(0, 1 - \frac{x}{1000}\right), \qquad \text{fcheap\_U1}(x) = \max\left(0, 1 - \frac{x}{2000}\right)$$

and

$$\text{fnew\_U1}(x) = \max\left(0, 1 - \frac{2005 - x}{20}\right)$$

giving us unary fuzzy predicates over resources (here in crisp binary relational encoding)

U1_Close		U1_Cheap		U1_New	
Hotel	close_score	Hotel	cheap_score	Hotel	new_score
H1	0.8	H1	0	H1	0.25
H2	0.7	H2	0.5	H2	0.7
H3	0.6	H3	0.75	H3	0.2
H4	0.5	H4	0.2	H4	0.9
H5	0	H5	0.625	H5	0.75

Step 2. *Combining attribute score to an overall score.* Using a fuzzy aggregation for degree of being close, cheap and new

$$@_{\text{U1}}(close, cheap, new) = \frac{2 * close + 3 * cheap + new}{6}$$

gives the fuzzy concept "a good hotel for vacation for user U1"

$$\{(H1; 0.31), (H2; 0.6), (H3; 0.61), (H4; 0.42), (H5; 0.44)\},$$

which reflect aggregation of all user's preferences, best is H3.

In what follows, we describe querying rule based systems enabling such computation for more complex (embedded) queries and rules.

3. Fuzzy predicate definite logic programming with fuzzy aggregation FDLP$_a$

We follow and extend our development of fuzzy definite logic programming as described in [26] and [15].

3.1. *Language*

Our language is a typed first-order predicate logic language $\mathcal{L}_{\mathrm{FDLP}_a}$. $\mathcal{L}_{\mathrm{FDLP}_a}$ contains a set of attributes $\mathcal{A}$, these coincide with relational attributes. For each attribute A there are variables $\mathcal{V}^A$ of type A, $\mathcal{V}^A = \{x^A, y^A, x_1^A, \ldots\}$ and a type function $type(x^A) = A$.

Function symbols $\mathcal{F} = \{\mathcal{F}^{\overline{A}}: \overline{A} \in {}^{n+1}\mathcal{A}, n > 0\}$, for an ordered list $\overline{A} = [A_0, \ldots, A_{n-1}, A_n]$ (with possible repetitions), *in-type*$(f) = [A_0, \ldots, A_{n-1}]$ and *out-type*$(f) = A_n$. Constants $\mathcal{C} = \{\mathcal{C}^A, A \in \mathcal{A}\}$ of all types. Terms are built from variables and constants inductively using function symbols and preserving all in/out types and for $t = f(t_1, \ldots, t_k)$ the *type*(t) = *out-type*(f) (we call them sometimes domain terms to distinguish from annotation terms).

Predicate symbols $\mathcal{P} = \{\mathcal{P}^{\overline{A}}: \overline{A} \in {}^n\mathcal{A}, n > 0\}$ for an ordered list of attributes $\overline{A} = [A_0, \ldots, A_{n-1}]$ (with possible repetitions). For $p \in \mathcal{P}^{\overline{A}}$ this corresponds to relational schema $p(A_0, \ldots, A_{n-1})$ in domain relational calculus.

The basic truth value set $\mathcal{T}$ is the unit interval of real numbers $[0, 1]$ with its natural ordering, although most of our results extend to arbitrary lattice or sometimes even partially ordered sets. Very often we work with some finite (recursively encoded) subset of $\mathcal{T}$ like rational numbers in the real unit interval $\mathcal{T}_Q = [0, 1] \cap Q$ or $\mathcal{T}_k = \{0, \frac{1}{k}, \ldots, \frac{k-1}{k}, 1\}$ for some natural number k (as we will see later, it is quite natural to work simultaneously with different $\mathcal{T}_{k_1}, \ldots, \mathcal{T}_{k_j}$). So in future we will use $\mathcal{T}$ or $[0, 1]$ as a variable denoting one of these truth value sets.

We assume finite sets of connectives: binary – conjunction symbols $\mathbb{C}$, disjunction symbols $\mathbb{D}$, implication symbols $\mathbb{I}$ and connectives of arbitrary arity – aggregation symbols $\mathbb{A}$.

We do not use negation here, because we are convinced that in web applications negation propagates downward to fuzzy predicates.

Formulas are built inductively from atoms using connectives and quantifiers. The main syntactic object of our language is a formula graded by a truth value constant

$$\varphi.c.$$

In practical application the specific truth value is not important, it is rather important the comparison of truth values for different assertions.

3.2. *Truth functions*

For a connective $\heartsuit$ the corresponding truth value function is denoted by $\heartsuit^\bullet$. We recognize several sorts of truth functions.

A function of two real variables $C : [0, 1]^2 \to [0, 1]$ (with mnemonic notation of variables $C(b, r)$, b for body and r for rule) is said to be a *conjunctor* if the following hold: C extends the two-valued conjunction (i.e. $C(0, 0) = C(0, 1) = C(1, 0) = 0$ and $C(1, 1) = 1$); C preserves ordering in both variables (i.e. $b_1 \leqslant b_2$ and $r_1 \leqslant r_2$ implies $C(b_1, r_1) \leqslant C(b_2, r_2)$): C is left continuous in the r variable (in sense of continuity of real functions) and $(\forall r > 0)(C(1, r) > 0)$.

A function of two real variables $I : [0, 1]^2 \to [0, 1]$ (with mnemonic notation of variables $I(b, h)$, b for body and h for head) is said to be an *implicator* if the following hold: I extends the two valued implication (i.e. $I(0, 0) = I(0, 1) = I(1, 1) = 1$

and $I(1,0) = 0$); I inverts ordering in the body variable and preserves ordering in the head variable (i.e. $b_1 \leqslant b_2$ implies $I(b_1,h) \geqslant I(b_2,h)$ and $h_1 \leqslant h_2$ implies $I(b,h_1) \leqslant I(b,h_2)$); I is right continuous in the h variable (in the sense of continuity of real functions) and $(\forall h < 1)(I(1,h) < 1)$.

A function of finitely many real variables $g : [0,1]^k \to [0,1]$ is a *fuzzy aggregation operator* if $g(0,\dots,0) = 0$, $g(1,\dots,1) = 1$; g is order preserving in all variables and left continuous in all variables.

We assume for every $\& \in \mathbb{C}$ the $\&^\bullet$ is a conjunctor, for every $\to \in \mathbb{I}$ the $\to^\bullet$ is an implicator, for every $\vee \in \mathbb{D}$ the $1 - \vee^\bullet(1-x, 1-y)$ is a conjunctor and for every $@ \in \mathbb{A}$ the $@^\bullet$ is a fuzzy aggregation operator.

Recall some classical connectives. *The Łukasiewicz logic connectives* $\&^\bullet_L(x,y) = \max(0, x+y-1)$, $\to^\bullet_L(x,y) = \min(1, 1-x+y)$, $\vee^\bullet_L(x,y) = \min(1, x+y)$. *The Gödel logic connectives* $\&^\bullet_G(x,y) = \min(x,y)$, $\to^\bullet_G(x,y) = y$ if $x > y$ else 1, $\vee^\bullet_G(x,y) = \max(x,y)$. *The product logic connectives* $\&^\bullet_P(x,y) = x.y$, $\to^\bullet_P(x,y) = \min(1, \frac{y}{x})$, $\vee^\bullet_P(x,y) = x+y-xy$.

Note that all conjunctors and disjunctors are also fuzzy aggregation operators.

DEFINITION 1. Let C and I be as above. We say C and I are residuated if for all b, h, $r \in [0,1]$

$$r \leqslant I(b,h) \quad \text{iff} \quad C(b,r) \leqslant h.$$

Note that our approach allows also noncommutative connectives. One such example from [23] is the material implication generated by the max–min connectives and $1-x$ negation and is used in fuzzy description logic FDL $\to^\bullet_{\mathrm{FDL}}(b,h) = \max(1-b, h)$ the corresponding residual conjunctor is $\wedge^\bullet_{\mathrm{FDL}}(b,r) = 0$ if $b + r \leqslant 1$ else r the disjunctor $\vee_{\mathrm{FDL}}$ is coupled with conjunctor by deMorgan law.

3.3. *Structures*

A structure $\mathfrak{M}$ of our language consists of attribute domains $D^A_{\mathfrak{M}}$. We consider evaluations of variables $e = \{e^A\colon A \in \mathcal{A}\}$ of all types, $e^A : \mathcal{V}^A \to D^A_{\mathfrak{M}}$.

Constants $\mathcal{C}_{\mathfrak{M}} = \{\mathcal{C}^A_{\mathfrak{M}}, A \in \mathcal{A}\}$ are elements of corresponding domains, i.e. $\mathcal{C}^A_{\mathfrak{M}} \subseteq D^A_{\mathfrak{M}}$.

Interpretations of function symbols $\mathcal{F}_{\mathfrak{M}} = \{\mathcal{F}^{\overline{A}}_{\mathfrak{M}}\}$ where for $\overline{A} = \{A_0, \dots, A_n\} \subseteq \mathcal{A}$ and $f \in \mathcal{F}^{\overline{A}}$ there is a function

$$f_{\mathfrak{M}} : \prod_{i=1}^{n-1} D^{A_i}_{\mathfrak{M}} \to D^{A_n}_{\mathfrak{M}}$$

inducing term evaluation $t[e]$.

Interpretation of predicate symbols $\mathcal{P}_{\mathfrak{M}} = \{\mathcal{P}^{\overline{A}}_{\mathfrak{M}}\}$, for $\overline{A} = \{A_0, \dots, A_{n-1}\} \subseteq \mathcal{A}$ and $p \in \mathcal{P}^{\overline{A}}$ there is a fuzzy predicate

$$p_{\mathfrak{M}} : \prod_{i=1}^{n-1} D^{A_i}_{\mathfrak{M}} \to [0,1]$$

typically for an evaluation e of variables with $p_{\mathfrak{M}}[e] \in [0, 1]$.

The truth value of a formula is calculated along the complexity of formula using truth value functions of connectives, infimum for universal quantifier. So a structure $\mathfrak{M}$ assigns to each formula φ and evaluation e the truth value $\mathfrak{M}^\bullet(\varphi)[e]$.

$\mathfrak{M}$ is a model of $\varphi.c$ if for all e we have

$$\mathfrak{M}^\bullet(\varphi)[e] \geqslant c.$$

Our logic is a fuzzy logic in a narrow sense according to P. Hájek [11].

3.4. *Fuzzy definite logic programs and declarative semantics*

Our fuzzy definite logic program (see [15]) consists of implicative rules (universally quantified) with positive atoms of the form

$$R = H \longleftarrow @(B_1, \ldots, B_m).r \tag{1}$$

where H is the head positive atom and B_i are positive body atoms and r is the truth value of the rule. Structure $\mathfrak{M}$ is a model of the rule (1) if for all evaluations of variables e

$$\mathfrak{M}^\bullet(R)[e] = \rightarrow^\bullet\big(@^\bullet\big((B_1)_{\mathfrak{M}}[e], \ldots, (B_m)_{\mathfrak{M}}[e]\big); H_{\mathfrak{M}}[e]\big) \geqslant r.$$

A fact is an atomary formula with assigned weight

$$H.c. \tag{2}$$

Structure $\mathfrak{M}$ is a model of the fact (2) if for all evaluations e

$$\mathfrak{M}^\bullet(H)[e] = H_{\mathfrak{M}}[e] \geqslant c.$$

A set of rules and facts is a fuzzy definite logic program P. $\mathfrak{M}$ is a model of P if it is a model of all rules and facts in P.

DEFINITION 2. Assume P is a fuzzy definite logic program, Q is an atom, θ is a substitution and x is a real number. We say that θ, x is a *correct* answer to query $? - Q$ with respect to program P if for all models $\mathfrak{M}$ of the program P we have

$$\mathfrak{M}^\bullet(Q\theta) \geqslant x.$$

Fuzzy Herbrand structures $\mathfrak{H}$ are based on the (crisp) Herbrand universe – a unique (syntactic) interpretation of domains by constant terms: each domain $D^A_{\mathfrak{H}}$ is the set of all constant terms with *type*$(t) = A$, constants $c^A_{\mathfrak{H}} = c^A$ and functions $f_{\mathfrak{H}}(t_1, \ldots, t_n) = f(t_1, \ldots, t_n)$ (the last is correct, because if t_is are constant terms of respective *in-type*(f) then $f(t_1, \ldots, t_n)$ is a constant term of type *out-type*(f)).

(Crisp) Herbrand base is the set $\mathcal{B}$ of all ground (constant) atoms. We represent fuzzy Herbrand structures by

$$\mathfrak{H} : \mathcal{B} \rightarrow [0, 1]$$

which corresponds to fuzzy interpretation of a predicate symbol $p_{\mathfrak{H}}(t_1, \ldots, t_n) = \mathfrak{H}^\bullet(p(t_1, \ldots, t_n))$.

Similarly as in classical case Herbrand structures suffice for modeling semantics of definite logic programs

THEOREM 1. *Assume P is definite fuzzy logic program and $\mathfrak{M} \models P$. Then there is a fuzzy Herbrand structure $\mathfrak{H}_{\mathfrak{M}}$ which is a model of P. Moreover, θ, x is a correct answer to $?-Q$ with respect to P iff for all Herbrand models $\mathfrak{H}$ of P we have $\mathfrak{H}^\bullet(Q\theta) \geqslant x$.*

PROOF. Let $\mathfrak{M}$ be a model of P. For a constant term $t = t_{\mathfrak{H}}$ which is equal to its fuzzy Herbrand interpretation, there is $t_{\mathfrak{M}}$ the corresponding interpretation in $\mathfrak{M}$ (note that two different constant terms can be interpreted as a same element in a domain of $\mathfrak{M}$).

Define the fuzzy Herbrand interpretation of ground atoms

$$\mathfrak{H}_{\mathfrak{M}}\big(p(t_1, \ldots, t_k)\big) = p_{\mathfrak{M}}\big((t_1)_{\mathfrak{M}}, \ldots, (t_k)_{\mathfrak{M}}\big).$$

Observe that evaluation of variables in $\mathfrak{H}$ corresponds to ground substitutions. So instead of calculating truth value in $\mathfrak{H}_{\mathfrak{M}}$ according to all evaluations of variables it is enough to calculate truth value of a ground instance of a rule or fact. Take a ground instance $R' = H \leftarrow @(B_1', \ldots, B_n').r$ of a program rule (1) with constant terms $t_1, \ldots, t_k$ substituted instead of variables $x_1, \ldots, x_k$. Then the truth value

$$\begin{aligned}
&\mathfrak{H}^\bullet_{\mathfrak{M}}\big(H \leftarrow @(B_1, \ldots, B_n)\big)[t_1, \ldots, t_k] \\
&\quad = \rightarrow^\bullet\big(@^\bullet\big(\mathfrak{H}^\bullet_{\mathfrak{M}}\big(B_1(t_1, \ldots, t_k)\big), \ldots, \mathfrak{H}^\bullet_{\mathfrak{M}}\big(B_m(t_1, \ldots)\big)\big); \mathfrak{H}^\bullet_{\mathfrak{M}}\big(H(t_1, \ldots)\big)\big) \\
&\quad = \rightarrow^\bullet\big(@^\bullet\big((B_1)_{\mathfrak{M}}\big((t_1)_{\mathfrak{M}}, \ldots, (t_k)_{\mathfrak{M}}\big), \ldots, (B_m)_{\mathfrak{M}}\big((t_1)_{\mathfrak{M}}, \ldots\big)\big); \\
&\qquad\qquad H_{\mathfrak{M}}\big((t_1)_{\mathfrak{M}}, \ldots\big)\big)
\end{aligned}$$

which corresponds to $\mathfrak{M}$ evaluation $e(x_i) = (t_i)_{\mathfrak{M}}$ and because $\mathfrak{M}$ is a model of P, this is greater than or equal r. We see that $\mathfrak{H}_{\mathfrak{M}}$ assigns ground rules and atoms same value as $\mathfrak{M}$ – the rest of theorem follows. □

3.5. *Procedural semantics of FDLP$_a$*

Procedural semantics of fuzzy logic programming we use here was described in [15] and is based on the many valued modus ponens and residual conjunctors C_i to rule implicators $\rightarrow^\bullet_i$

$$\frac{(B.b), (H \longleftarrow_i B.r)}{(H.C_i(b, r))}.$$

We know, that this is a sound rule, provided $\longleftarrow^\bullet_i$ and C_i are residuated [11]. Procedural semantics is based on the "backward usage of modus ponens". A computation looks like

$$\begin{gathered}
?Q \\
C_\rightarrow\big(@^\bullet(B_1, \ldots, B_m), r\big)\theta_1 \\
\ldots \\
C_\rightarrow\big(@^\bullet(b_1, \ldots, b_m), r\big)\theta_l
\end{gathered}$$

Namely, a query Q is (after a successful unification θ_1 with the head H of a rule, e.g., (1)) replaced by a mixed expression $C_i(@^\bullet(B_1, \ldots, B_m), r)$ – the initial segment of the annotation term calculating the truth value of the answer $C_\rightarrow(@^\bullet(\_), r)$ with embedded atoms from the body $B_1, \ldots, B_m$ not computed thus far. After all truth values $b_1, \ldots, b_m$ of all atoms $B_1, \ldots, B_m$ are calculated from rules and facts, we evaluate the whole expression. Note that the aggregation operator @ can be a disjunction, or at least not vanishing if some inputs are zero. For this case in all computational steps we can replace an atom by 0 (this acts also as a cut for blind computational branches).

DEFINITION 3. Assume P is a fuzzy definite logic program, Q is an atom, ϑ is a substitution and y is a real number. We say that ϑ, y is a *computed* answer to query $? - Q$ with respect to program P if there is a computation $Q_0, \ldots, Q_l$ such that $Q_0 = Q$, where the step Q_i, Q_{i+1} is either backward usage of modus ponens with residuated pair C, I for some program rule for a selected atom, or usage of a fact, or arbitrary insertion of zeros, Q_l does not contain any predicate symbols (hence it is a composition of conjunctors and aggregations with truth value constants), ϑ is the composition of all substitutions from computational steps and y is equal to the evaluation of $Q_l^\bullet$.

Note that $C_{\rightarrow}(@^\bullet(x_1, \ldots, x_n), r) = @^\bullet_{C,r}(x_1, \ldots, x_n)$ is again an aggregation operator (with hidden parameters C and r). Moreover $@^\bullet_{C,r}$ is again left continuous in all variables, because $C_{\rightarrow}$ was left continuous in the body variable.

Hence the rule (1) is computationally and declaratively equivalent to the rule (3)

$$H \longleftarrow_L @_{C,r}(B_1, \ldots, B_n).1 \tag{3}$$

where $\rightarrow_L$ is the Łukasiewicz implication with residual conjunctor $\&^\bullet_L = C_L$ fulfilling for all $b \in [0, 1]$ the condition $C_L(b, 1) = b$. Hence, from now on, we can assume all rules have truth value 1 (and we need not write it).

We can assume in our language there are all compositions of aggregation operators needed for computation and $@^\bullet_{C,r}$ as above.

From now on we assume our fuzzy definite logic program has rules of the form (3) with truth value 1, the only source of fuzziness are facts (and in practical applications we will see, that only fuzzy facts will come from user). Nevertheless aggregation operators can give fuzzy answers also from crisp data.

Now a computation looks like

$$\begin{gathered} ?Q \\ @^\bullet(B_1, \ldots, B_m)\theta_1 \\ \ldots \\ @^\bullet(b_1, \ldots, b_m)\theta_l \end{gathered}$$

which makes it easier to handle and optimize.

4. Pavelka completeness of fuzzy definite logic programs by fixpoint semantics

J. Pavelka in [19] showed that arbitrary fuzzy theory over his logic (which is in fact Łukasiewicz propositional logic with truth constants) the infimum of truth degree of a formula φ in all models of the theory equals supremum of degrees of graded proofs of φ in that theory (with a special set of logical axioms).

Pavelka has also shown that this does not hold in full generality for fuzzy theories over other logical systems generated by t-norms. (A presentation of Pavelka's results together with a simplification of his system can be found in [11], which is our standard citation for fuzzy logic.) The reason is that all residual implicators of t-norms except of Lukasiewicz are discontinuous at (0, 0). Nevertheless, formulas showing this are more complex and contain embedded implications. So there is a possibility to have Pavelka completeness for simpler theories.

We will show that fuzzy definite logic programming can have Pavelka completeness property under quite general conditions. We will develop the fixpoint semantics. Strategy is to follow classical logic programming development (see, e.g., [16]), take a complete lattice of Herbrand interpretations, define a continuous production (Datalog) operator, show that models correspond to prefixpoints of this operator, the minimal fixpoint is the minimal Herbrand model and completeness follows.

In fuzzy case we have a complete lattice of all fuzzy Herbrand interpretations. Problem will be with continuity. Correspondence of models and prefixpoints is guaranteed by residuation.

DEFINITION 4. Assume P is a fuzzy definite logic program and $\mathcal{F} = \mathcal{B}^{[0,1]}$ be the complete lattice of all fuzzy Herbrand interpretations (ordered coordinatewise). Then for $\mathfrak{H} \in \mathcal{F}$ and an ground atom H

$$\begin{aligned} & T_P(\mathfrak{H})(H) \\ & \quad = \max\bigl\{\sup\bigl\{\mathfrak{H}^\bullet\bigl(@(B_1,\ldots,B_m)\bigr)\colon \bigl(H \leftarrow_L @(B_1,\ldots,B_m)\bigr) \\ & \qquad\qquad \text{is a ground instance of a rule in the program } P\bigr\}, \\ & \qquad\qquad \sup\bigl\{c\colon (H.c) \text{ is a ground instance of a fact in the program } P\bigr\}\bigr\}. \end{aligned}$$

Continuity of the T_P operator means, it preserves all suprema of upwards directed sets in $\mathcal{F}$ (see [16]).

THEOREM 2. *Assume the truth value set is* $[0,1]$ *or its subset. The* T_P *operator is continuous for all* FDLP_a *programs* P *(in the lattice of Herbrand interpretations) if and only if all truth functions of fuzzy aggregation in body are left continuous (in the sense of functions of real numbers) in all variables.*

PROOF. Let $\mathcal{X} \subseteq \mathcal{F}$ be an upward directed set of fuzzy Herbrand interpretations. We need to check $T_P(\bigvee \mathcal{X}) \leqslant \bigvee\{T_P(\mathfrak{M})\colon \mathfrak{M} \in \mathcal{X}\}$ for all constant atoms H. Denote $\bigvee \mathcal{X} = \mathfrak{N}$. It is sufficient to show that for each $\varepsilon > 0$ there is an $\mathfrak{M} \in \mathcal{X}$ such that

$$T_P(\mathfrak{N})(H) - \varepsilon < T_P(\mathfrak{M})(H)$$

for this take a rule $H \longleftarrow @(B_1,\ldots,B_m)$ such that

$$@^\bullet\bigl(\mathfrak{N}(B_1),\ldots\bigr) > T_P(\mathfrak{N})(H) - \varepsilon.$$

From continuity of $@^\bullet$ there are $\delta_1,\ldots,\delta_m$ such that whenever $a_i > \mathfrak{N}(B_i) - \delta_i$ then $@^\bullet(a_1,\ldots,a_m) > T_P(\mathfrak{N})(H) - \varepsilon$. Now, because $\mathfrak{N}$ is supremum of all interpretations in $\mathcal{X}$ there are $\mathfrak{M}_i \in \mathcal{X}$ such that $\mathfrak{M}_i(B_i) > \mathfrak{N}(B_i) - \delta_i$. Now by upward directedness there is an $\mathfrak{M} \in \mathcal{X}$ dominating all $\mathfrak{M}_i$'s. □

To show opposite direction (necessity), assume $@^\bullet$ is left discontinuous at point $a_1,\ldots,a_m$ and $a_1^i,\ldots,a_m^i$, for $i \to \infty$ converges increasingly to $a_1,\ldots,a_m$ and

$$\lim_{i\to\infty} @^\bullet\bigl(a_1^i,\ldots,a_m^i\bigr) < @^\bullet(a_1,\ldots,a_m).$$

Put $\mathfrak{M}_i(B_j) = a^i_j$, then $\mathcal{X} = \{\mathfrak{M}_i\colon i \in \omega\}$ is upward directed because the sequence of a^i_j's is increasing and consider a fuzzy logic program P consisting of only one rule $H \longleftarrow_L @(B_1, \ldots, B_m)$ for some ground atoms $H, B_1, \ldots, B_m$. Then

$$\bigvee_{i\in\omega}\{T_P(\mathfrak{M}_i)(H)\} = \lim_{i\to\infty} @^\bullet(a^i_1, \ldots, a^i_m) < @^\bullet(a_1, \ldots, a_m) = T_P\Big(\bigvee \mathcal{X}\Big)(H).$$

In the case of truth value set being unit interval of real numbers (or its subset) we can prove even a stronger result: Assuming $@^\bullet$ is left discontinuous, then there is a program P such that $T^\omega_P(0)$ is not a minimal model of P.

Next step we need is a characterization of fuzzy models

THEOREM 3. *Assume P is a definite fuzzy logic program. A Herbrand structure $\mathfrak{H}$ is a model of P iff $T_P(\mathfrak{H}) \leqslant \mathfrak{H}$ (hence the minimal fixpoint of T_P is a minimal Herbrand model of P).*

PROOF. All our rules use same implication $\rightarrow^\bullet_L = I$ with the residual conjunctor C_L fulfilling $C_L(b, 1) = b$ for all b's. For a rule $R = H \longleftarrow @(B_1, \ldots, B_m)$ denote $R_B = @(B_1, \ldots, B_m)$. Now $\mathfrak{H}$ is a model of R if $I(\mathfrak{H}(R_B), \mathfrak{H}(H)) = 1$ and this is by residuation equivalent to

$$C_L\big(\mathfrak{H}(R_B), 1\big) = \mathfrak{H}(R_B) \leqslant \mathfrak{H}(H).$$

Now, $\mathfrak{H}(R_B)$ are all contribution to T_P value of H and theorem follows. □

THEOREM 4. *Assume our language contains only left continuous annotations, P is a fuzzy definite logic program, θ is a substitution and $x \in [0, 1]$. Then*

(soundness) *if θ, x is a computed answer to $?- Q$ with respect to P, then θ, x is a correct answer*

(approximate completeness) *if θ, x is a correct answer to $? - Q$ with respect to P, then for every $\varepsilon > 0$ there is a computed answer (y, ϑ) to $? - Q$ with respect to P such that $x - \varepsilon < y$ and $\vartheta = \theta\gamma$ (for some γ).*

PROOF. Soundness is based on soundness of fuzzy modus ponens.

Completeness follows from characterization of minimal fuzzy Herbrand model of P. As $T^\omega_P(0)(Q\vartheta) \geqslant x$, there is an n, such that $T^n_P(0)(Q\vartheta) \geqslant x - \varepsilon$ and the rest of the proof proceeds as in the classical case (see, e.g., [16]). □

5. A variant of Generalized Annotated Programs GAP^+_v

In [14] M. Kifer and V.S. Subrahmanian developed a theory of Generalized annotated programs GAP which has discontinuous restricted semantics with truth value set [0, 1]. In [15] we have developed a variant of GAP without constant annotations in body of rules and substitutional semantics with continuity property for [0, 1]. Here we give a brief description of an equivalent variant GAP^+_v, for details we refer to [15].

The language of GAP^+_v is the same as that of FDLP_a with some additional features and a little bit different terminology. We assume there are truth variables $\mathcal{V}^\mathcal{T}$ and a two-valued conjunction AND and a two-valued implication IF–THEN and the annotation symbol ":". GAP_v^+ rules are expressions of the form

$$H : @(x_1, \ldots, x_m) \text{ IF } B_1 : x_1 \text{ AND } \ldots \text{ AND } B_m : x_m \tag{4}$$

where the fuzzy aggregation @ is playing here the role of an annotation term and $x_1, \ldots, x_m \subseteq \mathcal{V}^{\mathcal{T}}$ and facts are

$$H : c$$

where $c \in [0, 1]$.

Structures for GAP_v^+ are the same as for FDLP_a.

What is different is the satisfaction, namely for a fuzzy Herbrand structure $\mathfrak{H}$ and a GAP_v^+ rule (4) we define

$$\mathfrak{H} \models_{\mathrm{GAP}_v^+} H : @(x_1, \ldots, x_m) \text{ IF } B_1 : x_1 \text{ AND } \ldots \text{ AND } B_m : x_m \tag{5}$$

if for every truth value assignment $e(x_i) \in [0, 1]$ the following (two-valued implication) is true:

$$\begin{aligned} &\text{IF } \mathfrak{H}(B_1) \geqslant e(x_1) \text{ AND } \ldots \text{ AND } \mathfrak{H}(B_m) \geqslant e(x_m) \\ &\qquad \text{THEN } \mathfrak{H}(H) \geqslant @^\bullet\big(e(x_1), \ldots, e(x_m)\big). \end{aligned}$$

For a fuzzy Herbrand structure $\mathfrak{H}$ we have

THEOREM 5. *Under previous notations and assumptions*

$$\mathfrak{H} \models_{\mathrm{FLP}_a} H \to_L @(B_1, \ldots, B_m)$$

iff

$$\mathfrak{H} \models_{\mathrm{GAP}_v^+} H : @(x_1, \ldots, x_m) \textit{ IF } B_1 : x_1 \textit{ AND } \ldots \textit{ AND } B_m : x_m.$$

PROOF. Put $a_i = \mathfrak{H}(B_i)$, then FDLP_a satisfaction says

$$\to^\bullet\big(@^\bullet(a_1, \ldots, a_m), \mathfrak{H}(H)\big) = 1$$

which is, as showed previously, by residuation equivalent to

$$@^\bullet(a_1, \ldots, a_m) \leqslant \mathfrak{H}(H).$$

The satisfaction of the body of GAP_v^+ two valued implication is true if all $e(x_i) \leqslant a_i$, hence by monotonicity of $@^\bullet$ we have

$$@^\bullet\big(e(x_1), \ldots, e(x_m)\big) \leqslant @^\bullet(a_1, \ldots, a_m) \leqslant \mathfrak{H}(H).$$

□

This shows an equivalence of fuzzy definite logic programming FDLP_a and GAP_v^+ programs and from now on we can develop both approaches simultaneously.

The advantage of substitutional procedural semantics is that computations in FDLP_a and in GAP_v^+ are literally the same, namely for the same query and rule translation

$$\begin{gathered} ?Q \\ @^\bullet(B_1, \ldots, B_m)\theta_1 \\ \ldots \\ @^\bullet(b_1, \ldots, b_m)\theta_l \end{gathered}$$

is identical as in FDLP_a.

6. Best/top-k answer semantics

In the semantic web it is very useful to have only most relevant, best, top-k answers to our query. We would like to formalize both declarative and procedural parts of semantics.

There are several problems. First the notion of correct answer is rather weak, e.g., 0 is a correct answer to every ground query. FDLP_a fixpoint showed there is a sort of best answer to a query Q, namely, the $T_P^\omega(0)(Q)$. Nevertheless this is not always computable (we have only approximate completeness), because a GAP_v^+ program (see [14])

$$p : \frac{1+2x}{4} \text{ IF } p : x$$
$$p : 0$$

has $T_P^\omega(0)(p) = \frac{1}{2}$ but no computation gives computed answer $\frac{1}{2}$.

So one solution is to work with programs without recursion and the second possibility is to study a notion of approximate best (top-k) answers. Moreover we are free to choose the additive or multiplicative approximation, we believe the additive version is more reasonable for practical applications.

We consider here all declarative versions, nevertheless the procedural version is elaborated only for nonrecursive programs.

DEFINITION 5. Assume P is a fuzzy definite logic program, Q is an atom of our language, θ_i are substitution to all free variables of Q (that is $Q\theta_i$ are ground atoms from the Herbrand base) and $a_i \in [0,1]$ are real numbers, $\varepsilon > 0$.

We say that the ordered sequence $[(\theta_1, a_1), \ldots, (\theta_k, a_k)]$ is the correct additive approximate ε-top-k answer to query Q with respect to the program P if

(a) $a_1 \geqslant a_2 \cdots \geqslant a_k$ and

(b) every (θ_i, a_i) is a correct answer to $?-Q$ with respect to P and

(c) for every substitution ϑ different from all θ_i there is a model $\mathfrak{H}$

$$\mathfrak{H}(Q\theta) \leqslant a_k + \varepsilon. \tag{c1}$$

The answer is multiplicative approximate ε-top-k answer if in (c1) there is

$$\mathfrak{H}(Q\theta) \leqslant (1+\varepsilon)a_k.$$

The answer is top-k answer if (c1) reads

$$\mathfrak{H}(Q\theta) \leqslant a_k.$$

Typically, if Q is the hotel_for_vacation predicate, then substitutions are hotel names, and top-k list, is the ordered list with top-k best hotels.

6.1. *Procedural semantics of best answer*

For fuzzy definite logic programs without recursion to compute best (top-k) we can always follow the backward modus ponens semantics, or the Datalog semantics of the T_P operator or some combination of both.

In a web environment the computational tree of a query is not very deep and does not branch very much, so there is a possibility of gluing together all aggregations along the computation and have the computed answer in the form

$$@^\bullet(B_1, \ldots, B_m)$$

which for different θ depends on truth value of facts $B_i\theta, b_i$.

Typically, most of B_is are unary ground fuzzy predicates like $cheap(H_2) = 0.5$, coming from RDF data as in our example.

In this limited case we can use heuristics like those in [7,8] and [10] on m separate lists of all ground facts of B_i ordered (indexed) by truth values.

7. Some additional aspects of our fuzzy rule system

One of advantages of our approach is we can learn the user aggregation function from her/his classification [27,12], in order it fits her/his requirements.

7.1. *Learning the aggregation function from user evaluation*

For example, user U2 does not formulate her/his particular requirements for vacation, rather, he evaluates a sample of hotels given to him, assigning them degrees from a scale of seven (so-called Likert scale 1-7, corresponds to linguistic expressions 7 = best, 6 = very good, 5 = good, 4 = neutral, 3 = weak reject, 2 = reject, 1 = strong reject), e.g., in our example $H2 = \frac{5}{6}$ = very good, $H4 = \frac{4}{6}$ = good, $H1 = H3 = \frac{1}{2}$ = neutral, $H5 = \frac{2}{6}$ = weak reject.

We assume a set of hotel $\mathcal{H}$ with m database attributes. Assume a user classifies (a representative sample of) hotels by a fuzzy classification $\mathcal{C}:\mathcal{H} \to [0,1]$. By a qualitative learning method (e.g., [21,4]) we can learn the order of her/his preferences according to ordering of particular attribute domains. Result is a vector of values "increasing", "decreasing", "first increasing then decreasing", ... which corresponds to fuzzy values "small", "medium" and "large", meaning what values user prefers the most. This corresponds to ordering induced by fixed set of fuzzy functions for notions "small", "medium" and "large" in all domains (medium can be parameterized) and induces a translation of the set of all hotels to a fuzzy subset $\mathcal{FH} \subseteq [0,1]^m$ and correspondingly we can translate the user classification $\mathcal{C}$ to

$$\mathcal{FC}:\mathcal{FH} \to [0,1].$$

Using methods of fuzzy inductive logic programming developed in [27] we can learn an aggregation function which preserves users preferences (ordering of sample hotels).

Namely, let us cut fuzzy classification $\mathcal{FC}$ to α-cuts, forming positive and negative examples $\mathcal{C}_\alpha = E_\alpha^+, E_\alpha^-$. From this we can learn by an classical ILP method (see [5]) rules guaranteeing the truth value at least α.

By this we get a part of the graph of the aggregation function with values bigger than or equal to α. Gluing these parts together (for details see [27]) we get the aggregation function following user's preferences.

There is one problem, namely, learning in classical ILP does not "know" our semantics is order preserving, a statement is true in degree α if in all structures it has truth value at least α.

On the side of example set $\mathcal{FC}$ we took upper cuts, for background knowledge B we have to add monotonicity axioms for each body predicate, e.g., $close(h, x) \leftarrow close(h, y)$ for all $x < y$.

Then learning takes into account "better" examples and does not violate ordering of classification.

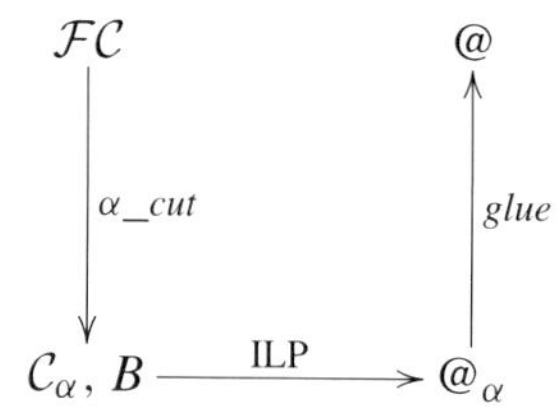

Note, adding axiom for each $x < y$, we have to work with some finite part of the truth value set [0, 1]. Also for translating our fuzzy ILP to multiple use of classical ILP by α cuts, we can run this only finitely many times. So our learning procedures forces us to work with some finite approximation, say with $\{0, \frac{1}{k}, \ldots, \frac{k-1}{k}, 1\}$ (our FDLP$_a$ system works also under this assumption).

7.2. *Discretization – finite domains*

To have decidable procedures it is sometimes necessary to work with finite domains (it can reduce the computational complexity too). Moreover we have seen that for induction we need a finite truth value set. Next feature of our system is that it works with finite truth value set via rounding and preserves order of computation up to some additive ε approximation.

It suffices to make following transformation. For a real number $x \in [0, 1]$ and a natural number n, denote

$$r_n(x) = \frac{\lceil n * x \rceil}{n}.$$

This corresponds to granulated perception Lotfi is speaking often about [29].

For every function $@(x_1, \ldots, x_n)$ and arbitrary natural numbers (rounding of inputs) $k_1, \ldots, k_m$ and j rounding of output, put

$$@^j_{k_1,\ldots,k_m}\big((x_1, \ldots, x_m)\big) = r_j\big(@\big(r_{k_1}(x_1), \ldots, r_{k_m}(x_m)\big)\big).$$

Note that r_n is a left continuous function and hence $@^j_{k_1,\ldots,k_m}$ is left continuous in all variables. Same with $(\&^\bullet_L)^j_{k_1,k_2}$ – only boundary condition is violated, so we skip this and finitize only @'s.

Note that rounding preserves ordering, namely if $x_i \leqslant y_i$ then $r_{k_i}(x_i) \leqslant r_{k_i}(y_i)$ and hence also $@^j_{k_1,\ldots,k_m}((x_1, \ldots, x_m)) \leqslant @^j_{k_1,\ldots,k_m}((y_1, \ldots, y_m))$.

Having this, we have to be aware of the fact that some computed results can be glued together, but sharp inequality between results cannot be changed to opposite sharp inequality. On the other side, for a query Q we can estimate an $\varepsilon > 0$ such that for two computed results x, y with a distance $|x - y| > \varepsilon$ these remain different and ε-additive top-k remains top-k.

7.3. *Similarities*

In RDF an URI usually contains a prefix, that has been assigned to a namespace URI followed by a local name, e.g.,

`<http://www.pv-FLSW-example.org/hotelname#"Hotel Crystal">`

in abbreviated form

`exhotelname:"Hotel Crystal"`

Now both hotel names can be spelled differently and in different countries accommodation facilities have different names and different ontologies (Europe is rich on languages), e.g.,

```
exhotelname:"Hotel Crystal"
ex2albergo:"Hotel Crystal"
exunterkunft:"Zimmer Frei"
exszállás:"Hotel Crystal"
exhotelname:"Salaš"
exschronisko:"Szalas"
```

In [17] we have developed a theory of fuzzy similarities for arbitrary multiadjoint lattices of truth values. For better computability properties we need to work with (small) finite domains. Our similarity theory can be extended to fit working with discretized (rounded) values.

The particular semantics given to our logic programming, based on the implicative rules, enables one to easily implement a version of fuzzy unification.

Namely, when extending arbitrary logic program by axioms of similarity

$$s(x, x).1$$

$$s(x, y) \rightarrow_L s(y, x)$$

$$s(x, y) \mathbin{\&} s(y, z) \rightarrow_L s(x, z)$$

$$p(x_1, \ldots, x_n) \mathbin{\&} s_1(x_1, y_1) \mathbin{\&} \cdots \mathbin{\&} s_n(x_n, y_n) \rightarrow_L p(y_1, \ldots, y_n)$$

$$s_1(x_1, y_1) \mathbin{\&} \cdots \mathbin{\&} s_n(x_n, y_n) \rightarrow_L s\big(f(x_1, \ldots, x_n), f(y_1, \ldots, y_n)\big)$$

(where & is an arbitrary conjunction) we get computed answers with similarity match in unification.

This is much more powerful, than max–min based similarity, usually used in systems. In [20] we have shown that similarity based Datalog can find most similar object in one step, provided similarity fulfills the transitivity property.

Tuning similarity transitivity. In order to have transitivity for some data downloaded from the web, e.g., for local names of hotels extending URI prefixes, having similarity degree (also in ontologies) like

s(hotel, albergo) = 0.9, s(hotel, szállás) = 0.5 and (albergo, szállás) = 0.3

we do not have transitivity according to the min, but we have transitivity, e.g., according to the Łukasiewicz conjunction.

Another feature of our system is, we have to find (one of) minimal conjunctors, for which our similarities fulfill transitivity. Then our similarity based Datalog [20] is effective and we have sound computation.

8. Extension of web rule languages

We propose an extension of Web rule languages (see, e.g., [25]) by the fuzzy aggregation construct, which can look for example like

```
<ex:imp>
  <ex:_body
    <ex:aggregation>User1_aggregation</ex:aggregation>
          <ex:individualPropertyAtom: Distance>
              <ex:var>_:Hotel_name</ex:var>
              <ex:var>_:Distance_value
    </ex:var></ex:individualPropertyAtom>
              ...
  </ex:_body
  <ex:_head
              ...
  </ex:_head
</ex:imp>
```

and user dependent fuzzy queries

```
<ex:query>
    <ex:var>_:user U1 </ex:var>
    <ex:var>_:hotel  </ex:var>
    <ex:fvar>_:close </ex:var>
    <ex:var >_:price#cheap </ex:var>
     ...
</ex:query>
```

It is a question of a further query processing, how to identify the user U1 with $@_{U1}$ and his fuzzy requirements. In principle our strategy is to have a rather limited number of fuzzy predicate for “small”, “medium” and “large” in each domain and tune the system by learning users fuzzy aggregation (or a group of users can have the same aggregation).

9. Conclusion

In this paper, we have presented an integration of several results into a system guided by a web application, especially multifeature querying. We have proposed several extensions of fuzzy and annotated definite logic programming deductive and inductive tasks and emphasized the necessity to extend some web modeling techniques with fuzzy aggregation feature.

References

[1] A. Bordiga, M. Lenzerini, R. Rosati, *Description logic for databases*, in: F. Baader, D. Calvanese, D. McGuinness, D. Nardi, P. Patel-Schneider (Eds.), The Description Logic Handbook – Theory, Implementation and Application, Cambridge University Press, 2003, pp. 462–484.

[2] A. Borgida, *On the relative expressiveness of description logics and predicate logics*, Artificial Intelligence **82** (12) (1996), 353–367.

[3] S. Borzsonyi, D. Kossmann, K. Stocker, *The skyline operator*, in: Proc. ICDE, Heidelberg, Germany, Apr 2001, pp. 421–430.

[4] I. Bratko, D. Šuc, *Learning qualitative models*, AI Magazine **24** (4) (2003), 107–119.

[5] S. Džeroski, N. Lavrač, *An introduction to inductive logic programming*, in: S. Džeroski, N. Lavrač (Eds.), Relational Data Mining, Springer, Berlin, 2001.

[6] T. Eiter, T. Lukasiewicz, R. Schindlauer, H. Tompits, *Combining answer set programming with description logics for the semantic web*, in: D. Dubois, et al. (Eds.), Proc. KR 2004, Morgan Kaufmann, 2004, pp. 141–151.

[7] R. Fagin, *Combining fuzzy information from multiple systems*, J. Comput. System Sci. **58** (1999), 83–99.

[8] R. Fagin, A. Lotem, M. Naor, *Optimal aggregation algorithms for middleware*, J. Comput. System Sci. **66** (2003), 614–656.

[9] B.N. Grosof, I. Horrocks, R. Volz, S. Decker, *Description logic programs: combining logic programs with description logic*, in: Proc. WWW 2003, ACM, 2003, pp. 48–57.

[10] U. Guentzer, W. Balke, W. Kiessling, *Optimizing multi-feature queries for image databases*, in: A. El Abbadi, et al. (Eds.), Proc. VLDB 2000, Morgan Kaufmann, 2000, pp. 419–428.

[11] P. Hájek, *Metamathematics of Fuzzy Logic*, Kluwer Academic Publishers, 1998.

[12] T. Horváth, F. Sudzina, P. Vojtáš, *Mining rules from monotone classification measuring impact of information systems on business competitiveness*, in: Proc. BASYS, Springer, Wien, 2004.

[13] I.F. Ilyas, R. Shah, W.G. Aref, J.S. Vitter, A.K. Elmagarmid, *Rank-aware query optimization*, in: SIGMOD 2004, ACM, 2004.

[14] M. Kifer, V.S. Subrahmanian, *Theory of generalized annotated logic programming and its applications*, J. Logic Programming **12** (1992), 335–367.

[15] S. Krajči, R. Lencses, P. Vojtáš, *A comparison of fuzzy and annotated logic programming*, Fuzzy Sets and Systems **144** (2004), 173–192.

[16] J.W. Lloyd, *Foundation of Logic Programming*, Springer, Berlin, 1987.

[17] J. Medina, M. Ojeda-Aciego, P. Vojtáš, *Similarity-based unification: a multi-adjoint approach*, Fuzzy Sets and Systems **146** (2004), 43–62.

[18] E. Naito, J. Ozawa, I. Hayashi, N. Wakami, *A proposal of a fuzzy connective with learning function*, in: P. Bosc, J. Kaczprzyk (Eds.), Fuzziness Database Management Systems, Physica-Verlag, 1995, pp. 345–364.

[19] J. Pavelka, *On fuzzy logic I, II, III*, Z. Math. Logik Grundl. Math. **25** (1979), 45–52, 119–134, 447–464.

[20] J. Pokorný, P. Vojtáš, *A data model for flexible querying*, in: Proc. ADBIS'01, in: Lecture Notes in Comput. Sci., vol. 2151, Springer-Verlag, Berlin, 2001, pp. 280–293.

[21] *QUIN and Qualitatively Faithful Numerical Learning*, http://ai.fri.uni-lj.si/dorian/q2/quin.htm.

[22] *RDF Semantics*, W3C Recommendation, www.w3c.org, 2004.

[23] U. Straccia, *Uncertainty in description logic programs*, TR-ISTI-CNR 1, Pisa, 2004.

[24] *Swoogle – search and metadata for semantic web*, http://swoogle.umbc.edu/.

[25] *The Rule Markup Initiative*, http://www.ruleml.org.

[26] P. Vojtáš, *Fuzzy logic programming*, Fuzzy Sets and Systems **124** (3) (2001), 361–370.

[27] P. Vojtáš, T. Horváth, S. Krajči, R. Lencses, *An ILP model for a monotone graded classification problem*, Kybernetika **40** (3) (2004), 317–332.

[28] L.A. Zadeh, *Fuzzy sets*, Inform. and Control **8** (3) (1965), 338–353.

[29] L.A. Zadeh, *Toward a theory of fuzzy information granulation and its centrality in human reasoning and fuzzy logic*, Fuzzy Sets and Systems **90** (2) (1997), 111–127.

[30] H.-J. Zimmermann, P. Zysno, *Latent connectives in human decision making*, Fuzzy Sets and Systems **4** (1980), 37–51.

Ontologies, Information Retrieval

CHAPTER 18

A Fuzzy Logic Approach to Information Retrieval Using an Ontology-based Representation of Documents

Mustapha Baziz, Mohand Boughanem and Henri Prade
IRIT, 118 Route de Narbonne, 31062 Toulouse Cedex 04, France
E-mail: {baziz,boughanem,prade}@irit.fr

Gabriella Pasi
Università degli Studi di Milano Bicocca, Via Bicocca degli Arcimboldi 8, 20126 Milano, Italy
E-mail: pasi@disco.unimib.it

Abstract

The paper proposes an approach to information retrieval based on the use of a fuzzy conceptual structure (ontology) that is used both for indexing document and expressing user queries. The conceptual structure is hierarchical and it encodes the knowledge of the topical domain of the considered documents. It is formally represented as a weighted tree. In this approach, the evaluation of conjunctive queries is based on the comparison of minimal sub-trees containing the two sets of nodes corresponding to the concepts expressed in the document and the query respectively. The comparison is based on the computation of a multiple-valued degree of inclusion. Some candidate implications are discussed on the basis of their respective semantics. The proposed approach generalizes standard fuzzy information retrieval and its evaluation on benchmark example is also presented.

1. Introduction

Textual Information Retrieval (IR) is based on keywords or expressions (generally associated with importance weights) extracted from documents and employed as the building blocks of both document and query representations: both are expressed in terms of (weighted) keywords. However, keywords may have different levels of granularity. For instance, "earth science" is a more general expression than "geology". Thus, some may

FUZZY LOGIC AND THE SEMANTIC WEB
Edited by Elie Sanchez
© 2006 Elsevier B.V. All rights reserved

refer to general topics while others are more specific descriptors. It may also happen that terms which can be used for describing the general topic(s) of a document are not so much present in the document. Still they are useful for classifying the document and referring to its contents.

Recently, an increasing number of approaches to IR have defined and designed IR models which are based on concepts rather than keywords, thus modeling document representations at a higher level of granularity, trying to describe the topical content and structure of documents. These efforts gave raise to the so called concept-based Information Retrieval, which aims at retrieving relevant documents on the basis of their meaning rather than their keywords. The main idea at the basis of conceptual IR is that the meaning of a text depends on conceptual relationships to objects in the world rather than to linguistic relations found in text or dictionaries [22]. To this aim, sets of words, phrases, names are related to the concepts they encode.

In this paper, a fuzzy set approach to concept-based Information Retrieval is proposed. Based on the existence of a conceptual hierarchical structure which encodes the contents of the domain to which the considered collection of documents belongs, both documents and queries are represented as weighted trees. The evaluation of a conjunctive query evaluation is then interpreted as computing a degree of inclusion between sub-trees.

The ontology-based description of the contents of the documents takes into account the semantic equivalences between expressions, as well as the basic principle stating that if a document explicitly heavily includes some terms, it also concerns to some extent more general concepts. This latter point is handled at the technical level by a completion procedure which assesses positive weights also to terms which do not appear directly in the documents. The possible completion of queries is also discussed.

The paper is organized into six main sections. In Section 2 a synthetic overview of some approaches to concept-based IR is presented. In Section 3 the proposed concept-based representations of documents and queries are introduced. In Section 4 the process of the conceptual query evaluation is described. Section 5 provides an illustrative example. Finally, the results of the experiments are reported and discussed in Section 6.

2. Concept-based information retrieval

The primary objective of the research in information retrieval is to define models, which allow designing effective IR systems, i.e. systems able to retrieve from the considered archive the documents concerned with the topics relevant to a user and expressed by a user query. The production of effective retrieval results depends on both subjective factors, such as the users' ability to express their information needs in a query, and the characteristics of the IR system. The indexing process plays a crucial role in determining the effectiveness of an IR system: in fact, to provide IR systems with powerful query languages or sophisticated retrieval mechanisms is not sufficient to achieve effective results if the representation of documents oversimplifies their information content. The indexing process has the aim of generating a formal representation of the contents of the information items (documents' surrogates). The most used automatic indexing procedures are based on term extraction and weighting: the documents are represented by means of a collection of index terms with associated weights (the index term weights); an index term weight expresses the degree of significance of the index term as a descriptor of the document information content. The automatic computation of the index term weight is based on the occurrence count of the term in the document and in the whole archive. In this case, a numeric weight is computed for each document d and each term t by means of an indexing function [21].

As a consequence, also query languages are usually based on (weighted) keywords, thus allowing the matching mechanism to compare two compatible representations. However, keyword-based retrieval models have several limitations; an important one is that they do not take into account the topical structure and content of documents, thus preventing concept-oriented document representation and query formulation.

Recently, some approaches have been proposed to concept-based Information Retrieval. In concept-based IR, sets of words, names, noun phrases are mapped into the concepts they encode [12]. By these models, a document is represented as a set of concepts: to this aim a crucial component is a conceptual structure for mapping concepts to document representations. Conceptual structures can be general or domain specific. In [12] an analysis of conceptual structures and their usage to improve retrieval is presented. Conceptual structures include dictionaries, thesauri and ontologies, and they can be either manually or automatically generated or they may pre-exist [11]. WordNet and EuroWordNet are examples of (thesaurus-based) ontologies widely employed to improve the effectiveness of IR systems.

A conceptual structure can be represented using distinct data structures: trees, semantic networks, conceptual graphs, etc. In [16] and [22], the use of conceptual graphs for representing documents and queries is discussed. The authors propose a method for measuring the similarity of phrases represented as conceptual graphs. In [10], the authors propose an indexing method based on the use of WordNet synsets: the vector space model is employed, by using synsets as indexing space instead of word forms. In a similar spirit in [3,5,14], the authors have advocated the use of possibilistic ontologies, distinguishing between specialization and approximate synonymy relations, in information querying.

In [6], a concept based information retrieval approach based on the use of a thesaurus is proposed. In [13], an approach to detect the topical structure of a set of documents is presented. Probabilistic latent semantic analysis has been employed as a mean to define conceptual structures; it applies an unsupervised learning technique to define a semantic (topic-based) language model [1,18].

In this paper, we do not face the problem of generating a conceptual structure, we take as a preliminary assumption the existence of a conceptual structure (more precisely an ontology) describing the contents of a considered document collection. This structure is a hierarchical tree, as it will be explained in the next section.

3. Concept-based representation of documents and queries

In the proposed approach, the query evaluation of textual documents is supposed to be mediated by an ontology made by a unique tree-like hierarchy H of concepts, which are supposed to be sufficient for describing the contents of the considered documents with an appropriate level of accuracy. Leaves in H can be thought as simple keywords, i.e. keywords expressing specialized concepts, while other nodes refer to keywords, which are labels of more general concepts. Edges in this hierarchy represent the classical *is-a* link.

Both documents and queries are supposed to be interpreted or expressed in terms of labels of nodes of H, possibly in association with weights.

Let d be a document. Each document d is identified by means of a set of pairs $R_d = \{(w_i, \alpha_i), i = 1, k(d)\}$ where w_i is a key word or phrase taken from d and α_i is its importance weight (index term weight) computed by an occurrence-based indexing function that is usually employed in IR [21], $k(d)$ is the number of terms in document d.

We then compute the projection H_d of d on H, in the following way:

(i) a weighted subset $N(H, d)$ of nodes of H, namely $N(H, d) = \{(n_j, \gamma_j), j = 1, m(d)\}$ where for any node n_j, there exists w_i in $d = \{(w_i, \alpha_i), i = 1, k(d)\}$ such that $n_j = w_i$, or n_j is known as a *strict equivalent* [9] of w_i in the conceptual structure H, and then we take $\gamma_j = \alpha_i$. $m(d)$ is the number of nodes in H that are equivalent to some terms in R_d. When several equivalent expressions w_i in H exist, it is assumed that they are counted as a unique expression.

(ii) H_d is the minimal sub-tree of H which contains $N(H, d)$, where the weights associated with the nodes are those obtained at step i if the nodes belong to $N(H, d)$ and are 0 otherwise.

N.B. If necessary, one may restrict $N(H, d)$ to those expressions for which γ_j is sufficiently high, thus specifying an acceptance threshold.

Let q be a query obtained by selecting a collection of labels (concepts) in H, with possibly an importance weighting, namely as a set $\{(l_k, \delta_k), k = 1, r(q)\}$. We assume that the query is viewed as a (weighted) conjunction of the concept labels. A query q is also modeled by a sub-tree H_q of H. Namely H_q is the minimal weighted sub-tree of H containing $\{(l_k, \delta_k), k = 1, r(q)\}$, keeping the weights δ_k, and putting 0 on the other nodes of H_q.

4. Query evaluation

Query evaluation is based on the comparison of two weighted subsets of nodes of H, one corresponding to the query and the other to the current document. First, a minimal sub-tree of H containing both subsets is determined, then an inclusion degree is computed for evaluating to what extent a document includes all the features of the query. The use of various implication connectives in the definition of the inclusion degree is discussed. Lastly, some possible procedures for completing the description of the document or extending the query, by propagating weights in H, are presented.

4.1. *Comparison based on the minimal common sub-tree*

Let H_E be the minimal non-weighted sub-tree which contains both H_d and H_q. Let H_d^* and H_q^* be the extensions of H_d and H_q on H_E putting zero weights on the nodes of $H_E - H_d$ and $H_E - H_q$ respectively.

The evaluation of a *conjunctive* query q with respect to a document d, is performed in terms of a degree of relevance $\text{rel}_c(d; q)$ of d with respect to q computed as degree of inclusion of H_q into H_d, namely

$$\text{rel}_c(d; q) = \min_{n \in H_E} \mu_{H_q^*}(n) \to \mu_{H_d^*}(n), \tag{1}$$

where $\mu_{H_d^*}(n)$ (resp. $\mu_{H_q^*}(n)$) is the weight associated with node n in H_d^* (resp. H_q^*), and $\to$ is a multiple-valued implication connective expressing that all the concepts of the query should appear in the description of the document.

4.2. *Choice of an implication connective*

Several choices can be considered for the implication $\rightarrow$ used in (1), depending on the intended semantics of the weights in the query. Among the possible choices we just consider the following ones, as they have clear semantics in a retrieval context (see, e.g., [7] in a database context and [17] in an IR context):

- using Dienes implication $a \rightarrow b = \max(1 - a, b)$ amounts to view the $\mu_{H_q^*}(n)$'s (antecedent a) as levels of importance in the following sense. The impact of having n absent in document $d(\mu_{H_d^*}(n) = 0)$ leads to an evaluation that will be upper-bounded by $1 - \mu_{H_q^*}(n)$, a bound which is higher as the importance $\mu_{H_q^*}(n)$ of n in the query is smaller. It is assumed that $\max_n \mu_{H_q^*}(n) = 1$ (at least one weight in the query is maximal),
- using Gödel implication $a \rightarrow b = 1$ if $a \leqslant b$, $a \rightarrow b = b$ if $a > b$, it amounts to use the weight $\mu_{H_q^*}(n)$ as a demanding threshold, since the corresponding term in conjunctive aggregation (1) is equal to 1 as soon as $\mu_{H_d^*}(n)$ is larger than $\mu_{H_q^*}(n)$.
- using Lukasiewicz implication $a \rightarrow b = \min(1, 1 - a + b)$, we mix the two above effects (namely $a \rightarrow b = 1$ if $a \leqslant b$ and $a \rightarrow b = 1 - a$ if $b = 0$).

As explained in [8], by delocalizing the weights and using the first implication, we can easily build an ordered weighted minimum (Owmin) aggregation, and model a query asking for the satisfaction of *most* of the terms of the query (rather than all).

REMARK 1. Note that a prototypical document can be directly used in this approach as a query. If the document is itself present in the collection, it will be retrieved with the maximal estimated degree of relevance 1. Thus, the approach is appropriate for handling a case-based querying process.

REMARK 2. The evaluation of a *disjunctive* query q by a degree $\mathrm{rel}_d(d; q)$ of a non-empty intersection between H_d and H_q can be computed as follows:

$$\mathrm{rel}_d(d; q) = \max_{n \in H_E} \min\big(\mu_{H_d^*}(n), \mu_{H_q^*}(n)\big). \tag{2}$$

This expresses that at least one of the important concepts of the query is somewhat relevant to the document.

REMARK 3. One may think of introducing equivalence connectives in place of implications in (1) for requiring that the topic of the document correspond exactly to the topic of the query. However, note that looking for exact matching may be dangerous: suppose we are looking for documents dealing with topic A $(q = A)$ but there does not exist any document dealing with A without B $(d = A, B)$; in such a case the exact matching strategy will give nothing. However, strict equivalence could be relaxed into approximate similarity by weakening the equivalence connective by means of a similarity relation.

Strict conjunctive evaluation as in (1) may be too requiring, and in information retrieval, "best matching" is usually preferred to exact matching. Therefore, a simple function that is known to allow best matching is the sum computed as follows, and that we will also evaluate in this paper:

$$\mathrm{rel}_d(d, q) = \sum_{n \in H_E} \mu_{H_q^*}(n) \rightarrow \mu_{H_d^*}(n). \tag{3}$$

4.3. *Completing the description of a document/enlarging the query*

In the above procedure, the weights both in H_d^* and H_q^* have been used as they are. However, it may be advisable to modify some of the zero weights both in the description of the document, and in the query, due to different reasons.

Indeed, regarding document d, if a node has a non-zero weight in H_d^*, we may think that a node which is an ancestor of the node in H is also somewhat relevant for the description of the document (even if its own weight in H_d^*, is zero or small). Then, we may think of "completing" H_d^* by computing updated weights in the following way. Let α_i^s and α_i^{s+1} denote weights at level s and $s+1$ in the hierarchy (the root is at level 0). The idea is to recursively update the weights of the nodes starting from the leaves by having the revised weights computed as:

$$\alpha_{i,\text{rev}}^s = \max\left(\alpha_i^s, \left(\max_i \alpha_{i,\text{rev}}^{s+1}\right)^* \text{disc}(s)\right),$$

where $\text{disc}(s)$ is a discounting factor possibly depending on level s. Indeed, if a document includes many instances of the word 'cat', it clearly deals with 'pets' (the "father" of 'cat'), but to a smaller extent if the word 'pets' (or its synonyms) do not appear as much in the document. In order to control the number of nodes to be added to document/query descriptions, only the common ancestors of couples or triples of co-occurring words in a same document might be considered in the completion procedure.

Regarding the queries, the completion procedure of the weights may be motivated by a potential enlargement of the query to less specific terms. Here, the use of a discounting factor will reflect the fact that documents dealing directly with the terms initially chosen should be preferred to more general documents. A similar idea has been used in [22] when dealing with fuzzy conceptual graphs for handling possibilistic information and fuzzy queries (however with a different interpretation for the weights in the fuzzy conceptual graphs leading to a different evaluation procedure).

REMARK 4. One might also think of enlarging the query by introducing children of nodes present in q. In fact, this makes the query more demanding (at least if we keep unchanged the levels of importance for the labels present in the original conjunctive query).

5. An illustrative example

Let consider an ontology H as illustrated by Figure 1, where each node, named n_i, represents a concept (keyword). Let d be a document indexed with the following weighted keywords $d = \{$(natural science, 0.1), (geology, 1), (geography, 0.5), (geophysics, 0.8)$\}$ and q be a query represented by the following weighted keywords $q = \{$(earth science, 1), (geography1)$\}$. This query means that the user is looking for documents containing both keywords.

The respective projection of the document d and the query q on H are:

$$N(H,d) = \{(n_4, 0.1), (n_{12}, 1), (n_{14}, 0.5), (n_{16}, 0.8)\},$$
$$N(H,q) = \{(n_9, 1), (n_{14}, 1)\}.$$

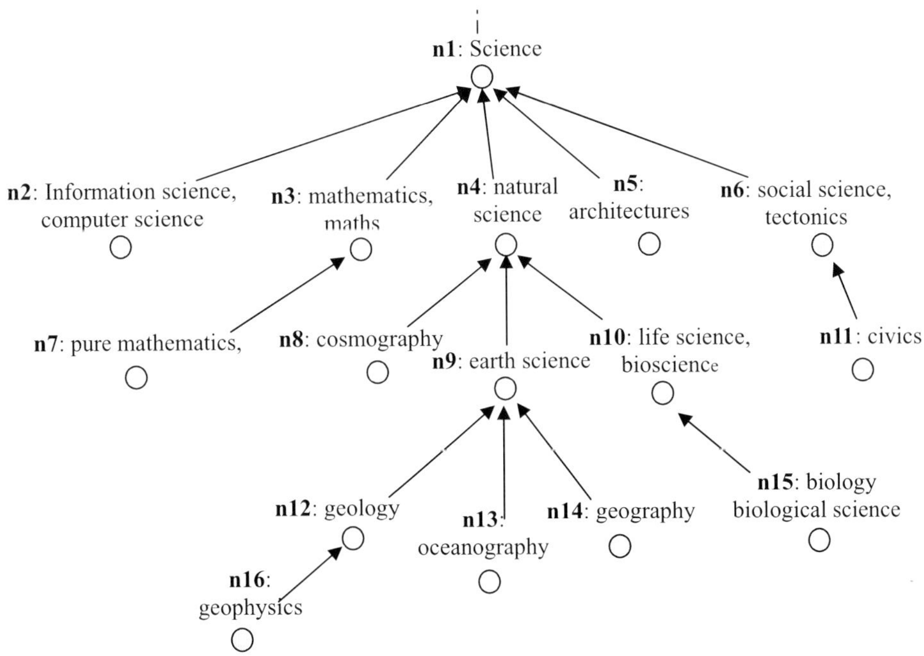

Fig. 1. An excerpt of ontology of science.

The minimal sub-trees H_d and H_q can be easily built from $N(H, d)$ and $N(H, q)$ as follows:

$$H_d = \{(n_4, 0.1), (n_9, 0), (n_{12}, 1), (n_{14}, 0.5), (n_{16}, 0.8)\},$$
$$H_q = \{(n_9, 1), (n_{14}, 1)\}.$$

Then, the minimal non-weighted sub-tree, H_E, containing H_d and H_q is represented by the following nodes: $H_E = \{n_4, n_9, n_{12}, n_{14}, n_{16}\}$.

In order to illustrate the effect of the completion, the query is evaluated, using Dienes implication, for both cases: without and with completion.

The extensions of H_d and H_q in H_E, in the case of no completion are given by:

$$H_d^* = \{(n_4, 0.1), (n_9, 0), (n_{12}, 1), (n_{14}, 0.5), (n_{16}, 0.8)\},$$
$$H_q^* = \{(n_4, 0), (n_9, 1), (n_{12}, 0), (n_{14}, 1), (n_{16}, 0)\}.$$

The evaluation of the conjunctive query gives:

$$\text{rel}_c(d, q) = \min\big(\max(1 - 0, 0.1), \max(1 - 1, 0), \max(1 - 0, 1), \max(1 - 1, 0.5), \max(1 - 0, 0.8)\big) = 0.$$

Note that the evaluation of the disjunctive counterpart of the query would give a non-zero value.

$$\text{rel}_d(d, q) = \max\big(\min(0, 0.1), \min(1, 0), \min(0, 1), \min(1, 0.5), \min(0, 0.8)\big) = 0.5.$$

Let us consider now the completion of the document with $\mathrm{disc}(s)$ as constant equal to 0.7, H_d^* and H_q^* are:

$$H_d^* = \{(n_4, 0.49), (n_9, 0, 7), (n_{12}, 1), (n_{14}, 0.5), (n_{16}, 0.8)\},$$
$$H_q^* = \{(n_4, 0), (n_9, 1), (n_{12}, 0), (n_{14}, 1), (n_{16}, 0)\}.$$

Then, $\mathrm{rel}_c(d, q) = 0.5$.

It can be noticed that the completion allows to better identify the concepts occurring in the document. In fact, in the above example, when no completion is considered, as the keyword "earth science" (the node n_9 in H) does not occur in the document d, the matching between the query and this document gives zero; the document is not selected. Whereas, it can be seen that, as the document d is about "geology", "geography" and "natural science", even though it does not contain the keyword "earth science", it deals with "earth science" topics. Thus, the completion procedure can take care of this problem, the node n_9 becomes non-zero in H_d^*, which then allows the retrieval of document d.

The example showed above illustrates the effect of the completion of the document tree only, which enlarges the explicit scope of the document and increases its potentiality to match a query. The completion could be also useful to be applied for both documents and queries, at least in the case of disjunctive queries. Indeed completing a conjunctive query makes it more demanding.

Let us consider an example with a disjunctive query. Let d_1 and d_2 be two documents represented by the following keywords $d_1 = \{(\text{natural science}, 0.1), (\text{geophysics}, 0.8)\}$ and $d_2 = \{(\text{natural science}, 0.1), (\text{earth science}, 1), (\text{geophysics}, 0.8)\}$ and q be a query, represented by $q = \{(\text{geography}, 1), (\text{geology}, 1)\}$.

The minimal sub trees H_{d_1}, H_{d_2} and H_q resulting from the projection of d_1, d_2 and q on H are:

$$H_{d_1} = \{(n_4, 0.1), (n_9, 0), (n_{12}, 0), (n_{16}, 0.8)\},$$
$$H_{d_2} = \{(n_4, 0.1), (n_9, 1), (n_{12}, 0), (n_{16}, 0.8)\},$$
$$H_q = \{(n_9, 0), (n_{12}, 1), (n_{14}, 1)\}.$$

The extension of H_{d_1} and H_q, and of H_{d_2} and H_q, in their corresponding non-weighted minimal sub-trees, in the case of *no* completion are given by:

$$H_{d_1}^* = \{(n_4, 0.1), (n_9, 0), (n_{12}, 0), (n_{14}, 0), (n_{16}, 0.8)\},$$
$$H_q^* = \{(n_4, 0), (n_9, 0), (n_{12}, 1), (n_{14}, 1), (n_{16}, 0)\},$$
$$H_{d_2}^* = \{(n_4, 0.1), (n_9, 1), (n_{12}, 0), (n_{14}, 0), (n_{16}, 0.8)\},$$
$$H_q^* = \{(n_4, 0), (n_9, 0), (n_{12}, 1), (n_{14}, 1), (n_{16}, 0)\}.$$

Note that in this particular example, the minimal non-weighted sub-trees containing H_{d_1} and H_q, and H_{d_2} and H_q are the same ones.

The evaluation of the disjunctive query on each document gives:

$$\mathrm{rel}_d(d_1, q) = \max(\min(0, 0.1), \min(0, 0), \min(1, 0), \min(1, 0), \min(0, 0.8)) = 0,$$
$$\mathrm{rel}_d(d_2, q) = \max(\min(0, 0.1), \min(0, 1), \min(1, 0), \min(1, 0), \min(0, 0.8)) = 0.$$

When the completion of document only is considered, with disc(s) equal to 0.7, $H^*_{d_1}$ and $H^*_{d_2}$ become:

$$H^*_{d_1} = \{(n_4, 0.27), (n_9, 0.39), (n_{12}, 0.56), (n_{14}, 0), (n_{16}, 0.8)\},$$
$$H^*_{d_2} = \{(n_4, 0.7), (n_9, 1), (n_{12}, 0.56), (n_{14}, 0), (n_{16}, 0.8)\}.$$

Then, $\text{rel}_d(d_1, q) = \text{rel}_d(d_2, q) = 0.56$.

Thus, we can conclude that d_1 and d_2 are both "equally" relevant to the query.

But, this result seems quite unsatisfactory. Indeed, we can easily notice that d_2 is topically more related to the query than d_1. In fact, d_2 contains all the keywords of d_1 *plus* "earth science" with a high weight equal to 1. Among all keywords of d_1 and d_2, "earth science" is the one that is conceptually the most related to those of the query (e.g., "geology" and "geography"), but as the query was not completed, the keyword "earth science" has no effect in the evaluation stage.

Let us consider now the query completion H^*_q:

$$H^*_q = \{(n_4, 0.49), (n_9, 0.7), (n_{12}, 1), (n_{14}, 1), (n_{16}, 0)\}.$$

The evaluation of d_1 and d_2 give:

$$\text{rel}_d(d_1, q) = 0.56,$$
$$\text{rel}_d(d_2, q) = 0.7.$$

The document d_2 is now better ranked than d_1. This result is more effective than the previous one.

6. Experiments

The aim of the experiments is to evaluate the effectiveness of the concept-based approach proposed here compared to classical IR approach. Especially two main contributions described in the paper are evaluated:

- How good is the concept-based approach compared to the classical one.
- How good is the completion of documents and/or queries compared to the classical?

The classical approach used in these experiments is based on a vector space model which is implemented in the Mercure system [4].

Below, we detail the experiments settings.

6.1. *Document collection*

The test collection we used in these experiments is issued from the MuchMore project[1] [5]. This collection contains 7823 documents (medical papers abstracts) obtained from the Springer Link web site, 25 topics from which the queries are extracted and a relevance judgment file which determines for each topic its set of relevant documents. These assessments were established by domain experts from Carnegie Mellon University, LT Institute.

[1] http://muchmore.dfki.de

Table 1
Average number of terms per document and query covered by WordNet.

	Average number of terms	
	Documents	Queries
All terms (classical)	56.01	4.05
WordNet terms only	48.89	3.1
%	87%	77%

6.2. *Ontology*

WordNet [15] is used as a general purpose ontology. In WordNet, concepts are organized into taxonomies where each node is a set of synonyms (called synset) representing a single sense. Several semantic relationships between nodes are defined, denoting generalization, specialization, composition links, etc. We used only the concept hierarchy identified by the *ISA* relation.

As the collection deals with the medical domain, the question of the suitability of WordNet for this kind of collections could be asked. Statistics carried out over the collection show that the vocabulary of the documents of the collection is almost covered by WordNet. Table 1 gives the cover rate. It can be seen that about 87% (respectively 77%) of terms used in documents (respectively queries) appear in WordNet.

6.3. *Evaluation methodology*

In order to evaluate our approach, two sets of experiments were carried out. The first set is based on classical indexing and the second one on the approach proposed in this paper. In the classical approach, the documents were first indexed using a classical term indexing. It consists in selecting single words occurring in the documents, and then stemming these words using Porter algorithm [19] and at the end removing stop-words according to a standard list [21]. A weight is then assigned to each term following a kind of BM25 TF.IDF formula [4]. The same process is applied to queries. A vector-based model [21] is then used to retrieve documents. We used this run as a baseline.

In the concept-based approach, the documents and the queries are indexed as follows. Keywords (single terms, phrases) are first extracted from each document (query) using an approach developed in [2]. The keywords are then projected onto WordNet. As these keywords could appear in different synsets, a disambiguation process [2] using the Resnik similarity measure [20], which selects for each keyword a unique synset; is then applied. The result of this stage is that each document (query) is represented by a set of weighted synsets (the weight of the keyword is assigned to the corresponding synset). Once nodes representing the documents (queries) are identified, the corresponding sub-trees, (H_d^*) and (H_q^*) are built by computing recursively the nodes subsuming the extracted ones until having a unique sub-tree covering all the concepts. It may arise that the highest nodes of the resulting sub-trees are located in the immediate vicinity of the root of ontology. As these nodes are abstract, they are not useful, so they are omitted using a pruning method. An illustration of this method is presented in the Appendix.

Once the document and the query sub-trees are built, the query evaluation is carried out. It is based on Lukasiewicz, Gödel and Dienes implications using three aggregation functions namely, min, max and sum described above in Equations (1), (2) and (3) respectively.

In order to measure the effectiveness of the concept-based approach two sets of results were built. The first set concerns the classical approach based on vector space model. The second set concerns the concept based results. In this latter, several sets of results were built by combining the three following factors: the implication, the aggregation function and the completion procedure.

The experimental method follows the TREC protocol [24]. For each query, the first 1000 retrieved documents are returned by the search engine and precisions are computed at different points.

6.4. *Results and discussion*

The most representative results are summarized in Table 2. It describes the P5, P10, P15 and P30 representing the mean precision values for the 20 used queries at the top 5, 10, 15 and 30 selected documents and MAP, the Mean Average Precision over the 20 queries.

The first run, labeled Classical, describes the results of the classical approach. The remaining runs, concern those combining Lukasiewicz, Dienes and Godel implications with min aggregation (for conjunctive query evaluation) labeled *CO_Lukas_conj*, *CO_Dienes_conj*, *CO_Godel_conj*, those using, max (for disjunctive query evaluation), labeled *CO_Lukas_dis*, *CO_Dienes_dis* and *CO_Godel_dis* and those using sum that are labeled *CO_Lukas_sum*, *CO_Dienes_sum* and *CO_Godel_sum*. The second column of the table indicates whether the completion is used for queries and/or documents.

First, it can be seen that the implication functions have no impact in the results. Indeed, Table 2 shows that the results obtained by the implications, for a given aggregation, are comparable. However, one notices that the aggregation type is crucial. Table 2 shows that the results of classical are strongly better than those of the conjunctive and disjunctive query evaluation at all considered precision levels. In the case of the conjunctive function, the reason is that only queries having all terms present in the relevant documents are evaluated (in our case 12 queries from the 20 used). In fact, this function which performs an exact matching is mostly adapted to Data Retrieval (a document is retrieved if and only if it contains all query terms), whilst best match is used in Information Retrieval [23]. The disjunctive function gives the worst results among those presented in Table 2. In fact, the disjunctive query evaluation is not suitable for rank-ordering. For instance, a document having query evaluation values (0.6, 0, 0, 0) is ranked better than another having much more common terms with the query but with lower weights (0.5, 0.5, 0.5, 0.5). When completion is used for queries and/or documents, the results of disjunctive and conjunctive are still also very low compared to the classical approach.

Now when the sum is used as aggregation function (*CO_Lukas_sum*, *CO_Dienes_sum* and *CO_Godel_sum*), the results look differently. It can be seen that the sum, when no completion is used, achieves better ranking than both disjunction and conjunction aggregation. The results are in the same class as the classical approach, with a slight advantage for the classical approach. However, when completion is used to queries and/or to documents the precisions improve in all considered cases with a real benefit when completion is used

Table 2
Comparison between classical and conceptual approaches.

		P5	P10	P15	P30	MAP
(1) *Classical*		**0.7400**	**0.6400**	**0.6233**	**0.4900**	**0.4171**
(2) *C0_Lukas_conj*	No completion	0.4200	0.3100	0.2333	0.1900	0.1447
	Completion of docs.	0.4400	0.3350	0.2567	0.2075	0.1532
	Completion of queries	0.0400	0.0400	0.0467	0.0500	0.0585
	Both docs and queries are completed	0.3800	0.2650	0.1967	0.1675	0.1297
(3) *C0_Dienes_conj*	No completion	0.4200	0.3100	0.2333	0.1433	0.1449
	Completion of docs.	0.4400	0.3400	0.2567	0.1517	0.1537
	Completion of queries	0.0400	0.0400	0.0467	0.0500	0.0585
	Both docs and queries are completed	0.3800	0.2650	0.1967	0.1300	0.1295
(4) *C0_Godel_conj*	No completion	0.4200	0.3100	0.2333	0.1433	0.1447
	Completion of docs.	0.4400	0.3400	0.2567	0.1517	0.1537
	Completion of queries	0.0400	0.0400	0.0467	0.0500	0.0585
	Both docs and queries are completed	0.3500	0.2500	0.1867	0.1183	0.1193
(5) *C0_Lukas_dis*	No completion	0.1600	0.1550	0.1367	0.1483	0.1432
	Completion of docs.	0.1600	0.1500	0.1367	0.1450	0.1407
	Completion of queries	0.2000	0.1833	0.1608	0.1618	0.1771
	Both docs and queries are completed	0.2000	0.1850	0.1600	0.1333	0.1436
(6) *CO_Dienes_dis*	No completion	0.1680	0.1560	0.1333	0.1333	0.1283
	Completion of docs.	0.1520	0.1520	0.1200	0.1267	0.1221
	Completion of queries	0.1900	0.2000	0.1967	0.1500	0.1476
	Both docs and queries are completed	0.3500	0.3550	0.3367	0.3200	0.2530
(7) *CO_Godel_dis*	No completion	0.1600	0.1550	0.1367	0.1483	0.1432
	Completion of docs.	0.1600	0.1500	0.1367	0.1450	0.1407
	Completion of queries	0.2000	0.2000	0.1667	0.1617	0.1559
	Both docs and queries are completed	0.2000	0.1850	0.1600	0.1333	0.1436
(8) *CO_Lukas_sum*	No completion	0.7000	0.6300	0.5767	0.4383	**0.3747**
	Completion of docs.	0.7200	0.5480	0.4853	0.3733	0.3179
	Completion of queries	0.7200	0.6550	0.6267	0.4733	0.3977
	Both docs and queries are completed	**0.7500**	**0.7100**	**0.6667**	**0.5150**	**0.4450**
(9) *CO_Dienes_sum*	No completion	0.6900	0.6250	0.5767	0.4383	0.3745
	Completion of docs.	0.7200	0.6400	0.5733	0.4333	0.3733
	Completion of queries	0.6500	0.5900	0.5333	0.4000	0.3495
	Both docs and queries are completed	**0.7400**	**0.6700**	**0.6233**	**0.5117**	**0.4286**
(10) *CO_Godel_sum*	No completion	0.7000	0.6300	0.5767	0.4383	0.3747
	Completion of docs.	0.7200	0.6400	0.5733	0.4350	0.3749
	Completion of queries	0.7200	0.6550	0.6267	0.4733	0.3977
	Both docs and queries are completed	**0.7500**	**0.7100**	**0.6733**	**0.5167**	**0.4464**

for both queries and documents. Indeed, in this latter, all precision cut-off are better than those obtained with the classical method.

The main interesting results that can be drawn from these experiments concern document completion. Indeed, most "of" IR "completion/expansion" approaches are used for query but never for documents. These preliminary experiments tend to indicate that completing documents and queries could be useful to identify important concepts that are not appearing explicitly in both document and query, providing thus, a contextual search capability.

7. Concluding remarks

In this paper an approach to concept-based IR has been proposed, which models both document and queries as tree-like ontologies where nodes are weighted. The query evaluation process uses fuzzy connectives.

The preliminary experiments carried out on an IR collection indicate that the proposed approach is viable in IR. The results showed that the concept-based approach outperforms the classical IR one based on vector space model. Future works will focus the evaluation of the concept-based on larger collection such the TREC collection [24].

Appendix: Pruning abstract nodes

In practice, when completing a document or a query, not all intermediate nodes are added to the document (query) representation. Indeed, nodes located in the high level of the hierarchy are removed as they represent abstract concepts. This additional stage, which consists in a pruning method, uses two pieces of information in order to decide whether a node could be added to a document (query) representation. The first one is the position in the hierarchy (*depth*) of the head node of the sub-tree containing original nodes and the second one is the *length* (number of nodes) of the current sub-tree branch containing the candidate intermediate nodes to be added. So, for a branch B_i of a given sub-tree, the number *Nb* of extra nodes to add, is given by the following formula:

$$Nb(B_i) = \min\bigl[(length(B_i) - 1 + depth)/2, length(B_i) - 1\bigr]. \tag{A.1}$$

A high value of *depth* means that a node is located near leaves and allows adding specific intermediate nodes, while a low value of *depth* permits to prune abstract nodes, as they are located in the immediate vicinity of the root.

EXAMPLE. In order to show how abstract intermediate nodes are pruned from the sub-tree, let us consider two concepts (nodes) extracted from a document (named Arthroskopie.00130047 in the used collection): the sense 1 of the name alternative (*alternative#n#1*) and the sense 1 of the name (*amount#n#1*). The sub-tree the two nodes belong to is as represented in Figure 2. *Depth* of the head node of the branches B_1 and B_2 equals 1 as the head node in this case represents the root of the hierarchy.

Let B_1 and B_2 the two branches as represented in Figure 2:

$$\begin{aligned} B_1 &= \text{alternative\#}n\text{\#1} \to \text{decision\_making\#}n\text{\#1} \\ &\quad \to \text{higher\_cognitive\_process\#}n\text{\#1} \to \text{process\#}n\text{\#3} \\ &\quad \to \text{cognition\#}n\text{\#1} \to \text{psychological\_feature\#}n\text{\#1}, \\ B_2 &= \text{amount\#}n\text{\#1} \to \text{magnitude\#}n\text{\#1} \\ &\quad \to \text{property\#}n\text{\#3} \to \text{attribute\#}n\text{\#2} \to \text{abstraction\#}n\text{\#6}. \end{aligned}$$

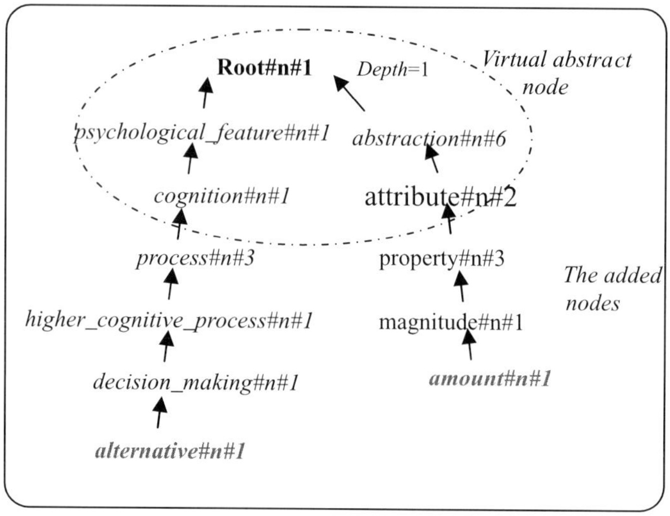

Fig. 2. A sub-tree containing the nodes labeled alternative#n#1 and amount#n#1.

Thus, $length(B_1) = 6$ and $length(B_2) = 5$.

So the number of nodes to be added (starting from the leaf) from the branch B_1 is: $Nb(B_1) = \min[(length(B_1) - 1 + depth)/2, length(B_1) - 1] = 3$ nodes which are: *decision_making*#n#1, *higher_cognitive_process*#n#1 and *process*#n#3.

And from the second branch B_2: $Nb(B_2) = \min[(length(B_2) - 1 + depth)/2, length(B_2) - 1] = 2$ nodes which are: magnitude#n#1 and property#n#3.

The remaining nodes are effectively more abstract and could be pruned from the document representation. They form with the Root node what we call a *virtual abstract node*.

References

[1] L. Azzopardi, M.L. Girolami, C.J. van Rijsbergen, *Topic based language models for ad hoc information retrieval*, in: The Proceedings of the International Joint Conference on Neural Networks, Budapest, Hungary, 2004.

[2] M. Baziz, M. Boughanem, N. Aussenac-Gilles, C. Chrisment, *Semantic cores for representing documents in IR*, in: Proceeding of the 2005 ACM Symposium on Applied Computing, Santa Fe, NM, USA, March 2005, vol. 2, pp. 1011–1017.

[3] G. Bordogna, G. Pasi, *Linguistic aggregation operators of selection criteria in fuzzy information retrieval*, Int. J. Intelligent Systems **10** (1995), 233–248.

[4] M. Boughanem, T. Dkaki, J. Mothe, C. Soulé-Dupuy, *Mercure at TREC-7*, in: Proceeding of Trec-7, 1998.

[5] P. Buitelaar, D. Steffen, M. Volk, D. Widdows, B. Sacaleanu, S. Vintar, S. Peters, H. Uszkoreit, *Evaluation resources for concept-based cross-lingual IR in the medical domain*, in: Proc. of LREC2004, Lissabon, Portugal, May 2004.

[6] H. Chen, K.J. Lynch, K. Basu, D.T. Ng, *Generating, integrating, activating thesauri for concept-based document retrieval*, IEEE EXPERT, Special Series on Artificial Intelligence in Text-based Information Systems **8** (2) (April 1993), 25–34.

[7] D. Dubois, M. Nakata, H. Prade, *Extended divisions for flexible queries in relational databases*, in: O. Pons, M.A. Vila, J. Kacprzyk (Eds.), Knowledge Management in Fuzzy Databases, Physica-Verlag, Heidelberg, 1999, pp. 105–121.

[8] D. Dubois, H. Prade, *Semantics of quotient operators in fuzzy relational databases*, Fuzzy Sets and Systems **78** (1996), 89–93.

[9] D. Dubois, H. Prade, *Bipolarity in flexible querying*, in: Flexible Query Answering Systems, 5th International Conference (FQAS'02), Copenhagen, Denmark, 27–29 Oct. 2002, Lecture Notes in Artificial Intelligence, vol. 2522, Springer-Verlag, Berlin, pp. 174–182.

[10] J. Gonzalo, F. Verdejo, I. Chugur, J. Cigarrán, *Indexing with WordNet synsets can improve text retrieval*, in: Proc. the COLING/ACL '98 Workshop on Usage of WordNet for Natural Language Processing, 1998.

[11] N. Guarino, C. Masolo, G. Vetere, *OntoSeek: content-based access to the Web*, IEEE Intelligent Systems **14** (3) (May/June 1999), 70–80.

[12] H.M. Haav, T. L. Lubi, *A survey of concept based information retrieval tools on the Web*, in: A. Caplinkas, J. Eder (Eds.), Advances in Databases and Information Systems, Proc. of 5th East-European Conference ADBIS*2001, vol. 2, Technika, Vilnius, 2001, pp. 29–41.

[13] M.Y. Kan, J.L. Klavans, K.R. McKeown, *Synthesizing composite topic structure trees for multiple domain specific documents*, Tech. Report CUCS-003-01, Columbia University, 2001.

[14] Y. Loiseau, H. Prade, M. Boughanem, *Qualitative pattern matching with linguistic terms*, AI Comm. **17** (1) (2004), 25–34.

[15] G. Miller, *Wordnet: a lexical database*, Comm. ACM **38** (11) (1995), 39–41.

[16] M. Montes-y-Gómez, A. López-López, A. Gelbukh, *Information retrieval with Conceptual Graph matching*, in: Proc. DEXA-2000, 11th Int. Conf. on Database and Expert Systems Applications, Greenwich, England, September 4–8, 2000, Lecture Notes in Comput. Sci., vol. 1873, Springer-Verlag, pp. 312–321.

[17] G. Pasi, *A logical formulation of the Boolean model and of weighted Boolean models*, Workshop on Logical and Uncertainty Models for Information Systems (LUMIS 99), University College London, 5–6 July 1999.

[18] J.M. Ponte, W.B. Croft, *A language modeling approach to information retrieval*, in: Proceedings of the Twenty First ACM-SIGIR, Melbourne, Australia, 1998, ACM Press, pp. 275–281.

[19] M. Porter, *An algorithm for suffix stripping*, Program **14** (3) (1980), 130–137.

[20] P. Resnik, *Using information content to evaluate semantic similarity in a taxonomy*, in: Proceedings of the 14th Intern. Joint Conference on Artificial Intelligence (IJCAI), 1995.

[21] G. Salton, M.J. McGill, *Introduction to Modern Information Retrieval*, McGraw-Hill, 1984.

[22] R. Thomopoulos, P. Buche, O. Haemmerlé, *Representation of weakly structured imprecise data for fuzzy querying*, Fuzzy Sets and Systems **140** (2003), 111–128.

[23] C.J. Van Rijsbergen, *Information Retrieval*, 2nd ed., Butterworths, London, 1979, http://www.dcs.glasgow.ac.uk/Keith/Preface.html.

[24] E.M. Vorhees, D.K. Harman, *Overview of the sixth Text REtrieval Conference (TREC-6)*, in: E.M. Vorhees, D.K. Karman (Eds.), Proceedings of the Sixth Text Retrieval Conference (TREC-6), 1998, pp. 83–91.

CHAPTER 19

Towards a Semantic Portal for Oncology Using a Description Logic with Fuzzy Concrete Domains

Mathieu d'Aquin, Jean Lieber and Amedeo Napoli

Orpailleur research group, LORIA (CNRS, INRIA, Nancy Universities), BP 239, 54506 Vandœuvre-lès-Nancy, France

E-mail: {daquin,lieber,napoli}@loria.fr

Abstract

This paper presents three systems that are fully implemented and a proposal for a fourth one. KASIMIR is a knowledge based-system using an ad hoc formalism similar to a simple description logic with concrete domains which is used for representing decision protocols in oncology. FUZZY-KASIMIR is an extension of KASIMIR with fuzzy concrete domains taking into account discontinuities in the decision that are due to numerical thresholds. Another extension of KASIMIR has led to embed it into a semantic portal for oncology, which has been motivated by the need to share knowledge for geographically distributed physicians and has led to change the ad hoc formalism to the standard OWL DL. A combination of these two extensions of KASIMIR is currently under implementation and will lead to a semantic portal for oncology with fuzzy datatypes.

Keywords

fuzzy description logics, semantic portal for oncology, OWL DL, fuzzy datatypes

1. Introduction

The KASIMIR project studies knowledge management and decision support in oncology [4]. It has led to the implementation of the KASIMIR system in which various medical decision protocols have been represented. The most complex protocol represented so far in the KASIMIR system is the one for breast cancer treatment. Besides protocol representation, tools for edition, visualization and maintenance of the knowledge contained in protocols have been implemented in the KASIMIR system [3]. KASIMIR uses an ad hoc

FUZZY LOGIC AND THE SEMANTIC WEB
Edited by Elie Sanchez
© 2006 Elsevier B.V. All rights reserved

representation formalism close to object-based representation formalisms (OBRFs [17]) and to description logics (DLs [1]).

Two further versions of KASIMIR have been realized in parallel. The first one is FUZZY-KASIMIR: a fuzzy extension of the formalism and the inferences of KASIMIR, that can be considered as a fuzzy DL. The second one relies on semantic Web principles and technologies. KASIMIR is embedded into a semantic portal, dedicated to the diffusion of knowledge and services for oncology. Particularly, protocols are now represented and shared thanks to the OWL DL formalism, and decision support relies on standard DL reasoning within this language.

Section 2 shows how medical decision protocols are expressed in KASIMIR and why this classical representation may be insufficient because of a threshold problem in concrete domains. Section 3 presents the FUZZY-KASIMIR system, that has been implemented in order to solve this problem. The current semantic portal for oncology has been developed using the OWL DL formalism which does not integrate fuzzy notions; Section 4 describes the main features of this portal and explains how it is planned to be extended in order to have a semantic portal for fuzzy decision protocols in oncology. Section 5 discusses our approach and compares it to related work, in particular to fuzzy DLs and similar formalisms; it discusses also how fuzziness can be introduced in OWL DL, taking into account the semantic Web origin of this formalism.

2. Decision protocol representation in KASIMIR

KASIMIR uses an ad hoc formalism for representing decision protocols, that is inspired by OBRFs and DLs, but for the sake of simplicity, the syntax of the $\mathcal{ALC}(\mathtt{D})$ formalism [1] is used in this section.

Let us consider the protocol of Figure 1. It has been represented by the knowledge base of Figure 2. The $\mathtt{C}_i$'s are *concepts* representing sets of individuals. For instance, $\mathtt{C}_0$ represents the women of Figure 1's context and $\mathtt{C}_1$ represents, among these women, the ones whose age is lower than 70. The axioms [A0] to [A8] define the concepts of patients $\mathtt{C}_0$ to $\mathtt{C}_8$ by necessary and sufficient conditions. The axioms [A9] to [A13] relate the concepts of patients to concepts of treatments: an axiom $\mathtt{Pat} \sqsubseteq \exists\mathtt{reco.Ttt}$ means that for each patient of the set of individuals represented by $\mathtt{Pat}$, there is a recommended treatment in the set of treatments represented by $\mathtt{Ttt}$. For example, [A11] means that for the women represented by the concept $\mathtt{C}_5$, a treatment by tamoxifen (a hormonotherapy drug) is recommended by the protocol. Let $\mathtt{KB}$ be a knowledge base: $\mathtt{KB}$ is a set of axioms.

A concept $\mathtt{C}$ represents a set and a role $\mathtt{r}$ represents a binary relation. An interpretation $\mathcal{I}$ is a pair $(\Delta_\mathcal{I}, \cdot^\mathcal{I})$ where $\Delta_\mathcal{I}$ is a non-empty set and $\cdot^\mathcal{I}$ maps a concept $\mathtt{C}$ (resp., a role $\mathtt{r}$) to a subset $\mathtt{C}^\mathcal{I}$ of $\Delta_\mathcal{I}$ (resp., to a subset $\mathtt{r}^\mathcal{I}$ of $\Delta_\mathcal{I} \times \Delta_\mathcal{I}$). A concept is either atomic (a concept name) or defined by an expression. Given two concepts $\mathtt{C}$ and $\mathtt{D}$ and a role $\mathtt{r}$, $\mathtt{C} \sqcap \mathtt{D}$ (conjunction) and $\exists\mathtt{r.C}$ (role restriction) are two new concepts such that, for every interpretation $\mathcal{I}$, $(\mathtt{C} \sqcap \mathtt{D})^\mathcal{I} = \mathtt{C}^\mathcal{I} \cap \mathtt{D}^\mathcal{I}$ and $(\exists\mathtt{r.C})^\mathcal{I}$ is the set of the $x \in \Delta_\mathcal{I}$ for which exists $y \in \mathtt{C}^\mathcal{I}$ with $(x, y) \in \mathtt{r}^\mathcal{I}$. The third KASIMIR constructor is concrete feature restriction and is based on concrete domains.

The definitions below are inspired from [15]. A *concrete domain* $\mathtt{D}$ is a pair $(\Delta_\mathtt{D}, \Phi_\mathtt{D})$ where $\Delta_\mathtt{D}$ is a set and $\Phi_\mathtt{D}$ is a set of predicate names on $\Delta_\mathtt{D}$. In the following, only unary

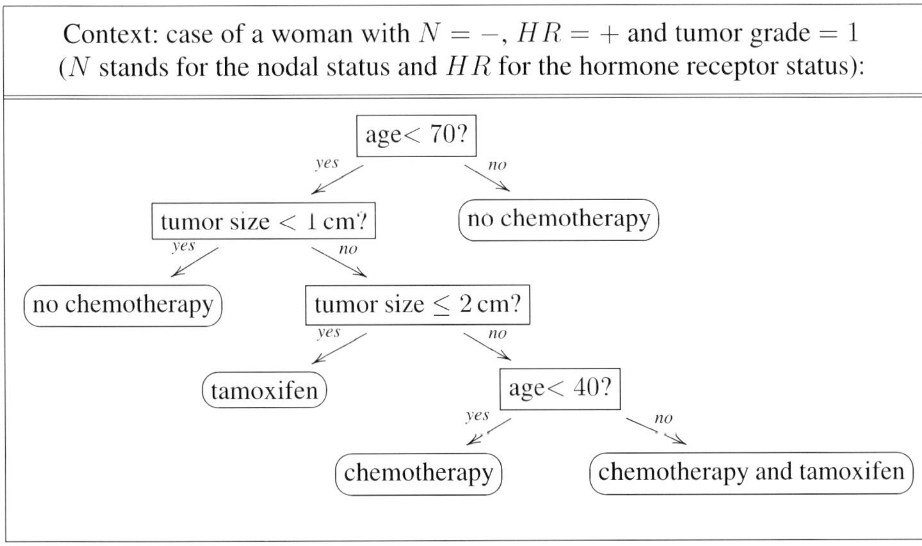

Fig. 1. A protocol represented by a decision tree (extracted and adapted from the decision protocol of breast cancer treatment described in www.oncolor.org/).

[A0] $\mathsf{C}_0 \equiv \exists\mathtt{sex.female} \sqcap \exists\mathtt{N.false} \sqcap \exists\mathtt{HR.true} \sqcap \exists\mathtt{tumor}.\exists\mathtt{grade}.=_1$

[A1] $\mathsf{C}_1 \equiv \mathsf{C}_0 \sqcap \exists\mathtt{age}.<_{70}$

[A2] $\mathsf{C}_2 \equiv \mathsf{C}_0 \sqcap \exists\mathtt{age}.\geq_{70}$

[A3] $\mathsf{C}_3 \equiv \mathsf{C}_1 \sqcap \exists\mathtt{tumor}.\exists\mathtt{size}.<_{1.0}$

[A4] $\mathsf{C}_4 \equiv \mathsf{C}_1 \sqcap \exists\mathtt{tumor}.\exists\mathtt{size}.\geq_{1.0}$

[A5] $\mathsf{C}_5 \equiv \mathsf{C}_4 \sqcap \exists\mathtt{tumor}.\exists\mathtt{size}.\leq_{2.0}$

[A6] $\mathsf{C}_6 \equiv \mathsf{C}_4 \sqcap \exists\mathtt{tumor}.\exists\mathtt{size}.>_{2.0}$

[A7] $\mathsf{C}_7 \equiv \mathsf{C}_6 \sqcap \exists\mathtt{age}.<_{40}$

[A8] $\mathsf{C}_8 \equiv \mathsf{C}_6 \sqcap \exists\mathtt{age}.\geq_{40}$

[A9] $\mathsf{C}_2 \sqsubseteq \exists\mathtt{reco.no\text{-}chemo}$

[A10] $\mathsf{C}_3 \sqsubseteq \exists\mathtt{reco.no\text{-}chemo}$

[A11] $\mathsf{C}_5 \sqsubseteq \exists\mathtt{reco.tamoxifen}$

[A12] $\mathsf{C}_7 \sqsubseteq \exists\mathtt{reco.chemotherapy}$

[A13] $\mathsf{C}_8 \sqsubseteq \exists\mathtt{reco.chemo\text{+}tam}$

Fig. 2. A knowledge base representing the protocol of Figure 1.

predicates are considered, i.e., $\varphi \in \Phi_{\mathrm{D}}$ is interpreted as a subset of Δ_{D} denoted by φ^{D}. Two concrete domains are represented in KASIMIR: $\mathsf{R} = (\Delta_{\mathsf{R}}, \Phi_{\mathsf{R}})$ and $\mathsf{Z} = (\Delta_{\mathsf{Z}}, \Phi_{\mathsf{Z}})$.

$\Delta_{\mathsf{R}} = \mathbb{R}$ is the set of real numbers and Φ_{R} is the set of the predicate names $\top_{\mathsf{R}}$, $\bot_{\mathsf{R}}$ and P_x where $P \in \{<, \leqslant, \geqslant, >\}$ and $x \in \Delta_{\mathsf{R}}$. $\top_{\mathsf{R}}$ is a name for Δ_{R} and $\bot_{\mathsf{R}}$, a name for $\emptyset$: $\top_{\mathsf{R}}^{\mathsf{R}} = \Delta_{\mathsf{R}}$, $\bot_{\mathsf{R}}^{\mathsf{R}} = \emptyset$. P_x represents the set P_x^{R} of $t \in \Delta_{\mathsf{R}}$ such that $t\ P\ x$. For example, $>_{1.}^{\mathsf{R}} =]1.; +\infty[$. More generally, P_x^{R} is an interval with one infinite bound.

$\Delta_{\mathsf{Z}} = \mathbb{Z}$ is the set of integers and Φ_{Z} is the set of the predicate names $\top_{\mathsf{Z}}$, $\bot_{\mathsf{Z}}$ and P_x where $P \in \{<, \leqslant, \geqslant, >\}$ and $x \in \Delta_{\mathsf{Z}}$. The interpretation of these predicates are $\top_{\mathsf{Z}}^{\mathsf{Z}} = \Delta_{\mathsf{Z}}$, $\bot_{\mathsf{Z}}^{\mathsf{Z}} = \emptyset$ and $P_x^{\mathsf{Z}} = \{n \in \Delta_{\mathsf{Z}} \mid n\ P\ x\}$ (e.g., $\geqslant_3^{\mathsf{Z}} = \{3, 4, 5, \ldots\}$).

Both R and Z are *admissible*, that is, for $\text{D} \in \{\text{R}, \text{Z}\}$: (1) Φ_D is closed under negation[1] and contains a name $\top_\text{D}$ for Δ_D; (2) the satisfiability of any conjunction of predicates from Φ_D is decidable.

Let $\text{D} = (\Delta_\text{D}, \Phi_\text{D})$ be a concrete domain, $\varphi \in \Phi_\text{D}$ and g be a D-concrete feature, i.e., a name interpreted by a mapping $g^\mathcal{I}$ from $\Delta_\mathcal{I}$ to Δ_D. The concept defined by $\exists g.\varphi$ represents the set of individuals $x \in \Delta_\mathcal{I}$ such that $g^\mathcal{I}(x) \in \varphi^\text{D}$. For example, if age is a Z-concrete feature, size is a R-concrete feature and tumour is a role, then $\exists\texttt{age}.{>_{70}}$ represents the set of individuals older than seventy and $\exists\texttt{tumour}.\exists\texttt{size}.{\leqslant_2}$. represents the set of individuals whose tumor size s is such that $s \leqslant 2$ cm.

Concepts of the form $\exists g.{=_x}$ are also used in the examples of this paper (for example, $\exists\texttt{grade}.{=_1}$ in Figure 2). In fact, it is not built on other predicate names, but it is the abbreviation of $\exists g.{\geqslant_x} \sqcap \exists g.{\leqslant_x}$.

Let $\mathcal{I} = (\Delta_\mathcal{I}, \cdot^\mathcal{I})$, an interpretation. $\mathcal{I}$ satisfies an axiom $\text{C} \equiv \text{D}$ if $\text{C}^\mathcal{I} = \text{D}^\mathcal{I}$. $\mathcal{I}$ satisfies an axiom $\text{C} \sqsubseteq \text{D}$ if $\text{C}^\mathcal{I} \subseteq \text{D}^\mathcal{I}$. A model of KB is an interpretation satisfying all its axioms.

The main inferences implemented in KASIMIR are the subsumption test and the classification. Given two concepts C and D, C is subsumed by D – denoted by $\text{C} \sqsubseteq \text{D}$ – if, for every model $\mathcal{I}$ of the knowledge base, $\text{C}^\mathcal{I} \subseteq \text{D}^\mathcal{I}$. Given a knowledge base KB and a concept C, classification highlights the *subsumers* of C, i.e., the concepts D of KB such that $\text{C} \sqsubseteq \text{D}$.

For example, let us consider the knowledge base of Figure 2 and a 39 year old woman with $N = -$, $HR = +$, a tumor grade of 1 and a tumor size of 3 cm. This medical case can be modeled by a concept MC (there is no mechanism for managing instances in KASIMIR):

$$\text{MC} = \text{C}_0 \sqcap \exists\texttt{age}.{=_{39}} \sqcap \exists\texttt{tumour}.\exists\texttt{size}.{=_3}.$$

The classification points out the concepts more general than MC in the knowledge base. Among them, there is the concept ∃reco.chemotherapy (cf. axioms [A0, 1, 4, 6, 7 and 12]). Thus, a chemotherapy is proposed for this patient.

Now, it can be remarked that the age 39 is close to the decision threshold 40 of the protocol. This raises a problem if the choice of this threshold is not assumed to be very accurate or if other factors, not expressed in the protocol, can be taken into account (e.g., another health problem). Indeed, if the woman was just one year older, this would lead to a different proposition of treatment (chemo+tam, cf. axioms [A0, 1, 4, 6, 8 and 13]). Therefore, it seems reasonable that the KASIMIR system proposes *both* treatments chemotherapy and chemo+tam (the latter is a treatment with a cure of chemotherapy and a cure of tamoxifen). The KASIMIR user, a physician, can use these propositions to build his/her own proposition that can be a choice between them or a combination of them (it may be the same chemotherapy and another hormonotherapy, with a lower dose of tamoxifen). To implement this mechanism, (classical) thresholds have been replaced by fuzzy thresholds; the extension of KASIMIR for managing them is presented in the next section.

3. FUZZY-KASIMIR: formalism and inferences

This section presents how fuzziness is introduced into the formalism of KASIMIR. First, some notions about fuzzy sets are reminded (for more information about fuzzy sets, see [7]).

[1] I.e., for each $\varphi \in \Phi_\text{D}$, exists $\psi \in \Phi_\text{D}$ such that $\psi^\text{D} = \Delta_\text{D} \backslash \varphi^\text{D}$.

Given a set X, called the *universe*, a fuzzy subset of X is a function $A: x \in X \mapsto A(x) \in [0; 1]$ (for the sake of simplicity, a fuzzy set and its membership function are not distinguished). Let A and B be two fuzzy subsets of X. A is a subset of B – denoted by $A \subseteq B$ – if, for every $x \in X$, $A(x) \leqslant B(x)$. $A \cap B$ is the fuzzy subset of X such that $(A \cap B)(x) = \min(A(x), B(x))$, for $x \in X$. Fuzzy logics are extensions of classical logic such that the truth values are elements of [0; 1] (0 stands for "false" and 1, for "true") and such that the logical connectors are extended on [0; 1]. We use in the following the Zadeh fuzzy logic for which, in particular, $\wedge$ is fuzzified in min, $\vee$ in max and $\neg$ in $n: x \mapsto 1 - x$. The implication $\Rightarrow$ can be fuzzified in different ways by a function $F_{\Rightarrow}: (x, y) \in [0; 1]^2 \mapsto F_{\Rightarrow}(x, y)$. In the following, the Łukasiewicz implication defined by $F_{\Rightarrow}(x, y) = \min(1, 1-x+y)$ is used (we could also have chosen the Gödel implication as in [5]). Based on $F_{\Rightarrow}$, the gradual inclusion between two fuzzy subsets A and B of X can be defined by $F_{\subseteq}(A, B) = \inf_{x \in X} F_{\Rightarrow}(A(x), B(x))$. Note that $F_{\subseteq}(A, B) = 1$ iff $A \subseteq B$. Given a fuzzy subset A of X and $\alpha \in]0; 1]$, A_α, the α-*cut* of A, is a classical set such that $x \in A_\alpha$ iff $A(x) \geqslant \alpha$. If $\alpha = 1$, A_α is called the *core* of A. The *support* of A is $\{x \in X \mid A(x) > 0\}$.

The formalism of FUZZY-KASIMIR extends the one of KASIMIR by the addition of new predicate names in the concrete domains $\mathtt{D} \in \{\mathsf{R}, \mathsf{Z}\}$. Such a new predicate name φ is interpreted as a fuzzy unary predicate $\varphi^{\mathtt{D}}$, i.e., as a fuzzy subset of $\Delta_{\mathtt{D}}$.

The new predicate names of the concrete domain R are $\leqslant_{a \pm b}$ and $\geqslant_{a \pm b}$ where $a, b \in \Delta_{\mathsf{R}}$ and $b > 0$. They are interpreted by:

$$\text{for } x \in \Delta_{\mathsf{R}}, \qquad \leqslant^{\mathsf{R}}_{a\pm b}(x) = \begin{cases} 1 & \text{if } x \leqslant a - b, \\ \dfrac{a + b - x}{2b} & \text{if } x \in [a - b; a + b], \\ 0 & \text{if } x \geqslant a + b], \end{cases}$$

$$\geqslant^{\mathsf{R}}_{a\pm b}(x) = n\big(\leqslant^{\mathsf{R}}_{a\pm b}(x)\big) = 1 - \leqslant^{\mathsf{R}}_{a\pm b}(x).$$

The new predicate names of Z are $\leqslant_{a \pm b}$ and $\geqslant_{a \pm b}$ with $a, b \in \Delta_{\mathsf{Z}}$ and $b > 0$. The interpretations of these predicate names are the same as for the concrete domain R, except that the universe is Δ_{Z}.

R and Z remain admissible with the addition of these new predicate names, if the conjunction and negation used in the definition of admissibility are min and n.

The semantics of a FUZZY-KASIMIR concept C is, for an interpretation $\mathcal{I}$, a fuzzy subset $\mathtt{C}^{\mathcal{I}}$ of $\Delta_{\mathcal{I}}$. Given two concepts C and D, a role r, a concrete domain $\mathtt{D} = (\Delta_{\mathtt{D}}, \Phi_{\mathtt{D}})$, a D-concrete feature g and $\varphi \in \Phi_{\mathtt{D}}$, $\cdot^{\mathcal{I}}$ maps the concept $\mathtt{C} \sqcap \mathtt{D}$ into the fuzzy set $\mathtt{C}^{\mathcal{I}} \cap \mathtt{D}^{\mathcal{I}}$, the concept $\exists \mathtt{r}.\mathtt{C}$ into the fuzzy set

$$(\exists \mathtt{r}.\mathtt{C})^{\mathcal{I}}: x \in \Delta_{\mathcal{I}} \mapsto \sup_{x \in \Delta_{\mathcal{I}}} \min\big(\mathtt{r}^{\mathcal{I}}(x, y), \mathtt{C}^{\mathcal{I}}(y)\big) \in [0; 1]$$

and the concept $\exists g.\varphi$ into the fuzzy set

$$(\exists g.\mathtt{C})^{\mathcal{I}}: x \in \Delta_{\mathcal{I}} \mapsto \varphi^{\mathtt{D}}\big(g^{\mathcal{I}}(x)\big) \in [0; 1].$$

The subsumption between concepts of FUZZY-KASIMIR can be defined as in KASIMIR: $\mathtt{C} \sqsubseteq \mathtt{D}$ if for every model of KB, $\mathtt{C}^{\mathcal{I}} \subseteq \mathtt{D}^{\mathcal{I}}$, the only difference is that here, $\subseteq$ stands for inclusion between fuzzy sets. Therefore, the hierarchical classification can be still applied. Another inference implemented in FUZZY-KASIMIR is the f*degree of subsumption* between two concepts C and D:

$$F_{\sqsubseteq}(\mathtt{C}, \mathtt{D}) = \inf\big\{F_{\subseteq}\big(\mathtt{C}^{\mathcal{I}}, \mathtt{D}^{\mathcal{I}}\big) \mid \mathcal{I}\text{: model of KB}\big\}.$$

It must be noticed that $F_{\sqsubseteq}(\mathtt{C}, \mathtt{D}) = 1$ iff $\mathtt{C} \sqsubseteq \mathtt{D}$. The fuzzy hierarchical classification highlights the concepts D of KB such that $F_{\sqsubseteq}(\mathtt{C}, \mathtt{D}) > 0$. An algorithm for fuzzy hierarchical classification is presented in [14].

For instance, let KB be the knowledge base obtained by substituting in the knowledge base of Figure 2, the axioms [A6], [A7] and [A8], respectively by

[A6′] $\mathtt{C}_6 \equiv \mathtt{C}_4 \sqcap \exists\mathtt{tumor}.\exists\mathtt{size}.\geqslant_{2.\pm 1}$.

[A7′] $\mathtt{C}_7 \equiv \mathtt{C}_6 \sqcap \exists\mathtt{age}.\leqslant_{40\pm 5}$

[A8′] $\mathtt{C}_8 \equiv \mathtt{C}_6 \sqcap \exists\mathtt{age}.\geqslant_{40\pm 5}$

Let us now consider the example of Section 2 with the new knowledge base. First, it can be noticed that MC is not subsumed with a degree 1 by any concept related to a treatment by an axiom. By contrast, it is subsumed with degrees of]0; 1[by ∃reco.chemotherapy and ∃reco.chemo+tam, with the respective degrees 0.6 and 0.4 (cf. axioms [A0, 1, 4, 6, 7′ and 12] and axioms [A0, 1, 4, 6, 8′ and 13]).

Therefore, the system recommends the two treatments chemotherapy and chemo+tam, with a preference for the former. Thus, the physician using FUZZY-KASIMIR decides either one of these treatments, or a combination of them (e.g., with a compromise in the doses of the chemotherapy and hormonotherapy drugs).

4. Towards a semantic portal for oncology based on fuzzy datatypes

One purpose of the KASIMIR project is the diffusion of standard knowledge in oncology, in order to improve health-care practice. KASIMIR has to provide a direct, updated and intelligent access to the current knowledge, for geographically distributed users. This leads to standardization and distribution requirements for both knowledge and software components within KASIMIR. For this reason, the KASIMIR system is currently embedded into a semantic portal for oncology. This semantic portal is a Web server used to supply and share knowledge and intelligent services for oncology, thanks to semantic Web principles and technologies. In particular, medical decision protocols are represented within the OWL DL formalism and supplied services are based on standard DL reasoning with this formalism. Note that this leads to an improved expressivity with respect to the ad hoc KASIMIR formalism, and that it allows KASIMIR to take advantage of the other knowledge sources and tools available on the semantic Web.

In the following, we focus on the representation of medical decision protocols within the OWL DL formalism, and particularly on the currently debated subject of concrete domain representation (Section 4.1). As explained in the previous section, representation of decision protocols leads us to introduce fuzziness in the representation formalism. Then, on the basis of lessons learnt from the development of FUZZY-KASIMIR, Section 4.2 discusses the introduction of fuzzy datatypes into OWL DL.

4.1. *Protocol representation in the current semantic portal*

The semantic portal has been developed within an architecture based on Web services dedicated to OWL reasoning. The KASIMIR user interface, used to associate a treatment to a patient description, has been replaced by a more general instance editor called EDHIBOU that uses these Web services. The principles of the translation of KASIMIR knowledge bases in OWL DL is presented hereafter.

The KASIMIR ad hoc syntax is closer to object-based representation formalisms than to DLs. A first work was to translate it in the DL syntax of $\mathcal{ALC}(\mathrm{D})$. This part of this translation is not detailed here: KASIMIR has already been described in the DL syntax in the previous sections.

The representation of concrete domains in OWL is based on *XML Schema datatypes*. A datatype corresponds to a (possibly infinite) set of values that are addressed through literals. A literal is a string that is interpreted as a value of a datatype. For example, `"40"^^xsd:integer` is a literal representing the integer `40`, `xsd:integer` being the XML Schema datatype for integers. There are two disjoint types of properties in OWL: object properties, that relate instances to other instances, and datatype properties, that relate instances to literals. Object properties correspond to roles and datatype properties to concrete features in classical DLs. Now, some concept constructors can use datatype properties and datatypes to build complex concepts. For example, the concept ∃`age.integer`, where `age` is a datatype property, corresponds to $\exists \texttt{age}.\top_{\mathbb{Z}}$ in the previous KASIMIR representation. In the XMLsyntax of OWLL, ∃`age.integer` can be written:

```
<owl:Restriction>
    <owl:onProperty rdf:resource="#age"/>
    <owl:someValuesFrom rdf:resource=
      "http://www.w3.org/2001/XMLSchema#integer"/>
</owl:Restriction>
```

XML Schema provides mechanisms to build user defined datatypes like *derived simple types*. In particular, restrictions on existing datatypes can be used to express basic unary predicates like the ones that are used for protocol representation in KASIMIR. For example, $\exists \texttt{age}.{<_{40}}$ in KASIMIR is expressed in OWL by ∃`age.integerUnder40`, with `integerUnder40` the derived type of the integers $n < 40$, defined by:

```
<xsd:simpleType name="integerUnder40">
    <xsd:restriction base="integer">
      <xsd:maxExclusive value="40"/>
    </xsd:retriction>
</xsd:simpleType>
```

Then, subsumption between concepts is performed thanks to the *subtype test* between XML Schema datatypes. For example, if `integerBetween20and30` is a subtype of `integerUnder40`, then ∃`age.integerBetween20and30` ⊑ ∃`age.integerUnder40`.

The problem is that derived types are not supported in the current OWL DL recommendation and that protocol representation for the KASIMIR portal uses it in a non-standard way. Actually, only *built-in* datatypes (`xsd:integer`, `xsd:float`, etc.) are taken into account, but current studies like OWL-E [18] aims at extending these recommendations for expressive datatype expressions. Thus, there is a reasonable hope that a future recommendation extending OWL DL will include user-defined or derived datatypes.

4.2. *The future semantic portal, with fuzzy datatypes*

An important preliminary remark must be done about OWL reasoning with datatypes. Datatype properties and object properties are disjoint in OWL, thus a reasoner on OWL can be composed of two distinct parts:

- The *concept reasoner* deals with concepts and, whenever it meets a datatype, calls the datatype reasoner.
- The *datatype reasoner* deals with datatypes and provides services (such as the subtype test) independently from the concepts of the DL.

This principle can be used to implement in a straightforward way two inferences on OWL with fuzzy datatypes: the subsumption ($\sqsubseteq$) and the α-subsumption ($\sqsubseteq_\alpha$). The implementation of the degree of subsumption ($F_\sqsubseteq$) is more complex and will be discussed.

REMARK. Many current DL reasoners use the method of semantic tableau to compute the subsumption test. In fact, the test computed by this method is the satisfiability of a concept and it calls the test of consistency of a datatype implemented in the datatype reasoner [19]. However, subsumption test between concepts (resp., subtype test between datatypes) can be computed thanks to concept satisfiability (resp., datatype consistency). Thus, in the following, only the subsumption test between concepts and the subtype test between datatypes are considered.

4.2.1. *Fuzzy datatypes*

A (classical) datatype represents a set of values. For example, the datatype `integerUnder40` represents the set {39, 38, 37, ...}, i.e. $<^{\mathrm{Z}}_{40}$ with the notation of Section 2. We introduce the notion of *fuzzy datatype*, which represents a *fuzzy set* of values. For example, let `integerUnder40±5` be the fuzzy datatype representing the fuzzy set $\leqslant^{\mathrm{Z}}_{40\pm5}$, with the notation of Section 3. More generally, it is assumed that a syntax is defined to introduce some fuzzy datatypes, in particular, the fuzzy datatypes `integerUnder`$a\pm b$ and `integerAbove`$a\pm b$ ($a, b \in \Delta_{\mathrm{Z}}$, $b > 0$), representing respectively the fuzzy sets $\leqslant^{\mathrm{Z}}_{a\pm b}$ and $\geqslant^{\mathrm{Z}}_{a\pm b}$. For example, we can imagine that a definition of the fuzzy datatype `integerUnder40±5` could be something like:

```
<xsd:simpleType name="integerUnder40pm5">
    <xsd:restriction base="integer">
      <xsd:maxInclusive value="40" plusOrMinus="5"/>
    </xsd:retriction>
</xsd:simpleType>
```

Another possibility is to follow the idea presented in [5] and to define this fuzzy datatype by two classical datatypes: `integerUnderEq35` representing the core of $\leqslant^{\mathrm{Z}}_{40\pm5}$ and `integerUnder45` representing its support. Then the definition of `integerUnder40±5` could be:

```
<fxsd:fuzzyType name="integerUnder40pm5">
    <fxsd:core value="integerUnderEq35"/>
    <fxsd:support value="integerUnder45"/>
</fxsd:fuzzyType>
```

where `fxsd` would refer to a new namespace for fuzzy XML Schema datatype definitions.

Let fd1 and fd2 be two fuzzy datatypes and A_1 and A_2, the corresponding fuzzy sets. Let $F_{\preccurlyeq}(\texttt{fd1}, \texttt{fd2}) = F_{\subseteq}(A_1, A_2)$, the so-called *subtype degree* of fd1 in fd2. For example:

$$F_{\preccurlyeq}(\texttt{integerUnder40}\pm\texttt{5}, \texttt{integerUnder42}\pm\texttt{8}) = 0.935.$$

For $\alpha \in]0; 1]$, let $\preccurlyeq_\alpha$ be the crisp relation between fuzzy datatypes defined by fd1 $\preccurlyeq_\alpha$ fd2 if $F_{\preccurlyeq}(\texttt{fd1}, \texttt{fd2}) \geqslant \alpha$. We call $\preccurlyeq_\alpha$ the α-subtype relation (it is the α-cut of $F_{\preccurlyeq}$). Finally, let $\preccurlyeq$ be the so-called *subtype relation* between fuzzy datatypes, i.e., fd1 $\preccurlyeq$ fd2 if $F_{\preccurlyeq}(\texttt{fd1}, \texttt{fd2}) = 1$ (in other words, $\preccurlyeq \,=\, \preccurlyeq_1$). $\preccurlyeq$ verifies fd1 $\preccurlyeq$ fd2 iff $A_1 \subseteq A_2$. It is assumed that, for the fuzzy datatypes that are introduced, $F_{\preccurlyeq}$ (and thus $\preccurlyeq_\alpha$ and $\preccurlyeq$) is computable. If the only fuzzy datatypes introduced are of the forms integerUnder$a\pm b$ and integerAbove$a\pm b$, this property of computability holds.[2]

The crisp and fuzzy relations $\preccurlyeq$, $\preccurlyeq_\alpha$ and $F_{\preccurlyeq}$ between fuzzy datatypes are used in the following to compute the relations $\sqsubseteq$, $\sqsubseteq_\alpha$ and $F_{\sqsubseteq}$ between concepts.

4.2.2. *Subsumption between concepts with fuzzy datatypes* ($\sqsubseteq$)

Let C and D be two concepts described in OWL DL, possibly using fuzzy datatypes. As mentioned above, the concept reasoner used to compute the test C $\sqsubseteq$ D uses the services of the datatype reasoner, in particular the subtype test. Thus, to compute the subsumption test, it is not necessary to modify the concept reasoner, but only the datatype reasoner, so that it can take into account fuzzy datatypes and the computation of $\preccurlyeq$.

Let us consider these two examples:

$$\exists\texttt{age.integerUnder40}\pm\texttt{5} \sqsubseteq \exists\texttt{age.integerUnder50}\pm\texttt{10}$$

$$\exists\texttt{age.integerUnder40}\pm\texttt{5} \not\sqsubseteq \exists\texttt{age.integerUnder42}\pm\texttt{8}$$

To compute these tests, the concept reasoner calls the datatype reasoner with the queries:

integerUnder40±5 $\preccurlyeq$ integerUnder50±10? (answer: *yes*)

integerUnder40±5 $\preccurlyeq$ integerUnder42±8? (answer: *no*)

4.2.3. *α-subsumption between concepts with fuzzy datatypes* ($\sqsubseteq_\alpha$)

Let $\alpha \in [0; 1]$ and $\sqsubseteq_\alpha$ be the binary relation between concepts defined by C $\sqsubseteq_\alpha$ D if $F_{\sqsubseteq}(\texttt{C}, \texttt{D}) \geqslant \alpha$. As it was the case for $\sqsubseteq$, the test $\texttt{C}_1 \sqsubseteq_\alpha \texttt{C}_2$ can be computed without modifying the concept reasoner. The principle of the implementation of $\sqsubseteq_\alpha$, as it is planned, is illustrated by Figure 3. Let us assume that a query C $\sqsubseteq_\alpha$ D is sent to the service for the α-subsumption. This service gives the parameter α to the fuzzy datatype reasoner, so that, each time a subtype test is requested for a pair (fd1, fd2) of fuzzy datatypes, it computes fd1 $\preccurlyeq_\alpha$ fd2. Therefore, the computation of the test C $\sqsubseteq_\alpha$ D amounts to the computation by the concept reasoner of C $\sqsubseteq$ D, which calls the fuzzy datatype reasoner parameterized by the value α.

[2] Indeed, in order to compute $F_{\preccurlyeq}(\texttt{fd1}, \texttt{fd2}) = \inf_x F_{\Rightarrow}(A_1(x), A_2(x))$, it is sufficient to compute this for $x \in \{a_1 - b_1, a_1 + b_1, a_2 - b_2, a_2 + b_2\}$, for fd$i$ = integerUnder$a_i \pm b_i$ or fd$i$ = integerAbove$a_i \pm b_i$, $i \in \{1, 2\}$.

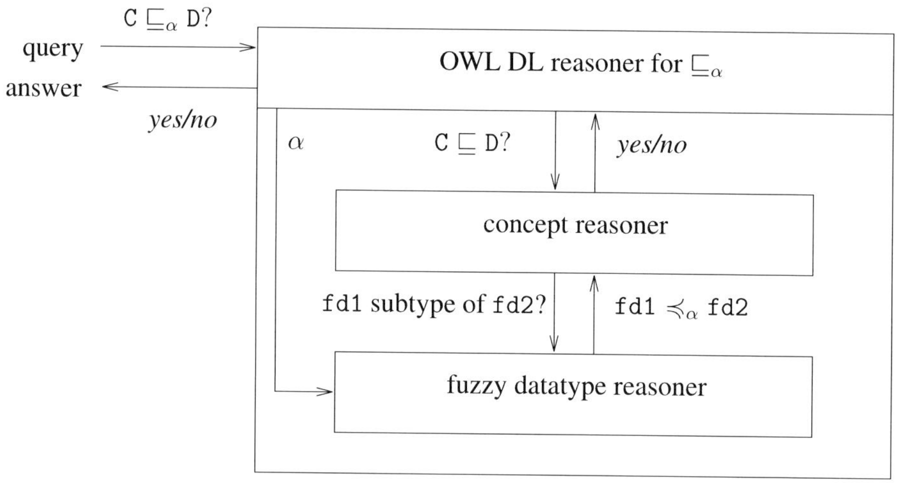

Fig. 3. Architecture of an OWL DL reasoner supporting the $\sqsubseteq_\alpha$ inference.

This service should give the following results on these examples:

$\exists$age.integerUnder40±5 $\sqsubseteq_\alpha$ $\exists$age.integerUnder50±10

for any $\alpha \in [0; 1]$

$\exists$age.integerUnder40±5 $\sqsubseteq_\alpha$ $\exists$age.integerUnder42±8

iff $\alpha \leqslant 0.935$

REMARK. For this computation it is assumed that the concept reasoner does not store the results of previous subsumption tests for future reuse, or that this storage can be reseted. Indeed, if the result of any subsumption test is stored, then, after a test $\mathtt{C} \sqsubseteq_\alpha \mathtt{D}$ has been performed with a positive answer, any test $\mathtt{C} \sqsubseteq_\beta \mathtt{D}$, with the same concepts C and D, will reuse this answer!

4.2.4. *Degree of subsumption between concepts with fuzzy datatypes ($F_\sqsubseteq$)*

What makes possible the implementation of $\sqsubseteq$ and $\sqsubseteq_\alpha$ without modifying the concept reasoner is that the result of these tests are binary. Now, a service implementing $F_\sqsubseteq$ would return a fuzzy truth value, thus using the output of the classical concept reasoner as the output of this service is not possible. However, it is possible to reuse – without changing it – the concept reasoner in order to implement $F_\sqsubseteq$. The principle is as follows. Let us assume that the α-subsumption has been implemented according to the architecture of Figure 3. The computation of $F_\sqsubseteq(\mathtt{C}, \mathtt{D})$ is based on the following equation:

$$F_\sqsubseteq(\mathtt{C}, \mathtt{D}) = \max\{\alpha \in [0; 1] \mid \mathtt{C} \sqsubseteq_\alpha \mathtt{D}\}. \tag{1}$$

Thus, to compute $F_\sqsubseteq(\mathtt{C}, \mathtt{D})$, it is sufficient to find this maximum value of α. This search can be done by dichotomy, given ε the precision of the required result (i.e., $|F_\sqsubseteq(\mathtt{C}, \mathtt{D}) - \alpha| \leqslant \varepsilon$, where α is the value that is computed by this search):

Table 1
Fuzzy DLs and related work, according to the way fuzziness is introduced in the formalism (F-K stands for FUZZY-KASIMIR).

	[26]	[25]	[16]	[22]	[23]	[9]	[24]	[20]	[11,10]	[13]	[6]	F-K
(FPCD)	×		×				×				×	×
(M)		×	×			×	×					
(FQ)								×				
(FAX)				×			×		×	×		
(FA)				×	×	×	×		×	×		

1. Let $a = 0$, $b = 1$ and $\alpha = \frac{1}{2}$.
2. While $b - a > \varepsilon$ do
   2.i If $\mathrm{C} \sqsubseteq_\alpha \mathrm{D}$ (cf. Section 4.2.3) then let $a = \alpha$ else let $b = \alpha$.
   2.ii Let $\alpha = \frac{a+b}{2}$.
3. Return α

This computation requires at most $\lceil -\log_2 \varepsilon \rceil$ queries $\mathrm{C} \sqsubseteq_\alpha \mathrm{D}$ (where $\lceil x \rceil$ is the lowest integer a such that $x \leqslant a$).

This algorithm can be modified to provide an exact result if the set of the fuzzy truth values explored is finite. This occurs if a finite scale of fuzzy truth values is used, as suggested in [5].

An ongoing study aims at finding an *exact* value of $F_\sqsubseteq(\mathrm{C}, \mathrm{D})$, even with the scale of truth values [0; 1]. The principle would be, given the concepts C and D, to make the fuzzy datatype reasoner generate a finite set of values *FTV*. Then, the dichotomy can be performed in *FTV*, which would require at most $1 + \lceil \log_2 |FTV| \rceil$ tests $\mathrm{C} \sqsubseteq_\alpha \mathrm{D}$. How this set can be efficiently generated is still an open question.

Another way to implement $F_\sqsubseteq$ could be to modify the concept reasoner so that it can manipulate fuzzy truth values, but designing and implementing a fuzzy concept reasoner which is as optimized as current (classical) DL reasoners is a much more complex task, from both theoretical and applied viewpoints.[3]

5. Discussion and related work

The contribution of this paper is twofold. First, a description of a simple representation language similar to a fuzzy DL – the FUZZY-KASIMIR formalism – has been described. Section 5.1 situates this language among the family of fuzzy DLs and related formalisms. The second contribution is our proposal of a semantic portal using the formalism OWL DL with fuzzy datatypes. OWL DL is not only a formalism equivalent to an expressive DL, it is also a language that has been designed for the semantic Web, which involves additional requirements. Section 5.2 shows how our proposal fulfills these requirements.

[3] For this purpose, Umberto Straccia has suggested to use techniques for mixed integer programming [24].

5.1. *Fuzzy DLs: how fuzziness is introduced into DLs*

The fuzzy DLs we have met in the literature are built following the same general principle: starting from a (classical) DL, such as $\mathcal{ALC}$, additional features are added to it in order to express fuzziness. In any case, the semantics consists in interpreting a concept by a fuzzy set and a role by a fuzzy binary relation. By contrast, there are different ways to introduce fuzziness into the DL formalism. To our knowledge, there are five of these ways:

(FPCD) Fuzzy predicates in concrete domains,
(M) Modifiers of concepts,
(FQ) Fuzzy quantifiers,
(FAX) Fuzzy axioms and
(FA) Fuzzy assertions.

Table 1 presents some fuzzy DLs and related formalisms, under the viewpoint of these five fuzziness introduction types. The related formalisms are either DLs taking into account some uncertainty (possibilistic logic [11,10] or probabilistic reasoning [13]), or object-based representation formalisms with fuzzy features (like [6] and FUZZY-KASIMIR).

(FPCD) is the approach used in FUZZY-KASIMIR. Sub-categories of (FPCD) can be defined by precising what are the concrete domains involved and what are the families of fuzzy predicate names that are supported.

(M) is based on the constructors $m\mathrm{C}$, where m is a modifier (also called a manipulator or a hedge) and C is a concept. A modifier is a function $m : [0; 1] \to [0; 1]$. For example, the modifier *very* is often defined by $very(t) = t^2$. Given an interpretation $\mathcal{I}$, $m\mathrm{C}$ is interpreted by $(m\mathrm{C})^{\mathcal{I}}(x) = m(\mathrm{C}^{\mathcal{I}}(x))$ for $x \in \Delta_{\mathcal{I}}$. Different families of modifiers give birth to different sub-categories of (M).

(FQ) consists in extending the classical quantifiers of DLs ($\exists$ and $\forall$) with fuzzy quantifiers. The language of the chosen fuzzy quantifiers characterizes the fuzzy extension of the DL.

(FAX), contrasting with the previous ways of introducing fuzziness, does not consist in defining new concept constructors, but it consists in associating to axioms some pieces of informations about them, such as constraints on their fuzzy truth values.

(FA) is based on the same idea as (FAX), except that it concerns assertions (i.e., specifications on particular instances) and not concepts.

Finally, as Table 1 shows, combinations of the ways of introducing fuzziness into DLs can be done.

5.2. *Introduction of fuzziness into OWL DL*

OWL DL can be seen as an expressive DL designed to be the standard language for representing ontologies in the semantic Web. It is a "semantic Web DL" which involves some particular features that must be taken into account in the building of a fuzzy extension for this formalism:

- OWL is standard. It has the status of a W3C recommendation.
- There exist open, stable and efficient systems that are actually used for reasoning with OWL.
- OWL is divided into sub-languages (Lite, DL and Full) that differ in their level of complexity, and in the expected usage of the included constructors in practical applications.

The syntax and the semantics of OWL are defined in a recommendation of the W3C (World Wide Web Consortium) [2]. This means that it is a standard language and so, that an OWL ontology must be considered in the same way, i.e. leads to the same behavior, in different OWL implementations. Being a standard also constraints the language to be stable in time: the extension of a standard technology has to be compatible with the extended technology (principle of backward compatibility). Considering a fuzzy extension of OWL, any ontology described in the classical OWL formalism must also be an ontology in the fuzzy OWL formalism and so, must lead to the same inferences in the two languages. This constraint is fulfilled by the fuzzy extension proposed in this paper. Indeed, on the basis of XML Schema datatypes, datatype definitions are extended for introducing fuzziness in our framework. Since the OWL recommendation is not directly concerned with datatype definitions, the actual syntax of the OWL formalism remains the same. Moreover, the semantics introduced to take into account fuzzy datatypes in OWL DL is fully compatible with ontologies made to be interpreted in a classical way, using the classical semantics. For example, if C and D are two concepts of an ontology that does not use any fuzzy datatype, then C is subsumed by D in the classical sense, iff C is subsumed by D in the fuzzy sense.

Several systems have been developed for reasoning on $\mathcal{SHOIN}(\mathrm{D})$, the OWL DL underlying description logic (e.g., FaCT, Pellet and Racer [12,21,8]). Since these systems are open, stable and efficient, the development of a reasoner for OWL DL with fuzzy datatypes can benefit from reusing them. As said in Section 4.2, datatype reasoning and concept reasoning are two distinct parts of an OWL DL reasoner. In this way, the reasoner for OWL DL with fuzzy datatypes can be built on top of an existing concept reasoner without modifying it and so, can have benefit of its efficiency.

Furthermore, this layered architecture comes in complement of the existing layers of OWL (Lite, DL and Full), and fuzzy datatypes can be used in any sub-language of OWL. Moreover, in the same way that the constructors included in OWL Lite provide "a minimal useful subset of language features, that are relatively straightforward for tool developers to support" [2], using fuzzy datatypes seems to be a simple and useful way of introducing fuzziness in OWL. Indeed, in our application for oncology, this choice is motivated by requirements concerning the representation of medical decision protocols. In many other applications, the need to define fuzzy knowledge elements also comes from thresholds on concrete data (most often numeric data). For this reason, the proposed extension of OWL with fuzzy datatypes and the future associated reasoner, should provide a simple answer to the need for fuzzy representation in many semantic Web applications.

6. Conclusion

This paper presents three systems that are fully implemented (including user interfaces), and a proposal for a fourth one. Kasimir is a knowledge-based system using an ad hoc formalism similar to a simple description logic with concrete domains which is used for representing decision protocols in oncology. Fuzzy-Kasimir is an extension of Kasimir with fuzzy concrete domains taking into account discontinuities in the decision that are due to numerical thresholds. Another extension of Kasimir has led to embed it into a semantic portal for oncology, which has been motivated by the need to share knowledge for geographically distributed physicians and has led to change the ad hoc formalism to the standard OWL DL. A combination of these two extensions of Kasimir (Fuzzy-Kasimir and the semantic portal for oncology) is currently under implementation and will lead to a semantic portal for oncology with fuzzy datatypes. The implementation of this new portal follows the principles of the semantic Web (in particular, backward compatibility) and is based on the decomposition of an OWL DL reasoner into a concept reasoner and a datatype reasoner. The inferences that are studied are the subsumption, the α-subsumption and the degree of subsumption between OWL DL concepts with fuzzy datatypes.

References

[1] F. Baader, D. Calvanese, D. McGuinness, D. Nardi, P. Patel-Schneider (Eds.), The Description Logic Handbook, Cambridge University Press, Cambridge, UK, 2003.

[2] S. Bechhofer, F. van Harmelen, J. Hendler, I. Horrocks, D.L. McGuinness, P.F. Patel-Schneider, L.A. Stein, *OWL Web Ontology Language Reference*, www.w3.org/TR/owl-ref. Last consultation: February 2005.

[3] M. d'Aquin, C. Bouthier, S. Brachais, J. Lieber, A. Napoli, *Knowledge edition and maintenance tools for a semantic portal in oncology*, Internat. J. Human Computer Studies **62** (5) (2005), 545–552.

[4] M. d'Aquin, S. Brachais, J. Lieber, A. Napoli, *Decision support and knowledge management in oncology using hierarchical classification*, in: K. Kaiser, S. Miksch, S.W. Tu (Eds.), Proceedings of the Symposium on Computerized Guidelines and Protocols (CGP-2004), Studies in Health Technology and Informatics, vol. 101, IOS Press, 2004, pp. 16–30.

[5] D. Dubois, J. Mengin, H. Prade, *Possibilistic uncertainly and fuzzy features in description logic, a preliminary discussion*, in: E. Sanchez (Ed.), Proceedings of the Workshop Fuzzy Logic and the Semantic Web, Marseille, France, 2005, pp. 5–7.

[6] D. Dubois, H. Prade, J.-P. Rossazza, *Vagueness, typicality and uncertainty in class hierarchies*, Internat. J. Intelligent Systems **6** (1991), 167–183.

[7] D. Dubois, H. Prade (Eds.), Fundamentals of Fuzzy Sets, The Handbooks of Fuzzy Sets, vol. 7, Kluwer Academic Publishers, 2000.

[8] V. Haarslev, R. Möller, *Description of the racer system and its applications*, in: C. Goble, D.L. McGuinness, R. Möller, P.F. Patel-Schneider (Eds.), Proceedings of the 2001 International Description Logics Workshop, 2001, pp. 46–55.

[9] S. Hölldobler, T.D. Khang, H.P. Störr, *A fuzzy description logic with hedges as concept modifiers*, in: N.H. Phuong, H.T. Nguyen, N.C. Ho, P. Santiprabhob (Eds.), Proceedings of the Joint Third International Conference on Intelligent Technologies and Third Vietnam–Japan Symposium on Fuzzy Systems and Applications (InTech/VJFuzzy'2002), Hanoi, Vietnam, 2002, pp. 25–34.

[10] B. Hollunder, *An alternative proof method for possibilistic logic and its application to terminological logics*, Research Report RR-93-01, DFKI, 1993.

[11] B. Hollunder, *An alternative proof method for possibilistic logic and its application to terminological logics*, in: R. Lopez de Mantaras, D. Poole (Eds.), Proceedings of the 10th Conference on Uncertainty in Artificial Intelligence, Morgan Kaufmann Publishers, San Francisco, CA, 1994, pp. 327–335.

[12] I. Horrocks, *Using an expressive description logic: FaCT or fiction?* in: A.G. Cohn, L. Schubert, S.C. Shapiro (Eds.), Principles of Knowledge Representation and Reasoning: Proceedings of the Sixth International Conference (KR'98), Morgan Kaufmann Publishers, June 1998, pp. 636–647.

[13] M. Jaeger, *Probabilistic reasoning in terminological logics*, in: J. Doyle, E. Sandewall, P. Torasso (Eds.), KR'94: Principles of Knowledge Representation and Reasoning, Morgan Kaufmann, 1994, pp. 305–316.

[14] J. Lieber, *Strong, fuzzy and smooth hierarchical classification for cae-based problem solving*, in: F. van Harmelen (Ed.), Proceedings of the 15th European Conference on Artificial Intelligence (ECAI-02), Lyon, France, IOS Press, Amsterdam, 2002, pp. 81–85.

[15] C. Lutz, *Description logics with concrete domains – a survey*, in: Advances in Modal Logics, vol. 4, King's College Publications, 2003.

[16] R. Molitor, C.B. Tresp, *Extending description logics to vague knowledge in medicine*, in: P. Szczepaniak, P.J.G. Lisboa, S. Tsumoto (Eds.), Fuzzy Systems in Medicine, Studies in Fuzziness and Soft Computing, vol. 41, Springer-Verlag, 2000, pp. 617–635.

[17] A. Napoli, C. Laurenço, R. Ducournau, *An object-based representation system for organic synthesis planning*, Internat. J. Human–Computer Studies **41** (1/2) (1994), 5–32.

[18] J.Z. Pan, *Reasoning Support for OWL-E (extended abstract)*, in: Proc. of Doctoral Programme in the 2004 International Joint Conference of Automated Reasoning (IJCAR2004), July 2004.

[19] J.Z. Pan, I. Horrocks, *Web ontology reasoning with datatype groups*, in: D. Fensel, K. Sycara, J. Mylopoulos (Eds.), Proceedings of the 2nd International Semantic Web Conference (ISWC2003), 2003.

[20] D. Sánchez, A.G.B. Tettamanzi, *Reasoning and quantification in fuzzy description logics*, in: Proceedings of the Workshop Fuzzy Logic and the Semantic Web, Marseille, France, 2005, pp. 15–17.

[21] E. Sirin, B. Parsia, *Pellet: an OWL DL reasoner, porter*, in: Third International Semantic Web Conference (ISWC2004), November 2004.

[22] U. Straccia, *A fuzzy description logic*, in: Proceedings of the 15th National Conference on Artificial Intelligence (AAAI-98) and of the 10th Conference on Innovative Applications of Artificial Intelligence (IAAI-98), AAAI Press, 1998, pp. 594–599.

[23] U. Straccia, *Reasoning within fuzzy description logics*, J. Artificial Intelligence Research **14** (2001), 137–166.

[24] U. Straccia, Fuzzy description logics with concrete domains, Technical Report 2005-TR-03, Istituto di Scienza e Tecnologie dell'Informazione, Consiglio Nazionale delle Ricerche, Pisa, Italy, 2005.

[25] C.B. Tresp, R. Molitor, *A description logic for vague knowledge*, in: Proceedings of the 13th European Conference on Artificial Intelligence (ECAI'98), Wiley, 1998, pp. 361–365.

[26] J. Yen, *Generalizing term subsumption languages to fuzzy logic*, in: Proceedings of the 12th International Joint Conference on Artificial Intelligence (IJCAI'91), Sidney, August 1991, pp. 472–477.

CHAPTER 20

Fuzzy Relational Ontological Model in Information Search Systems

Rachel Pereira, Ivan Ricarte and Fernando Gomide

Faculty of Electrical and Computer Engineering, State University of Campinas, Campinas 13083-852, Brazil
E-mail: {rpereira,ricarte,gomide}@dca.fee.unicamp.br

Abstract

Ontology is an essential ingredient to improve information search efficiency and success. Ontology can be used to provide semantic-based access to the Web documents and extract meaningful information from texts. This chapter presents an information search model based on ontology encoded by fuzzy relations. The model uses the principles of fuzzy set theory and approximate reasoning for knowledge representation and information search. Two query algorithms are developed emphasizing, without loss of generality, document search. Experimental results show that the fuzzy relational ontological model performs better when compared with two alternative approaches based on thesauri and fuzzy conceptual network.

Keywords

information search, fuzzy relational ontology, search algorithms

1. Introduction

Currently, document search in the World Wide Web is a common task and may represent a complex activity for many users. The considerable growth of documents in both, amount and variety, increases the need to develop specialized knowledge-based tools to identify and extract semantic information from unstructured documents in databases, particularly in the Web [1]. The Semantic Web is a major ingredient to develop new generation of search systems since it aims encoding information in a structured way and with well-defined meaning. This is especially convenient to enable computer and people to work in cooperation [10,16]. In this context, ontologies play an essential role to guarantee success

FUZZY LOGIC AND THE SEMANTIC WEB
Edited by Elie Sanchez
© 2006 Elsevier B.V. All rights reserved

and dissemination [6,9]. Basically, an ontology defines the knowledge about a domain of interest through a set of concepts and their relations. Ontologies can be used to describe complex, mixed content containing text, images, audio (voice, music), video, etc. Here, however, we emphasize, without loss of generality, the development of relational ontologies using a database of documents. In general, however, a knowledge base may encode ontologies, thesauri or conceptual networks [4,7,8,11,17,18], but here ontology is used as a synonym to knowledge base.

This chapter presents a model for knowledge-based fuzzy information search (KBIS), namely, the fuzzy relational ontological model [15], whose aim is to improve efficiency in information search systems. Differently from alternative approaches developed in the literature, the idea here is to encode knowledge in a fuzzy relational ontology, a fuzzy relation defined in the space of words and categories.

2. Knowledge based fuzzy information search models

Many information search models and systems have been developed recently [2]. In this section we emphasize two particular models, the ones introduced by Ogawa [11] and Horng [7], once they constitute important knowledge-based fuzzy information search models addressed in the literature.

The model proposed by Ogawa assumes a fuzzy document retrieval system using a keyword connection matrix (fuzzy thesauri) to represent similarities between keywords. The matrix elements represent keywords co-occurrence in a collection of documents, and their values are defined as

$$W_{ij} = \begin{cases} \dfrac{N_{ij}}{N_i + N_j - N_{ij}}, & i \neq j, \\ 1, & i = j \end{cases}$$

where W_{ij} is the relationship value between the jth and the ith keywords in the keyword connection matrix; N_{ij} is the number of documents containing both ith and jth keywords; N_i is the number of documents that contain the ith keyword, and N_j is the number of documents that contain the jth keyword.

After the keyword connection matrix is created, fuzzy indices are generated for each index term in the collection of documents.

Three steps must be followed to determine the relevant documents for a query:

1 – Generation of fuzzy indices.

The fuzzy index value determines the degree of relevance of an index term and a document in the collection.

2 – Computation of relevance values for each sub-query.

The user query contains some keywords and possibly logical operators such as AND (conjunction), OR (disjunction) and NOT (negation). The query in this model is converted into conjunctive normal form, consisting of sub-queries including only OR and NOT.

3 – Computation of overall relevance value.

After that relevance values for all the sub-queries are determined and the overall relevance value computed. This step finds which documents are more relevant for the user query. The retrieved documents are shown to the user in decreasing order of relevance.

Figure 1 summarizes the documents retrieval process of Ogawa model.

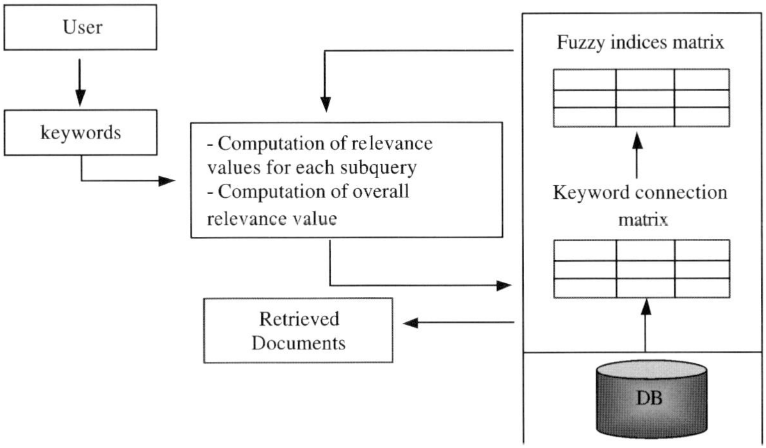

Fig. 1. Documents retrieval process – Ogawa model.

In the Horng model [7], the knowledge base is a fuzzy conceptual network that defines the relationships and the respective relevance degrees between concepts. Conceptual networks use four types of fuzzy relationships: fuzzy generalization, fuzzy specialization, fuzzy positive association, and fuzzy negative association. A concept is regarded as a fuzzy generalization of another concept if the first includes all of meanings of the second. The fuzzy specialization relationship is the inverse of the fuzzy generalization. Fuzzy positive association relates synonymous concepts or concepts that have a similar meaning in some context. On the other hand, the fuzzy negative association relates complementary or antonymous concepts.

The fuzzy conceptual network is represented by a matrix U_r as follows:

$$U_r = \begin{array}{c} \\ c_1 \\ c_2 \\ \vdots \\ c_y \end{array} \begin{array}{c} \begin{array}{cccc} c_1 & c_2 & \cdots & c_y \end{array} \\ \begin{bmatrix} u_{11} & u_{12} & \cdots & u_{1y} \\ u_{21} & u_{22} & \cdots & u_{2y} \\ \vdots & \vdots & \ddots & \vdots \\ u_{y1} & u_{y2} & \cdots & u_{yy} \end{bmatrix} \end{array}$$

where U_r is a relational relevance matrix, $u_{ij} \in [0, 1]$ is the relevance degree between concepts c_i and c_j based on relationship r; $r \in \{P, N, G, S\}$ where "P" stands for "fuzzy positive association", "N" stands for "fuzzy negative association", "G" stands for "fuzzy generalization" and "S" stands for "fuzzy specialization". The detailed steps to create the matrix U_r can be found in the original Horng work [7].

Horng model enables the retrieval of documents that are not directly related to the user query. In addition, an algorithm, namely the *query vector expansion algorithm*, is introduced to add new concepts of the fuzzy conceptual network to the user query. After this algorithm is executed, the satisfaction degree relating the user query and the documents collection is computed [7]. As a result, more relevant documents are retrieved. Figure 2 shows the documents retrieval process of Horng model.

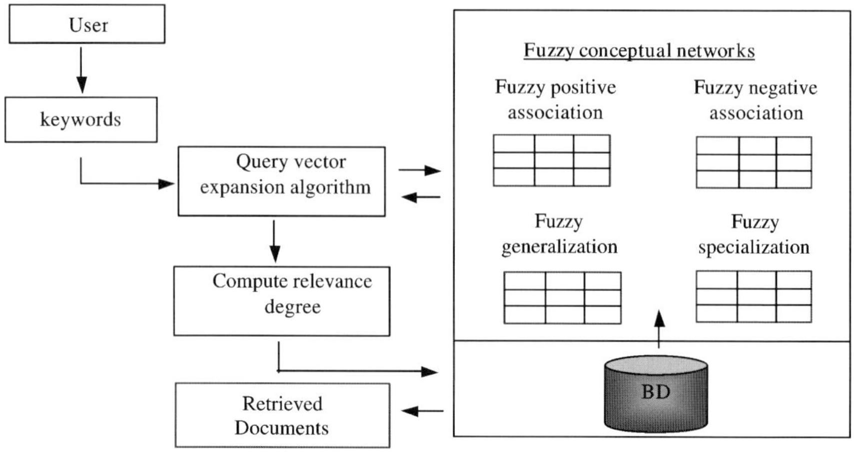

Fig. 2. Documents retrieval process – Horng model.

3. Relational ontological model

3.1. *Fuzzy relational ontology*

The fuzzy ontological model uses an ontology comprising words and categories to conceptually represent the content of a documents collection. A system designer should define the relationship degree between words and categories. This relationship is a fuzzy relation [14] with values in [0, 1], where 0 indicates no relationship between word and category and 1 indicates the strongest possible relationship. The system designer is also responsible to select the words and categories related with each collection (documents indexing process) and to establish the relationship degree between words and documents, and between category and document as well. We can find, in the literature, many mechanisms to index documents whose purpose is to facilitate the information search process. The most traditional, the Boolean indexing, is based on presence (1) or absence (0) of a word in a document [11]. Other traditional mechanism, an evolution of Boolean indexing mechanism, select only the most frequent words in documents [7]. In both situations, if a word specified in the user query does not appear in the document, then it will never be retrieved. A way to optimize the indexing process is suggested by soft computing techniques [12]. An interesting approach uses fuzzy techniques to personalize the indexing procedure. This procedure takes into account the user indication to interpret the document content and to build its synthesis on the basis of this interpretation [3,13]. Other fuzzy set-based technique [5] uses FIS-CRM (Fuzzy Interrelations and Synonymy Concept Representation Model) to represent the indexed documents.

Despite the relative success of many techniques, none is able to exactly identify the words that define the document content fully. Because of this, we adopt an indexing process based on analysis developed by the system designer.

Two layers compose the fuzzy relational ontology. The first layer contains category names whereas the second layer contains keywords related with the category names of the first layer. Figure 3 shows an example of a fuzzy relational ontology with two cate-

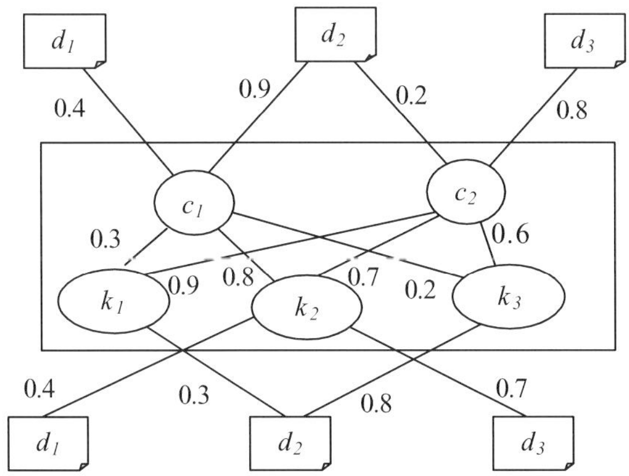

Fig. 3. Example of a fuzzy relational ontology.

gories, c_1 and c_2, and three keywords, k_1, k_2 and k_3. The figure shows documents d_1, d_2 and d_3 connected with the fuzzy relational ontology.

A relevance matrix R defines a fuzzy relational ontology as follows:

$$R = \begin{array}{c} \\ k_1 \\ k_2 \\ \vdots \\ k_n \end{array} \begin{array}{c} \begin{array}{cccc} c_1 & c_2 & \ldots & c_m \end{array} \\ \begin{bmatrix} r_{11} & r_{12} & \cdots & r_{1m} \\ r_{21} & r_{22} & \cdots & r_{2m} \\ \vdots & \vdots & \ddots & \vdots \\ r_{n1} & r_{n2} & \cdots & r_{nm} \end{bmatrix} \end{array}$$

where $1 \leqslant i \leqslant n$, n is the number of keywords of the second layer; $1 \leqslant j \leqslant m$, m is the number of categories of the first layer, and $r_{ij} \in [0, 1]$ is the relevance degree between k_i and c_j.

3.2. *Characteristics of the ontological model*

Architecture. The architecture of the information search system proposed in this work uses the fuzzy relational ontological model as illustrated in Figure 4.

Query representation. Let $Q = \{k_i, c_j\}$ the set of words and categories in the user query, eventually linked by operators AND or OR (in case of a query with more than one category and/or keyword). The query Q is represented by vectors $x = [x_1, x_2, \ldots, x_i, \ldots, x_n]$, $1 \leqslant i \leqslant n$, and $y = [y_1, y_2, \ldots, y_j, \ldots, y_m]$, $1 \leqslant j \leqslant m$, such that:

$$x_i = \begin{cases} 1 & \text{if } k_i \in Q, \\ 0 & \text{otherwise,} \end{cases}$$

$$y_j = \begin{cases} 1 & \text{if } c_j \in Q, \\ 0 & \text{otherwise.} \end{cases}$$

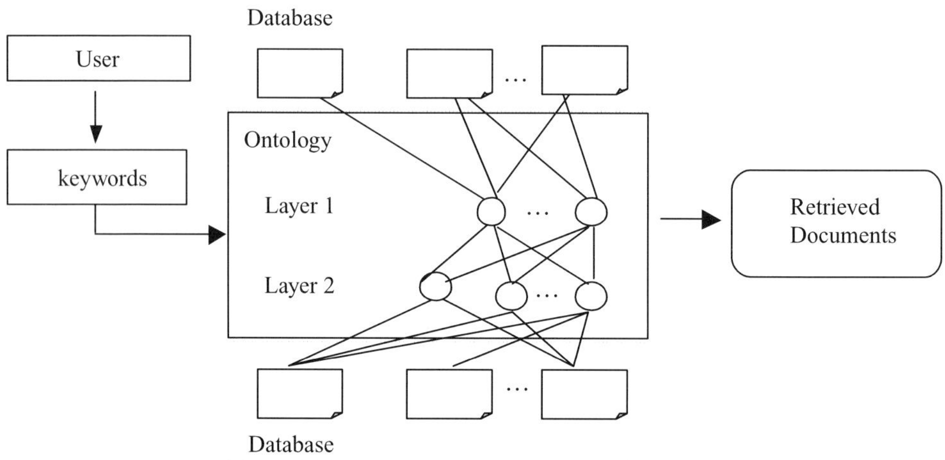

Fig. 4. System architecture.

Documents representation. Let D the set (collection) of documents, $D = \{d_1, d_2, \ldots, d_{\mathrm{doc}}, \ldots, d_u\}$, K the set of keywords, $K = \{k_1, k_2, \ldots, k_i, \ldots, k_n\}$ and C the set of categories, $C = \{c_1, c_2, \ldots, c_j, \ldots, c_m\}$. Documents are represented by matrix T_k as follows,

$$T_k = \begin{array}{c} \\ d_1 \\ d_2 \\ \vdots \\ d_u \end{array} \begin{array}{c} \begin{array}{cccc} k_1 & k_2 & \ldots & k_n \end{array} \\ \begin{bmatrix} \alpha_{11} & \alpha_{12} & \cdots & \alpha_{1n} \\ \alpha_{21} & \alpha_{22} & \cdots & \alpha_{2n} \\ \vdots & \vdots & \ddots & \vdots \\ \alpha_{u1} & \alpha_{u2} & \cdots & \alpha_{un} \end{bmatrix} \end{array} \tag{3.1}$$

where $\alpha_{\mathrm{doc},i} \in [0, 1]$ is the relevance degree between the document d_{doc} and the keyword k_i, $1 \leqslant \mathrm{doc} \leqslant u$, $1 \leqslant i \leqslant n$; and by a T_c matrix,

$$T_c = \begin{array}{c} \\ d_1 \\ d_2 \\ \vdots \\ d_u \end{array} \begin{array}{c} \begin{array}{cccc} c_1 & c_2 & \ldots & c_m \end{array} \\ \begin{bmatrix} \beta_{11} & \beta_{12} & \cdots & \beta_{1m} \\ \beta_{21} & \beta_{22} & \cdots & \beta_{2m} \\ \vdots & \vdots & \ddots & \vdots \\ \beta_{u1} & \beta_{u2} & \cdots & \beta_{um} \end{bmatrix} \end{array} \tag{3.2}$$

where $\beta_{\mathrm{doc},j} \in [0, 1]$ is the relevance degree between the document d_{doc} and the category c_j, $1 \leqslant \mathrm{doc} \leqslant u$, $1 \leqslant j \leqslant m$.

Information retrieval with relational ontological model

DEFINITION 3.1. Let x be the vector of query keywords and let R be a fuzzy relational ontology. The composition of x and R is the fuzzy set G_c that represents the categories associated the query terms. That is,

$$G_c = x \circ R \tag{3.3}$$

where $\circ$ denotes the max–min composition [8].

DEFINITION 3.2. Let y be the vector of query categories and let R be a fuzzy relational ontology. The composition of R and y is the fuzzy set G_k that represents the keywords associated the query terms. That is,

$$G_k = R \circ y. \tag{3.4}$$

Three situations are considered to find relevant documents:
1 – Q contains elements of set K only;
2 – Q contains elements of set C only;
3 – Q contains elements of both sets K and C.

Document ranking. Let $F_k = [f_{k_1}, f_{k_2}, \ldots, f_{k_i}, \ldots, f_{k_n}]$, $1 \leqslant i \leqslant n$, a fuzzy set derived from G_k (3.4) and let $F_c = [f_{c_1}, f_{c_2}, \ldots, f_{c_j}, \ldots, f_{c_m}]$, $1 \leqslant j \leqslant m$, a fuzzy set derived from G_c (3.3), such that:

$$f_{k_i} = \begin{cases} g_{k_i} & \text{if } g_{k_i} > z_1, \\ 0 & \text{otherwise,} \end{cases}$$

$$f_{c_j} = \begin{cases} g_{c_j} & \text{if } g_{c_j} > z_1, \\ 0 & \text{otherwise} \end{cases}$$

where g_{k_i} is the ith component of G_k, g_{c_j} is the jth component of G_c, and z_1 is a threshold defined by the system designer.

DEFINITION 3.3. The composition of F_k and T_k is the fuzzy set V_{DK} that represents the relevance degree between collection documents and the keyword k_i. That is,

$$V_{DK} = F_k \circ T_k. \tag{3.5}$$

DEFINITION 3.4. The composition of F_c and T_c is the fuzzy set V_{DC} that represents the relevance degree between collection documents and the category c_j. That is,

$$V_{DC} = T_c \circ F_c' \tag{3.6}$$

where $'$ denotes transpose.

3.3. *Information search algorithms*

Two algorithms were developed to process user's queries considering the above three situations. The documents search process used by both algorithms is based on max–min composition [8] to find the categories (3.3) and words (3.4) related with the user's query. The max–min composition is also used for compute the relevance degree between the documents and the user's query (3.5), (3.6). The z_1 threshold is used to select the categories and words related with the user's query, and the z_2 threshold is used to select the documents that will be presented to user (the most relevant documents). The details of each algorithm are as follows. Each situation is illustrated using an example based on the fuzzy relational ontology of Figure 3 assuming:

$$R = \begin{bmatrix} 0.3 & 0.9 \\ 0.8 & 0.7 \\ 0.2 & 0.6 \end{bmatrix}, \qquad T_c = \begin{bmatrix} 0.4 & 0 \\ 0.9 & 0.2 \\ 0 & 0.8 \end{bmatrix}, \qquad T_k = \begin{bmatrix} 0 & 0.4 & 0 \\ 0.3 & 0 & 0.8 \\ 0 & 0.7 & 0 \end{bmatrix}.$$

The first situation applies when $Q = \{k_i\}$, $1 \leqslant i \leqslant n$, and $1 \leqslant |Q| \leqslant n$. In this case, the following steps are performed for the first algorithm:

(a) Compute $G_c = [g_{c_j}]$ as indicated in (3.3),
(b) Select categories $c_j \in G_c$ with $g_{cj} > z_1$, where z_1 is a threshold selected by the system designer,
(c) Get the database documents related with the categories of step (b),
(d) Compute V_{DC} as indicated in (3.6),
(e) Rank the vector V_{DC} in decreasing order,
(f) Show the list of documents of step (c) following the order defined by vector V_{DC} (step (e)) for all $v_{DC} > z_2$, where z_2 is a threshold chosen by the system designer.

Consider the case in which $Q = \{k_1\}$, $z_1 = 0.8$, $z_2 = 0.3$. Then, in the first step, G_c is computed:

$$G_c = x \circ R = [\,1 \quad 0 \quad 0\,] \circ \begin{bmatrix} 0.3 & 0.9 \\ 0.8 & 0.7 \\ 0.2 & 0.6 \end{bmatrix} = [\,0.3 \quad 0.9\,].$$

In the second step, the only category satisfying the condition $g_{c_j} > 0.8$ is c_2. Documents related with the category c_2 (step (c)) are d_2 and d_3. The next step is to compute V_{DC}:

$$V_{DC} = T_c \circ F_c' = \begin{bmatrix} 0.4 & 0 \\ 0.9 & 0.2 \\ 0 & 0.8 \end{bmatrix} \circ \begin{bmatrix} 0 \\ 0.9 \end{bmatrix} = \begin{bmatrix} 0 \\ 0.2 \\ 0.8 \end{bmatrix}$$

which should be ranked in decreasing order (step (e)):

$$V_{DC} \text{ in decreasing order } = \begin{bmatrix} 0.8 \\ 0.2 \\ 0 \end{bmatrix}.$$

The final result is obtained for the documents for which $v_{DC} > 0.3$, which in this case yields document d_3.

The second situation applies when $Q = \{c_j\}$, $1 \leqslant j \leqslant m$ and $1 \leqslant |Q| \leqslant m$. In this case, the same algorithm is applied with the following steps:

(a) Compute $G_k = [g_{k_i}]$ as indicated in (3.4),
(b) Select the keywords $k_i \in G_k$ with $g_{k_i} > z_1$, where z_1 is a threshold selected by the system designer,
(c) Get the database documents related with the keywords of step (b),
(d) Compute V_{DK} as indicated in (3.5),
(e) Rank the vector V_{DK} in decreasing order,
(f) Show the list of documents of step (c) following the order defined by vector V_{DK} (step (e)) for all $v_{DK} > z_2$, where z_2 is a threshold chosen by the system designer.

Consider the example with $Q = \{c_2\}$, $z_1 = 0.7$, $z_2 = 0.3$. The first step is to compute G_k,

$$G_k = R \circ y = \begin{bmatrix} 0.3 & 0.9 \\ 0.8 & 0.7 \\ 0.2 & 0.6 \end{bmatrix} \circ \begin{bmatrix} 0 \\ 1 \end{bmatrix} = \begin{bmatrix} 0.9 \\ 0.7 \\ 0.6 \end{bmatrix}.$$

The only keyword for which $g_{k_i} > 0.7$ is k_1 (step (b)), and the only document related to it is d_2 (step (c)). Therefore, in the next step V_{DK} is computed as

$$V_{DK} = F_k \circ T_k = [\,0.9 \quad 0 \quad 0\,] \circ \begin{bmatrix} 0 & 0.4 & 0 \\ 0.3 & 0 & 0.8 \\ 0 & 0.7 & 0 \end{bmatrix} = [\,0 \quad 0.4 \quad 0\,]$$

which, in step (e), yields

$$V_{DK} \text{ in decreasing order} = [\,0.4 \quad 0 \quad 0\,].$$

Finally, the result yields the only document for which $v_{DK} > 0.3$, which in this case is d_2.

The third and last situation for which the first algorithm applies happens when $Q = \{k_i, c_j\}$, $1 \leqslant i \leqslant n$, $1 \leqslant j \leqslant m$. In this case, the algorithm steps are:

(a) Split Q in two subsets: $Q_1 = \{p_i\}$ and $Q_2 = \{c_j\}$,
(b) Compute $G_c = [g_{c_j}]$ as in (3.3),
(c) Compute $G_k = [g_{k_i}]$ as in (3.4),
(d) Select categories $c_j \in G_c$ (step (b)) with $g_{c_j} > z_1$, where z_1 is a threshold,
(e) Select keywords $k_i \in G_k$ (step (c)) with $g_{k_i} > z_1$, where z_1 is a threshold,
(f) Get database documents related with the categories of step (d),
(g) Get database documents related with the keywords of step (e),
(h) Select database documents related with the categories presented in Q_2,
(i) Select database documents related with the keywords presented in Q_1,
(j) If the elements in the user query are connected by operator AND, then select the intersection of documents obtained in steps (h) and (i); if the operator used is OR, then get the union,
(k) Compute V_{DK} as indicated in (3.5),
(l) Compute V_{DC} as indicated in (3.6),
(m) Compare vectors V_{DK} and V_{DC} retrieving the highest values of relevance degrees,
(n) Rank the result of step (m) in decreasing order,
(o) Show the list of documents of step (j) following the same order as defined in step (n) with values higher than z_2, where z_2 is a threshold.

Consider the example with $Q = \{k_2 \text{ AND } c_1\}$, $z_1 = 0.6$, $z_2 = 0.6$. In the first step, $Q_1 = \{k_2\}$, $Q_2 = \{c_1\}$. Then G_c is computed as

$$G_c = x \circ R = [\,0 \quad 1 \quad 0\,] \circ \begin{bmatrix} 0.3 & 0.9 \\ 0.8 & 0.7 \\ 0.2 & 0.6 \end{bmatrix} = [\,0.8 \quad 0.7\,]$$

and G_k as

$$G_k = R \circ y = \begin{bmatrix} 0.3 & 0.9 \\ 0.8 & 0.7 \\ 0.2 & 0.6 \end{bmatrix} \circ \begin{bmatrix} 1 \\ 0 \end{bmatrix} = \begin{bmatrix} 0.3 \\ 0.8 \\ 0.2 \end{bmatrix}.$$

Thus, c_1 and c_2 are the categories with $g_{c_j} > 0.6$ (step (d)) and k_2 is the keyword with $g_{k_i} > 0.6$ (step (e)). The next steps get the documents related with the categories c_1 and c_2, which in this case are d_1, d_2, and d_3 (step (f)), and related with the keyword k_2, which are d_1 and d_3 (step (g)). Similarly, in the next two steps are obtained the documents related with the categories of set Q_2 (d_1 and d_2) and with the keywords of set Q_1 (d_1 and d_3). As the query has the logical operator AND, d_1 is the only document resulting in step (j). Then,

$$V_{DK} = T_k \circ F_k = \begin{bmatrix} 0 & 0.4 & 0 \\ 0.3 & 0 & 0.8 \\ 0 & 0.7 & 0 \end{bmatrix} \circ \begin{bmatrix} 0 \\ 0.8 \\ 0 \end{bmatrix} = \begin{bmatrix} 0.4 \\ 0 \\ 0.7 \end{bmatrix}$$

and

$$V_{DC} = T_c \circ F_c' = \begin{bmatrix} 0.4 & 0 \\ 0.9 & 0.2 \\ 0 & 0.8 \end{bmatrix} \circ \begin{bmatrix} 0.8 \\ 0.7 \end{bmatrix} = \begin{bmatrix} 0.4 \\ 0.8 \\ 0.7 \end{bmatrix}.$$

Then, in step (m) both vectors are compared and combined, resulting:

$$V_{DK,DC} \text{ in decreasing order} = \begin{bmatrix} 0.8 \\ 0.7 \\ 0.4 \end{bmatrix}.$$

Thus, given the threshold $v_{DK,DC} > 0.6$, the resulting documents are d_2 and d_3.

The other algorithm is a particular case of the one presented above. Similarly, it can be applied in three situations, which are described next.

The first situation applies when $Q = \{k_i\}$, $1 \leqslant i \leqslant n$ and $1 \leqslant |Q| \leqslant n$. In this case, the second algorithm steps are:

(a) Compute $G_c = [g_{c_j}]$ as in (3.3),
(b) Select categories $c_j \in G_c$ with $g_{c_j} > z_1$ and put them in the set J_c, where z_1 is a threshold defined by the system designer,
(c) For each category in set J_c create a new query Q^*, such that $Q^* = c_p$ AND k_i, where p is the pth category in J_c and set *size* $= |J_c|$, the cardinality of J_c.
(d) Execute the following iteration:
for ($p = 1$ to size)
  for ($i = 1$ to n)
    Process Q^* similarly as situation 3 of method 1
    Put the retrieved documents on the list L
(e) Get all documents of the list L (step (d)) without repetitions,
(f) Compute V_{DC} as in (3.6),
(g) Rank the vector V_{DC} in decreasing order,
(h) Show the list of documents of step (e) in the same order as defined by V_{DC} with $v_{DC} > z_2$, where z_2 is threshold.

Consider the example when $Q = \{k_2\}$, $z_1 = 0.75$, $z_2 = 0.6$. In the first step,

$$G_c = x \circ R = [\,0 \quad 1 \quad 0\,] \circ \begin{bmatrix} 0.3 & 0.9 \\ 0.8 & 0.7 \\ 0.2 & 0.6 \end{bmatrix} = [\,0.8 \quad 0.7\,]$$

which yields c_1 as the only category for which $g_{c_j} > 0.75$. Then, in step (c), "$Q^* = k_1$ AND c_1", "$Q^* = k_2$ AND c_1", "$Q^* = k_3$ AND c_1". The next step includes $d_3 - d_2 - d_3 - d_2$ in list L; thus, in step (e), the documents in L are d_2 and d_3. V_{DC} is then computed as

$$V_{DC} = T_c \circ F_c' = \begin{bmatrix} 0.4 & 0 \\ 0.9 & 0.2 \\ 0 & 0.8 \end{bmatrix} \circ \begin{bmatrix} 0.8 \\ 0 \end{bmatrix} = \begin{bmatrix} 0.4 \\ 0.8 \\ 0 \end{bmatrix}$$

and in step (g)

$$V_{DC} \text{ in decreasing order} = \begin{bmatrix} 0.8 \\ 0.4 \\ 0 \end{bmatrix}.$$

Therefore, the only resulting document for which $v_{DC} > 0.6$ is d_2.

The second situation occurs when $Q = \{c_j\}$, $1 \leqslant j \leqslant m$ and $1 \leqslant |Q| \leqslant m$. In this case, the steps for the algorithm are performed as follows:

(a) Compute $G_k = [g_{k_i}]$ as indicated in (3.4),

(b) Select keywords $k_i \in G_k$ with $g_{ki} > z_1$ and put them in set J_k, where z_1 is a threshold defined by the system designer,

(c) For each keyword in set J_k create a new query Q^*, where "$Q^* = k_p$ AND c_j", p is the pth keyword in J_k.

(d) Execute the iteration:
  for ($p = 1$ to size)
    for ($i = 1$ to m)
      Execute Q^* similarly as situation 3 of method 1
      Put the retrieved documents on the list L

(e) Get all documents of list L (step (d)) with no repetition,

(f) Compute V_{DK} as indicated in (3.5),

(g) Rank the vector V_{DK} in decreasing order,

(h) Show the list of documents of step (e) in order defined by vector V_{DK} with $v_{DK} > z_2$, where z_2 is threshold.

For example, consider the case with $Q = \{c_1\}$, $z_1 = 0.75$, $z_2 = 0.6$. In the first step,

$$G_k = R \circ y = \begin{bmatrix} 0.3 & 0.9 \\ 0.8 & 0.7 \\ 0.2 & 0.6 \end{bmatrix} \circ \begin{bmatrix} 1 \\ 0 \end{bmatrix} = \begin{bmatrix} 0.3 \\ 0.8 \\ 0.2 \end{bmatrix}.$$

The only keyword with $g_{k_i} > 0.75$ is k_2. In step (c), "$Q^* = k_2$ AND c_1", "$Q^* = k_2$ AND c_2". The iteration in the next step yields $L = d_2 - d_3$, and thus the documents in the list L are d_2 and d_3. Thus, in step (f) V_{DK} is computed as

$$V_{DK} = T_k \circ F_k = V_{DC} = T_c \circ F_c' = \begin{bmatrix} 0 & 0.4 & 0 \\ 0.3 & 0 & 0.8 \\ 0 & 0.7 & 0 \end{bmatrix} \circ \begin{bmatrix} 0 \\ 0.8 \\ 0 \end{bmatrix} = \begin{bmatrix} 0.4 \\ 0 \\ 0.7 \end{bmatrix}$$

which, in step (g), yields

$$V_{DK} \text{ in decreasing order} = \begin{bmatrix} 0.7 \\ 0.4 \\ 0 \end{bmatrix}.$$

Therefore, the result contains the document for which $v_{DC} > 0.75$; in this case, document d_3.

For the third situation ($Q = \{k_i, c_j\}$, $1 \leqslant i \leqslant n$, $1 \leqslant j \leqslant m$), this algorithm performs exactly as the first algorithm.

4. Experiments

4.1. *Evaluation procedure*

Performance of information retrieval systems is normally measured scoring recall (R) and precision (P) defined as follows:

$$R = \frac{Number(Relevant\_documents\_in\_the\_result)}{Number(All\_relevant\_documents\_in\_the\_database)},$$

$$P = \frac{Number(Relevant\_documents\_in\_the\_result)}{Number(Documents\_in\_the\_result)}.$$

Recall indicates how many of the relevant documents are retrieved, whereas precision indicates how many of the retrieved documents are relevant.

Frequently, retrieval algorithms are evaluated by running several distinct queries. As suggested by Baeza-Yates [2], to evaluate the retrieval performance of an algorithm over a set of test queries means to compute the average precision at each recall level as follows:

$$\overline{P}(r) = \sum_{i=1}^{N_q} \frac{P_i(r)}{N_q}$$

where $\overline{P}(r)$ is the average precision at the recall level r, N_q is the number of queries, and $P_i(r)$ is the precision at recall level r for the ith query.

A precision versus recall plot is normally based on an 11 recall levels standard, from 0 to 100% in steps of 10%. As recall levels for each query might be distinct from these 11 levels, an interpolation procedure is necessary. This procedure, which is also used to determine the 0% recall level, is defined as follows.

Let r_j, $j \in \{0, 1, 2, \ldots, 10\}$, be a reference to the jth standard recall level (i.e., r_6 is a reference to the recall level 60%). Then,

$$P(r_j) = \max_{r_j \leqslant r \leqslant r_j+1} P(r)$$

which states that the interpolated precision at the jth standard recall level is the maximum known precision at any recall level between the jth recall level and $(j+1)$th recall level.

4.2. *Experimental results*

The information search system using the fuzzy relational ontological model was implemented on JSP (Java Server Pages) and servlets using the Java programming language, on a Unix workstation using the Tomcat HTTP server and PostgreSQL relational database server.

To validate the model, a database with 100 scientific articles about *Computational Intelligence* was assembled to evaluate of the retrieval algorithms. A total of 61 words were extracted from the documents, being 6 category names and 55 keywords. These words were used to create the fuzzy relational ontology for the proposed ontological model, the keywords connection matrix of Ogawa model, and the fuzzy conceptual network of Horng model.

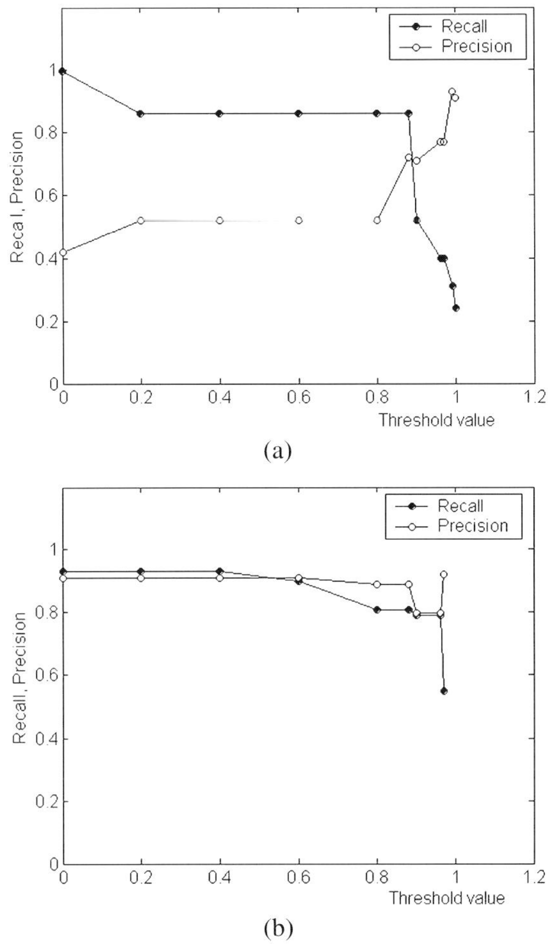

Fig. 5. Threshold × recall and precision.

Figure 5 shows the relationship between threshold values, recall, and precision for a query executed for algorithm 1 (Figure 5(a)) and algorithm 2 (Figure 5(b)). Both algorithms perform poorly for high threshold (z_2) values.

Figures 6, 7 and 8, based on the 11 recall levels standard, compare algorithm 1 with algorithm 2(a) and algorithm 1(b) and 2(c) with the Ogawa and Horng models, respectively. Figure 6 shows the result for five distinct queries with a single category name, *Information Retrieval*, *Genetic Algorithm*, *Fuzzy Logic*, *Agent*, and *Neural Network* assuming $z_2 = 0.75$.

Figure 7 shows the results of four distinct queries with a single keyword *fuzzy set*, *web agent*, *search engine*, *hybrid system* with $z_2 = 0.75$.

Figure 8 shows the result of five distinct composite queries that use operator AND, that is, *Agent* AND *Information Retrieval*, *Fuzzy Logic* AND *Information Retrieval*, *Information Retrieval* AND *Search Engine*, *Agent* AND *Genetic Algorithm*, *Genetic Algorithm* AND *Hybrid System* with $z_2 = 0.75$.

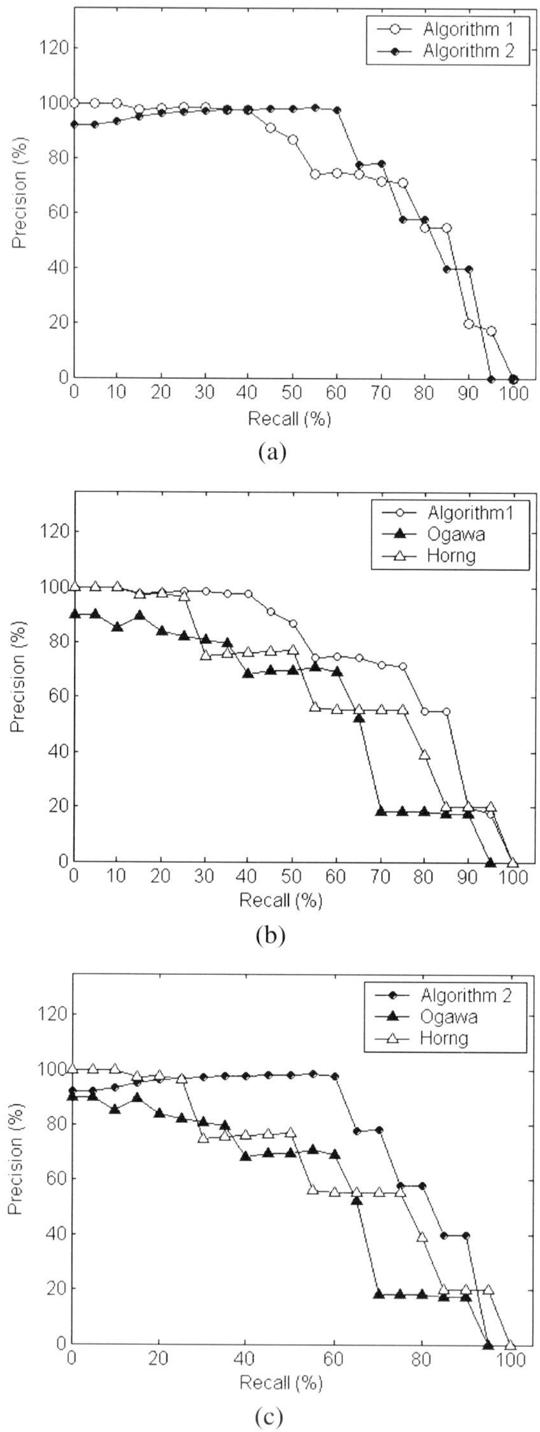

Fig. 6. Recall × precision, five single category queries.

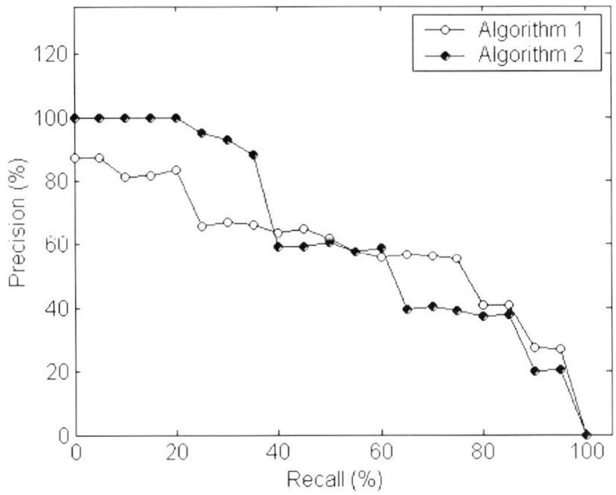

(a)

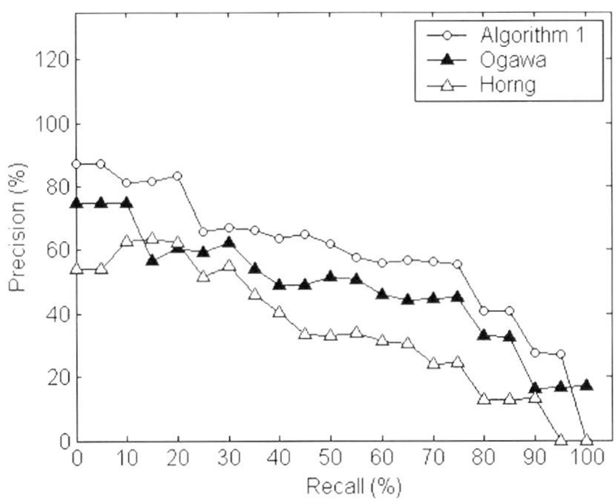

(b)

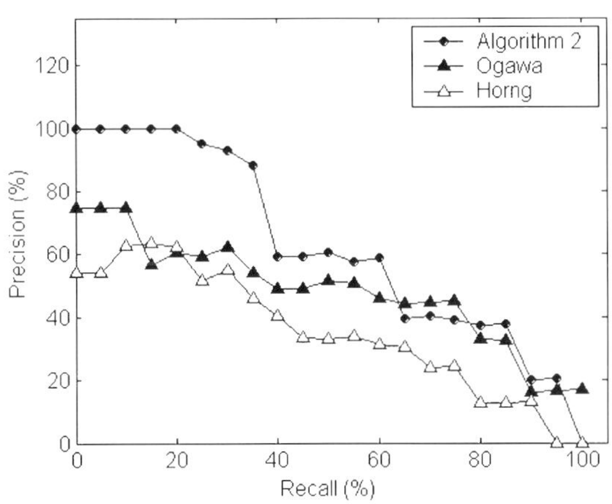

(c)

Fig. 7. Recall × precision, four single keyword queries.

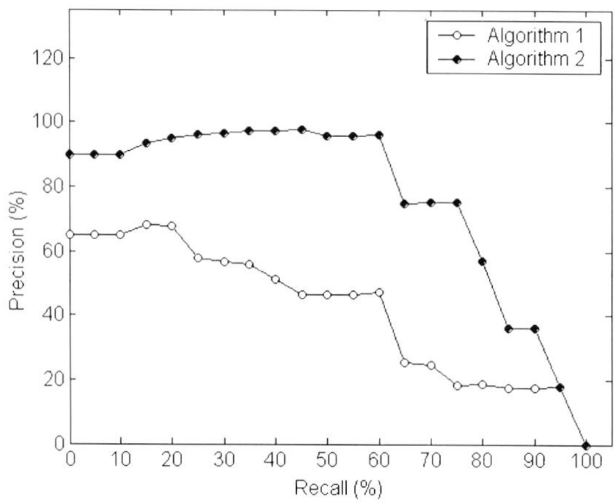

(a)

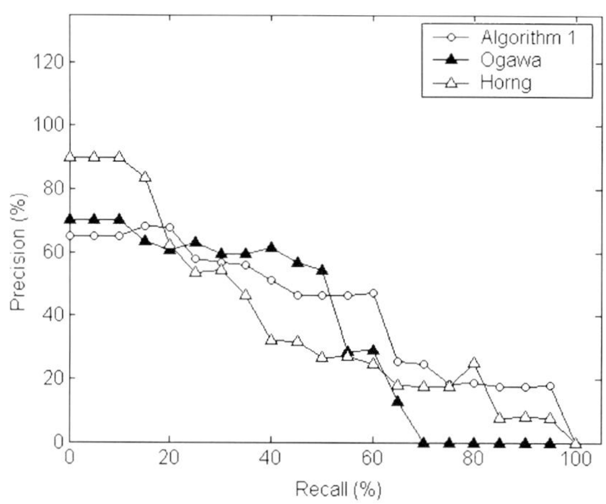

(b)

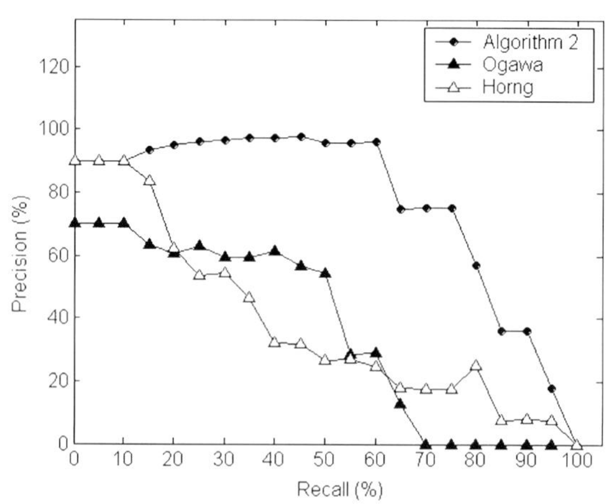

(c)

Fig. 8. Recall × precision, five composite queries.

Observe that the fuzzy ontological model performs better than the ones based on thesauri and conceptual networks once it retrieves more relevant documents. In other works, the fuzzy ontological model achieves, most of the time, higher recall and precision values.

5. Conclusions

The goal of the Semantic Web is to create a new form of Web content meaningful to computers. One way to represent this type of data in the Web is through of ontologies. This chapter has presented a fuzzy model based on a relational ontology for information search systems, with emphasis on search algorithms. Experimental results showed that the fuzzy relational ontological model achieves better performance when compared with Ogawa and Horng models. Among the algorithms described, the second algorithm has shown better performance than the first.

Although in this chapter the presentation of the fuzzy ontological models were restricted to text, the same principles can be applied to audio and image databases.

Acknowledgements

The first author thanks CAPES of the Brazilian Ministry of Education for a fellowship, and the last is grateful to CNPq, Brazilian National Research Council, for its support via grants 304299/03-0 and 475295/03-0.

References

[1] H. Alani, S. Kim, D.E. Millard, M.J. Weal, W. Hall, P.H. Lewis, N.R. Shadbolt, *Automatic ontology-based knowledge extraction from Web documents*, IEEE Intelligent Systems **18** (1) (2003), 14–21.

[2] R. Baeza-Yates, B. Ribeiro-Neto, *Modern Information Retrieval*, ACM Press, Addison-Wesley, 1999.

[3] G. Bordogna, P. Carrara, G. Pasi, *A fuzzy document representation supporting user adaptation in information retrieval*, in: Proceeding of the 2th IEEE International Conference on Fuzzy Systems, 1993, pp. 974–979.

[4] S.M. Chen, J.Y. Wang, *Document retrieval using knowledge-based fuzzy information retrieval techniques*, IEEE Trans. on Systems Man Cybernet. **25** (5) (1995), 1–6.

[5] P.J. Garcés, J.A. Olivas, F.P. Romero, *Conceptual matching in web search using FIS-CRM for representing documents*, in: Third Conference of the European Society for Fuzzy Logic and Technology, 2003, pp. 63–66.

[6] A. Gómez-Pérez, O. Corcho, *Ontology languages for the Semantic Web*, IEEE Intelligent Systems **17** (1) (2002), 54–60.

[7] Y.-J. Horng, S.-M. Chen, C.-H. Lee, *Automatically constructing multi-relationship fuzzy concept in fuzzy information retrieval systems*, in: IEEE International Fuzzy Systems Conference, 2001, pp. 606–609.

[8] G.J. Klir, B. Yuan, *Fuzzy Sets and Fuzzy Logic*, Prentice-Hall, Upper Saddle River, 1995.

[9] A. Maedche, S. Staab, *Ontology learning for the Semantic Web*, IEEE Intelligent Systems **16** (2) (2001), 72–79.

[10] R. Mihalcea, S.I. Mihalcea, *Word semantics for information retrieval: moving one step closer to the Semantic Web*, in: International Conference on Tools in Artificial Intelligence, 2001, pp. 280–287.

[11] Y. Ogawa, T. Morita, K. Kobayashi, *A fuzzy document retrieval system using the keyword connection matrix and a learning method*, Fuzzy Sets and Systems **39** (1991), 163–179.

[12] S.K. Pal, V. Talwar, P. Mitra, *Web mining in soft computing framework relevance, state of the art and future directions*, IEEE Trans. on Neural Network **13** (5) (2002), 1163–1177.

[13] G. Pasi, *Flexible information retrieval: some research trends*, Mathware and Soft Computing **9** (2002), 107–121.

[14] W. Pedrycz, F. Gomide, *An Introduction to Fuzzy Sets: Analysis and Design*, MIT Press, Cambridge, MA, 1998.

[15] R.C. Pereira, *Fuzzy relational ontological model in textual information retrieval systems*, MSc Thesis (in Portuguese), Faculty of Electrical and Computer Engineering, State University of Campinas, Campinas, São Paulo, Brazil, 2004.

[16] A. Sheth, C. Bertram, D. Avant, B. Hammond, K. Kochut, Y. Warke, *Managing semantic content for the Web*, IEEE Internet Computing **6** (4) (2002), 80–87.

[17] T. Takagi, K. Kawase, *A trial for data retrieval using conceptual fuzzy sets*, IEEE Trans. on Fuzzy Systems **9** (4) (2001), 497–505.

[18] D.H. Widyantoro, J. Yen, *A fuzzy ontology-based abstract search engine and its user studies*, in: Proceedings of the 10th IEEE International Conference on Fuzzy Systems, 2001, pp. 1291–1294.

Soft Computing in Various Domains

CHAPTER 21

Evolving Ontologies for Intelligent Decision Support

Paulo Gottgtroy, Nikola Kasabov and Stephen MacDonell

Auckland University of Technology, School of Computer and Information Sciences, Knowledge Engineering and Discovery Research Institute, KEDRI–AUT Tech Park, 581-585 Great South Road, Ronald Trotter House, Private Bag 92006, Auckland 1020, New Zealand

E-mail: {Paulo.gottgtroy,nkasabov,stephen.macdonell}@aut.ac.nz

Abstract

The explosive growth in volumes of data and the growing number of disparate data sources are exposing researchers to a 'new' challenge – how to acquire, maintain and share knowledge from large and distributed databases in the context of rapidly evolving research. Our approach addresses such questions by integrating soft computing techniques and ontology engineering. The primary outcome of this work is a process that facilitates the creation of evolving ontologies that are able to represent the dynamic and uncertain nature of domains in order to provide constant and ongoing support for the decision making process over time.

Keywords

ontologies, knowledge discovery, uncertainty, evolving connectionist systems, soft computing

1. Introduction

The explosive growth in volumes of data and the growing number of disparate data sources are exposing researchers to a 'new' challenge – how to acquire, maintain and share knowledge from large and distributed databases in the context of rapidly evolving research. In the biomedical domain, for instance, the challenge of knowledge discovery from diverse and heterogeneous biomedical data in order to make knowledge and concepts sharable over applications and reusable for several purposes is both complex and crucial. It is central to the support of decision making in medical practice as well as to enable comprehensive knowledge acquisition by medical research communities and molecular biologists involved in advancing biomedical understanding.

FUZZY LOGIC AND THE SEMANTIC WEB
Edited by Elie Sanchez
© 2006 Elsevier B.V. All rights reserved

The work described here addresses these three main tasks: knowledge acquisition, knowledge maintenance and knowledge sharing, in the context of ontologies for intelligent decision support. The primary outcome aim is a process that facilitates the creation of ontologies that are able to represent the dynamic and uncertain nature of domains in order to provide constant and ongoing support for the decision making process over time. These ontologies are based mainly on evolving processes that are *developing, changing over time in a continuous manner* [20].

In recent years ontologies [13] have been increasingly used to provide a common reference framework across disparate systems. Ontology is defined in the artificial intelligence literature as a specification of a conceptualization [16]. Ontology specifies at a higher level the classes of concepts that are relevant to an application domain and the classes of relations that exist between these classes. However in many domains relationships, concept understanding, classifications and so on are evolving over time. Thus ontology specifications need to be able to represent the dynamics and changes that occur in the real world. Otherwise, there is a danger that building and using an 'inflexible' ontology, that is, an ontology that has a fixed structure, may fossilize the current knowledge representation in a particular paradigm or point of view, so that only information that fits that paradigm is accessible by the users [6].

In the context of the Semantic Web, defined as an extension of the current web that provides a common framework enabling data to be shared and reused across applications and better enabling computers and people to work in cooperation, ontologies play a central role as a new formalism. The semantic web brings fundamental questions to the research community. For instance, semantic web languages are mainly logic based with facts either being true or not true; however the world is in some sense more uncertain that this, so how can we represent things which are not so discrete and are not amenable to such bivalent classifications? In other words, how can we represent knowledge that evolves over time, changing its structure and/or values?

Our approach addresses such questions. It is different from those that treat the dynamic and uncertain nature of domains through the improvement of the representation formalism itself, using nonclassical logics, the integration of fuzzy concepts in description logics, and fuzzy formal concept analysis, in that our approach is founded on the integration of soft computing techniques and ontology engineering.

Evolving ontologies deal with modification, adaptation and polymorphism representing meta-knowledge and are able to cope with different perspectives, uncertainty and changes in a domain. Evolving ontologies can be argued as a new attempt to solve an old problem. The knowledge acquisition bottleneck, for instance, has been investigated since the early stages of artificial intelligence research [11], mainly by advocates of expert systems. The need for knowledge maintenance, adaptation and generalization were among the driving forces for the development of connectionist systems. Evolving ontologies address the knowledge acquisition problem through different learning techniques inherited from the data mining discipline and its integration with ontology engineering. Adaptation is enabled by the Evolving Connectionist Systems (ECOS) [20] paradigm and its methods. Uncertain knowledge is represented through the ontology meta-model and can be programmatically acquired by the ECOS' modeling techniques.

This chapter is organized in four interrelated parts. In the first part we discuss ontologies, their applications, some philosophical aspects and define the characteristics required to represent dynamism and uncertainty in an ontological representation, giving a short definition to evolving ontologies.

The second part of the chapter presents the framework and the methodology used in our research. This part addresses ontological engineering aspects of the research, its integration with data mining techniques and the life cycle adopted for the knowledge discovery process.

The third part of the chapter explores a biomedical example of an evolving ontology and shows an ontological representation of the domain.

The final part presents future directions and results of the integration of our data mining workbench, Neucom, with an ontology tool. The chapter concludes with some final remarks.

2. Evolving ontologies and semantic web languages

Humans have been concerned with the description of things since the ancient Greeks. Essence, existence, universals, among others underlies the work on ontology, in the context of philosophy. As Kant [29] states "Socrates said he was the midwife to his listeners, i.e., he made them reflect better concerning that which they already knew and become better conscious of it". Ontology is about making things explicit; it is concerned with the description of concepts and relationships among concepts in a domain.

Although we have inherited the principles of ontological knowledge from philosophy, the artificial intelligence community has been working for many years on bringing knowledge forth and making it accessible in a machine-readable format. There are many definitions for ontology, some of them are more related to its formality, others related to expressivity. All reflect the respective background of their researchers.

Gruber, the most cited author in the ontology field, states that, in the context of knowledge sharing, *an ontology is a description (like a formal specification of a program) of the concepts and relationships that can exist for an agent or a community of agents* [16].

The formal ontology community defines ontology as an explicit formal specification of how to represent the objects, concepts and other entities that are assumed to exist in some area of interest and the relationships that hold among them.

There is value in both perspectives. In 1998 Studer and colleagues [31] attempted to merge the two in the following definition:

> An ontology is a formal, explicit specification of a shared conceptualization. Conceptualization refers to an abstract model of some phenomenon in the world by having identified the relevant concepts of that phenomenon. Explicit means that the type of concepts used, and the constraints on their use are explicitly defined. Formal refers to the fact that the ontology should be machine-readable. Shared reflects the notion that an ontology captures consensual knowledge, that is, it is not private of some individual, but accepted by a group.

A large number of ontologies have been developed by different research groups, under different approaches and with different methods and techniques [13]. In the context of ontologies and knowledge discovery integration we are concerned with two main characteristics: purpose and completeness. Our understanding takes into account the views of Guarino and colleagues [18] in relation to the notion of the *partial account of a conceptualization* and Gruber's notion of *treat* among people [17], agents, and users with some *common motive in sharing*. Ontology in our research is an agreement that represents a partial shared conceptualization of the world. Partial means incomplete knowledge, being constructed, dynamic, changing, and evolving. Shared means consensual knowledge, maintained through knowledge management.

There also exists in the literature another group of definitions that center on the techniques used to build ontology [33] and its intended purpose. Our understanding of ontology is influenced by the knowledge discovery in databases field and takes into account the goals of knowledge acquisition, maintenance and sharing.

In the context of ontology driven knowledge discovery, then, we define evolving ontology as follows:

Evolving ontology is a partial shared conceptualization that supports knowledge discovery from databases.

Evolving ontologies describe a process rather than a static model. As a consensual and partial conceptualization, ontology is an object of social pressures, different perspectives, negotiation, mental model interpretation and so on. These requirements make imperative the ability to change structure and content over time as new knowledge is made available.

The evolving characteristic of an ontology is achieved through the use of learning techniques inherited from data mining and through its meta-knowledge representation. For example, the hierarchical structure of an evolving ontology can be determined using the Evolving Clustering Method [28], which permits an instance to be part of more than one cluster. At the same time, evolving ontology uses its meta-knowledge representation to cope with this multiple clustering requirement.

It is important to note that we do not make any explicit distinction between formal and informal ontologies in our research. However, we do make the following comments in order to clarify our research position. The informal parts of a specification help explain the knowledge in a human-accessible fashion, which makes ontologies easier to understand and powerful in terms of expressivity. The formal parts make ontologies unambiguous, allowing some automated analysis and reasoning. We have been using both formal and informal languages, such as Frames and Description Logics, in our experiments. We believe that the most effective representation is often a mix of formal and informal, and is dependent on the methods used to build the ontology and the knowledge being represented. Description logic [2] in its current form, for example, is not able to cope with uncertainty and soft semantics. Different approaches have been taken to give more flexibility to the semantic web languages [24,26]. In the context of knowledge discovery in databases neither formal nor informal languages have proved dominant in supporting the ontology building process through soft computing techniques. In this investigation, then, the learning process is enacted using evolving connectionist systems and the resulting knowledge is represented in the ontology formalism.

2.1. *Ontology and knowledge discovery*

From a philosophical point of view, discovery can be defined as "*the act of becoming aware of something previously existing but unknown*" [23]. This broad definition includes both kinds of scientific discovery: factual and conceptual. The former typically happens during the investigation of current "known" facts or data. The latter emerges from different points of view of "unknown" facts or data.

Ontologies are used to facilitate both forms of scientific discovery in providing a common framework of understanding for several systems and problem solving methods, and, at the same time, connecting different facts and points of view.

The integration of ontologies and knowledge discovery in databases is aimed at the extraction of knowledge from data. Thus it is important to define and note the differences between data, information and knowledge. To this end Devlin [8] reported the following:

- Data – Recorded facts
- Information = Data + meaning
- Knowledge = Internalized information + ability to utilize the information

Considering the above we extend the previous definition of discovery to one that is more strongly related to the discovery *of knowledge* – "the act of adding meaning to data in order to share knowledge and be aware of the "unknown" ".

Although this definition helps to give meaning to a process that is able to leverage the advantages of ontology engineering and knowledge discovery, it cannot help in the evaluation of the usefulness and intelligent support that the integration of ontologies and knowledge discovery in databases (KDD) brings to the decision making process. Thus we use Dhar's intelligence density framework [9] as a guideline to allow us to measure the level of intelligent support provided by the ontology driven knowledge discovery process.

Adding value to data enables an organization, a decision maker, a decision system, or even a software agent to "know more" about a domain – its environment, circumstances under which the decision is made, and so on. Increased intelligence in a KDD process means improved KDD results in terms of transforming data into knowledge.

Intelligence density is defined by Dhar as a measure of productivity. It is the amount of useful decision support information that a decision maker gets from using the output of an analytic system. Intelligence density supports a decision through the transformation of data into knowledge in the following steps:

- Access;
- Scrub;
- Integrate;
- Transform;
- Discover;
- Learn.

By adding value to data from the Access step up to the Learning step we can transform data into knowledge. Thus to evaluate a system we should measure the level of support that it can give to such a decision process.

A data warehouse application, for instance, is able to provide Access, Scrub, Integrate and Transform functionality, but it cannot in itself Discover. To increase the intelligence and discover useful information we must use additional tools and techniques in conjunction with data warehouses. For example, when using on-line analytical processing (OLAP) tools to analyze multi-dimensional models represented by a data warehouse (Figure 1) the system can discover unknown information; however it cannot learn or acquire knowledge because it is not mining or storing the meaning of the discovered facts in any repository, such as an ontology or knowledge base.

The knowledge discovery definition presented here – "*the act of adding meaning to data in order to share knowledge and be aware of the "unknown"*" – encompasses both the intelligence density framework and KDD process.

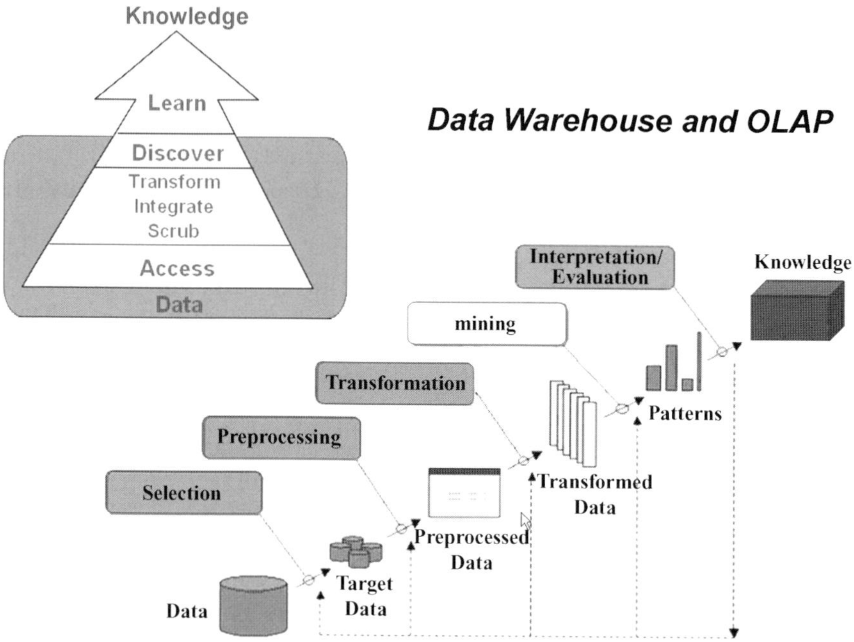

Fig. 1. An intelligence density representation of a data warehouse and OLAP application aligned to a KDD process.

3. Ontology driven knowledge discovery – ODKD

Knowledge Discovery in Databases (KDD) is an iterative process based on the analysis of current facts or data, pre-processing to clean and transform that data, application of mining algorithms, and deployment using the mining results on new data. Computer scientists have been working on various aspects of the KDD process for many years. Their objectives include: building simpler and more comprehensible models, improving data mining performance, and helping to prepare, to clean, and to understand data. In largely parallel research, ontologies have been widely used by the artificial intelligence community to represent domain knowledge and to integrate different database models.

The current interest in ontologies is the latest version of artificial intelligence's alternation of focus between content theories and mechanism theories [7]. Sometimes, a mechanism (such as rule systems, frame languages, neural nets, fuzzy logic) excites the artificial intelligence community, generating extensive research activity. At other times, the lack of a good knowledge representation is argued as the reason for the failure of AI-based systems and knowledge discovery tasks. In our investigation, we realize that, however wonderful the mechanism, it cannot do much without a good content theory of the domain on which it is to work, and vice-versa. Moreover, we recognize that the integration of good content and different techniques is the key to improving the knowledge discovery process. Once good content theory is available, many different techniques might be used equally well to implement effective systems all using and improving essentially the same content.

The common background underlying database and ontology research is well known. Although ontologists and data modelers have been working together to bridge both areas, for example, in topics like conceptual modeling, database integration and metadata representation, less work has been undertaken in the integration of ontologies and KDD that includes learning from schemata and learning from instances.

There are three approaches being investigated in this emergent topic research: Onto4KDD, KDD4Onto, and one approach that integrates both previous approaches, named here Onto4KDD4Onto.

Onto4KDD is defined as the application of ontologies in order to improve the KDD process. For example, domain ontologies can be used to improve the understanding of a problem and to support hypothesis-driven analysis and discovery approaches. Onto4KDD has been employed extensively in the biomedical informatics research area [4] where a huge amount of DNA sequence information is made available daily and thousands of projects need to be annotated. At the same time, millions of facts are being accumulated from a variety of sources in seemingly unrelated fields and are being made available either immediately on the Internet or published in hundreds of journals. Biological knowledge is evolving so rapidly that it is difficult for most scientists to assimilate. Gene Ontology (GO) [12], for example, has been used to "produce a controlled vocabulary that can be applied to all organisms even if knowledge of genes and proteins is changing". GO is the basis for classification systems that address the problem of linking biology knowledge and evolving literature, such as GO-KDS [34] and DiscoveryInsight [3].

In contrast, KDD4Onto approaches are focused on the application of mining techniques in order to automatically or semi-automatically acquire knowledge from data. Currently, ontologies are mostly learnt from text [1] by an integration of lexicons, taxonomies, and other ontologies, in the natural language processing and computational linguistics areas. Another incentive for this development comes from the Semantic Web research where ontology acquisition from text is a significant issue and has been well documented in recent years.

Although some researchers are addressing Onto4KDD or KDD4Onto, rare is the research that encompasses both perspectives. Our work is an attempt to integrate both approaches. It bridges the gap between ontological engineering and knowledge discovery in databases in order to improve both processes. In particular, it analyzes how data mining, and in particular Evolving Connectionist Systems (ECOS), can assist in efficient and effective large-volume data analysis in order to build a sharable and evolving knowledge repository, while at the same time leveraging the semantic content of ontologies to give intelligent support and to improve knowledge discovery in complex and dynamic domains.

3.1. *Life cycle of ODKD*

Several different knowledge discovery life cycles have been proposed in the literature. Most reflect the background of their proponents, originating from the database community [10], the artificial intelligence community [27], the decision support community [30], or the information systems community [40]. We take into consideration different aspects of these life cycles in order to establish a life cycle that integrates both ontological engineering and data mining requirements.

In ODKD we are concerned with the abstract level of the data – the concepts. We consider data as a representation of the knowledge within a domain. In this way we are looking at the data from a data modeling perspective where the database is a formal representation of the domain knowledge, containing useful data and metadata within its model. This perspective aligns with the artificial intelligence perception in which concepts are the most basic units of thought. Concepts are thus building blocks and these and their relationships can represent knowledge about a domain.

Moreover, when analyzing real and complex problems, the knowledge discovered in one KDD task can be used as a data input for another KDD task, and this chain might configure a decision-making workflow. For example, the knowledge extracted from a microarray about the expression of a gene in a tissue can be used as a data input for a drug discovery investigation for a completely different disease. Thus we do not presume that a 'complete' dataset is available and sufficient at the time a new data mining application is begun. On the contrary, at a conceptual level we believe that new domains will generate new types of variables and knowledge about them that can improve our understanding about both the specific problem and the wider domain.

Furthermore, we also believe that data mining is not simply the one-time application of a program to a new dataset. In our own work, data mining is an ongoing hypothesis-checking exercise. It frequently starts with small pilot studies and manual bias space search, including feature construction. With preliminary confirmation that the process can find some interesting relationships, more data and greater expectations are introduced leading to another round of investigation, utilizing a larger amount of data or at least a better understanding of the problem.

In order to consider this dynamic behavior, we are adopting some steps of the CRoss-Industry Standard Process for Data Mining (CRISP-DM) [40] in the ODKD life cycle. CRISP-DM is the most widespread methodology describing the steps of the KDD process in the business area.

The CRISP life cycle consists of six phases:

- Business Understanding;
- Data Understanding;
- Data Preparation;
- Modeling;
- Evaluation;
- Deployment.

The sequence of the phases is not strict. The dynamic behavior is represented by the cyclic nature of data mining. Moving back and forth between different phases is always required. The outcome of each phase determines which phase, or which particular task of a phase, is to be performed next.

ODKD is based on a hybrid approach that integrates content theory (ontologies) and mechanism theory (KDD). It defines an ontology driven knowledge discovery process framework (Figure 2) that combines KDD and ontological engineering aspects.

There are five steps in the ontology driven knowledge discovery process: ontology preparation, ontology population, instance selection, ontology mining, and ontology refining.

3.2. *ODKD process*

3.2.1. *Ontology preparation*

Ontology preparation is the process of analyzing data models and/or ontology models in order to gain domain understanding and select the target knowledge for the KDD task. In

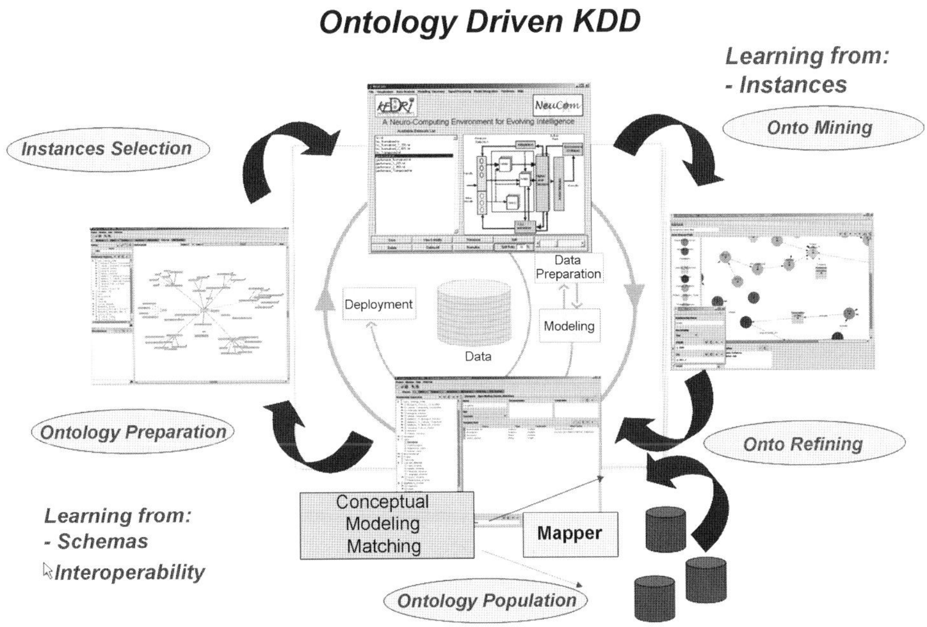

Fig. 2. The ontology driven knowledge discovery process.

this phase ontologies are utilized to guide conceptual matching with data models and to integrate different sources of related information.

This initial phase focuses on understanding the problem domain and the problem requirements, and then converting this knowledge into an ontological representation capable of providing a preliminary definition of the problem and the required KDD task.

This phase is subdivided into two main pipelines, ontology preparation and data preparation, and then in the selection of the target knowledge for the problem. The ontology pipeline is composed of the following sub-tasks: ontology candidates' selection; ontology merging/alignment; and ontology learning. The data preparation pipeline is composed of the following sub-tasks: data selection; data understanding; and conceptual matching.

There are two different approaches to ontology learning in this phase:

- Ontology learning from schemata;
- Ontology learning for interoperability.

Ontologies are used in this phase to integrate different databases and to merge with other available ontologies. Integrating databases enables a shared conceptualization about the domain that at the same time improves the quality of the 'input data source'.

3.2.2. *Ontology population*

Ontology population is the act of populating the knowledge base with the instances acquired from the databases. This is the next natural step after the conceptual model matching. There are different approaches to integrate and populate ontologies and databases [14, 25,19]. In brief, after two concepts have been identified as similar the instances of the ontology and/or the records of the database are stored in the ontology to build the knowledge base.

3.2.3. *Instance selection*

Instance selection is the final step before the choice and application of a data mining technique. In the KDD process this is one of the most important steps. In gene expression analysis, for example, the number of samples may be small and the number of genes can be incredibly huge, thus any support for gene selection should help to identify those that are significant in the problem analysis, that reduce the amount of computing time needed, and that give optimum meaning in the results. Instance selection in this case plays an important role in acquiring previous knowledge that can help the bioinformatician in the selection of such features.

Instance selection from the ontology requires explicit knowledge about the domain. Opportunities arise from the simple act of connecting different facts and points of view that have been created for one purpose, but in light of subsequent information, can be reused in a quite different context, to form new concepts or hypotheses. Ontology visualization techniques can be used to navigate through the hierarchy and through the relationships in order to increase the intelligence density, resulting in increased understanding of the domain.

On the other hand, KDD techniques can help ontologists to identify new relationships among concepts when using data analysis techniques, such as principal component analysis (PCA), independent component analysis (ICA), and when visualizing the data through 2D, 3D and Surface Plotting.

3.2.4. *Ontology mining*

Onto-mining is mainly recognized as ontology learning from instances. Most of the research in this area is concentrated on learning from textual and semi-structured resources in the process of building an ontology. However less has been said about the use of ontologies as an inference mechanism. The reasoning within the links and relationships of an ontology can help the mining techniques to produce better results. Ontologies can help mining techniques in the same sense as the application of mining techniques in a good data warehouse can lead to better results than when applied to raw data.

Data mining techniques can help to identify new relationships among concepts within an ontology, in turn refining or updating the ontology model. There are many mining techniques and cross-disciplinary methods unexplored, such as FCA, neural networks, and statistical methods for ontology mining.

Our view here is that both the ontology engineering and data mining fields have much to contribute to each other and a broad vision is needed to integrate them. In the Future Work section we present a brief introduction about our ongoing research to integrate ontologies and Neucom [44] – a Neurocomputing Environment for Data Mining, Knowledge Discovery and Intelligent Decision Support Systems.

3.2.5. *Ontology refining*

The main task in this phase is to acquire the knowledge discovered by the mining technique and represent it in the ontology formalism. Depending on the learning technique selected the resulting knowledge must be analyzed and mapped and/or translated into the formalism used by the ontology.

Some of the research in this area includes merging and mapping techniques utilized in ontology integration, such as FCA [32]. This can be undertaken according to similarities resulting from the insertion of new structures in an ontology and the integration of different ontologies.

Although we do not emphasize the difference between formal and informal ontology languages, we note that the formal approaches can exploit the advantages of logics to enable automatic reasoning when new knowledge or a new instance is classified by the onto mining step. Description logic, for example, is able to automatically classify new knowledge extracted through its reasoning mechanisms. However that knowledge must be first converted to the logic formalism in order to enable the reasoning to occur.

This step is the most crucial in the ODKD process. There remain many open questions in this research area. The semantic web research community is dealing with most of the aspects related to formal ontologies when using Ontology Web Language – OWL. Active groups are establishing patterns [45] that aim to address some of the critical points in this area. However, expressing and reasoning with imprecise knowledge is an active area where nonclassical logics play a key role.

In our research, the uncertainty is extracted by ECOS and translated to the core of the meta-knowledge representation by integrating relevant fuzzy rules in the ontological representation – Figure 2. In contrast to the text mining requirements where the boundaries of domain knowledge are well defined and the challenge is to extract semantics from text, knowledge discovery in database is concerned more with meta-knowledge. The main challenge is to extract knowledge from unrelated data samples and match it with a knowledge base; for instance, a column labeled 'sex' whose instances are filled with values 1 or 0 does not enable you to guess the type of animal represented by the attribute or even whether it represents hermaphrodite animals. A database is already a model of a partial conceptualization which should be mapped to another partial model using the values represented in tables, which are not expressed linguistically, and using previous knowledge about the domain expressed by other domain ontologies and its links.

Considering the KDD process, we have been using mapping techniques that are able to cope with the requirements of translating a data model into an ontological language. These mappings enable us to perform formalism translations, such as from a data mining workbench format to a frames formalism. We are also investigating the insertion of the fuzzy rules extracted by ECOS' methods in the ontology language using SweetRules [39], such as translating DENFIS [22] rules into JESS [36].

The onto refining step is responsible for closing one full iteration of the ontology driven knowledge discovery (ODKD) process. The knowledge extracted is inserted into the ontology beginning a new cycle of analysis; or it can be used by external applications to support a decision making process.

4. Biomedical evolving ontology case studies

Biomedical discovery is an intrinsically complex and risky process. One of the most challenging aspects of the biomedical discovery process is its necessarily iterative nature in terms of analyzing existing facts or data to support or refute current hypotheses and to generate new ones [5].

Integrating discovery of outputs from different experimentation techniques with extracted appropriate biological and medical knowledge could facilitate the discovery of new metabolic pathways and modeling of the metabolic and regulatory networks in living organisms, and ultimately a more complete understanding of the pathogenesis of various diseases. The biomedical community (in its widest sense) is suffering from communication problems and is lacking the ability to use resources to search the vast sources of information more effectively, to extract appropriate meanings. Many researchers and databases use (at least partially) idiosyncratic terms and concepts for representing biological information. It has been observed that often, terms and definitions differ between groups, with different groups not infrequently using identical terms with different meanings. The concept 'gene', for example, is used with different semantics by the major international genomic databases.

Biomedical ontologies provide a conceptual framework and factual knowledge that is necessary to deal critically with the rapidly changing science of biology. Ontology encloses the development of a semantic repository of systematically ordered relevant concepts in molecular biology. Such repositories can be used, for example, to bridge the different notions in various databases by explicitly specifying the meaning of and relation between fundamental concepts. Thus in general, ontology permits researchers to define and share domain-specific vocabularies.

However, in addition to this existing biological knowledge, the amount of potentially valuable data produced daily by medical information systems and medical decision support systems is growing at a staggering rate. We must consider that scientific biomedical information can include information stored in the genetic code, but also can include results from various experiments and databases, including patient statistics and clinical data. Large amounts of information and knowledge in medicine are also available on the WWW. Making medical knowledge and medical concepts sharable over applications and reusable for different purposes is crucial.

The search for biomedical knowledge brings several challenges to ontological representation. For example, it must provide conceptual links between data from seemingly disparate fields, which might include, for example, information collected in clinical patient data for clinical trial design, geographical and demographic data, epidemiological data, drugs, and therapeutic data, as well as from different perspectives such as those collected by nurses, doctors, laboratory experts, experimental researchers and so on. Consequently, any comprehensive biomedical ontology should reuse and integrate as many different specific ontologies as possible. The ontologies should integrate terminologies, such as Unified Medical Language System (UMLS) [42], with domain specific ontologies, such as particular disease ontologies and Gene Ontology.

At the same time, biomedical ontologies must be able to evolve their structure and content in order to cope with the rapid development of the area and introduction of novel techniques. They must also be able to represent uncertain knowledge and evaluate the quality of the knowledge discovered through different referenced measures, such as the quality of the annotation, number of related publications and so on.

Furthermore, to leverage the maximum power of biomedical ontologies, they must support domain understanding as well as the data preparation phase of knowledge discovery as the basis for a "semantic preparation phase" that will allow us to facilitate both forms of scientific discovery, factual and conceptual.

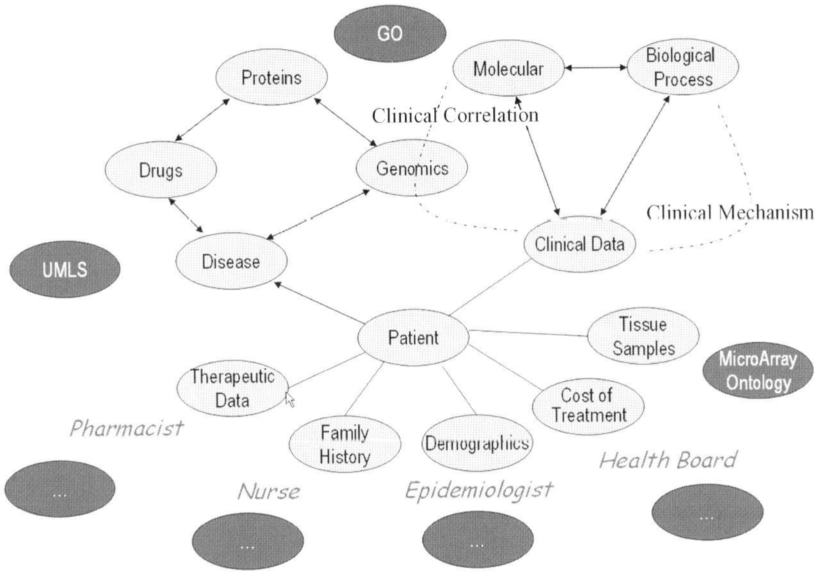

Fig. 3. Infogene map scope.

4.1. *Evolving infogene map*

Evolving Infogene Map (EIMp) aims to build multi-dimensional biomedical informatics ontologies (Figure 3), able to share knowledge from different experiments undertaken across aligned research communities in order to connect areas of science seemingly unrelated to the area of immediate interest. An EIMp for a leukemia case study, for example, is not only useful in modeling gene regulatory networks but the ontology framework also produces functional insight towards the understanding of hematopoietic cell differentiation, apoptosis and proliferation.

4.1.1. *Infogene map ontologies*

There are currently six ontologies included in or under development for the Evolving Infogene Map. Each represents a specific domain in the biomedical informatics area and the meta-knowledge that allows the ontology to evolve over time.

Concept metadata. The Concept Metadata entity enables a flexible knowledge representation for any concept present in the specific ontologies to be defined. This entity contains schemes to represent the following knowledge:

- Type Scheme – allows us to import and represent various types of information, such as image and text. The user can define a new type of information and constraints related to the insertion of instances.
- Spatial Scheme – represents the geographic knowledge.
- Temporal Scheme – represents the time notion in the ontologies.
- Language Scheme – allows us to acquire concepts in a language and maintain links with synonyms in other languages.

- Source Scheme – gives the flexibility to acquire information from different sources, such as UMLS, Internet sites, or clinical data, and maintain its independence of the original source. It is also concerned with the acquisition of data from external databases.
- Relationship Scheme – represents known relationships, such as *part_of* and *responsible_for*, and permits the creation of new relationships acquired from experts or through data mining activities. This meta-knowledge can define attributes that represent important characteristics of a relationship, such as its uncertainty and strength.

Biomedical domain. This entity represents the biomedical knowledge. It includes abstracts concepts, such as organism, and more concrete concepts, such as a disease and its instances.

Each biomedical concept uses the concept metadata to define its source of information and other abstract dimensions needed to represent its knowledge.

Biomedical informatics domain. The Biomedical Informatics domain represents the common knowledge between the biomedical and bioinformatics domains. Each subclass of this entity, such as oncogene, inherits characteristics from its domain and properties related with the biomedical informatics domain.

Clinical domain. Clinical domain classes represent the clinical knowledge contained in laboratories' results regarding treatments, drugs and so on.

The subclasses are mainly multi-inherited from the biomedical domain and its instances are directly updated from databases.

Gene ontology. Gene ontology (GO) represents the bioinformatics knowledge in the Evolving Infogene Map. This entity is directly imported to the main ontology and its instances are included through annotation tools.

In the current stage we maintain GO as part of the main ontology and use it without altering its knowledge representation.

In order to keep our ontology aligned with ongoing research projects around the world, every gene represented in the ontology is annotated with GO. At the same time, we are updating the included GO project monthly.

Disease gene map. This ontology is the core of our specific case studies. It enables us to build the gene/disease relationship and identify relationships among molecular weights, chromosome positions and so on. Each instance of this ontology represents experimental research and is traceable through a query language that allow us, for example, to answer questions such as "Which genes are related to the occurrence of Leukemia?"

Each map represents relationships graphically (Figure 4) in a way that enables visualization of existing knowledge and the creation of new relationships. Additional properties, such as uncertainty and importance, can be inferred and added by experts or programmatically using the evolving connectionist systems (ECOS) [20]. This approach enables us to evolve the maps as new knowledge is discovered through the use of the data mining techniques. In this sense the maps act as knowledge acquisition tools allowing the expert to annotate new knowledge acquired from external databases and creating instances linking it to external information.

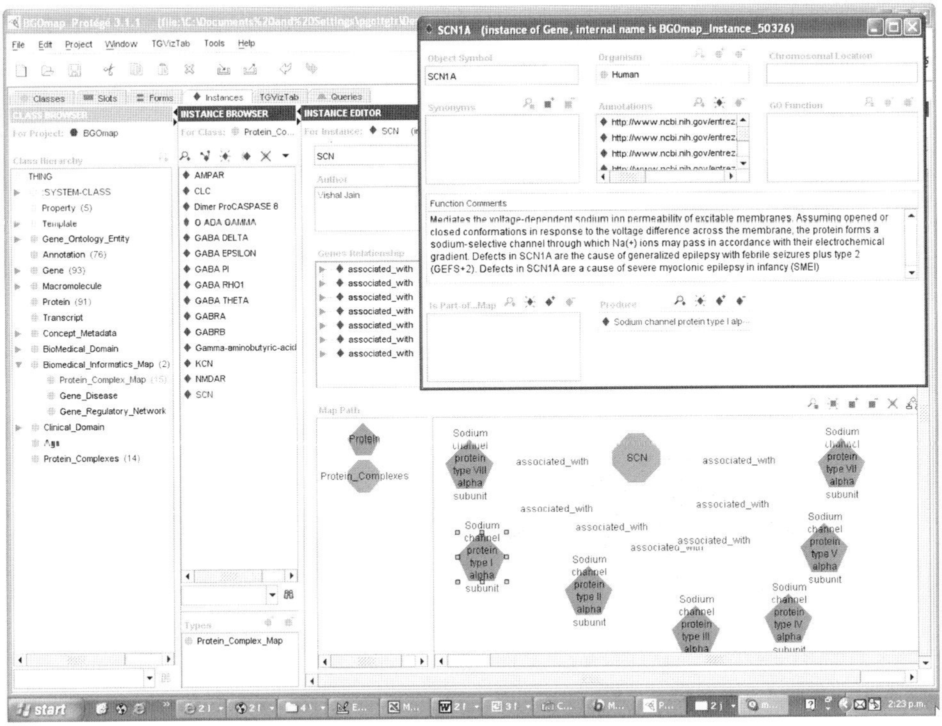

Fig. 4. Gene/disease map example.

Different experts can work on the same project or share different projects using the tools available in the Protégé environment [37]. In this sense, the system works as a knowledge management tool storing knowledge from diverse projects and sharing the knowledge discovered in each experiment. The maps can also be shown in different visualizations supporting the instance selection step of the ontology driven knowledge discovery process.

4.1.2. *Case studies*

It is well accepted in the ontology engineering community that reuse is a key factor in projects that aim to integrate different domains or different sources of information under the umbrella of ontology. Our Evolving Infogene Maps include two of the most used ontologies, Gene Ontology and Unified Medical Language System (UMLS), to represent respectively genes and biomedical knowledge as shown in Figure 5.

Our biomedical ontology is projected to be generic enough to integrate different sources of information and types of information. We represent biomedical concepts based on the UMLS semantic network, the UMLS metathesaurus, the knowledge acquired from domain experts as well as knowledge acquired directly from clinical databases.

Furthermore, any Infogene Map created in this representation is automatically linked to other projects. This characteristic enhances the broader understanding of the domain and reveals new relationships that were not linked individually in each project.

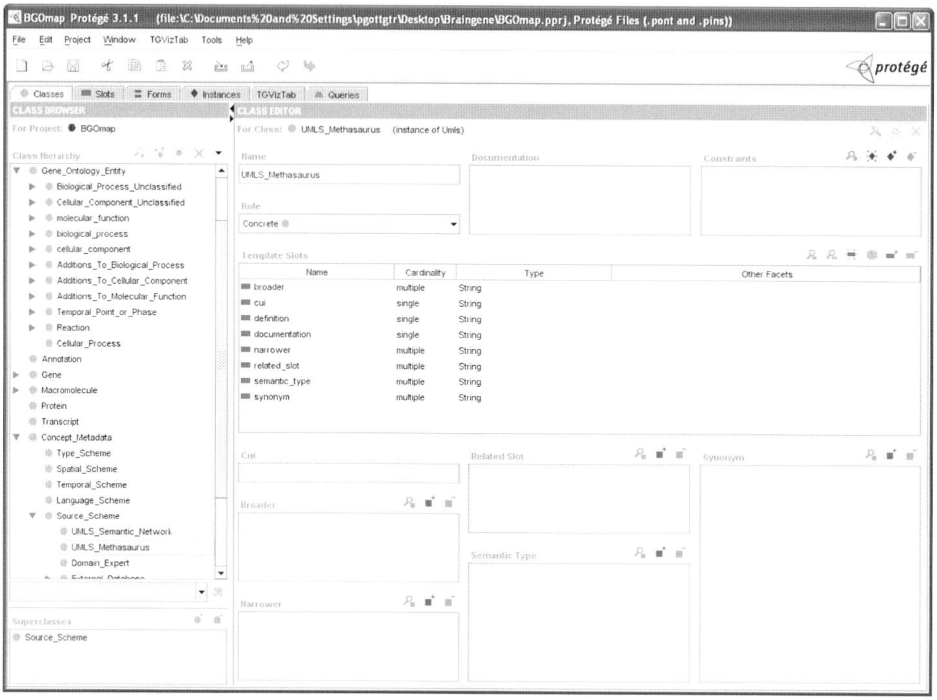

Fig. 5. Ontology.

The leukemia gene regulatory network map. In medical research it has been shown that human macrophage-like cell lines spontaneously produce a suppressor factor that inhibits production of interleukin 2 (IL 2) by human blood T lymphocytes. Considering regulatory genes and proteins from such cell lines an Infogene Map is obtained and is integrated with the apoptotic pathway.

The Infogene Maps in this case represent oncogenes, suppressor genes and genes that are related to leukemia in a network that helps to explain the genetic pathway of the disease – Figure 6. These maps are interconnected with other maps that express graphically the complex proteins and their roles in different diseases. Myc, for example, which is a powerful inducer of apoptosis, is a strategic controller of cell proliferation that acts pleiotropically to coordinate both cell growth and concomitant progression through the cell cycle. Myc is connected through the leukemia infogene map and the Subway Map of cancer pathways [41]. It enables investigators, for instance, to answer the question "In which cancers is Myc involved?"

Each relationship among genes/proteins represents inhibitor or activator behavior which is qualified by a degree of uncertainty (Figure 6) based on the evidence given by the expert or by the number of supporting publications found in the literature.

Brain disease. Complex interactions between genes and proteins in neurons affect the dynamics of the whole neural network. Gene and protein expression values may change due to the internal dynamics of the gene regulatory (interaction) network, initial conditions of the genes and external conditions. It is observed that different initial gene conditions can

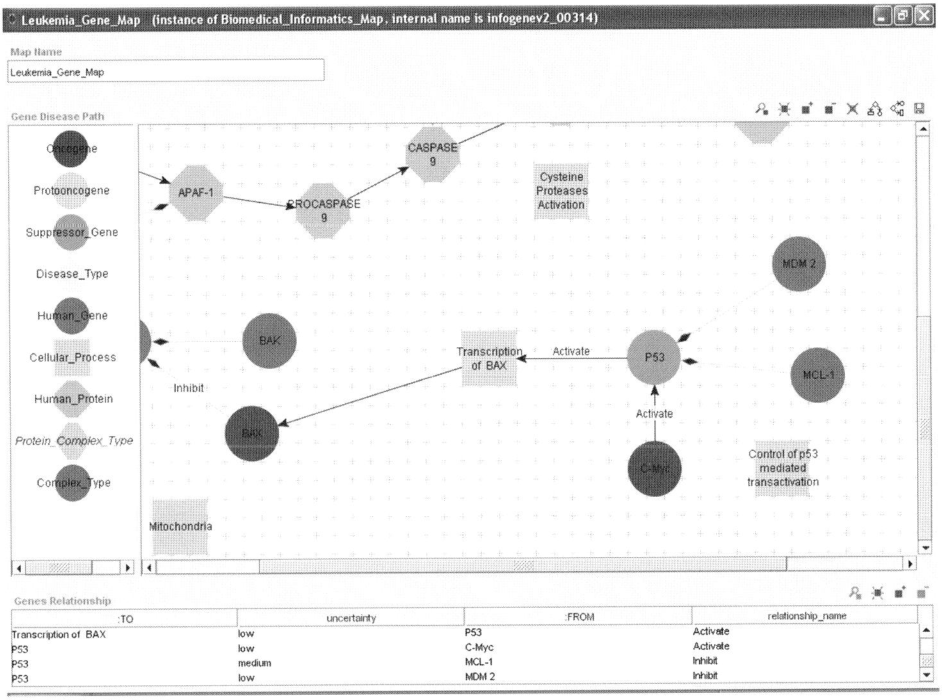

Fig. 6. Leukemia gene disease map.

lead to the same outcome in terms of neuronal activity. On the other hand, in the diseased brain, either altered initial conditions, mutated genes and/or altered interactions within the GN lead to abnormalities in network activity. Complete understanding through ontological representation about biological properties like protein length, molecular weight, location of genes on a chromosome, mutation and so on of these interacting elements enhances the chances of correctly tuning the gene interaction network and in assignment of initial gene/protein expression values through which different states of the neural network operation can be achieved and the desired behavior of the neural network to simulate certain brain states.

Infogene Maps not only directly facilitate the neurogenetic modeling [21] that integrates dynamic gene networks within neurons but also produce functional insight towards the understanding of gene regulatory networks.

5. Ontology driven knowledge discovery applications

This section describes two applications of the ODKD process, using different methods to address real problems. The first illustration follows the ODKD process in the integration of a data mining environment and an ontology editor. The second is an application of the on-line analytical process for ontological analysis in order to enhance the ODKD tasks.

5.1. *ODKD and Neucom*

A tool that gradually accumulates knowledge from databases is appropriate for and applicable to knowledge discovery from data because KDD is an iterative process where any change in one of the source databases should represent an input to a new knowledge discovery process. Our approach enables us to achieve this behavior through the integration of ontologies and the Evolving Connectionist Systems (ECOS) paradigm. ECOS is used to find new relationships and patterns in data and to extract fuzzy rules to update the ontology content and structure. The resulting ontology is used as a knowledge visualization tool for the whole data mining task.

Our ongoing research has been focused on the integration of Neucom [44] and Protégé. Our aim is to develop and test Neucom as an ontology-learning tool, and, at the same time, integrate it with an ontology editor in order to add semantics to Neucom and to build an ontology editor environment.

Neucom, a Neurocomputing Environment for Data Mining, Knowledge Discovery and Intelligent Decision Support Systems, is a self-programmable learning and reasoning computer environment based on connectionist (neurocomputing) modules. The Neucom modules can adapt to new incoming data in an on-line, incremental, life-long learning mode, and can extract meaningful rules that aid the discovery of new knowledge in multiple fields.

Protégé is an ontology editor and a knowledge-base editor developed at Stanford. It has plug-ins that extend the environment, enabling inference and reasoning, and visualization. Protégé was selected considering its features and its already available plug-in support for visualization, merging, and semantic web integration. We have developed other plug-ins to integrate the ontology environment with the data mining workbench in order to accomplish the ontology driven knowledge discovery tasks.

Our approach takes advantage of ECOS' evolving characteristic, accumulating knowledge from databases as new inputs are added to the model. In the symmetric direction, the results acquired from ECOS are represented in an ontology adding meaning to it, and allowing reuse of the knowledge discovered. This ongoing work aims to build a complete evolving environment that incorporates both state-of-the-art knowledge discovery and ontology engineering. Currently we are applying some of Neucom's features in order to learn from ontology instances. We have been testing this methodology in different projects [15].

5.1.1. *Evolving clustering method for hierarchal ontology learning*

The Evolving Clustering Method (ECM) is a fast, one-pass algorithm for dynamic clustering of an input stream of data. It is a distance-based clustering method in which the cluster centers are represented by evolved nodes in an on-line mode. The ECM was specially designed for the Dynamic Evolving Neural Fuzzy Inference System (DENFIS) [22], and in this case, the ECM dynamically performs scatter partitioning of the input space for the purpose of creating fuzzy inference rules and evolving a fuzzy system.

In the clustering process, the data samples come from a data stream and the process begins with an empty set of clusters. When a new cluster is created, its cluster center, Cc, is located and its cluster radius, Ru, is initially set with a value 0. With following samples presented one after another, some already created clusters will be updated through changing their centers' positions and increasing their cluster radiuses. Decisions regarding which cluster should be updated and how it should be changed depend on the position of

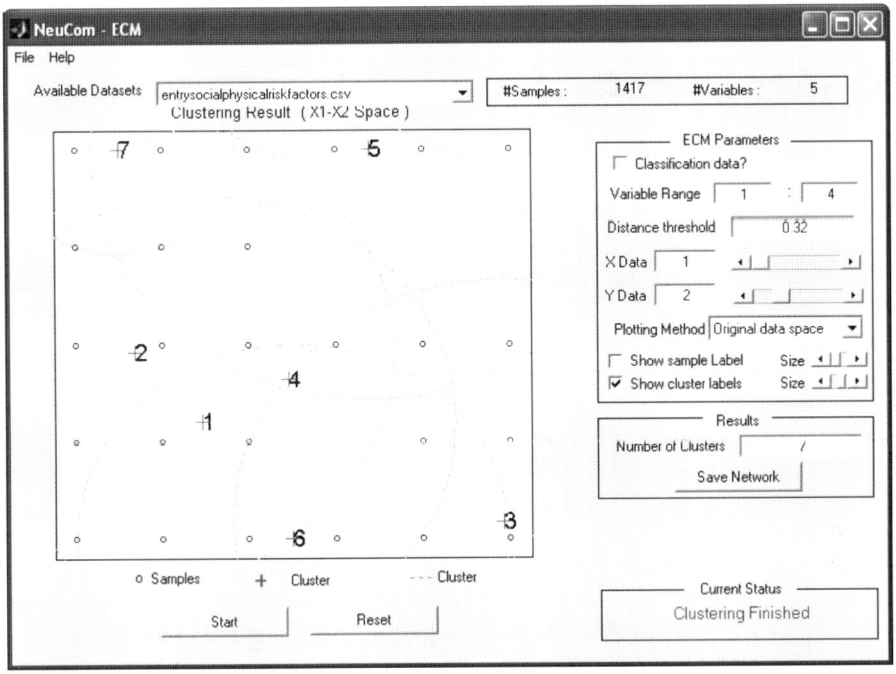

Fig. 7. ECM patient clusters based on social and physical characteristics of the Stulong dataset.

the current data sample. A cluster will not be updated any further when its cluster radius, R_u, has reached a specific value that is, usually, equal to a threshold value D_{thr}.

ECM has been used to learn the hierarchal structure of the ontology. It identifies new classes and sub-classes based on data samples and can be used to validate the domain structure developed by the ontology engineer.

We have been using ECM in medical analysis within the Stulong project [35]. ECM helps us to check the actual hierarchical structures in the database and identify clusters of patients that are likely to develop heart disease in the future – Figure 7. Instead of considering the crisp classification of patients in "risk-group", "normal group" and "pathological group", ECM allows us to represent each patient in a soft classification in which one patient can belong to more than one cluster at the same time. The resulting soft clusters are analyzed through a 20-year longitudinal study in order to identify relevant features that were not considered in the experimental study but which could help in the discovery of new relationships associated with the patients' lifestyle and the outcome of the diseases. For example, we are investigating why a patient previously classified as "normal" developed heart disease, and why a "risky" patient did not develop it.

5.2. *Ontology on-line analytical pre-processing (OOLAP)*

On-line analysis has been widely used in data warehousing in order to explore the huge amounts of data now available. The use of visualization techniques in conjunction with simple tabular data enable users, mainly decision makers, to work with tools that are familiar to them in the form of spreadsheets and graphs, resulting in the billion-dollar industry of business intelligence. The success of these tools is constrained, however, by the limited effectiveness of the data warehouse model, driven as it is by the static characteristics of an underlying data model. Any query, investigation or even insight desired that is out of the range of the multi-dimensional model is difficult and expensive to obtain. The ontology driven knowledge discovery approach avoids this restriction by adding a flexible semantic layer (evolving ontologies) to the data warehouse schema.

Ontology On-line Analytical Pre-processing (OOLAP) enables detailed and human-driven analysis of the domain. It converts any highly-dimensional ontology space into a simple tabular representation that can be translated via semantic mapping into a multi-dimensional format to be used by current business intelligence tools as well as to be further explored by data mining workbenches. Additionally, it interacts with available Online Analytical Processing (OLAP) and visualization tools to facilitate ontology exploration and analysis in order to refine the ontology model and improve the data and domain understanding. In particular, OOLAP supports the ontology preparation, population and instance selection activities of the ontology driven knowledge discovery process.

Our instance of the OOLAP process is currently implemented using a variety of tools for ontology engineering, data warehousing, OLAP, and data mining, namely: Protégé, MS-Analysis Service 2000, Excel spreadsheets and Neucom. We selected these tools in order to avoid any extra learning curve by our experts and users. However it is possible to use other tools for any of the required tasks.

5.2.1. *OOLAP alternative pipeline*

OOLAP encompasses three steps of the ontology driven knowledge discovery process: ontology preparation, ontology population and instance selection. It can be considered as an alternative pipeline that is akin to the CRISP-DM process. OOLAP adds an ontology analysis capability as well as an ontology integration tool able to integrate data warehouse models.

Ontology visualization. Ontology visualization supports the CRISP business understanding step (domain understanding). We utilize different visualization techniques developed in the Protégé environment, such as Jambalaya [43] and Ontoviz [38], in order to gain an understanding of the domain and to select the best concepts for analysis. The different ontology visualizations support user investigation and enable exploration of the knowledge base from different perspectives. Furthermore, the environment incorporates a query language that can speed up any data search or concept understanding effort.

Additionally, we explore Protégé's knowledge acquisition feature to enable the user to update the knowledge base on the fly when necessary. Visualization can facilitate the identification of important missing concepts or relationships that should be further explored in the data understanding step.

Data understanding. After the initial domain understanding, concepts in the ontology must be aligned with the data available. Our methodology makes it possible to import data from almost any available data source. It can import databases, ontologies, flat files, XML files, and other specific data mining workbench formats. The data imported is included in the knowledge base and made available for further analysis or for the conceptual model

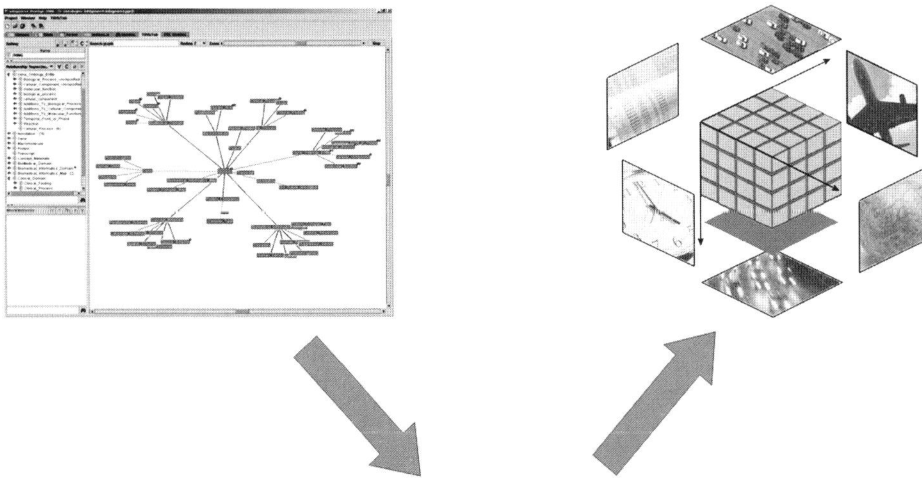

	A	B	C	D
1	Mr. Science	Editor	60000	
2	Chief Honcho	Editor	150000	
3	Sports Nut	Editor	100000	
4	Ms Gardiner	Editor	45000	
5	Larry Tennis-nut	Reporter	80000	
6	Joe Schmo	Reporter	120000	
7	Anne	Reporter	50000	
8				

Fig. 8. Ontology/cube mapping.

mapping phase. The main benefits of this effort are the identification of missing attributes and enrichment of the data model.

Using the highly-dimensional ontology model the user can identify relationships that were not present in the data model but are relevant for their analysis or they can request the inclusion of new knowledge in the ontology.

Conceptual model mapping. This stage is concerned with the translation of the ontology format into a simple tabular format used in the OLAP cube and thus able to be processed by the OLAP tool – Figure 8. Once they have a level of domain understanding the user can navigate in the ontology and select the scope of his/her analysis. This is performed using visualization techniques or through the query language. The selected data is exported and translated to a local cube format.

We translate the knowledge base into a text specific format to build the local cube. However the translation can be done for any cube format depending on the platform and OLAP tool(s) selected. One of the main characteristics of this approach is the total independence of the possibly complex database environments, such as SQL Server, MYSQL or Oracle.

OLAP tool. After exploring the multi-dimensional, application-independent and semantically powerful ontology environment and translating it to the cube format, the next step exploits all the advantages of well-developed OLAP techniques to analyze the target problem.

This step introduces human-driven analysis in the data preparation step of the CRISP-DM and ODKD processes. The data selected is made available for the decision maker/analyst through the local cube that enables him/her to use OLAP tools in their analysis. We have been using mainly MS-Excel as our OLAP tool. However we have also tested the cubes with MS-Data Analyser.

5.2.2. *OOLAP biomedical informatics application*

We tested our approach using the evolving brain disease ontology, which represents knowledge of genes and proteins that are related to specific brain related disorders like epilepsy and schizophrenia, as described in the related previous section. We focused on the crucial neuronal parameters that are in some way thought to control the phenomenon of epileptic seizures through their direct or indirect interactions with other genes/proteins, adding different dimensions in order to visualize their importance in the disease.

The OOLAP methodology helped us to identify missing concepts, new relationships and select the best candidates for our modeling tasks. Discovery of these interacting elements enhanced our ability to correctly tune the gene interaction network and in the assignment of initial gene/protein expression values through which different states of the neural network operation could be achieved. This OOLAP biomedical informatics case study was developed as part of a wider ontological research effort in neurogenetic modeling that integrates dynamic gene networks within neurons [21].

6. Conclusions and future work

As researchers we are faced with many complex and dynamic problems. The amounts of data produced daily by information systems are growing at a staggering rate. Terabyte and even petabyte databases, once unthinkable, are now becoming a reality in a variety of domains, including marketing, sales, finance, healthcare, molecular biology (e.g., the human genome project), and various government applications. This explosion of data and the growing number of disparate data sources are exposing us to a 'new' category of problem – how to acquire, maintain and share knowledge from these large and distributed databases in the context of this dynamic and evolving process.

Evolving processes are difficult to model because some of their parameters may not be known *a priori*. The acquisition of new knowledge can demand the development of new models and the embracing of different perspectives in order to understand the problem. Decision makers needs tools that can intelligently assist them in the decision making process. Thus, modeling of such processes is a challenging but potentially fruitful task with many practical applications in life sciences and business.

In biological systems, for example, everything is interconnected, and ostensibly unrelated fields are related – the separation of biology into different disciplines is in some respects artificial. Conceptual research can encompass many fields without limitation. So what is needed is a way to manage the context of the search, so that concepts having different meaning in different contexts can be represented appropriately. We also need ways to enable scientists to cross disciplines and search in areas outside their expertise, so that they can extract information critical for new discoveries.

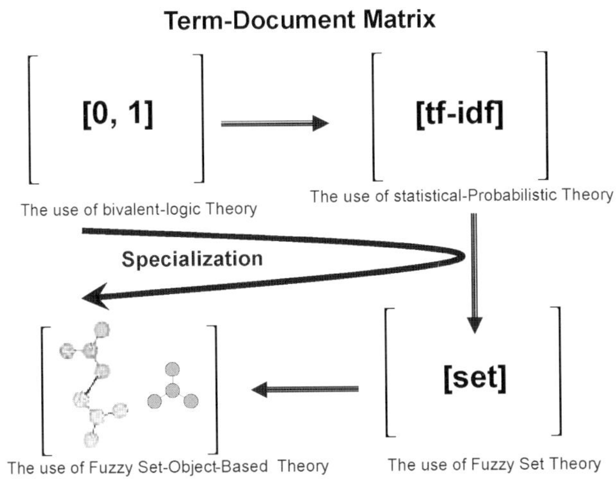

Fig. 2. Evolution of term–document matrix representation.

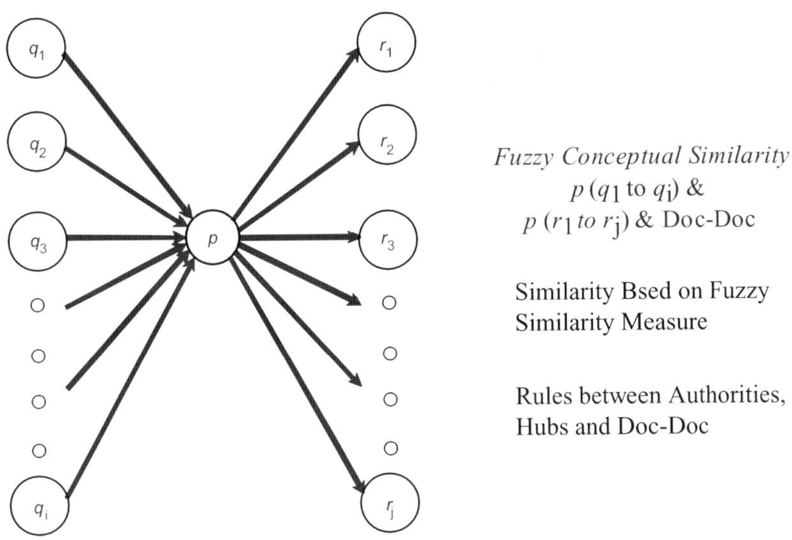

Fig. 3. Fuzzy conceptual similarity.

of the current search engines such as Google™, Teoma, Yahoo!, and MSN use this technique to store the term–document information. One can extend this model by the use of ontology and other similarity measures. This is the core idea that we will use to extend this technique to FCM. In this case, the existence of any keyword that does not exist directly in the document will be decided through the connection weight to other keywords that exist in this document. For example consider the followings:

- if the connection weight (based on the ontology) between term "i" (i.e. Automobile) and term "j" (i.e. Transportation) is 0.7; and the connection weight between term "j" (i.e. Transportation) and term "k" (i.e. City) is 0.3;

Webpages

RX' = *Webpages*

1 (Text_Sim, In_Link, Out_Link, Rules, Concept)	**2** (...)	**0** (...)	**0** (...)
0 (Text_Sim, In_Link, Out_Link, Rules, Concept)	**1** (...)	**1** (...)	**6** (...)
2 (Text_Sim, In_Link, Out_Link, Rules, Concept)	**0** (...)	**5** (...)	**4** (...)
0 (Text_Sim, In_Link, Out_Link, Rules, Concept)	**1** (...)	**4** (...)	**0** (...)

Text_sim: Based on Conceptual Term-Doc Matrix; It is a Fuzzy Set

In_Link & Out_Link: Based on the Conceptual Links which include actual links and virtual links; It is a Fuzzy Set

Rules: Fuzzy rules extracted from data or provided by user

Concept: Precisiated Natural Language definitions extracted from data or provided by user

Fig. 4. Matrix representation of fuzzy conceptual similarity model.

  - $w_{ij} = 0.7$,
  - $w_{jk} = 0.3$
- and term "i" doesn't exists in document "I", term "j" exists in document "I", and term "k" does not exists in document I
  - $T_i D_I = 0$,
  - $T_i D_I = 1$,
  - $T_k D_I = 0$.
- Given the above observations and a threshold of 0.5, one can modify the term–document matrix as follows:
  - $T_i D'_I = 1$,
  - $T_i D'_I = 1$,
  - $T_k D'_I = 0$.
- In general one can use a simple statistical model such as the concurrence matrix to calculate w_{ij} [14,15].

An alternative to the use of [1, 0] representation is the use of the tf–idf (term frequency–inverse document frequency) model. In this case, each term gets a weight given its frequency in individual documents (tf, frequency of term in each document) and its frequency in all documents (idf, inverse document frequency). There are many ways to create a tf–idf matrix [18]. In FCM, our main focus is the use of fuzzy set theory to find the association between terms and documents. Given such association, one can represent each entry in the term–document matrix with a set rather then either [0, 1] or tf–idf. The use of fuzzy-tf–idf is an alternative to the use of the conventional tf–idf. In this case, the original tf–idf weighting values will be replaced by a fuzzy set representing the original crisp value of that specific term. To construct such value, both ontology and similarity measure can be used. To develop ontology and similarity, one can use the conventional Latent Semantic Indexing (LSI) or Fuzzy-LSI [18]. Given this concept (FCM), one can also modify the link analysis (Figure 3) and in general Webpage – Webpage similarly (Figure 4). More information about this project and also a Java version of Fuzzy Search Tool (FST) that uses the

FCM model is available at http://www.cs.berkeley.edu/~nikraves/fst//SFT and a series of papers by the author at Nikravesh, Zadeh and Kacprzyk [15]. Currently, we are extending our work to use the graph theory to represent the term–document matrix instead of the use of fuzzy set. While, each step increases the complexity and the cost to develop the FCM model, we believe this will increase the performance of the model given our understanding based on the results that we have analyzed so far. Therefore, our target is to develop a more specialized and personalized model with better performance rather than a general, less personalized model with less accuracy. In this case, the cost and complexity will be justified.

2.1. *Fuzzy conceptual match and search engine*

The main problem with conventional information retrieval and search such as vector space representation of term–document vectors are that (1) there is no real theoretical basis for the assumption of a term and document space and (2) terms and documents are not really orthogonal dimensions. These techniques are used more for visualization and most similarity measures work about the same regardless of model. In addition, terms are not independent of all other terms. With regards to probabilistic models, important indicators of relevance may not be term – though terms only are usually used. Regarding Boolean model, complex query syntax is often misunderstood and problems of null output and Information overload exist. One solution to these problems is to use extended Boolean model or fuzzy logic. In this case, one can add a fuzzy quantifier to each term or concept. In addition, one can interpret the AND as fuzzy-MIN and OR as fuzzy-MAX functions. Alternatively, one can add agents in the user interface and assign certain tasks to them or use machine learning to learn user behavior or preferences to improve performance. This technique is useful when past behavior is a useful predictor of the future and wide variety of behaviors amongst users exist. In our perspective, we define this framework as *Fuzzy Conceptual Matching based on Human Mental Model*. The Conceptual Fuzzy Search (CFS) model will be used for intelligent information and knowledge retrieval through conceptual matching of both text and images (here defined as "Concept"). The selected query doesn't need to match the decision criteria exactly, which gives the system a more human-like behavior. The CFS can also be used for constructing fuzzy ontology or terms related to the context of search or query to resolve the ambiguity. It is intended to combine the expert knowledge with soft computing tool. Expert knowledge needs to be partially converted into artificial intelligence that can better handle the huge information stream. In addition, sophisticated management work-flow need to be designed to make optimal use of this information. The new model can execute conceptual matching dealing with context-dependent word ambiguity and produce results in a format that permits the user to interact dynamically to customize and personalized its search strategy.

2.2. *From search engine to Q/A systems: the need for new tools*

> Construction of Q/A systems has a long history in AI. Interest in Q/A systems peaked in the seventies and eighties, and began to decline when it became obvious that the available tools were not adequate for construction of systems having significant question-answering capabilities. However, Q/A systems in the form of domain-restricted expert systems have proved to be of value, and are

growing in versatility, visibility and importance. Upgrading a search engine to a Q/A system is a complex, effort-intensive, open-ended problem. Semantic Web and related systems for upgrading quality of search may be viewed as steps in this direction. But what may be argued, as is done in the following, is that existing tools, based as they are on bivalent logic and probability theory, have intrinsic limitations. The principal obstacle is the nature of world knowledge. Dealing with world knowledge needs new tools. A new tool which is suggested for this purpose is the fuzzy-logic-based method of computing with words and perceptions (CWP [20–23]), with the understanding that perceptions are described in a natural language [14,5,18,8,9]. A concept which plays a key role in CWP is that of Precisiated Natural Language (PNL). It is this language that is the centerpiece of our approach to reasoning and decision-making with world knowledge. The main thrust of the fuzzy-logic-based approach to question-answering which is outlined here, is that to achieve significant question-answering capability it is necessary to develop methods of dealing with the reality that much of world knowledge – and especially knowledge about underlying probabilities is perception-based. Dealing with perception-based information is more complex and more effort-intensive than dealing with measurement-based information. In this instance, as in many others, complexity is the price that has to be paid to achieve superior performance. (Extracted text from Prof. Zadeh's presentation and abstracts; Nikravesh et al., Web Intelligence: Conceptual-Based Model, Memorandum No. UCB/ERL M03/19, 5 June 2003)

2.3. *NeuSearch*™

There are two types of search engine that we are interested and are dominating the Internet. First, the most popular search engines that are mainly for unstructured data such as Google™, Yahoo, MSN, and Teoma which are based on the concept of Authorities and Hubs. Second, search engines that are task specifics such as (1) *Yahoo*!: manually-pre-classified, (2) *NorthernLight*: Classification, (3) *Vivisimo*: Clustering, (4) *Self-organizing Map*: Clustering + Visualization and (5) *AskJeeves*: Natural Languages-based Search; Human Expert. Google uses the PageRank and Teoma uses HITS for the Ranking. To develop such models, state-of-the-art computational intelligence techniques are needed [6,12,15, 17]. Figures 5 and 8 show, how neuro-science and PNL can be used to develop the next generation of the search engine. These include and are not limited to:

- Latent-Semantic Indexing and SVD for preprocessing,
- Radial-Basis Function Network to develop concepts,
- Support Vector Machine (SVM) for supervised classification,
- fuzzy/neuro-fuzzy clustering for unsupervised classification based on both conventional learning techniques and Genetic and Reinforcement learning,
- non-linear aggregation operators for data/text fusion,
- automatic recognition using fuzzy measures and a fuzzy integral approach,
- self-organization map and graph theory for building community and clusters,
- both genetic algorithm and reinforcement learning to learn the preferences,
- fuzzy-integration-based aggregation technique and hybrid fuzzy logic-genetic algorithm for decision analysis, resource allocation, multi-criteria decision-making and multi-attribute optimization,
- text analysis: next generation of the Text, Image Retrieval and concept recognition based on soft computing technique and in particular Conceptual Search Model (CSM). This includes
  - understanding textual content by retrieval of relevant texts or paragraphs using CSM followed by clustering analysis,
  - hierarchical model for CSM,

- integration of Text and Images based on CSM,
- CSM Scalability, and
- the use of CSM for development of Ontology, Query Refinement and Ambiguity Resolution, Clarification Dialog, and Personalization-User Profiling

Figure 5 shows a unified framework for the development of a search engine based on conceptual semantic indexing. This model will be used to develop the NeuSearch model based on the Neuroscience and PNL approach. As explained previously and represented by Figures 2 through 4 with respect to the development of FCM, the first step will be to represent the term–document matrix. Tf–idf is the starting point for our model. As explained earlier, there are several ways to create the tf–idf matrix [18]. One such attractive idea is to use the term ranking algorithm based on evolutionary computing, as it is presented in Figure 5, GA-GP context-based tf–idf or ranked tf–idf. The first version of this work has been developed for Book Indexing. The current version uses a file format such as PDF, Word, LaTex, PS, and HTML files. In this study for each chapter of the book we calculated the frequencies of words for:

- Title of Chapters (or article) + Book's Title (if the intension is not to create the Book Title)
- Section Titles
- Abstracts
- Conclusions
- References
- Body

Once these frequencies are created, an expert defines the series of the keywords that he/she feels are important and should have the highest value (say, 50–100 keywords). Given 50–100 keywords the model will adjust the weight for each set of terms in the title, section, abstract, conclusion, reference and body of the paper based on the following different criteria:

- GA should learn the weights such that
  - The 50–100 keywords gets the highest rank (no matter how they are distributed)
  - The 50–100 keywords, appear into the first 500 keywords
  - The 50–100 keywords, appear into the first 500 keywords, but it should be maximize in 50–100 keywords

This technique has been applied to select the keywords to be used to index the book entitled "Soft Computing and Intelligent Data Analysis in Oil Exploration" by the publisher (Elsevier).

One can also use this technique to do the reveres engineering such as:

- Given a sets of weights for each of the 5 attributes (Title, Section, Abstract, Conclusion, and References), and the information provided in Chapter, try to predict a New Title
- Given a sets of weights for each of the 5 attributes (Title, Section, Abstract, Conclusion, and References), and the information provided in Chapter, try to predict a Abstract
- Given a sets of weights for each of the 5 attributes (Title, Section, Abstract, Conclusion, and References), and the information provided in Chapter, try to predict a Conclusion

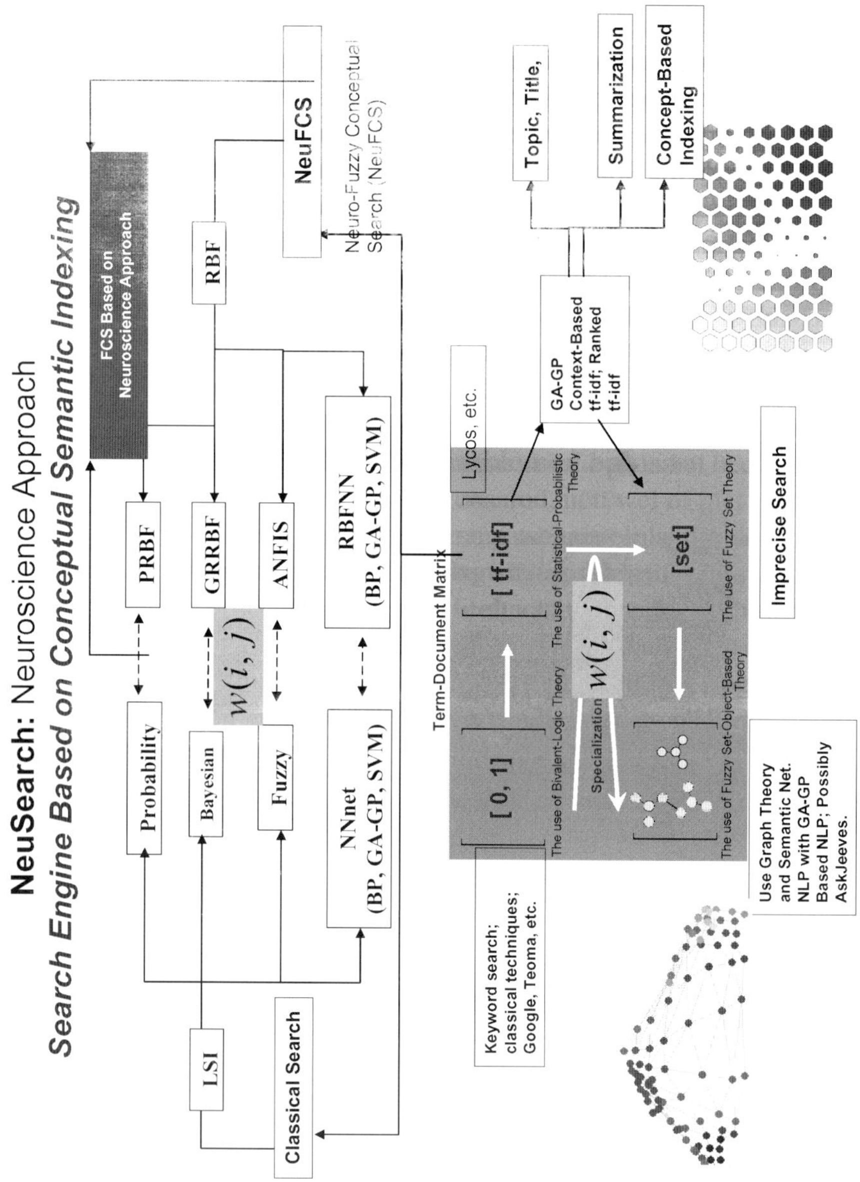

Fig. 5. Search engine based on conceptual semantic indexing.

- Given a sets of weights for each of the 5 attributes (Title, Section, Abstract, Conclusion, References, and body), and the information provided in Chapter, try to predict a Section Titles.

Another potentials use of such techniques are for book indexing so as create indexes based on multiple keywords or to organize the keywords. Also, one can explore the possibility of multiple keywords from the beginning of the GA learning process. Alternatively, one can create the summary using the top ranked keywords. The author has created such summary for promotional materials for the book entitled "Soft Computing and Intelligent Data Analysis in Oil Exploration", published by Elsevier in 2003. The author currently is working to extend this idea for non-technical document, such as NEWS and unstructured documents such as WebPages. In this case, a fuzzy transition state maker will be defined that will partition the document to a series of sections. In this case, we will assign a series of weight to each section. One can also use the tags (the bold keywords that appear in middle or certain locations of text document in box) as a transition point to define the boundary of sections within the text.

Once fuzzy tf–idf (term–document matrix) is created, the next step will be to use such indexing to develop the search and information retrieval mechanism. As shown in Figure 5, *there* are two alternatives (1) classical search models such as LSI-based approach and (2) NeuSearch model. The LSI based models include (1) Probability based-LSI, Bayesian-LSI, Fuzzy-LSI, and NNNet-based-LSI models. It is interesting that one can find an alternative to each of such LSI-based models using Radial Basis Function (RBF). For example, one can use Probabilistic RBF equivalent to Probability based-LSI, Generalized RBF as an equivalent to Bayesian-LSI, ANFIS as an equivalent to Fuzzy-LSI, and RBF function neural network (RBFNN) as an equivalent to NNnet-LSI. RBF model is the basis of the NeuSearch model (Neuro-Fuzzy Conceptual Search – NeuFCS). Given the NeuSearch model, one needs to calculate the $w(i, j)$ which will be defined in next section (the network weights). Depends on the model and framework used (Probability, Bayesian, Fuzzy, or NNnet model), the interpretation of the weights will be different. Figure 6 shows the typical Neuro-Fuzzy Conceptual Search (NeuFCS) model input–output. The main idea with respect to NeuSearch is concept-context-based search. For example, one can search for word "Java" in the context of "Computer" or in the context of "Coffee". One can also search for "apple" in the context of "Fruit" or in the context of "Computer". Therefore, before we relate the terms to the document, we first extend the keyword, given the existing concepts-contexts and then we relate that term to the documents. Therefore, there will be always a two-step process, based on NeuSearch model as shown below:

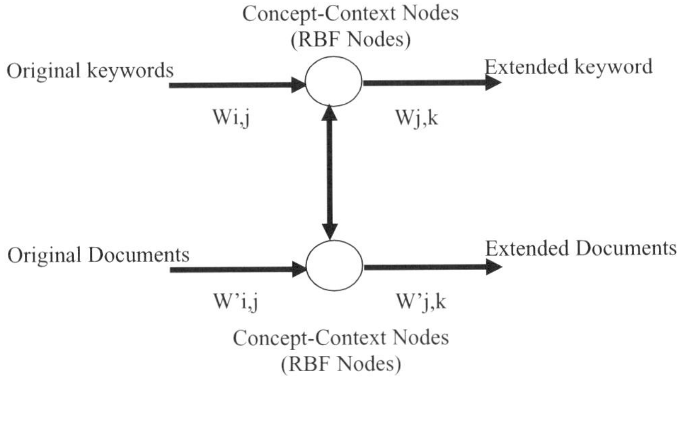

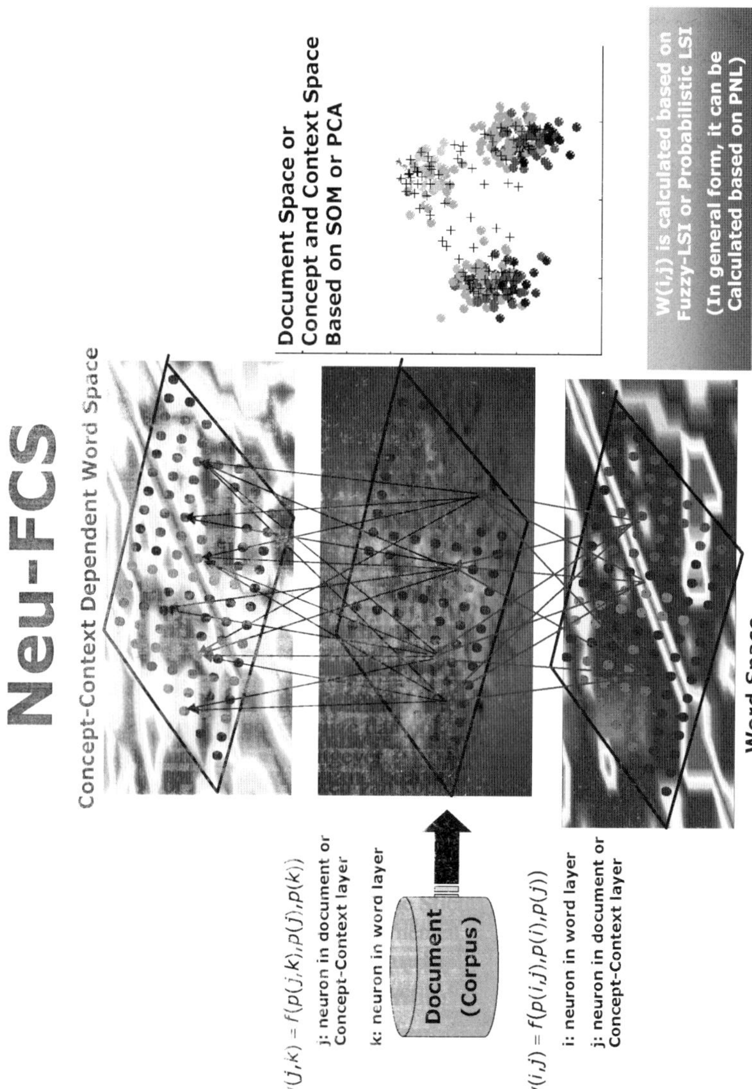

Fig. 6. Neuro-fuzzy conceptual search (NeuFCS).

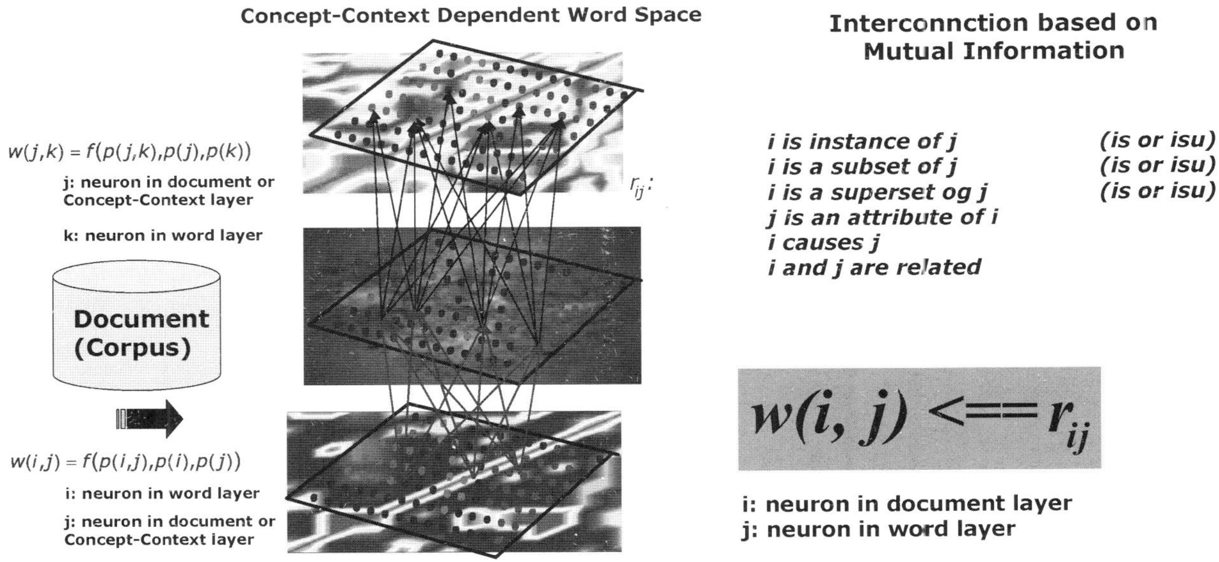

Fig. 7. PNL-based conceptual fuzzy search using brain science.

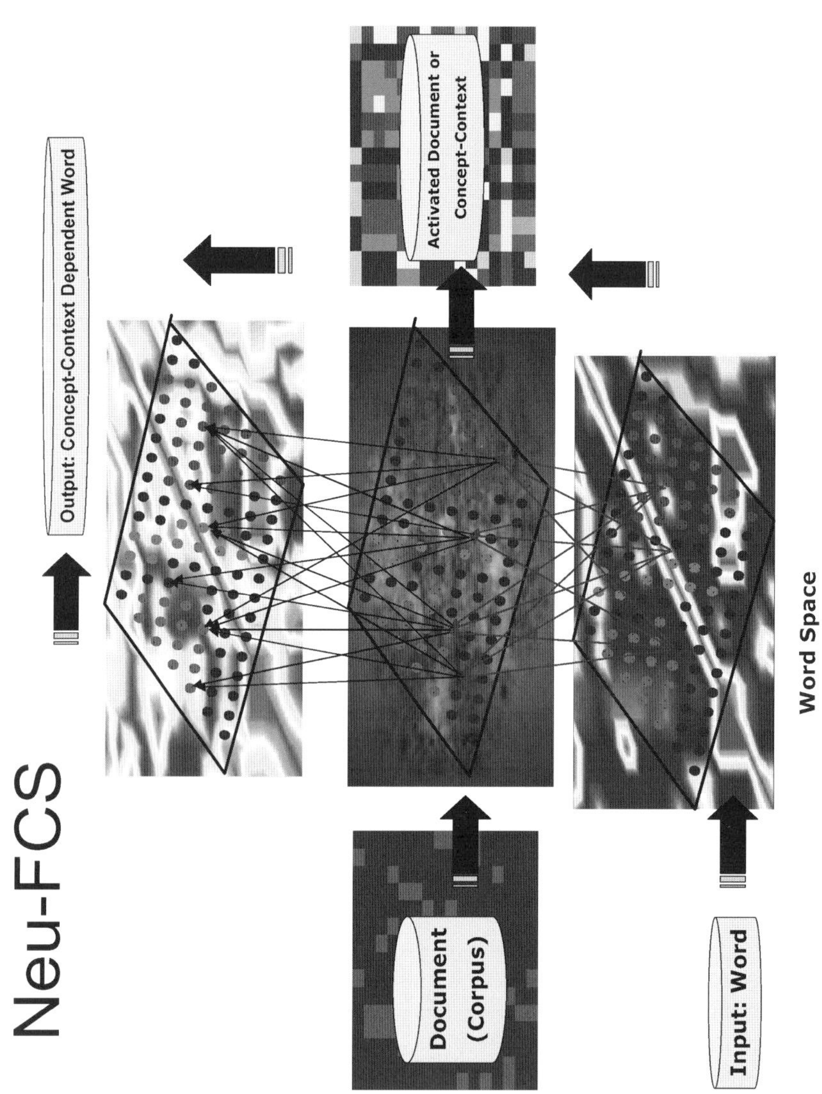

Fig. 8. Typical example of NeuFCS.

In general, $w(i, j)$ is a function of $p(i, j)$, $p(i)$, and $p(j)$, where p represent the probability. Therefore, one can use the probabilistic-LSI or PRBF given the NeuSearch framework. If the probabilities are not known, which is often times is the case, one can use Fuzzy-LSI model or ANFIS model or NNnet-LSI or RBFNN model using the NeuSearch model. In general, the PNL model can be used as unified framework as shown in Figure 7. Figure 7 shows PNL-Based Conceptual Fuzzy Search Using Brain Science model and concept presented based on Figures 2 through 6.

Based on PNL approach, $w(i, j)$ is defined based on $r(i, j)$ as follows:

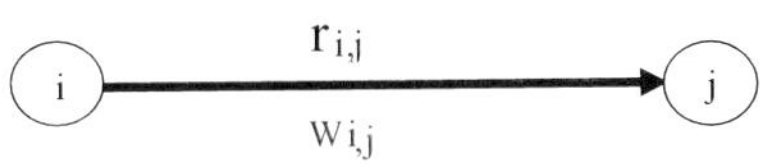

Where $w(i, j)$ is granular strength of association between i and j, $r(i, j)$ is epistemic lexicon, $w(i, j) \Longleftarrow r(i, j)$, and $r(i, j)$ is defined as follows:

$r(i, j)$: *i is an instance of j* (*is or isu*)
i is a subset of j (*is or isu*)
i is a superset of j (*is or isu*)
j is an attribute of i
i causes j (*or usually*)
i and j are related

Figure 8 shows a typical example of Neu-FCS result. Both term–document matrix and term-term document matrix/document-document matrix are reconstructed. For example, the original document corpus is very sparse as one can see in the figure; however, after processing through concept node, the document corpus is less spare and smoothens. Also, the original keywords are expanded given the concept-context nodes. While the nodes in the hidden space (Concept-Context nodes) are based on RBF (NueSearch model) and it is one-dimensional nodes, one can use SOM to rearrange the nodes given the similarity of the concept (in this case the RBF in each node) and present the 1-D nodes into 2-D model as shown in Figure 8.

3. Intelligent information systems in enterprise

To build any semantic web-based system, it will become necessary to construct a powerful logical language for making inferences and reasoning such that the system to become expressive enough to help users in a wide range of situations. This section will try to develop a framework to address this issue specially when we are dealing with structured data using web-based interface. Most of the existing search systems' 'software' are modeled using crisp logic and queries. We introduce fuzzy querying and ranking as a flexible tool allowing approximation where the selected objects do not need to match exactly the decision criteria resembling natural human behavior. The model consists of five major modules: the Fuzzy Search Engine (FSE), the Application Templates (AT), the User Interface (UI), the Database (DB) and the Evolutionary Computing (EC). We developed the software with many essential features. It is built as a web-based software system that users can access and use over the Internet. The system is designed to be generic so that it can run different application domains. To this end, the Application Template module provides information of a specific application as attributes and properties, and serves as a guideline structure for building a new application. The Fuzzy Search Engine (FSE) is the core module of the system. It has been developed to be generic so that it would fit any application. The main FSE component is the query structure, which utilizes membership functions, similarity functions and aggregators. Through the user interface a user can enter and save his profile, input criteria for a new query, run different queries and display results.

The user can manually eliminate the results he disapproves of or change the ranking according to his preferences. The Evolutionary Computing (EC) module monitors ranking preferences of the user's queries. It learns to adjust to the intended meaning of the users' preferences.

3.1. *BISC decision support system*

The BISC (Berkeley Initiative in Soft Computing) Decision Support System Components include:

- Data Management
  - database(s) which contains relevant data for the decision process
- User Interface
  - users and DSS communication
- Model Management and Data Mining
  - includes software with quantitative and fuzzy models including aggregation process, query, ranking, and fitness evaluation
- Knowledge Management and Expert System
  - model representation including linguistic formulation
- Evolutionary Kernel and Learning Process
- Data Visualization and Visual Interactive Decision Making

3.1.1. *Model framework*

The DSS system starts by loading the application template, which consists of various configuration files for a specific application and initializing the database for the application, before handling a user's requests, see Figure 9. Once the DSS system is initialized, users can enter their own profiles in the user interface or make a search with their preferences. These requests are handled by the control unit of the system. The control unit converts user input into data objects that are recognized by the DSS system. Based on the request types, it forwards them to the appropriate modules. If the user wants to create a profile, the control unit will send the profile data directly to the database module which stores the data in the database for the application. If the user wants to query the system, the control unit will direct the user's preferences to the Fuzzy Search Engine which queries the database. The query results will be sent back to the control unit and displayed to the users.

3.1.2. *Fuzzy engine*

Fuzzy Query, search and Ranking: To support generic queries, the fuzzy engine has been designed to have a tree structure. There are two types of nodes in the tree, category nodes and attribute nodes. While multiple category levels are not necessary, they are designed

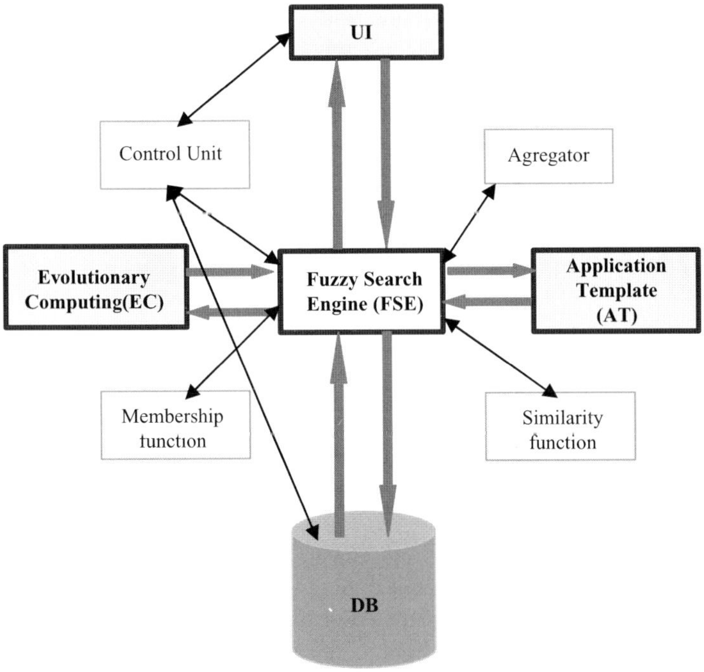

Fig. 9. The BISC-DSS general framework.

to allow various refinements of the query through the use of the type of aggregation of the children. Categories only act to aggregate the lower levels. The attribute nodes contain all the important information about a query. They contain the membership functions for the fuzzy comparison as well as the use of the various aggregation methods to compare two values. The attribute nodes handle the compare method slightly differently than the category nodes. There are two different ways attributes may be compared. The attribute nodes contain a list of membership functions comprising the fuzzy set. The degrees of membership for this set are passed to the similarity comparator object, which currently has a variety of different methods to calculate the similarity between the two membership vectors. In the other method, the membership vector created by having full membership to a single membership function specified in the fuzzy data object, but no membership value for the other functions. Membership function: Currently there are three membership functions implemented for the Fuzzy Engine. A generic interface has been created to allow several different types of membership functions to be added to the system. The three types of membership functions in the system are: Gaussian, Triangular and Trapezoidal. These functions have three main points, for the lower bound, upper bound and the point of maximum membership. For other functions, optional extra points may be used to define the shape (an extra point is required for the trapezoidal form).

3.1.3. *Application template*

The DSS system is designed to work with different application domains. The application template is a format for any new application we build; it contains data of different categories, attributes and membership functions of that application. The application template module consists of two parts: (1) the application template data file specifies all the membership functions, attributes and categories of an application. We can consider it as a configuration data file for an application. It contains the definition of membership functions, attributes and the relationship between them; (2) the application template logic parses and caches data from the data file so that other modules in the system can have faster access to definitions of membership functions, attributes and categories. It also creates a tree data structure for the fuzzy search engine to transverse.

3.1.4. *User interface*

It is difficult to design a generic user interface that suits different kind of applications for all the fields. For example, we may want to have different layouts for user interfaces for different applications. To make the DSS system generic while preserving the user friendliness of the interfaces for different applications, we developed the user interfaces into two parts.

First, we designed a specific HTML interface for each application we developed. Users can input their own profiles; make queries by specifying preferences for different attributes. Details for the DSS system are encapsulated from the HTML interface so that the HTML interface design would not be constrained by the DSS system. The second part of our user interface module is a mapping between the parameters in the HTML files and the attributes in the application template module for the application. The input mapping specifies the attribute names to which each parameter in the HTML interface corresponds. With this input mapping, a user interface designer can use input methods and parameter names freely.

3.1.5. *Database (DB)*

The database module is responsible for all the transactions between the DSS system and the database. This module handles all queries or user profile creations from the Fuzzy Engine and the Control Unit respectively. For queries from the Fuzzy Search Engine, it retrieves data from the database and returns it in a data object form. Usually queries are sets of attribute values and their associated weights. The database module returns the matching records in a format that can be manipulated by the user, such as eliminating one or more record or changing their order. To create a user profile, it takes data objects from the Control Unit and stores it in the database. There are three components in the DB module.

The DB Manager is accountable for two things: (1) setting up database connections and allocating database connections to DB Accessor objects when needed. It also supplies information to the database for authentication purposes (e.g., username, password, path to the database, etc.); (2) the DB Accessor Factory creates DB Accessor objects for a specific application. For example, if the system is running the date matching application, DB Accessor Factory will create DB Accessor objects for the date matching application. The existence of this class serves the purpose of using a generic Fuzzy Search Engine; (3) the DB Accessor is responsible for storing and getting user profiles to and from the database. It is the component that queries the database and wrap result from the database into data objects that are recognized by our application framework.

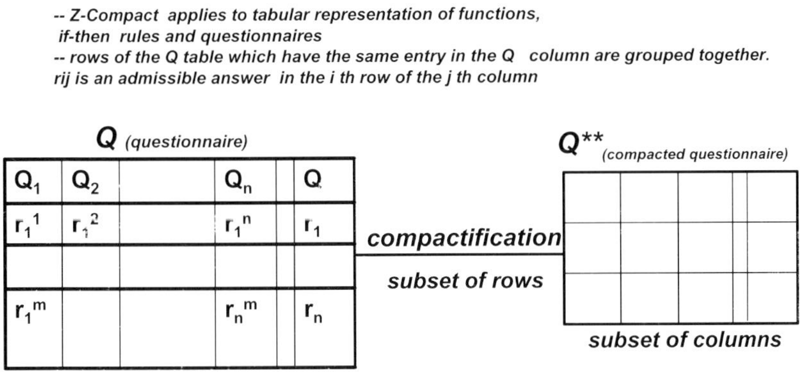

Fig. 10. Schematic Z-Compact algorithm.

3.1.6. *Measure of association and fuzzy similarity*

As in crisp query and ranking, an important concept in fuzzy query and ranking applications is the measure of association or similarity between two objects in consideration. For example, in a fuzzy query application, a measure of similarity between two queries and a document, or between two documents, provides a basis for determining the optimal response from the system. In fuzzy ranking applications, a measure of similarity between a new object and a known preferred (or non-preferred) object can be used to define the relative goodness of the new object. Most of the measures of fuzzy association and similarity are simply extensions from their crisp counterparts. However, because of the use of perception based and fuzzy information, the computation in the fuzzy domain can be more powerful and more complex.

Various definitions of similarity exist in the classical, crisp domain, and many of them can be easily extended to the fuzzy domain. However, unlike in the crisp case, in the fuzzy case the similarity is defined on two fuzzy sets. Suppose we have two fuzzy sets A and B with membership functions $\mu_A(x)$ and $\mu_B(x)$, respectively. The arithmetic operators involved in the fuzzy similarity measures can be treated using their usual definitions while the union and the intersection operators need to be treated specially. It is important for these operator pairs to have the following properties: (1) conservation, (2) monotonicity, (3) commutativity, and (4) associativity. It can be verified that the triangular norm (T-norm) and triangular co-norm (T-conorm) [1,2,7,10] conform to these properties and can be applied here. A detailed survey of some commonly used T-norm and T-conorm pairs along with other aggregation operators can be find at Nikravesh and Azvine [16].

In many situations, the controlling parameters, including the similarity metric, the type of T-norm/conorm, the type of aggregation operator and associated weights, can all be specified based on the domain knowledge of a particular application. However, in some other cases, it may be difficult to specify a priori an optimal set of parameters. In those cases, various machine learning methods can be employed to automatically "discover" a suitable set of parameters using a supervised or unsupervised approach. For example, the Genetic Algorithm (GA) and DNA-based computing, as described in later sections, can be quite effective. Another important and unique component of our system is compactification algorithm or Z-Compact [13,19,23]. Z-Compact algorithm developed by Lotfi A. Zadeh

Group 1(initial)

Q_1	Q_2	Q_3		Q	
a_1^1	a_2^2	a_1^3		a_1	
a_2^1	a_2^2	a_1^3		a_1	
a_3^1	a_2^2	a_1^3		a_1	
a_3^1	a_1^2	a_2^3		a_1	
a_1^1	a_2^2	a_2^3		a_1	
a_2^1	a_2^2	a_2^3		a_1	
a_3^1	a_2^2	a_2^3		a_1	
*	a_2^2	a_1^3		a_1	**Pass (1)**
*	a_2^2	a_2^3		a_1	
a_3^1	*	a_2^3		a_1	**Pass (2)**
a_1^1	a_2^2	*		a_1	
a_2^1	a_2^2	*		a_1	
a_3^1	a_2^2	*		a_1	**Pass (3)**
*	a_2^2	*		a_1	

Group 2 (initial)

Q_1	Q_2	Q_3		Q	
a_2^1	a_1^2	a_1^3		a_2	
a_2^1	a_1^2	a_2^3		a_2	
a_2^1	a_1^2	*		a_2	*Pass (3)*

Final Solution

Q_1	Q_2	Q_3		Q
a_3^1	*	a_2^3		a_1
*	a_2^2	*		a_1
a_2^1	a_1^2	*		a_2

Fig. 11. Z-Compact algorithm, example.

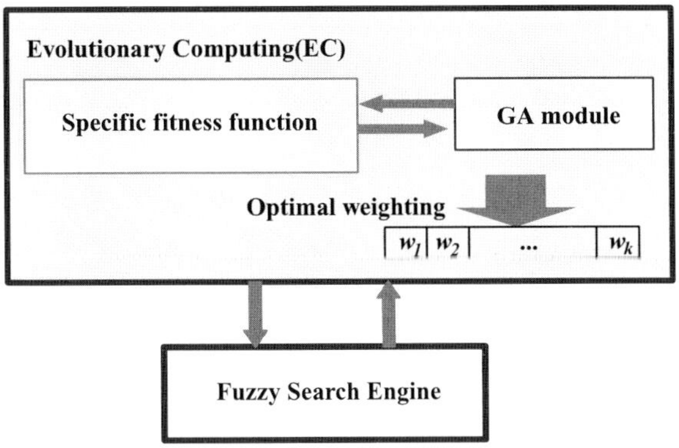

Fig. 12. Evolutionary computing module: preferences learning.

[19] and it has been implemented for the first time as part of BISC-DSS for automatons multi-agents modeling as part of ONR project [23]. The algorithm has been extended and currently is part of the BISC-DSS software and it has been applied in several applications [23] and [13]. Figures 10 and 11 show step by step how this algorithm works and how can be implemented [13,19,23].

3.1.7. *Evolutionary computing*

In the Evolutionary Computing (EC) module of the BISC Decision Support System, our purpose is to use an evolutionary method to allow automatic adjusting of the user's preferences. These preferences can be seen as parameters of the fuzzy logic model in form of weighting of the used variables. These preferences are then represented by a weight vector and genetic algorithms will be used to fix them. In the Evolutionary Computation approach, Genetic Programming, which is an extension of Genetic Algorithms, is the closest technique to our purpose. It allows us to learn a tree structure, which represents the combination of aggregators. The selection of these aggregators is included to the learning process using the Genetic Programming. In this section, we describe the GA [3] and GP [4] application to our problem [3,4,11]. Our aim is learning fuzzy-DSS parameters which are the weight vectors representing the user preferences associated to the variables that have to be aggregated on the one hand, and the adequate decision tree representing the combination of the aggregation operators that have to be used on the other hand. Weight vector being a linear structure, can be represented by a binary string in which weight values are converted to binary numbers. This binary string corresponds to the individual's DNA in the GA learning process. The goal is to find the optimal weighting of the variables. A general GA module can be used by defining a specific fitness function for each application as shown in Figure 12.

Aggregators can be combined in the form of a tree structure which can be built using a Genetic Programming learning module (Figure 13). It consists in evolving a population of individuals represented by tree structures. The evolution principle remains the same as in a conventional GP module but the DNA encoding needs to be defined according to the considered problem. We propose to define an encoding for aggregation trees which is more complex than for classical trees and which is common to all considered applications. As

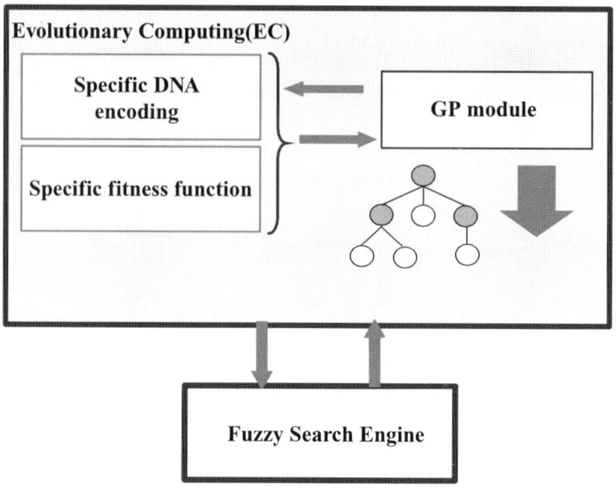

Fig. 13. Evolutionary computing module: aggregation tree learning.

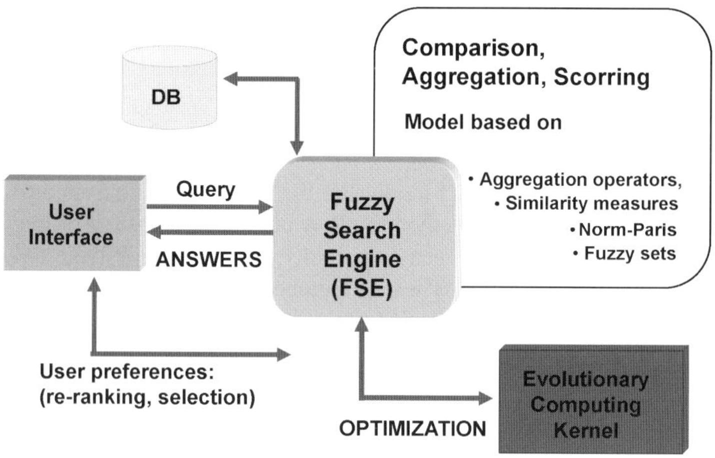

Fig. 14. Interaction and optimization.

shown in Figure 12, we need to define a specific encoding, in addition to the fitness function specification. Tree structures are generated randomly as in the conventional GP. But, since these trees are augmented according the properties defined above, the generation process has to be updated. So, we decided to randomly generate the number of arguments when choosing an aggregator as a node in the tree structure. And for the weights, we chose to generate them randomly for each node during its creation.

Concerning the fitness function, it is based on performing the aggregation operation and the root node of the tree that has to be evaluated. For the university admissions application, the result of the root execution corresponds to the score that has to be computed for each value vector in the training data set. The fitness function, as in the GA learning of the user preferences, consists in simple or combined similarity measures. In addition, we can include to the fitness function a complementary measure that represents the individual's size which has to be minimized in order to avoid over-sized trees. We have described the use of evolutionary computation methods for optimization problems in the BISC decision. BISC-DSS is also a strategy and framework for combining fuzzy logic, machine learning and evolutionary computation. Figure 14 shows how these components interact with each other.

4. Conclusions

Intelligent search engines with growing complexity and technological challenges are currently being developed. This requires new technology in terms of understanding, development, engineering design and visualization. While the technological expertise of each component becomes increasingly complex, there is a need for better integration of each component into a global model adequately capturing the imprecision and deduction capabilities. In addition, intelligent models can mine the Internet to conceptually match and rank homepages based on predefined linguistic formulations and rules defined by experts or based on a set of known homepages. The FCM model can be used as a framework for intelligent information and knowledge retrieval through conceptual matching of both text and images (here defined as "Concept"). The FCM can also be used for constructing fuzzy ontology or terms related to the context of the query and search to resolve the ambiguity. This model can be used to calculate conceptually the degree of match to the object or query.

Acknowledgements

Funding for this research was provided by the British Telecommunication (BT), the BISC Program of UC Berkeley, and Imaging and Informatics group at Lawrence Berkeley National Lab.

References

[1] R. Fagin, *Fuzzy queries in multimedia database systems*, in: Proc. ACM Symposium on Principles of Database Systems, 1998, pp. 1–10.

[2] R. Fagin, *Combining fuzzy information from multiple systems*, J. Comput. System Sci. **58** (1999), 83–99.

[3] J.H. Holland, *Adaptation in Natural and Artificial Systems: An Introductory Analysis with Applications to Biology, Control and Artificial Intelligence*, MIT Press, 1992, First Published by University of Michigan Press, 1975.

[4] J.R. Koza, *Genetic Programming: On the Programming of Computers by Means of Natural Selection*, MIT Press, Cambridge, MA, 1992, 819 pages.

[5] V. Loia, M. Nikravesh, L.A. Zadeh, J. Soft Computing, Special Issue, Fuzzy Logic and the Internet **6** (5) (August 2002).

[6] V. Loia, M. Nikravesh, L.A. Zadeh, *Fuzzy Logic and the Internet*, Studies in Fuzziness and Soft Computing, vol. 137, Physica-Verlag, Springer, 2004.

[7] M. Mizumoto, *Pictorial representations of fuzzy connectives, Part I: cases of T-norms, T-conorms and averaging operators*, Fuzzy Sets and Systems **31** (1989), 217–242.

[8] M. Nikravesh, *Fuzzy logic and Internet: perception based information processing and retrieval*, Berkeley Initiative in Soft Computing, Report No. 2001-2-SI-BT, September 2001.

[9] M. Nikravesh, *BISC and the new millennium, perception-based information processing*, Berkeley Initiative in Soft Computing, Report No. 2001-1-SI, September 2001.

[10] M. Nikravesh, *Perception-based information processing and retrieval: application to user profiling*, 2001 Research Summary, EECS, ERL, University of California, Berkeley, BT-BISC Project, http://zadeh.cs.berkeley.edu/, http://www.cs.berkeley.edu/~nikraves/, http://www-bisc.cs.berkeley.edu/.

[11] M. Nikravesh, *Credit scoring for billions of financing decisions*, Joint 9th IFSA World Congress and 20th NAFIPS International Conference. IFSA/NAFIPS 2001 "Fuzziness and Soft Computing in the New Millenium", Vancouver, Canada, July 25–28, 2001.

[12] M. Nikravesh, *Fuzzy conceptual-based search engine using conceptual semantic indexing*, NAFIPS-FLINT 2002, June 27–29, New Orleans, LA, USA.

[13] M. Nikravesh, *Evolutionary-based intelligent information and decision systems*, FuzzIEEE, Reno-Nevada, May 22–25, 2005 (Invited Talk).

[14] M. Nikravesh, B. Azvine, *New directions in enhancing the power of the Internet*, in: Proc. of the 2001 BISC Int. Workshop, University of California, Berkeley, Report: UCB/ERL M01/28, August 2001.

[15] M. Nikravesh, B. Azvine, *Fuzzy queries, search, and decision support system*, Internat. J. Soft Computing, Special Issue in Fuzzy Logic and the Internet, 2002.

[16] M. Nikravesh, B. Azvine, *Fuzzy queries, search, and decision support system*, J. Soft Computing **6** (5) (August 2002).

[17] M. Nikravesh, V. Loia, B. Azvine, *Fuzzy logic and the Internet (FLINT), Internet, World Wide Web, and Search Engines*, Internat. J. Soft Computing, Special Issue in Fuzzy Logic and the Internet, 2002.

[18] M. Nikravesh, et al., *Enhancing the Power of the Internet*, Studies in Fuzziness and Soft Computing, vol. 139, Physica-Verlag, Springer, 2004.

[19] L.A. Zadeh, *A fuzzy-algorithmic approach to the definition of complex or imprecise concepts*, Int. J. Man–Machine Studies **8** (1976), 249–291.

[20] L.A. Zadeh, *From computing with numbers to computing with words – from manipulation of measurements to manipulation of perceptions*, IEEE Trans. on Circuits and Systems **45** (1999), 105–119.

[21] L.A. Zadeh, *A new direction in AI: towards a computational theory of perceptions*, AI Magazine **22** (2001), 73–84.

[22] L.A. Zadeh, *Toward a perception-based theory of probabilistic reasoning with imprecise probabilities*, J. Statist. Plann. Inference **105** (2002), 233–264.

[23] L.A. Zadeh, M. Nikravesh, *Perception-based intelligent decision systems*, Office of Naval Research, Summer 2002 Program Review, Covel Commons, University of California, Los Angeles, July 30–August 1, 2002.

Author Index

Roman numbers refer to pages on which the author (or his/her work) is mentioned. Italic numbers refer to reference pages. Numbers between brackets are the reference numbers. No distinction is made between first and co-author(s).

Subject Index